Lecture Notes in Computer Science 15441

Founding Editors

Gerhard Goos

Juris Hartmanis

AF172956

The series Lecture Notes in Computer Science (LNCS), including its subseries Lecture Notes in Artificial Intelligence (LNAI) and Lecture Notes in Bioinformatics (LNBI), has established itself as a medium for the publication of new developments in computer science and information technology research, teaching, and education.

LNCS enjoys close cooperation with the computer science R & D community, the series counts many renowned academics among its volume editors and paper authors, and collaborates with prestigious societies. Its mission is to serve this international community by providing an invaluable service, mainly focused on the publication of conference and workshop proceedings and postproceedings. LNCS commenced publication in 1973.

Jun Zhao · Weizhi Meng

Editors

Science of Cyber Security

6th International Conference, SciSec 2024
Copenhagen, Denmark, August 14–16, 2024
Proceedings

 Springer

Editors
Jun Zhao 🆔
Nanyang Technological University
Singapore, Singapore

Weizhi Meng 🆔
Technical University of Denmark
Kongens Lyngby, Denmark

ISSN 0302-9743   ISSN 1611-3349 (electronic)
Lecture Notes in Computer Science
ISBN 978-981-96-2416-4   ISBN 978-981-96-2417-1 (eBook)
https://doi.org/10.1007/978-981-96-2417-1

This Springer imprint is published by the registered company Springer Nature Singapore Pte Ltd.
The registered company address is: 152 Beach Road, #21-01/04 Gateway East, Singapore 189721, Singapore

If disposing of this product, please recycle the paper.

Preface

This volume contains the papers presented at the 6th International Conference on Science of Cyber Security (SciSec 2024), which was held as a hybrid conference during 14–16 August, 2024, in Copenhagen, Denmark. SciSec is an annual international forum for researchers and industry experts to present and discuss the latest research, trends, breakthroughs, and challenges in the domain of cybersecurity. This new forum was initiated in 2018 and aims to catalyze the research collaborations between the relevant communities and disciplines that can work together to deepen our understanding of, and build a firm foundation for, the emerging Science of Cyber Security.

This year we received 79 submissions from around the world. Each paper was reviewed by at least three Program Committee members in a single-blind process. At last, we accepted 25 full papers with an acceptance rate of 31.6%. The conference program also included two invited keynote talks: the first keynote titled "AI and Cybersecurity" from Audun J, and the second keynote titled "Who was behind the Camera?" from Jeff Yan.

We would like to thank all of the authors of the submitted papers for their interest in SciSec 2024. We would like to express our heartfelt thanks to the reviewers, keynote speakers, and participants for their invaluable contributions to the success of SciSec 2024. We are also deeply grateful to the Program Committee, the Publicity Committee, and the Organizing Committee for their dedicated efforts in preparing and managing the event. We also extend our thanks to Proceeding Assistant Mr. Chang Liu. We hope you find the conference proceedings inspiring and that they will aid you in discovering future research opportunities.

August 2024

Jun Zhao
Weizhi Meng

Organization

General Chairs

Weizhi Meng	Lancaster University, UK
Feng Liu	Institute of Information Engineering, Chinese Academy of Sciences, China
Sokratis Katsikas	Norwegian University of Science and Technology, Norway

Steering Committee

Shouhuai Xu	University of Colorado Springs, USA
Feng Liu	Institute of Information Engineering, Chinese Academy of Sciences, China
Guoping Jiang	Nanjing University of Posts and Telecommunications, China
Moti Yung	Google and Columbia University, USA

Program Committee Chairs

Jun Zhao	Nanyang Technological University, Singapore
Christian D. Jensen	Aarhus University, Denmark

Publicity Chairs

Ziyao Liu	Nanyang Technological University, Singapore
Xiaoning (Maggie) Liu	RMIT University, Australia
Huangxun Chen	Hong Kong University of Science and Technology (Guangzhou), China
Qiang Tang	Luxembourg Institute of Science and Technology, Luxembourg
Mir Pritom	Tennessee Tech University, USA
Habtamu Abie	Norwegian Computing Center, Norway

Proceedings Committee

Jun Zhao	Nanyang Technological University, Singapore
Weizhi Meng	Lancaster University, UK
Chang Liu	Nanyang Technological University, Singapore

Program Committee

Habtamu Abie	Norwegian Computing Centre, Norway
Milad Taleby Ahvanooey	Nanyang Technological University, Singapore
Nicola Bena	University of Milan, Italy
Alessandro Brighente	University of Padua, Italy
Michele Carminati	Politecnico di Milano, Italy
Bo Chen	Michigan Technological University, USA
Huangxun Chen	Hong Kong University of Science and Technology, China
Hung-Yu Chien	National Chi Nan University, Taiwan
Michal Choras	Bydgoszcz University of Science and Technology, Poland
Nora Cuppens-Boulahia	Polytechnique Montréal, Canada
Paolo D'Arco	University di Salerno, Italy
Ali Dehghantanha	University of Guelph, Canada
Francesco Flammini	Mälardalen University, Sweden
Joaquin Garcia-Alfaro	Institut Polytechnique de Paris, France
Mehdi Gheisari	Islamic Azad University, Iran
Yong Guan	Iowa State University, USA
Yujuan Han	Shanghai Maritime University, China
Paul Haskell-Dowland	Edith Cowan University, Australia
Daojing He	East China Normal University, China
Julio Hernandez	University of Kent, UK
Huawei Huang	Sun Yat-sen University, China
Wei Huo	Institute of Information Engineering, Chinese Academy of Sciences, China
Pedro Inácio	Universidade da Beira Interior, Portugal
Christian-D. Jensen	Aarhus University, Denmark
Taeho Jung	University of Notre Dame, USA
Jan Jürjens	Fraunhofer Institute for Software & Systems Engineering ISST and University of Koblenz-Landau, Germany
Zbigniew Kalbarczyk	University of Illinois Urbana-Champaign, USA
Georgios Kambourakis	University of the Aegean, Greece

Sokratis Katsikas	Norwegian University of Science and Technology, Norway
Dimitris Kavallieros	Centre for Research & Technology Hellas, Greece
Elisavet Konstantinou	University of the Aegean, Greece
Panayiotis Kotzanikolaou	University of Piraeus, Greece
Kwok Yan Lam	Nanyang Technological University, Singapore
George Lazaridis	Centre for Research and Technology Hellas, Greece
Wenjuan Li	Education University of Hong Kong, China
Chang Liu	Nanyang Technological University, Singapore
Feng Liu	Institute of Information Engineering, Chinese Academy of Sciences, China
Xiaoning Liu	RMIT University, Australia
Xiwei Liu	Tongji University, China
Ziyao Liu	Nanyang Technological University, Singapore
Zhuo Lu	University of South Florida, USA
Xiapu Luo	Hong Kong Polytechnic University, China
Vasileios Mavroeidis	University of Oslo, Norway
Weizhi Meng	Lancaster University, UK
Andrew Odlyzko	University of Minnesota, USA
Irdin Pekaric	University of Liechtenstein, Liechtenstein
Sandeep Pirbhulal	Norwegian Computing Center, Norway
Joachim Posegga	University of Passau, Germany
Mir Mehedi Pritom	University of Texas at San Antonio, USA
Liangxin Qian	Nanyang Technological University, Singapore
Kouichi Sakurai	Kyushu University, Japan
Reijo Savola	University of Jyväskylä, Finland
Qingni Shen	Peking University, China
Gang Wang	Nanyang Technological University, Singapore
Weizheng Wang	City University of Hong Kong, China
Zefan Wang	Nanyang Technological University, Singapore
Jia Xu	Nanjing University of Posts and Telecommunications, China
Maochao Xu	Illinois State University, USA
Shouhuai Xu	University of Colorado Springs, USA
Lei Xue	Sun Yat-sen University, China
Toshihiro Yamauchi	Okayama University, Japan
Fei Yan	Wuhan University, China
Guanhua Yan	Binghamton University, State University of New York, USA
Wenhan Yu	Nanyang Technological University, Singapore
Fan Zhang	Zhejiang University, China

Leo Yu Zhang	Griffith University, Australia
Yuan Zhang	Nanjing University, China
Jun Zhao	Nanyang Technological University, Singapore
Cliff Zou	University of Central Florida, USA

Additional Reviewers

Aggelopoulos, Konstantinos
Berger, Christian
Chatzoglou, Efstratios
Chen, Huashan
Chi, Chihung
Han, Zhaoyang
Kardara, Antonia
Koirala, Nirajan

Modir Rousta, Mohammadhossein
Pöhls, Henrich C.
Qipeng, Xie
Sha, Kailun
Shi, Haichao
Smiliotopoulos, Christos
Tang, Wenyi

Contents

A Novel Scoring Algorithm Against HID Attacks Based on Static Text Feature Matching

Haiyang Li[1,2]([✉]), Zhiqiang Lv[1,2]([✉]), Yixin Zhang[1,2], and Yanan Xue[3]

[1] Institute of Information Engineering, Chinese Academy of Sciences, Beijing, China
{lihaiyang,lvzhiqiang}@iie.ac.cn
[2] School of Cyber Security, University of Chinese Academy of Sciences, Beijing, China
[3] North Automatic Control Technology Institute, Taiyuan, China

Abstract. In recent years, the proliferation of HID attacks has raised significant security concerns. Previous research has primarily focused on detecting these attacks by analyzing behavior features extracted from keystroke timestamps. However, the sophisticated HID attack could highly imitate user behavior features with the rising trend of countermeasures to overcome the known flaws of existing studies. In this paper, we propose a novel scoring algorithm that relies exclusively on keystroke text analysis. It depends on a HID keystroke text feature space, within which each feature matching result of a text segment is scored and their sum is calculated. Once the total score surpasses a predefined threshold, the text segment will be flagged as indicative of an unsafe HID interaction. In confrontational testing, our algorithm demonstrates a precision rate exceeding 95% in detecting HID attacks.

Keywords: HID attack · Keystroke text feature · Attack detection

1 Introduction

The USB specification is a universal serial bus transmission protocol. It has been developed from the initial version 1.0 to current 4.0. And the HID specification [1] is a sub-specification of the USB 2.0, it regulates human interface devices. It's application scenarios usually do not require high data speed, but need the timeliness of data transmission. A variety of USB attacks have emerged over years [2], and HID attack is one of them. HID attacks exploit a malicious USB device that masquerades as a legitimate USB keyboard when connected to the victim's computer. This allows the attacker to gain user access and imitate human keystrokes, facilitating the injection of malicious payloads. The well-known ones are BadUSB [3], USB Rubber Ducky [4] and Teensy [5].

A USB keyboard will generate keystroke data including timestamps and text. Consequently, detection mechanisms fall into two categories: those based on keystroke timestamps and those based on keystroke text. Research in the former category has yielded numerous effective detection methods for the early

J. Zhao and W. Meng (Eds.): SciSec 2024, LNCS 15441, pp. 1–18, 2025.
https://doi.org/10.1007/978-981-96-2417-1_1

HID attacks [6–15]. However, as the confrontational dynamics between attack and detection have evolved, attack's capacity to mimic behavior features has improved, potentially evading detection [16]. There are fewer results in the latter category. One approach for detection involves keyword matching, but it is easily bypassed in other ways.

The keystroke timestamps are the data that can characterize the behavior features, but the keystroke text is the actual carrier of the keystroke content. As shown in the Fig. 1, both the trusted device and the attacking device generate keystroke text. The user keystroke text is the content typed, but the HID attack text is the malicious payloads, which contain attack commands or attack data. Although the HID attacks can imitate some user's behavior features, it would not imitate the user's keystroke text. Because it aims at accomplishing a specific attack purpose, which can only be achieved through a well-designed keystroke sequence.

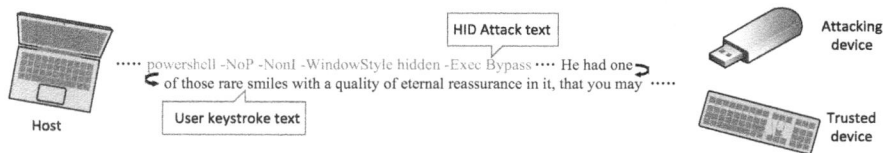

Fig. 1. USB HID keystroke injection attack.

In this paper, we propose a scoring algorithm based on HID text feature matching to realize the HID attack detection task. Though comparative analysis of user keystroke text and HID attack text, we extract features by text mining, select features that can express textual differences, and combine the features to form a text feature space. For a detection sample, its features are extracted within the scope of the feature space, and the matching results of each feature are scored and summed. When the total score is greater than the predefined threshold, it will infer that the sample is generated by HID attacks. Through confrontation test of the attacks and our detection, the detection precision reaches more than 95%.

The main contributions of this paper are threefold:

- Summarize the relevant research results, and raise the problems or deficiencies of the detection mechanisms.
- Study HID attacks from a purely textual perspective, and construct a HID text feature space through comparative analysis.
- Propose a scoring algorithm based on HID text feature matching to realize the detection task, and the experimental precision reaches over 95%.

2   Related Work and Problems

In this section, we summarize the existing detection mechanisms against HID attacks, and point out the problems or deficiencies.

The first category is to determine whether a device is of risk on the enumeration information. During the enumeration, descriptors are requested and parsed to identify the PID, VID, etc., especially the interface combination. There is a mapping relationship between the interface class and the functional device. When the parsed result does not match its actual appearance or the user's expected function, the device will be considered a dangerous device. Typically, USBWall [14] uses this method; Similarly, USBGuard [8] sets the black and white list of USB devices; USBsec [17] authenticates device's identity information; GoodUSB [7] presets expectations by the user. However, firmware of a USB device can easily be tampered with or custom designed to look like normal user devices. In fact, the USB device mainly performs related configuration work in enumeration, and cannot launch malicious attacks. So, only several devices of risk can be found at this stage.

The second category is to identify whether an attack has occurred with the data flow. In the transmission stage, a malicious USB device can launch a HID attack that generates abnormal traffic. The data of the HID attacks includes keystroke timestamps and keystroke text. The most notable feature of the initial HID attacks is that the keystroke injection speed is significantly higher, which is beyond the reach of humans. At present, there are some detection mechanisms that extract behavior features from timestamps. Typically, USBlock [13] directly uses the keystroke speed; Similarly, DuckHunt [18] uses DDT (down-down time, the time interval between pressing a key and pressing the second consecutive key); Literature [10] uses an eigenvector $F = [\mu,\ \sigma,\ \eta,\ \text{n, e, s}]$. However, HID attacks don't always inject quickly, and it is capable of injecting attack payloads at the keystroke rhythm of a human when the victim user is absent. There are two feasible ways as follows:

(i) The attacker uses tools such as Keylogger [19] to monitor the user's keystrokes, resulting in the leakage of the user's keystroke timestamps that can equip the HID attack with the same keystroke behavior features. The Malboard [16] launches a HID attack in a similar way.

(ii) The attacker extensively collects keystroke data from some human users who have similar occupation, keystroke proficiency and other factors as the victim user. Then imitate the similar keystroke behavior features by machine learning, and use them for attack.

There are fewer detection mechanisms based on keystroke text, and a conceivable one is keyword matching. The WHID Defense [15] detects attacks for Windows by matching the keywords that start Command Prompt or Run Box. But the keywords are easily bypassed and it may lead to frequent false alarm for that the user also types the same keywords. Additionally, in the analysis of the HID attack scripts, we found that there are two countermeasures:

(i) Evade detection mechanisms by adopting uppercase and lowercase letters, such as '*SYsTEM.io.cOmpressION.coMPRESsiOnmode*', which can escape the keywords of system API.

(ii) Bypass keywords in another way to achieve the same operation. For example, start powershell without the full 'powershell' by *'cmd /c "p^Owe%ALLUS ERSPROFILE:~7,1%Shell /NoPr /Nonin /wind hidD"'*.

Overall, the existing detection mechanisms are not capable of dealing with various HID attack methods due to the problems or deficiencies.

3   Method and Design

In this section, we study the HID attack detection task from the perspective of keystroke text, and design a scoring algorithm based on static feature matching.

First of all, we make a normative definition of keystroke text. It is generated during keystrokes of human users or imitated keystrokes of HID attacks, and consists of a continuous string of printable characters. Our research object is a text segment intercepted from the HID text buffer. Compared with the text in news, novels, etc., keystroke text contains significantly more noise. User keystroke text often has a lot of redundant characters because of type errors, and HID attack text is content of the attack commands or attack data that require expertise to read.

3.1   Method of Detection

In research, we find that HID attacks can imitate the keystroke behavior features of human users to evade detection mechanisms, but generally they would not imitate the user keystroke text. Because the attack keystroke text is the actual carrier of the attack payloads, and it has to inject a well-crafted keystroke sequence into the victim computer to achieve a specific attack purpose.

So, we propose a new detection method based on keystroke text. However, relying solely on a single text feature from an HID attack for detection is not advisable, as such feature may also appear in user keystroke text, albeit with a low probability of occurrence. In subsequent research, we have discovered that a text segment from an HID attack typically exhibits multiple text features, but segments from user keystroke text have a significantly lower likelihood of sharing similar text features. Based on this, our detection method is conceived as shown in the Fig. 2.

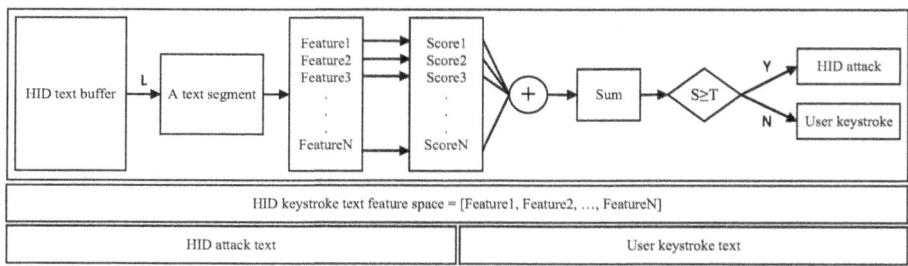

Fig. 2. Method of scoring algorithm based on static feature matching.

Firstly, a HID keystroke text feature space (HID-KTFS) is the basis of our detection method. It is a result of feature extraction, feature selection and feature combination on text. We need to process HID attack text and user keystroke text to construct the static HID-KTFS. The construction is according to the statistical and comparative analysis of keystroke text. When a text feature is very different on the two text categories, it will be included in the feature space. On the contrary, when a text feature has little difference, then it will not be included.

Secondly, a scoring algorithm is proposed based on keystroke text feature matching. There are two hyperparameters need to be preset, one is the detection sample length **L**, and another is the score threshold **T**. We intercept a text segment of length L from the HID keystroke text buffer through a sliding window as a detection sample, and mine various features of the sample. Within the range of the HID-KTFS, the matching result of a single feature is scored, and then all the scores are added for a total score. Finally, referring to the score threshold T, when the total score is greater than T, we consider that the text segment may be generated by HID attacks, which means that the USB device is of risk. On the contrary, we see it user keystroke.

The challenges of our mechanism are as follows: Firstly, how to select text features to build an efficient HID-KTFS; Secondly, how to design the scoring rules so that the scoring results are clearly distinguishable; Thirdly, how to configure parameters T and L to get a higher accuracy.

3.2 Design of Scoring Algorithm

In this subsection, we describe the design of our scoring algorithm in detail. The panorama of the detection is shown in the Fig. 3.

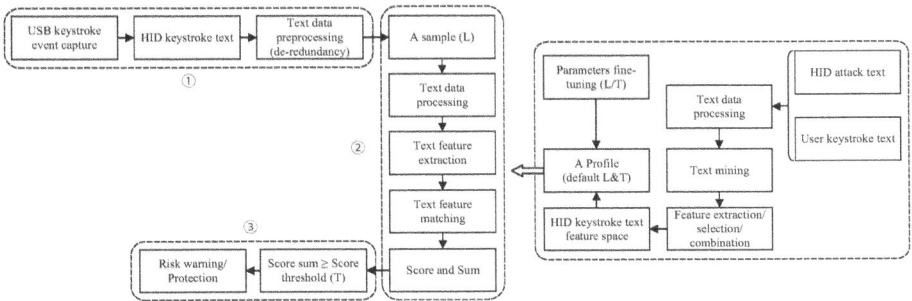

Fig. 3. Detection design of scoring algorithm against HID attacks.

A profile, as prior knowledge, should be originally built on the HID-KTFS. First, we process HID attack text and user keystroke text to extract various text features by text mining skills. Second, we select features that are significantly different in the two categories of text. Third, these text features are combined to construct the static HID-KTFS. In the profile, L and T are set to default values and can also be fine-tuned by user.

The scoring algorithm judges whether a text segment has the risk of HID attacks according to its overall performance in the HID-KTFS. It mainly contains three steps: ① First, capture the keystroke event and obtain the keystroke text. ② Second, preprocess the text and intercept a text segment as a detection sample. Within the HID-KTFS, extract the text features from the sample, and score each feature and calculate the sum. ③ Last, when the total score is greater than the score threshold, it will be inferred that an attack risk exists. Once a possible attack is detected, a risk warning or protective measures will be triggered.

3.3   HID Keystroke Text Feature Space

In this subsection, a HID-KTFS is constructed. Text mining skills are adopted to extract potential features that can be used for the detection task. Some features are directly extracted from the original text, while others are extracted from the preprocessed text. The text preprocessing mainly includes text cleaning (removing noise, punctuation marks, stop words, etc.), word segmentation, lemmatization, etc. By comparative analysis of keystroke text features, there are six kinds of features selected as shown in Fig. 4. They are significantly different between HID attack text segments and user keystroke text segments.

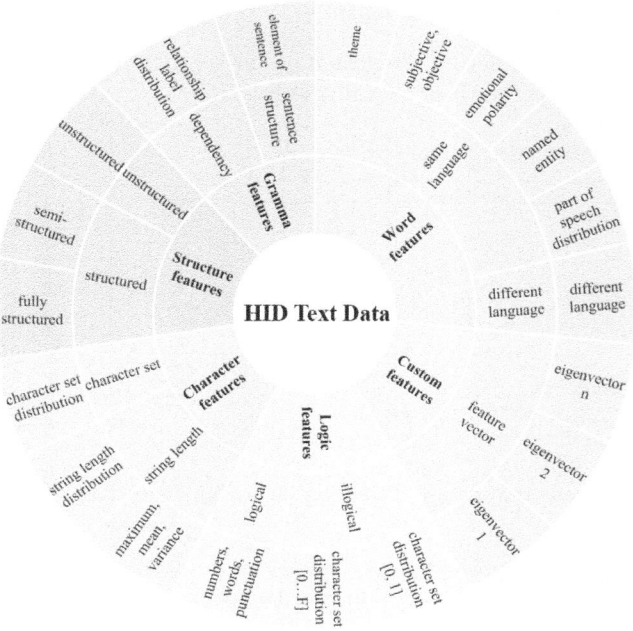

Fig. 4. Keystroke text features in the HID-KTFS.

Structure Features. User keystroke text is always unstructured, but the HID attack text may have some structured or semi-structured text in order to complete its malicious attack purpose. In addition, the HID attack text may also contain some malicious program blocks, which include preprocessing instructions, programming language keywords, runtime instructions. We take whether the keystroke text has a certain format or organizational pattern and code blocks as structure features.

Character Features. Longer strings appear frequently in the HID attack text and are tightly organized. On the contrary, the words are always separated by a space symbol when the user enters a sentence, and a long character string typed continuously is few. Secondly, there are some differences in the character distribution, the most notable is the wide distribution of special characters in HID attack text. We define the mean, variance, maximum of continuous string lengths and the character distribution as character features.

Logic Features. Generally, the user keystroke text is readable, and the basic elements are words, numbers, punctuation marks, etc. But the HID attack text is composed of attack commands and other payloads, and basic computer expertise is required to read. Additionally, there is an unusual HID attack method that injects forged data or Trojan files into the victim computer. Generally, such keystroke text is very long and the characters are concentrated. For example, if the data size is 1 KB, injecting it in binary would result in a keystroke text length of 8K with a character set of $[0, 1]$. Conversely, if injected in hexadecimal, the text length would be 2K, with the character set of $[0, \ldots, F]$. The integrity of forged data or Trojan files requires the correctness and continuity of attack text. But the user's keystroke text always has some typing errors, and its character distribution is richer. These features are combined into logic features.

Word Features. First, the user keystroke text is always composed of the vocabulary of the user's native language, but the words of the HID attack text are often consistent with the language of the victim computer. Because the attack needs to interact with the operating system and call the API to assist the malicious actions. Second, there are also obvious differences in the part-of-speech distribution. Although the HID attack text is dominated by verbs and nouns, adverbs and pronouns rarely appear, modal verbs and interjections hardly appear. However, the vocabulary in user keystroke text is very rich, and its part-of-speech distribution is wider. Third, the HID attack text presents an attack process with impersonal expression, which is objective. But the user keystroke text is often mixed with more personal opinions or judgments, which is subjective. Fourth, HID attack text often has no obvious emotional statements, and words that can express emotional polarity are rare. But the user keystroke text is often positive or negative emotional expression. Fifth, the named entity set in HID attack text is a subset of the named entity set, but the statistical result shows that the probability of hitting them in user keystroke text is very low. Lastly, the themes of user keystroke text are rich, involving various fields such as economy, culture, and politics, while these of HID attack text only revolve around the limited number of attack purpose.

Grammar Features. The grammatical relationships between words in a text segment are identified. By counting the number of all dependency tags, we obtain the distribution that is significantly different in both text categories. For example, the proportion of the tag 'compound' is very high in HID attack text segment, and its number is generally higher than the sum of the second and the third. However, there are always not many 'compound' tags in a user keystroke text segment, and its number is usually one of the last fewer ones. In addition, from the perspective of sentence structure, a HID attack text segment is always composed of many short sentences, and most of them are verb-object phrases. However, the components of a text segment in user keystroke data are diverse, typically featuring complete structures with subjects, predicates, and objects. Additionally, they often include noun clauses, definite clauses, or adverbial clauses.

Custom Features. Lastly, through manual observation of keystroke text, we find that some text fragments with dozens of characters frequently appear in HID attack text, but hardly appear in user keystroke text. Such as data encoding, changing the registry, network connection, sending and receiving emails, calling system API, etc. For example, to configure Windows Defender through *"HKLM:\SOFTWARE\Policies\Microsoft\Windows Defender\..."*, or to send an email by API *"Net.Mail.SmtpClient..."*. We choose these text fragments and define them as text eigenvectors.

The selected features are used to jointly construct a HID-KTFS. It is a static text feature set as shown in Fig. 4. Based on this, a profile will be generated for our scoring algorithm. Each selected text feature is based on statistical comparison results, so it will ensure the generalization performance of the algorithm.

3.4   Scoring Principles

The details of the main scoring principles for features in the HID-KTFS are described in Table 1. The P1 is the basis for making the scoring results differentiated. The P2 and P3 further make the scoring results more accurate.

3.5   Parameters Setting

The details of default setting and fine-tuning for the detection sample length and the score threshold are described below.

Table 1. The main scoring principles of the scoring algorithm.

No	Principle	Description
P1	Bonus Points or Minus Points	For the unique features of HID attack text, bonus points strategy is adopted for scoring. Conversely, for the unique features of user keystroke text, minus points strategy is adopted.
P2	Scoring Value of A Feature	The scoring value of each feature represents the weight of the feature in the HID-KTFS. The score of a single feature must be referenced to the scores of other features. When the probability of occurrence of a text feature is higher in one kind of text and is lower in another, the absolute value of the score will be larger.
P3	Ladder Scoring	For a quantifiable text feature, a staged scoring strategy is adopted. The larger the absolute value of the eigenvalue is, the bigger the absolute value of the score is.

Hyperparameter L. The larger L is, the more features a text sample may have and the higher sensitivity of the algorithm is. But, L should not be too large. If the length of the keystroke text generated by a complete attack is less than the length of a sample, it may cause a delay of the detection result. This will lead to a loss of real-time performance and missed protection opportunities. Conversely, it is worth noting that the L should not be set too small, because the amount of text feature that can be carried in a short text segment is fewer, which is not conducive to the detection task. In subsequent experiments, L is set based on the statistical results of the length of HID attack keystroke text.

Hyperparameter T. It is proportional to the absolute value of a single feature score. When L is set to a constant, the smaller T is, the higher the sensitivity of the algorithm is. In subsequent experiments, we set T according to the intersection point of probability density curves of total scores.

Parameters Fine-Tuning. There will be a default setting for L and T in the profile, but they can be fine-tuned by user. Their values should be positively correlated, and the smaller values lead to a more sensitive algorithm. Parameter fine-tuning is to cope with complex application scenarios and diverse HID attack methods.

4   Experiments and Results

In this section, a HID keystroke text dataset is constructed and a set of validation experiments are performed. We also make a preliminary assessment of the verification results.

4.1   Keystroke Text Dataset

Since there are no public keystroke text datasets for our study, the dataset is constructed first. We extensively collect HID attack scripts from BadUSB [3], USB Rubber Ducky [4], WHID [15], etc., and construct a HID attack payload library. The user keystroke text is captured from 10 users' typing. They are graduate students, company employees, journalists, etc. The typing content includes interactions in various scenarios such as work, study, and entertainment, which makes the dataset representative. Keystroke text samples are used to conduct confrontation tests, including ordinary user text, developer user text and HID attack text. Developer users are tested separately because of the particularity of their work.

Part of the attack text source from USB Rubber Ducky is presented in Table 2. It covers a variety of HID attack methods which will ensure the generalization of the algorithm.

Table 2. The composition of attack text from USB Rubber Ducky.

Attack Methods	Attack Scripts	Victim OS
Credentials	12	Linux/Windows
Execution	15	Linux/Windows
Exfiltration	38	Linux/Windows
Mobile	5	Android/iOS
Remote Access	16	Linux/Windows/Android
Prank	46	Windows/Linux/Mac
General	23	Windows/Linux

4.2   Experimental Results

According to statistics on attack scripts, taking a value for L within the range of [60, 160] effectively encompasses the text length typically encountered in HID attacks. So, in experiments, L is set as 64, 96, 128, and 160 respectively, and the probability density curves of the total score are obtained as shown in Fig. 5.

In the HID-KTFS, there are more features that bonus points and fewer features that minus points, which results in the distribution of the total scores. It can be seen that the scoring algorithm can well distinguish the HID attack text from the user keystroke text.

To balance False Accept Rate (FAR) and False Rejection Rate (FRR), the score thresholds are set to the curve intersections of HID attack and ordinary user in Fig. 5. The experimental results of the algorithm are shown in Tables 3, 4, 5 and 6. The accuracies on all samples are 0.918, 0.930, 0.949, and 0.956. The larger L is, the higher the overall accuracy is.

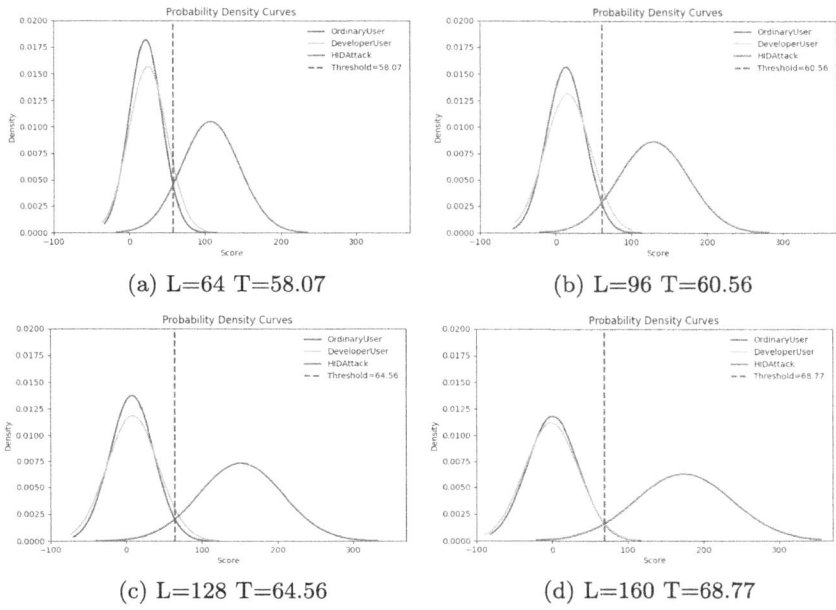

Fig. 5. Probability density curves of the scoring results (The left are the Ordinary User with the higher peak and the Developer User; the right is the HID Attack).

Table 3. The experimental verification results ($L = 64$, $T = 58.07$)

Predicted	Actual		
	HID Attack	*Ordinary User*	*Developer User*
HID Attack	1273	136	72
User	104	1696	511
Accuracy	0.924	0.926	0.877
		0.918	

Table 4. The experimental verification results ($L = 96$, $T = 60.56$)

Predicted	Actual		
	HID Attack	*Ordinary User*	*Developer User*
HID Attack	873	85	46
User	44	1136	343
Accuracy	0.952	0.930	0.882
		0.930	

Table 5. The experimental verification results (L = 128, T = 62.56)

Predicted	Actual		
	HID Attack	*Ordinary User*	*Developer User*
HID Attack	845	55	32
User	36	1117	341
Accuracy	0.959	0.953	0.914
		0.949	

Table 6. The experimental verification results (L = 160, T = 68.77)

Predicted	Actual		
	HID Attack	*Ordinary User*	*Developer User*
HID Attack	660	42	14
User	28	873	277
Accuracy	0.959	0.954	0.952
		0.956	

We also find that no matter what the value of the score threshold is, the precision of ordinary users is slightly higher than that of developer users, which is in line with acts. Because developer users are often engaged in software code programming, and frequently interact with the computer by commands. This increases the probability that the developer user text can match the features of HID attack text.

The evaluation metrics (L = 160, T = 68.77) are shown in Table 7, FAR is 0.0457 and FRR is 0.0407. The experimental results verify that the scoring algorithm has a good detection effect on known HID attacks. We preset L = 160 and T = 68.77 by default in the profile, and the detection of complex scenarios or possible unknown HID attacks may require proper fine-tuning of the parameters to improve the detection sensitivity.

Table 7. The evaluation metrics of our scoring algorithm (L = 160, T = 68.77)

Accuracy	Precision	Recall	F1-Score	FAR	FRR
0.9556	0.9218	0.9593	0.9402	0.0457	0.0407

5   Evaluation

In this section, we conduct practical attack detection and user input experiments to evaluate the performance of our scoring algorithm. There will yield comparative experimental results between representative detection mechanisms and a generalization of the scoring algorithm against different HID attacks.

5.1 Experimental Setup

The attack detection experiment employs a host of Windows 11 as the victim, utilizing the USB Rubber Ducky (DuckyScript 3.0), BadUSB (a U disk with tampered firmware) and Teensy (a Teensy 4.1 development board) as the HID attack tools shown in Fig. 6. For USB Rubber Ducky, it involves the extraction of 100 attack scripts, each of which is configured to enable Jitter. Jitter is a feature which varies the cadence, or delay, between individual key presses. Each deployment of a jitter-enabled payload will produce different results. The attack injection file is compiled by PayloadStudio maintained by Hak5, and copied to the storage of USB Rubber Ducky. BadUSB and Teensy respectively extract their own 100 scripts compiled and burned in Arduino. Dedicated delays are also added for both of them. Overall, the HID attack text will be typed continuously with a modulated delay between each imitated keystrokes, which allows the HID attacks to mimic the behavioral features of a human user's keystrokes.

Fig. 6. The experimental attack tools (From left to right: USB Rubber Ducky, BadUSB and Teensy).

The user input test experiment includes another 10 participants, comprising 5 ordinary users and 5 developer users. The input content is derived from the context of human-computer interaction in their professional and educational activities.

There are 5 tests as shown in Table 8. DuckHunt (DDT = 20 ms), WHID Defense and our scoring algorithm (default L = 160, T = 68.77) are deployed respectively for attack detection and user identification. DuckHunt focuses on keystroke timestamps, and WHID Defense focuses on keystroke text. T1 and T2 serve as experimental comparison benchmarks for evaluating the performance of T3. T4 and T5 are used for further test of the generalization performance.

Table 8. The experimental setting.

Test	Host	Installed Decision	Attack Tool	User Input
T1	PC-A	DuckHunt [18]	USB Rubber Ducky	Users
T2	PC-B	WHID Defense [15]	USB Rubber Ducky	Users
T3	PC-C	Scoring Algorithm	USB Rubber Ducky	Users
T4	PC-C	Scoring Algorithm	BadUSB	——
T5	PC-C	Scoring Algorithm	Teensy	——

5.2 Test Results

For 100 attack scripts, each script is executed 10 injection attacks. Additionally, 10 users each perform 100 consecutive keyboard inputs. That results in a total of 1000 HID attacks and 1000 normal user inputs for a test. The attack detection and user identification are carried out with the statistical number of correct inferences presented in Table 9. Overall-1 is a overall number of attack detection and user identification for every detection mechanism, while Overall-2 is the overall number on these three attack tools for our mechanism.

Table 9. The experimental results in deployment tests.

Test	Attack Detection[a]	User Identification[a]	Overall-1[a]	Overall-2[a]
T1	541	996	1537	——
T2	880	923	1803	——
T3	970	958	1928	2910
T4	980	——	——	
T5	960	——	——	

[a] The statistical base numbers of tests are 1000, 2000 and 3000 respectively.

Firstly, in comparing T1, T2, and T3, the accuracies of the detection mechanism can be clearly seen in Fig. 7. Our detection mechanism achieves the highest attack detection accuracy as well as Overall-1 accuracy. DuckHunt has the highest user identification accuracy because of that the DDT threshold is set to 20 ms which is generally difficult for users to reach at normal keystroke speed. But it has the lowest attack detection accuracy.

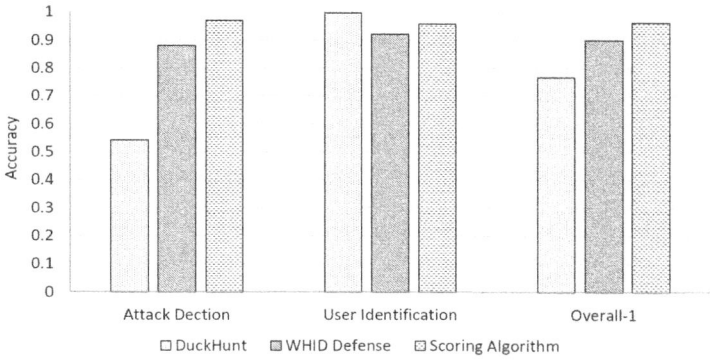

Fig. 7. Comparison of experimental results of representative detection mechanisms (T1, T2 and T3).

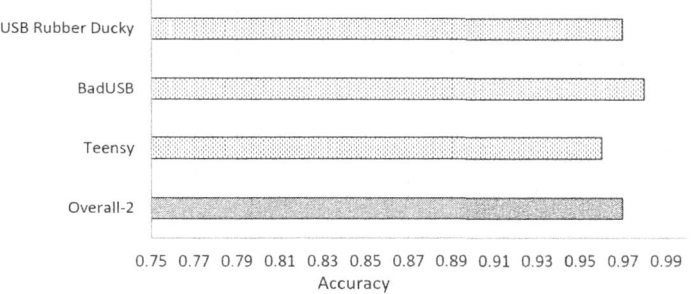

Fig. 8. Comparison of experimental results of different attack tools (T3, T4 and T5).

Secondly, in comparing T3, T4, and T5, the detection accuracy of our scoring algorithm is represented in Fig. 8. Our detection mechanism has effectiveness against various attack tools, yielding balanced results and an accuracy of 97% for Overall-2 that meets expectations.

These two comparison results demonstrate the superior performance of our scoring algorithm.

6   Analysis

6.1   Analysis of Detection Mechanisms

In this subsection, we analysis the representative detection mechanisms in Table 10. Firstly, the victim's keystroke behavior features could be imitated through technologies such as deliberate delays, enabled Jitter or the future AI modules. Secondly, for HID attack methods without malicious component implantation, there is no difference in side-channel analysis. Since none of the three attack tools in the evaluation experiment are hardware-implanted attacks, this mechanism was not used as the experimental benchmark. Thirdly, the keystroke text keywords could also be bypassed in other ways, and there are doubts about the detection effect of keyword matching on unknown HID attacks.

So, DuckHunt and WHID Defense did not perform well in the experiments, and theoretically Malboard cannot play a detection role. But, our scoring algorithm based on keystroke text feature matching is difficult to break through for that the HID attack text cannot avoid the HID-KTFS.

Table 10. A brief comparison of the representative detection mechanisms.

Mechanism	Method	Detection Basics	Source
DuckHunt [18]	Keystroke behavior features	Abnormal keystroke speed/DDT	Keystroke timestamps
Malboard [16]	Side-channel analysis	Higher power consumption/Vocal keystroke delay	Power consumption/Keystroke' sound
WHID Defense [15]	Keystroke text keyword matching	Launched by Command Prompt/Run Box	Keystroke text
Our Scoring Algorithm	Keystroke text feature matching	HID-KTFS	Keystroke text

6.2   Comparison of Mechanisms on Text

Both WHID Defense and our scoring algorithm are keystroke text-based detection mechanisms. So, we make a more detailed comparison in Table 11. For WHID Defense, the detection keywords are also typed occasionally by normal users, especially the key combination "Win+r" is frequently used by developer users. This will result in a lower accuracy of user identification. And the attacks that open a shell by "Win+x"+"i" or "Win"+"powershell" cannot be detected. This mechanism can only work for Windows.

But the HID-KTFS is result of statistical analysis of HID keystroke text, so our scoring algorithm is not susceptible evolved attacks. Therefore it has better robustness against varied HID attacks. In addition, it has good portability and can work as a lightweight software detection tool on hosts of Windows, Linux or Mac OS.

Table 11. Comparison of the detection mechanisms on keystroke text.

Mechanism	Detailed Basics	Results	System
WHID Defense [15]	"Win"+"cmd"/ "Win"+"run"/ "Win+r"	Worse user identification/ attack detection	Windows
Our Scoring Algorithm	Statistical features of text differences	Balanced evaluation metrics in Table 7 and Fig. 7	Windows/Linux/ Mac OS

6.3   Limitation of Scoring Algorithm

In verification and evaluation experiments, our detection mechanism achieved better performance, but it also has a limitation. In analyzing the samples where the algorithm inferred incorrectly, we found that the text fragments possess few features in the HID-KTFS. They do not have precise security properties and may appear in both HID attack text and user text. Our detection algorithm is incompetent, and it is not easy or even impossible to have correct detection results by professional researchers. So it might have to audit through user judgment whether the text is his or her own typing. But fortunately, the proportion of such samples is low, and it does not impact the overall accuracy of the scoring algorithm.

7   Conclusion

In this paper, we propose a scoring algorithm based on static text feature matching to achieve the detection of keystroke injection attacks. It has a strong capability to deal with various HID attack methods. In the future, a richer HID-KTFS and more detailed scoring rules will help improving detection accuracy. And the deployment combined with methods based on significant keystroke behavior features such as keystroke speed variations, and user-specific typing patterns will further enhance the detection capabilities. Our study applies keystroke text only for attack detection and there will be no in-depth analysis, because it may involve legal and ethical issues.

But there is a newly emerging and unconventional HID attack that exfiltrates data from a Trojan lurking on the victim host to a malicious HID device with custom HID reports. There is a covert HID interface used as a data backdoor. It does not generate keystroke data, and requires new methods for detection in the future.

Acknowledgement. We would like to thank the anonymous reviewers for their constructive comments. This work was partially supported by the Program of Key Laboratory of Network Assessment Technology, the Chinese Academy of Sciences, the Program of Beijing Key Laboratory of Network Security and Protection Technology, the National Key Research and Development Program of China (No. 2021YFB2910109), and the National Natural Science Foundation of China (No. 62202465).

References

1. Device class definition for Human Interface Devices (HID) 1.11, USB IF Std. (2020). https://www.usb.org/sites/default/files/hid1_11.pdf
2. Nissim, N., Yahalom, R., Elovici, Y.: USB-based attacks. Comput. Secur. **70**, 675–688 (2017)
3. Nohl, K., Lell, J.: BadUSB-on accessories that turn evil. Black Hat USA **1**(9), 1–22 (2014)
4. Hak5. USB rubber ducky. https://shop.hak5.org/products/usb-rubber-ducky. Accessed 14 June 2023
5. PJRC. Teensy USB development board. https://www.pjrc.com/teensy/. Accessed 22 June 2023
6. Fu, J., Huang, J., Zhang, L.: Curtain: keep your hosts away from USB attacks. In: Nguyen, P., Zhou, J. (eds.) Information Security, ISC 2017. LNCS, vol. 10599. Springer, Cham (2017). https://doi.org/10.1007/978-3-319-69659-1_25
7. Tian, D.J., Bates, A., Butler, K.: Defending against malicious USB firmware with GoodUSB. In: Proceedings of the 31st Annual Computer Security Applications Conference, pp. 261–270 (2015)
8. GitHub. USBGuard. https://github.com/USBGuard/usbguard. Accessed 14 June 2023
9. Loe, E.L., Hsiao, H.-C., Kim, T.H.-J., Lee, S.-C., Cheng, S.-M., SandUSB: an installation-free sandbox for USB peripherals. In: IEEE 3rd World Forum on Internet of Things (WF-IoT), pp. 621–626. IEEE (2016)
10. Jiang, J., Chang, Z., Lü, Z., Zhang, N.: Research on USB hid attack detection technology. Chin. J. Comput. **42**(5), 1018–1030 (2019)
11. Griscioli, F., Pizzonia, M.: USBCaptchain: preventing (un)conventional attacks from promiscuously used USB devices in industrial control systems. J. Comput. Secur. **29**(1), 51–76 (2021)
12. Griscioli, F., Pizzonia, M., Sacchetti, M.: USBCheckIn: preventing BadUSB attacks by forcing human-device interaction. In: 2016 14th Annual Conference on Privacy, Security and Trust (PST), pp. 493–496. IEEE (2016)
13. Neuner, S., Voyiatzis, A.G., Fotopoulos, S., Mulliner, C., Weippl, E.R.: USBlock: blocking USB-based keypress injection attacks. In: Kerschbaum, F., Paraboschi, S. (eds.) DBSec 2018. LNCS, vol. 10980, pp. 278–295. Springer, Cham (2018). https://doi.org/10.1007/978-3-319-95729-6_18
14. Kang, M., Saiedian, H.: USBWall: a novel security mechanism to protect against maliciously reprogrammed USB devices. Inf. Secur. J. A Global Perspect. **26**(4), 166–185 (2017)
15. Lv, Z., Xue, Y., Zhang, N., Feng, Z., Jin, Z.: Whid defense: detection and protection technology for USB hid attack. J. Cyber Secur. **6**, 110–128 (2021)
16. Farhi, N., Nissim, N., Elovici, Y.: Malboard: a novel user keystroke impersonation attack and trusted detection framework based on side-channel analysis. Comput. Secur. **85**, 240–269 (2019)
17. Wang, Z., Johnson, R., Stavrou, A.: Attestation & authentication for USB communications. In: 2012 IEEE Sixth International Conference on Software Security and Reliability Companion, pp. 43–44. IEEE (2012)
18. GitHub. duckhunt:prevent rubberducky attacks. https://github.com/pmsosa/duckhunt. Accessed 14 June 2023
19. Keelog. USB-keylogger. https://www.keelog.com/usb-keylogger/. Accessed 18 June 2023

Identifying Ransomware Functions Through Microarchitectural Side-Channel Analysis

Connor Startzel, Dane Brown(ID), T. Owens Walker III(ID),
and Jennie E. Hill$^{(\boxtimes)}$(ID)

United States Naval Academy, Annapolis, MD 21412, USA
jehill@usna.edu

Abstract. Ransomware continues to be an effective and lucrative means to extort large sums of money from organizations which depend on reliable access to data to meet their objectives. Often, the public embarrassment associated with falling victim to a ransomware attack is an effective motivator to pay and avoid the damaging headlines. The goal of this research is to improve the ability of security teams to identify ransomware samples in real-time. This type of determination has traditionally been difficult as attackers have been able to make small changes to malware to create variants which still infect systems but are not recognized by existing analysis tools. The work presented here uniquely leverages leaked source code which has been used by malicious actors to create custom ransomware variants for real ransomware campaigns. By isolating individual options for a build, we are able to determine ground truth and fingerprint its behavior using hardware-based side channel data collected by a CPU performance monitoring tool. This approach allowed us to successfully identify the Hardware Performance Counters and associated function calls which are directly correlated with specific capabilities inherent to the prolific Lockbit 3.0 ransomware. The specific function calls identified were those used to terminate Windows Defender protection and to search for shared network resources to encrypt. The methodology presented will help analysts detect ransomware samples for which there is no existing signature and narrow the scope of follow-on analysis, thus saving valuable time.

Keywords: Ransomware · Ransomware Builders · Hardware Performance Counters · Micro-architectural Side-channel · Reverse Engineering

1 Introduction

In the summer of 2023, MGM casino in Las Vegas, Nevada was digitally taken hostage through the use of ransomware [1], severely disrupting business operations and revenue. Beginning with the social engineering of a help desk employee,

J. Zhao and W. Meng (Eds.): SciSec 2024, LNCS 15441, pp. 19–36, 2025.
https://doi.org/10.1007/978-981-96-2417-1_2

the ALPHV ransomware group was able to gain valid credentials within the MGM network and use that access to install their BlackCat ransomware on the network [2]. This caused significant disruptions of various hotel and casino operations and is expected to cost the company an estimated $100 million U.S. dollars [3]. Unfortunately, this kind of incident is becoming far too common. Cases of ransomware and the associated losses have been on an exponential rise over recent years and are expected to cause total damages of $265 billion by 2031 [4].

The lucrative nature of this illicit activity presents a model where malicious actors can make incredible profits while taking relatively little risk. To make matters worse, criminals with little to no cybersecurity expertise are able to carry out these attacks by using Ransomware-as-a-Service (RaaS). Under this model, experienced hackers write software that is capable of creating custom ransomware builds which a client can deploy in a victim environment [5]. One notorious vendor of RaaS is the Lockbit group. Their platform has been used to attack Continental Automotive, UK Royal Mail, Foxsemicon, Taiwan Semiconductor, and the city of Oakland, California [6–8].

The frequency of these attacks as well as the disruption they cause has drawn concerted attention from legal authorities. In February of 2024, the group behind the Lockbit ransomware family was shut down through a stealthy, coordinated effort between American and British law enforcement agencies [9–11]. The research presented here adds another tool which system defenders and law enforcement can use in their battles against unrelenting ransomware attacks. Our case study focuses on the Lockbit ransomware building kit, which is one of several such kits which have been leaked on the public internet by a client, whether intentionally or unintentionally [12]. By using the Lockbit builder, we are able to enable and disable various features to assess their exact impact on the resultant custom variant. This allows us to find the similarities and differences between various builds of Lockbit and identify them based on runtime behaviour, rather than static signatures.

This work leverages a ransomware builder to create custom ransomware samples that incrementally isolate key ransomware features. These features can then be correlated to specific hardware performance counter events. Here, we compare the version of Lockbit with all features enabled against variants where we have selectively disabled two features: terminating (or killing) Windows Defender and searching for network shares. We submit that this process can be utilized to identify an Hardware Performance Counter (HPC) based runtime signature for any desired ransomware feature.

To this effect, this paper makes the following contributions:

1. A method for using ransomware builders to isolate and analyze key ransomware features, and
2. A method for using hardware performance counters to correlate microarchitectural fingerprints to specific functions in a ransomware build.

These contributions lay the groundwork for the implementation of efficient and accurate ransomware analysis processes.

The remainder of this work is organized as follows. Terms and concepts which are relevant to the presented concepts are discussed in Sect. 2. Previous work that provided a foundation for this research is discussed in Sect. 3. The ransomware variant building procedure is found in Sect. 4. Section 5 discusses the method for analyzing our custom variants using HPCs. The results and conclusion of this study are then detailed in Sect. 6.

2   Background

Reverse engineering various ransomware samples which have been made available to the public has shown that these ransomware variants retain many similar features. Recent works have focused on using HPCs for ransomware detection. This works because HPCs are special-purpose registers built into processors that store the counts of software and hardware related events, such as cache misses and instructions committed [13]. These values are used to create a library of HPC counts collected for known ransomware, which can then be used to compare the run-time HPCs to the library of known ransomware HPCs.

It is evident that ransomware is a serious threat to both individuals and organizations which place value on their data. Ransomware was seen as early as 1989 with the AIDS Trojan which was distributed via floppy disks. It encrypted filenames on the victim's computer, demanding a ransom to be sent to a P.O. Box in Panama for the decryption key. By the early 2000s, ransomware attacks were still infrequent and used simple symmetric encryption algorithms, demanding payment via conventional methods like wire transfer or prepaid cards. The modern era of ransomware began in 2012 with the emergence of CryptoLocker. This ransomware used robust asymmetric encryption and demanded payment via Bitcoin, which provided a degree of anonymity to the attackers. CryptoLocker marked a significant shift in the ransomware landscape, demonstrating the potential for significant financial gain with a reduced risk of facing legal consequences [14].

In recent years, ransomware attacks have become more targeted and devastating. High-profile incidents include the attacks on the Colonial Pipeline (2021) and JBS Foods (2021), which highlighted the vulnerabilities in critical infrastructure and supply chains. The LockBit ransomware group, known for its RaaS model, has been involved in several high-profile attacks, demonstrating the effectiveness and reach of RaaS platforms. The rise of double extortion, where attackers steal sensitive data before encrypting it and threaten to release the data publicly if the ransom is not paid, has added a new dimension to ransomware attacks.

Traditional antivirus, such as Windows Defender, attempt to detect threats through signatures and heuristics. Signatures are useful if the same code base is seen multiple times, however modern malware creators tend to use techniques to obfuscate commonalities across samples. This may include using packers, encrypting code segments, staging the code, or changing functionality between variants as done with ransomware builders. Heuristic analysis may perform better against modern malware as it examines the actual behavior. However, there is quite a bit of uncertainty associated with these heuristics as many malicious behaviors are also used in completely benign applications. For example, ransomware performs file encryption, but so would an email client when a user is trying to maintain confidentiality. This duality where behaviors are sometimes benign and other times malicious leads to many false positives and false negatives when using heuristic approaches. Trustworthy detection requires analysis of how those functions are used differently by benign and malicious programs. This is the analysis we perform in our research using HPCs.

3   Related Work

This paper presents novel methods for identifying the functions performed by ransomware which could be key to implementing effective detection. These methods build off of important works in recent years which have attempted to recognize and mitigate ransomware threats using a variety of techniques.

Some research has evaluated the use of on-board sensors to detect ransomware. Computers are manufactured with sensors that receive and report information about temperature, power, light, battery, fans, and more. Careful analysis of these sensor reports can indicate ransomware activity [15]. These types of sensors are known as side channels, since they are indirectly affected by the executing malware. Additional research efforts have attempted to detect ransomware quickly based on these side channels while differentiating it from other systems loads [16,17]. There has even been research into detecting ransomware through monitoring processor and disk usage [18]. Others have explored the frequency of systems events, applying Fast Fourier Transforms and analyzing the ransomware through Deep Neural Networks [19]. Machine Learning continued to gain popularity for being able to classify ransomware activity using one-class and two-class models analyzing low-level system information [20].

Our work continues to bring these techniques to the forefront with a specific focus on using readily available HPC information to classify ransomware functionality. These counters are not monitoring a single process but the whole system, however it has been shown that HPCs are effective at verifying the expected functionality of a particular program [13]. A representative illustration of performance counter research utilized machine learning to analyze runtime HPC collections and then classify them as either ransomware or benign [21]. This classification was done utilizing machine learning models that were trained using labeled ransomware and non-ransomware files. Similarly, another approach is able to differentiate between applications like antivirus, file search, and file encryption applications with ransomware executables [22]. The RAPPER project introduced a two step detection framework that involved training an Artificial Neural Network on HPC events to prevent ransomware from performing disk encryption operations [23]. The RanStop project similarly used HPCs to observe microarchitectural event sets in order to detect known and unknown ransomware at run-time [24]. These projects made important contributions that showed researchers the viability of HPC analysis for ransomware classification and detection, however they mostly perform their analysis from the safety of a virtualized system which can leave doubt as to whether the techniques would successfully transfer to a standard, bare metal hardware setup. This was recently proved via the analysis of ransomware using HPC on a non-virtualized system [25].

A Virtual Machine (VM) environment is a software-based emulation of a computer system that runs an operating system and applications as if it were a physical computer. Each VM has its own virtualized hardware, including CPU, memory, storage, and network interfaces. They are incredibly useful during malware analysis as they provide isolation from the host system and other VM guests, snapshots of their state along with the ability to roll back to a snapshot following infection, and faithful replication of nearly any environment a researcher would require for testing [26,27]. Another option for safe isolation is a sandboxed network, which is a controlled and isolated network environment designed to safely execute and observe the behavior of software, particularly potentially malicious code. This environment prevents the software from spreading to other systems or networks. A sandbox network also provides controlled interactions for monitoring network traffic and system calls, as well as the ability to simulate realistic environments implementing any network topology or configuration that may not be implemented in a virtual environment [28]. This research makes exclusive use of these isolation strategies in order to responsibly contain any malicious code from attempting to spread and to ensure experiments could be reproduced starting from a trustworthy, clean system.

In this paper, we explore ransomware builders which enable malicious actors to create ransomware that run undetected by deployed malware signatures since they customize select features in the ransomware builds. A ransomware builder provides a user-friendly interface to allow desired ransomware features to be selected or deselected and also allows a custom ransom message to be shown to the eventual victim(s). Once these customizations are selected, the builder will compile a unique variant of the ransomware which will have no known detection signature. Such a builder is highly valuable to a threat actor who does not have the skill or desire to create ransomware that can bypass detection [29,30]. Our group has acquired actual builders of ransomware that have been used in real-world campaigns. They were leaked and then archived an online malware research repository named `vx-underground` [31–33]; some may have been purposely leaked by threat actors who purchased the builder, while others may have been unintentionally leaked or stolen. Either way, these samples set our work apart from theoretical analysis of what ransomware might do and allows us to establish ground truth based on what deployed ransomware is actually doing.

The research presented in this study is unique in that it examines incremental changes among ransomware variants and correlates specific functions in the ransomware to specific hardware performance counters. Since ransomware builders make it trivial to bypass signatures for a particular ransomware build, it is important to develop methods that looks at specific, suspicious behaviors being performed. Several projects have attempted to analyze the types of behaviors exhibited by ransomware, either explicitly [34,35] or through machine learning [36,37]. This paper will focus on analyzing which HPCs have a significant change as a result of certain functionality being enabled or disabled in a particular build of ransomware. The performance monitoring software, VTune, additionally provides insight regarding which procedure in the code was executing when that counter was recorded. This is invaluable insight as it can drastically decrease the time and effort required by a malware analyst to triage the code; the given function address directly correlates the HPCs of interest with the code that performs the specific functionality of that variant.

At this point, the analyst can open the ransomware sample with Reverse Engineering tools in a properly contained environment. One of the most popular free Reverse Engineering tools is Ghidra, a disassembler and decompiler which was open-sourced by the United States National Security Agency in 2019 [38]. Contemporary work in ransomware Reverse Engineering has focused on using the structure of previously studied ransomware samples to interpolate the structure of new ransomware [39]. This paper takes a different approach and explains how hardware performance counters can be used to isolate the specific functions which implement features of interest in new Ransomware samples, thus significantly reducing the scope and time required to perform static code analysis on a malicious application.

4   Ransomware Builder Study

This section describes the first of two studies leveraging the microarchitectural side-channel to analyze function-level behavior of custom ransomware. This first software-based study explores the use of ransomware builders to create custom ransomware in a virtual environment, while the second study (Sect. 5) is a proof-of-concept hardware-based collection and analysis of several variants of the popular LockBit ransomware in a non-virtualized environment. The test apparatus for this study of ransomware builders was a sandboxed Linux virtual environment running a Windows VM. This first study consisted of a ransomware builder exploration (Sect. 4.1), ransomware variant creation (Sect. 4.2), ransomware process analysis (Sect. 4.3), and ransomware function analysis (Sect. 4.4), as described below.

4.1   Ransomware Builders

Ransomware builders were acquired from vx-underground and installed in a Windows VM for six different ransomware families: Yashma, ChaosE, ChaosD, LockBit 3.0, Hakunana Matata, and Thanos. Yashma, ChaosE, and ChaosD are related ransomware families, with Yashma being the newest iteration and ChaosD being the oldest iteration. Of the installed builders, all were fully functional except for Thanos, which we found to be corrupted.

Figure 1 shows the Yashma ransomware builder. It includes options for different features - such as disabling windows task manager or leaving a unique message on the wall paper - as well as the ability to leave a ransom note from the malicious actor on the victim's device. These features provide the ability to tailor the ransomware for a specific attack in order to evade detection services. Other benefits of a builder include creating a variant that is less likely to attributed to a specific group. Of note, these builders often include the ability to target features not commonly utilized, such as deleting shadow copies or disabling windows recovery mode. In this research, ransomware builders were leveraged to create variants with a high degree of control over variables so that slight changes in the ransomware functionality could be examined. Ransomware variant creation is described in Sect. 4.2.

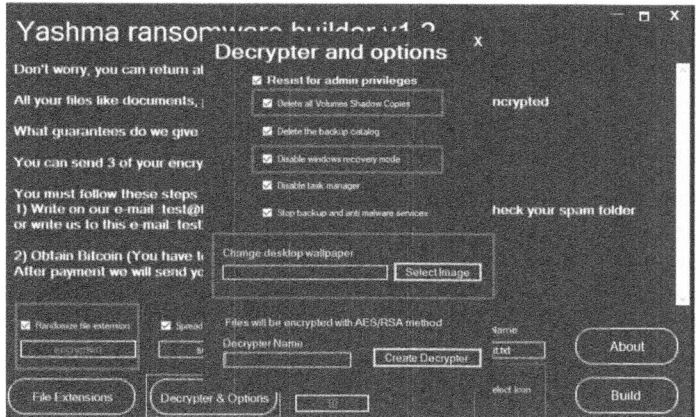

Fig. 1. Ransomware builder GUI

4.2   Ransomware Variant Creation

A single variant using the preset options was created for each ransomware to run a system check on each builder, followed by variants with all options disabled, and a single option disabled. Figure 2 shows the builder options in the left column while the names of the variants created are in the top row. A green check indicates the option was enabled, while a red X indicates options that were disabled in the respective variant.

Options of interest in all of the builders were identified and ransomware variants were created based on these options (and the related functions). Accordingly, ransomware variants were produced that were identical in all regards except for the inclusion or exclusion of a single, specific, targeted function. This enabled the collection of HPCs for only that one function by effectively subtracting the HPCs for the exclusion variant from those of the inclusion variant. All ransomware variants were executed on the test platform to ensure proper functionality.

Option \ Variant	Default Options	No Options	No Kill Defender	No Network Shares	No Self Destruct
impersonation	✓	✗	✓	✓	✓
local_disks	✓	✗	✓	✓	✓
network_shares	✓	✗	✓	✗	✓
kill_processes	✓	✗	✓	✓	✓
kill_services	✓	✗	✓	✓	✓
running_one	✓	✗	✓	✓	✓
print_note	✓	✗	✓	✓	✓
set_wallpaper	✓	✗	✓	✓	✓
set_icons	✓	✗	✓	✓	✓
self_destruct	✓	✗	✓	✓	✗
kill_defender	✓	✗	✗	✓	✓
gpo_netspread	✓	✗	✓	✓	✓
gpo_ps_update	✓	✗	✓	✓	✓
delete_eventlogs	✓	✗	✓	✓	✓

Fig. 2. Options within LockBit 3.0 builder (Color figure online)

4.3 Ransomware Process Analysis

Process monitor was used to identify and analyze the processes called during the execution of the individual ransomware variants. Figure 2 shows the results of a Process Monitor capture during the execution of the LockBit variant utilizing the default options within the builder. Of note, three operations seen in the figure were frequently called: (1) CreateFile, (2) QueryBasicInformation, and (3) CloseFile (Fig. 3).

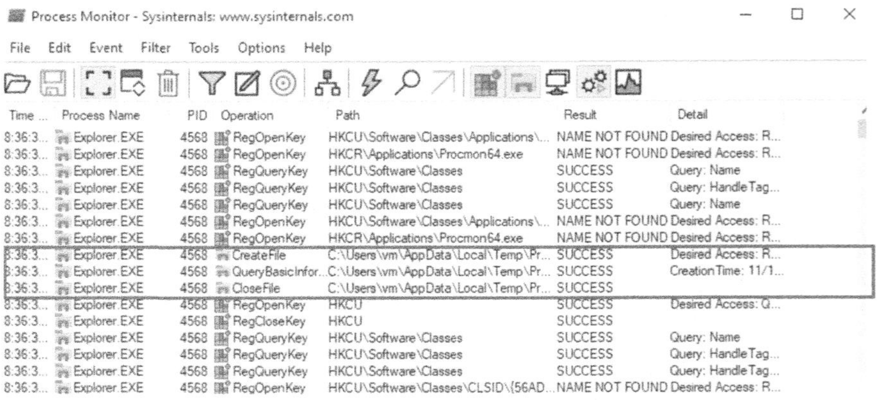
Process Monitor - Sysinternals: www.sysinternals.com

File Edit Event Filter Tools Options Help

Time ...	Process Name	PID	Operation	Path	Result	Detail
8:36:3...	Explorer.EXE	4568	RegOpenKey	HKCU\Software\Classes\Applications\...	NAME NOT FOUND	Desired Access: R...
8:36:3...	Explorer.EXE	4568	RegOpenKey	HKCR\Applications\Procmon64.exe	NAME NOT FOUND	Desired Access: R...
8:36:3...	Explorer.EXE	4568	RegQueryKey	HKCU\Software\Classes	SUCCESS	Query: Name
8:36:3...	Explorer.EXE	4568	RegQueryKey	HKCU\Software\Classes	SUCCESS	Query: HandleTag...
8:36:3...	Explorer.EXE	4568	RegQueryKey	HKCU\Software\Classes	SUCCESS	Query: Name
8:36:3...	Explorer.EXE	4568	RegOpenKey	HKCU\Software\Classes\Applications\...	NAME NOT FOUND	Desired Access: R...
8:36:3...	Explorer.EXE	4568	RegOpenKey	HKCR\Applications\Procmon64.exe	NAME NOT FOUND	Desired Access: R...
8:36:3...	Explorer.EXE	4568	CreateFile	C:\Users\vm\AppData\Local\Temp\Pr...	SUCCESS	Desired Access: R...
8:36:3...	Explorer.EXE	4568	QueryBasicInfor...	C:\Users\vm\AppData\Local\Temp\Pr...	SUCCESS	CreationTime: 11/1...
8:36:3...	Explorer.EXE	4568	CloseFile	C:\Users\vm\AppData\Local\Temp\Pr...	SUCCESS	
8:36:3...	Explorer.EXE	4568	RegOpenKey	HKCU	SUCCESS	Desired Access: Q...
8:36:3...	Explorer.EXE	4568	RegCloseKey	HKCU	SUCCESS	
8:36:3...	Explorer.EXE	4568	RegQueryKey	HKCU\Software\Classes	SUCCESS	Query: Name
8:36:3...	Explorer.EXE	4568	RegQueryKey	HKCU\Software\Classes	SUCCESS	Query: HandleTag...
8:36:3...	Explorer.EXE	4568	RegQueryKey	HKCU\Software\Classes	SUCCESS	Query: HandleTag...
8:36:3...	Explorer.EXE	4568	RegOpenKey	HKCU\Software\Classes\CLSID\{56AD...	NAME NOT FOUND	Desired Access: R...

Fig. 3. Process Monitor of LockBit

Process monitor revealed that in the first two minutes of execution, LockBit caused nearly 3 million operations to occur. In comparison, Ghidra, another resource-intensive but non-ransomware operation, only triggered 200 thousand operations over the same startup period. This demonstrated that HPC metrics could be useful to capture since ransomware variants require a high number of system operations during early stages of execution.

4.4   Ransomware Function Analysis

Figure 4 provides a grouping of common features identified across the ransomware builders. The leftmost column contains a description of the capabilities, while the top row contains the name of the ransomware. Blank spots indicate that no builder option was identified to implement the described capability.

Capability \ Ransomware	Lockbit	Yashma	ChaosE	ChaosD	Hakuna Matata
Network Spreads	✓	✓	✓	✓	✓
Set Wallpaper	✓	✓	✓	✓	✓
Dropped File	✓	✓	✓	✓	✓
Disable recovery mode	✓	✓	✓	✓	✓
Delete the backup catalog		✓	✓	✓	✓
Delete shadow copies		✓	✓	✓	✓
Disable task manager		✓	✓	✓	

Fig. 4. Family independent ransomware capabilities

These features are of note given their appearance across unrelated ransomware. This is important as it indicates that they are potential signs of ransomware and essential capabilities in an effective ransomware campaign. Disable recovery mode and delete shadow copies are both of note given that they are operations not commonly utilized and seem to occur before encryption. Additionally, while the set wallpaper feature occurs post encryption, it may still be useful in preventing network spreads.

The consistent appearance of these options means it can be reasonably assumed that any future ransomware created will include one or more of these capabilities. Thus, by linking the source code of these functions with the corresponding HPCs, a ransomware detector could be created that scans files for blocks of code previously tied to ransomware through HPCs.

As was shown in Fig. 4, there are certain functions found across different ransomware builders that accomplish the same result. One of these functions is the DisableTaskManager feature in Yashma. It is clear that these functions utilize uncommon capabilities and occur in ways that a standard user would not call them. It is illogical for a normal Windows user to disable their startup manager, making variants based on capabilities like this potentially good candidates for HPC capture.

5   Ransomware Variant Analysis Using Hardware Performance Counters

This section describes the proof-of-concept micro-architectural side-channel-based study of a single set of ransomware variants created in Sect. 4. For this study, variants of LockBit ransomware were selected due to its prevalence and popularity amongst ransomware gangs [40]. Section 5.1 describes the hardware set up and initial analysis is detailed in Sect. 5.2.

5.1   Hardware Test Apparatus

To enable accurate collection of HPCs, a hardware-based test apparatus is highly desirable [41]. The testbed computer is a Dell Precision 7920 Tower with Intel Xeon Silver 4208 8-core 2.10 GHz System and Cascade Lake microarchitecture with 32 GB DDR4 memory running Windows 10 Pro Build 10.0.19044.1586 Version 21H1 Experience Pack 120.2212.551.0. The system largely reflects an off-the-shelf configuration, with the exception of disabling Windows Defender by setting the "DisableAntiSpyware" flag to 1 through the Registry Editor, to allow ransomware to execute.

The microarchitectural side-channel is accessed through VTune Performance Analyzer 2022.3.0, which was used to collect a subset of 90 HPCs. The test setup and HPC selection used in this proof-of-concept study was derived from [25], with the variants of LockBit created in Sect. 4.2 added to the Python trial collection script and the hardware restore image. The selected LockBit variants are (1) LockBit (default), (2) LockBit No Kill Defender, and (3) LockBit No Network Shares and are described in more detail in Sect. 5.2.

Data Collection Procedure. A python test script was used to automate HPC collection for individual ransomware trials. The test script initiated VTune Performance Analyzer to collect selected performance counters and then triggered the selected LockBit ransomware variant after 15 s. HPC data for ten trials of each of the three LockBit variants were collected in randomized order at various times of day to minimize the impact of outside factors on results, resulting in thirty total trials. After each ransomware trial, the VTune database was moved to a firewall-protected data server before initiating a safe restore procedure to return the hardware to its pre-encrypted state.

To align ransomware execution times in the collected data, VTune start-up time was measured to determine if there was a delay initialization of the software after it was called. By utilizing a timer, the VTune start-up delay was found to be within a 2% tolerance of 5.5 s over 20 trials. This start up delay resulted in a data collection timeline that was offset by approximately 5.5 s from the test script, which can be observed in Fig. 5.

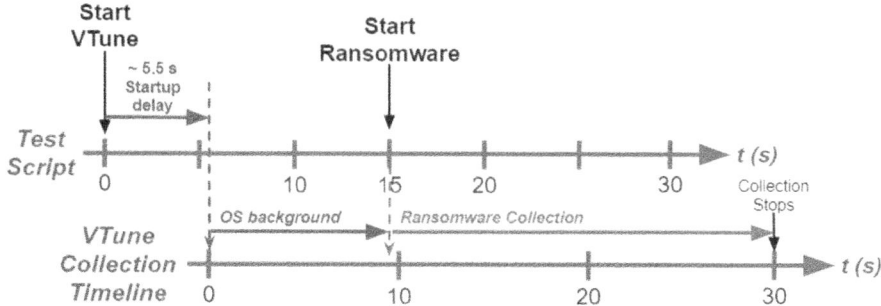

Fig. 5. Test Script and VTune HPC Collection Timeline

5.2 LockBit Variant Analysis

For this proof-of-concept LockBit variant analysis, the default builder version of LockBit with all options enabled was compared to two additional variants, as described below [42].

1. LockBit (default). The default version of LockBit ransomware with all options enabled, as shown in Fig. 2.
2. LockBit NKD (No Kill Defender). This variant of LockBit was created using the builder with all default features enabled except for killing Microsoft's Windows Defender antivirus software, which was disabled.
3. LockBit NNS (No Network Shares). This variant of LockBit disables encryption of network shares, but leaves all other options listed in Fig. 2 enabled.

The three LockBit versions were analyzed using: (1) Total HPC Count, (2) HPC Count Per Function, and (3) Time Domain analysis, which are described in the following sections.

Total HPC Count Analysis. A high-level analysis using the total HPC count was performed by averaging the total event count for each of the ten 30 s trials for each of the 90 selected HPCs collected. Figure 6 shows the average count over 10 trials for each LockBit variant for ten of the 90 HPCs collected. The default LockBit version is shown in blue, while the LockBit No Kill Defender variant is orange, and LockBit No Network Shares is green.

The subset of ten HPCs plotted in Fig. 6 were selected because they displayed an appreciable difference in average event count, and are listed below with brief descriptions adapted from [43].

1. `FP_ARITH_INST_RETIRED.SCALAR_DOUBLE` counts Single Instruction/Multiple Data (SIMD) scalar computational double precision floating-point instructions retired, where each count represents 1 computational operation (on least significant element).

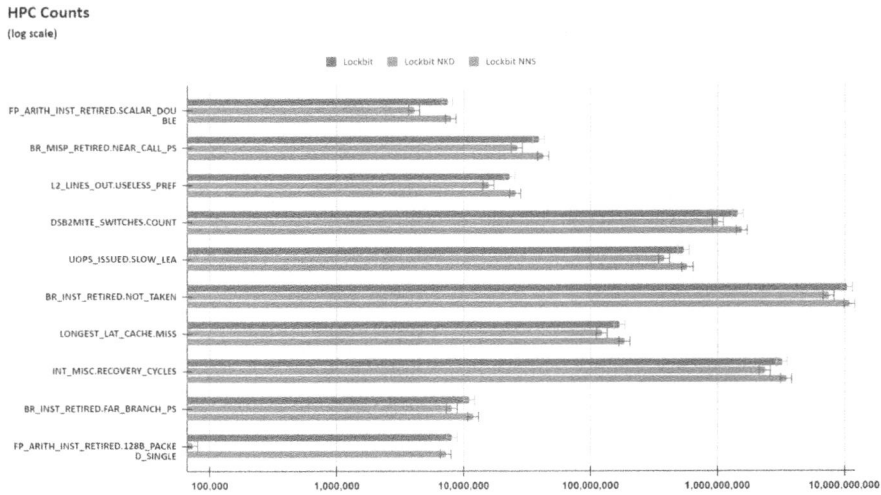

Fig. 6. Counts for LockBit, LockBit (No Kill Defender), and LockBit (No Network Shares) HPCs, logarithmic scale (Color figure online)

2. `BR_MISP_RETIRED.NEAR_CALL_PS` counts retired mispredicted direct and indirect (register and memory) near calls both taken and not taken, using Precise Event Based Sampling (PEBS) [44].
3. `L2_LINES_OUT.USELESS_PREF` counts the number of lines that were prefetched by the L2 cache, not used, and then evicted.
4. `DSB2MITE_SWITCHES.COUNT` counts the number of the Decode Stream Buffer (DSB)-to-Micro-instruction Translation Engine (MITE) legacy decode pipeline switches [45].
5. `UOPS_ISSUED.SLOW_LEA` Number of slow Load Effective Address (LEA) uops being allocated.
6. `BR_INST_RETIRED.NOT_TAKEN` counts not taken branch instructions retired.
7. `LONGEST_LAT_CACHE.MISS` counts core-originated cacheable requests that miss the L3 (Longest Latency) cache, such as data reads, speculative accesses, and hardware prefetches from L1 and L2 cache.
8. `INST_MISC.RECOVERY_CYCLES` counts core cycles the Resource allocator was stalled due to recovery from an earlier branch misprediction or machine clear event.
9. `BR_INST_RETIRED.FAR_BRANCH_PS` counts far branch instructions retired using PEBS.
10. `FP_ARITH_INST_RETIRED.128B_PACKED_SINGLE` counts SIMD 128-bit packed computational single precision floating-point instructions retired, where each count represents 4 computation operations (one for each element, computed in parallel).

For all HPCs, the event counts for LockBit default and LockBit NNS were within one standard deviation, indicated by the error bars in Fig. 6. This is expected in a stand-alone sand-boxed (non-networked) hardware collection environment, where disabling encryption of network shares would have little effect.

The minimal difference between LockBit and its NNS variant stands in stark contrast to the LockBit NKD variant. The FP_ARITH_INST_RETIRED.128B_PACKED_SINGLE counter displays the strongest indicator of whether LockBit (blue) or LockBit without the Kill Defender feature (orange) is executing. Apparent through the large difference in magnitude between LockBit (nearly 10M events) and LockBit NKD (less than 100k events) on the logarithmic scale, the number of 128-bit packed computational single precision floating-point instructions retired provides the largest relative difference in HPC counts. For all ten HPCs displayed, the magnitude of the LockBit NKD event count is at least two standard deviations less than the default LockBit ransomware. This suggests HPC may be a viable method of differentiating between certain LockBit variants.

HPC Count Per Function Analysis. Next, HPC counts were analyzed per function call. The "drill down" feature of the VTune Performance Analyzer GUI provided the event count of all 90 HPCs for each function called during the 30 s data capture. The HPC signature for each of the three LockBit variants was found by comparing the HPC counts per function call for the NKD and NNS variants to the default LockBit variant. An example subset of event count by function is shown in Fig. 7, with the red outline indicating the total number of events for each HPC indicated by the column header for the function at address 0x401072.

Function / Call Stac	INST_RETI	CPU_CLK_	BR_INST_	BR_INST_	BR_INST_	BR_INST_	BR_INST_	BR_INST_	BR_MISP_	BR_MISP_
func@0x40105c	5636672	6000006	2731300	598	277680	278226	0	0	0	0
func@0x401072	45751603	76000076	7226180	7176	853008	854438	2817100	4804826	260988	235092
func@0x40108c	6455782	9000009	0	0	0	0	0	0	0	0
func@0x4010bc	5703594	37000037	137280	312	48334	48698	63128	25974	338	312
func@0x401180	8.21E+08	3E+08	61666228	2418	2821754	2823912	66575912	2416804	2395822	2395484
func@0x4011c4	26856783	22000022	0	0	0	0	3882164	3314584	132522	115830
func@0x401264	17323858	20000020	0	0	0	0	0	0	0	0

Fig. 7. HPC Counts per function call for LockBit No Network Shares

The per function analysis of the default LockBit version showed that the FP_ARITH_INST_RETIRED.128B_PACKED_SINGLE count was zero for every function call for the average of the 10 trials except for the function at memory location 0x004020ac. However, in the NKD variant, the FP_ARITH_INST_RETIRED.128B_PACKED_SINGLE count was 0 for every function, including the location 0x004020ac function. This observation suggests a correlation between LockBit disabling Windows Defender and the FP_ARITH_INST_RETIRED.128B_PACKED_

SINGLE count, and that a spike in the count of this HPC could indicate Windows Defender is being targeted. This was reinforced by analyzing the FP_ARITH_INST_RETIRED.SCALAR_DOUBLE value at the memory address 0x004020ac. With Kill Defender enabled the value was routinely in the tens of thousands range, whereas with Kill Defender disabled this value was only in the thousands range.

Function / Call Stack	INST_RETIRED.ANY	CPU_CLK_ UNHALTED.REF_TSC	BR_INST_RETIRED. CONDITIONAL_PS	BR_INST_RETIRED. FAR_BRANCH_PS	BR_INST_RETIRED. NEAR_CALL_PS
func@0x401072	45751603	76000076	7226180	7176	853008
func@0x401072	52549634	72000072	9684714	9074	1129128
% CHANGE:	14.8586	-5.2632	34.0226	26.4493	32.3702
func@0x40de78	28348117	52000052	3220672	3094	298402
func@0x40de78	37937937	54000054	1229826	1274	137592
% CHANGE:	33.8288	3.8462	-61.8146	-58.8235	-53.8904
func@0x40f21c	6406335	6000006	2642822	2002	79898
func@0x40f21c	7219889	6000006	0	0	0
% CHANGE:	12.6992	0.0000	-100.0000	-100.0000	-100.0000

Fig. 8. Percent Change Between LockBit (black) and LockBit No Network Shares (blue) (Color figure online)

This approach was refined for the LockBit NNS variant by taking the individual percent change for each HPC count by function for the average of the 10 trials for the default version of LockBit and the NNS variant, as shown in Fig. 8. This highlighted BR_INST_RETIRED.NEAR_CALL_PS, shown in the right column of Fig. 8, as the HPC with the greatest percent change between variants for the functions at memory locations 0x401072, 0x40de78, 0x40f21c, which suggests these functions are possible locations of the network shares feature.

Narrowing in on a handful of functions in this manner significantly reduces the scope of Reverse Engineering work for a malware analyst. A typical ransomware sample may have hundreds to thousands of functions to analyze which may take a considerable amount of time for an analyst to comb through and determine specific functionality. This process can pinpoint locations in code that would likely contain the malicious functionality in question, allowing the analyst to work faster and be more productive. Figure 9 depicts Ghidra with a view of both the disassembly and the decompiled approximation of function 0x40de78, which is implied by the results of Fig. 8 to be one of 3 functions which may implement the network spread capability of Lockbit.

Time Domain Analysis. Plots of the event count as a function of time over the duration of the 30 s trial for each LockBit variant were generated with the VTune GUI. Figure 10 shows the event count versus time for the BR_MISP_RETIRED.NEAR_CALL_PS HPC for a representative trial of the default LockBit (top row), the NKD variant (middle), and the NNS variant (bottom).

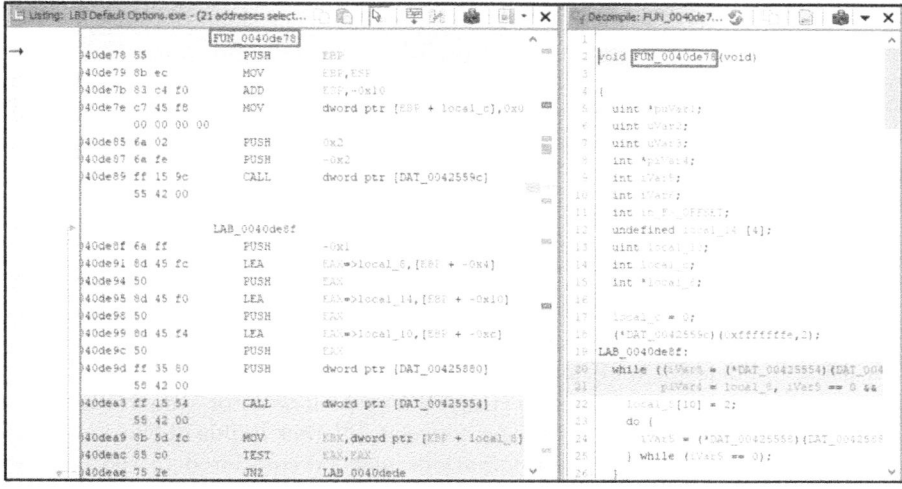

Fig. 9. Ghidra analysis of Function 0x40de78 in Lockbit

Of note, due to the VTune start-up delay discussed in Sect. 5.1, the first approximately 10 s of this plot shows HPC event count for background operating system activity before each LockBit ransomware variant is executed.

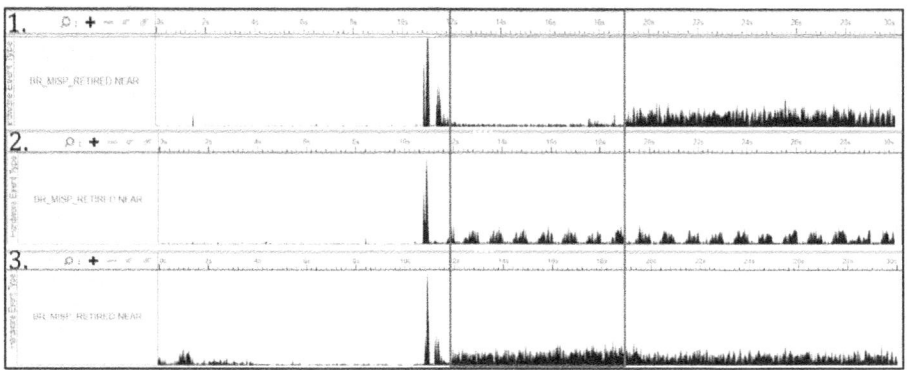

Fig. 10. BR_MISP_RETIRED.NEAR_CALL_PS event count over 30 s trial duration for 1. LockBit (default, top), 2. LockBit No Kill Defender (NKD, middle), and 3. LockBit No Network Shares (NNS, bottom)

The time plots show clear differences between the three variants. After the ransomware is executed around the 10 s mark, there is a noticeable drop in activity for the default version of LockBit. In contrast, the NKD variant shows a periodic surge of retired mispredicted branch instructions every half second, while the NNS variant maintains increased retired mispredicted branch instructions over the same time period (indicated by a red outline). Since the retired branch misprediction activity, indicated by the BR_MISP_RETIRED.NEAR_CALL_PS, occurs ten seconds into VTune capture, it can be clearly attributed to the different behavior of the ransomware variants as opposed to the OS only activity of the first ten seconds.

6   Conclusion

This work presents novel methods for identifying the functions performed by ransomware which could be key to implementing effective detection. It began with using builders for real ransomware to create custom variants which could be analyzed during runtime by collecting HPCs which monitor system hardware. These metrics showed clear variations between different builds of the same ransomware, which infers that those variations may be correlated to the features differences between those builds. Lockbit 3.0 ransomware showed notable distinctions between default, No Kill Defender, and No Network Share builds. Further, the HPC data delineated the functions in the executable code where these distinctions occurred, further bolstering the insight that these measurements provide potentially valuable data for ransomware detection and code analysis. For example, FP_ARITH_INST_RETIRED.128B_PACKED_SINGLE may be indicative of a ransomware process that is trying to protect itself by disabling the Windows Defender service.

This exploratory analysis consisted of just 3 variants of one ransomware family, with the results of 10 trials of each variant averaged before analysis. While this work is already showing encouraging results, additional analysis is required to determine if these averaged results are closely representative of individual trial results over a larger sample set. Immediate next steps will incorporate additional variants as well as adding new variants from other ransomware families. Finally, extensive Reverse Engineering analysis will be required to confirm the implied correlations between HPC measurements and specific ransomware capabilities.

References

1. Brill, A., Thompson, E.: Ransomware, a tool and opportunity for terrorist financing and cyberwarfare. Defence Against Terror. Rev. **12** (2019)
2. Team ZCySec. Mgm resorts data breach FAQ: what happened, who was affected, what was the impact? (2023)
3. Siddiqui, Z.: MGM casino expects $100 million hit from hack that led to data breach. Insurance J. (2023). Accessed 10 May 2024
4. Braue, D.: Global ransomware damage costs predicted to exceed $265 billion by 2031 (2021)
5. Chauhan, P.S., Kshetri, N.: Ransomware as a service kit: a novel cybercrime strategy to monetize victims' data. Computer **56**(10), 102–106 (2023)
6. Gatlan, S.: CISA: lockbit ransomware extorted $91 million in 1,700 U.S. attacks. BleepingComputer (2023). Accessed 10 May 2024

7. Kovacs, E.: Law enforcement hacks lockbit ransomware, delivers major blow to operation. SecurityWeek (2023). Accessed 10 May 2024
8. The biggest ransomware attacks of 2023. Kaspersky Official Blog (2023). Accessed 10 May 2024
9. CrowdStrike. U.S. and U.K. law enforcement seize lockbit ransomware infrastructure. CrowdStrike Blog (2024). Accessed 10 May 2024
10. O'Donnell-Welch, L.: Europol, FBI announce lockbit ransomware crackdown. Decipher - Duo Security (2024). Accessed 10 May 2024
11. Office of Public Affairs. U.S. and U.K. disrupt lockbit ransomware variant. United States Department of Justice (2024). Accessed 10 May 2024
12. Alzahrani, S., Xiao, Y., Sun, W.: An analysis of conti ransomware leaked source codes. IEEE Access **10**, 100178–100193 (2022)
13. Malone, C., Zahran, M., Karri, R.: Are hardware performance counters a cost effective way for integrity checking of programs. In: Proceedings of the Sixth ACM Workshop on Scalable Trusted Computing, STC 2011, New York, NY, USA, pp. 71–76. Association for Computing Machinery (2011)
14. Gonzalez, D., Hayajneh, T.: Detection and prevention of crypto-ransomware. In: 2017 IEEE 8th Annual Ubiquitous Computing, Electronics and Mobile Communication Conference (UEMCON), pp. 472–478. IEEE (2017)
15. Taylor, M.A., Smith, K.N., Thornton, M.A.: Sensor-based ransomware detection. In: Future Technologies Conference, pp. 794–801 (2017)
16. Taylor, M.A., Larson, E.C., Thornton, M.A.: Rapid ransomware detection through side channel exploitation. In: 2021 IEEE International Conference on Cyber Security and Resilience (CSR), pp. 47–54. IEEE (2021)
17. Taylor, M.A., Larson, E.C., Thornton, M.A.: General process detection through physical side channel characterization. In: 2022 IEEE International Systems Conference (SysCon), pp. 1–8. IEEE (2022)
18. Thummapudi, K., Lama, P., Boppana, R.V.: Detection of ransomware attacks using processor and disk usage data. IEEE Access (2023)
19. Alam, M., Bhattacharya, S., Dutta, S., Sinha, S., Mukhopadhyay, D., Chattopadhyay, A.: RATAFIA: ransomware analysis using time and frequency informed autoencoders. In: 2019 IEEE International Symposium on Hardware Oriented Security and Trust (HOST), pp. 218–227. IEEE (2019)
20. Woralert, C., Liu, C., Blasingame, Z., Yang, Z.: A comparison of one-class and two-class models for ransomware detection via low-level hardware information. In: 2023 Asian Hardware Oriented Security and Trust Symposium (AsianHOST), pp. 1–6. IEEE (2023)
21. Aurangzeb, S., Rais, R.N.B., Aleem, M., Islam, M.A., Iqbal, M.A.: On the classification of microsoft-windows ransomware using hardware profile. PeerJ Comput. Sci. **7**, e361 (2021)
22. Anand, P.M., Charan, P.V.S., Shukla, S.K.: Hiper-early detection of a ransomware attack using hardware performance counters. Digit. Threats: Res. Pract. **4**(3), 1–24 (2023)
23. Sinha, S., Alam, M., Bhattacharya, S., Mukhopadhyay, D., Chattopadhyay, A., Dutta, S.: RAPPER: ransomware prevention via performance counters. In: Kangacrypt 2018, Adelaide, Australia (2018)
24. Pundir, N., Tehranipoor, M., Rahman, F.: RanStop: a hardware-assisted runtime crypto-ransomware detection technique. arXiv preprint arXiv:2011.12248 (2020)
25. Hill, J.E., Walker, T.O., Blanco, J.A., Ives, R.W., Rakvic, R., Jacob, B.: Ransomware classification using hardware performance counters on a non-virtualized system. IEEE Access (2024)

26. Dinaburg, A., Royal, P., Sharif, M., Lee, W.: Ether: malware analysis via hardware virtualization extensions. In: Proceedings of the 15th ACM Conference on Computer and Communications Security, pp. 51–62 (2008)
27. Sierra-Arriaga, F., Branco, R., Lee, B.: Security issues and challenges for virtualization technologies. ACM Comput. Surv. (CSUR) **53**(2), 1–37 (2020)
28. Or-Meir, O., Nissim, N., Elovici, Y., Rokach, L.: Dynamic malware analysis in the modern era-a state of the art survey. ACM Comput. Surv. (CSUR) **52**(5), 1–48 (2019)
29. Arora, A., Hasan, R., Warner, G.: Obsolete ransomware: a comprehensive study of the continued threat to users. Int. J. Internet Technol. Secur. Trans. **7** (2020)
30. Davidson, R.: The fight against malware as a service. Netw. Secur. **2021**(8), 7–11 (2021)
31. Mahmoud, R.-V., Anagnostopoulos, M., Pastrana, S., Pedersen, J.M.: Enhancing analysis through sysmon and ELK integration. IEEE Access, Redefining Malware Sandboxing (2024)
32. Grelot, F., Larinier, S., Salmon, M.: Automation of binary analysis: from open source collection to threat intelligence. In: Proceedings of the 28th C&ESAR, vol. 41 (2021)
33. Dasgupta, P., Osman, Z.: A comparison of state-of-the-art techniques for generating adversarial malware binaries. arXiv preprint arXiv:2111.11487 (2021)
34. Kharraz, M., Arshad, S., Kirda, E.: UNVEIL: a large-scale, automated approach to detecting ransomware. In: 25th USENIX Security Symposium (USENIX Security 2016), Austin, TX, pp. 757–772 (2016)
35. McIntosh, T., Kayes, A., Chen, Y.-P., Ng, A., Watters, P.: Ransomware mitigation in the modern era: a comprehensive review, research challenges, and future directions. ACM Comput. Surv. (CSUR) **54**(9), 1–36 (2021)
36. Wan, Y.-L., Chang, J.-C., Chen, R.-J., Wang, S.-J.: Feature-selection-based ransomware detection with machine learning of data analysis. In: 2018 3rd international conference on computer and communication systems (ICCCS), pp. 85–88. IEEE (2018)
37. Sun, R., et al.: Mate! Are you really aware? An explainability-guided testing framework for robustness of malware detectors. In: Proceedings of the 31st ACM Joint European Software Engineering Conference and Symposium on the Foundations of Software Engineering, pp. 1573–1585 (2023)
38. Rohleder, R.: Hands-on Ghidra-a tutorial about the software reverse engineering framework. In: Proceedings of the 3rd ACM Workshop on Software Protection, pp. 77–78 (2019)
39. Zimba, A., Simukonda, L., Chishimba, M.: Demystifying ransomware attacks: reverse engineering and dynamic malware analysis of wannacry for network and information security. Zambia ICT J. **1**(1), 35–40 (2017)
40. Zugec, M.: Bitdefender threat debrief (2024). Accessed 11 May 2024
41. Das, S., Werner, J., Antonakakis, M., Polychronakis, M., Monrose, F.: SoK: the challenges, pitfalls, and perils of using hardware performance counters for security. In: 2019 IEEE Symposium on Security and Privacy (SP), pp. 20–38. IEEE (2019)
42. Ovalle, E., Figurelli, F., Souza, C., Muñoz, A.: Revisiting the lockbit 3.0 builder files (2024). Accessed 30 May 2024
43. 2nd generation intel xeon processor scalable family based on cascade lake product. Accessed 30 May 2024
44. Intel® VTune^{TM} profiler user guide. Accessed 30 May 2024
45. Intel VTune amplifier XE and intel VTune amplifier for systems help: DSB switches. Accessed 30 May 2024

STARMAP: Multi-machine Malware Analysis System for Lateral Movement Observation

Shota Fujii[1]([⊠])(iD), Yoichi Tsuzuki[2], Takanori Okamoto[2], Yu Tamura[1], and Takayuki Sato[1]

[1] Hitachi, Ltd., Tokyo, Japan
shota.fujii.xh@hitachi.com
[2] FFRI Security, Inc., Tokyo, Japan

Abstract. One of the functions of malware used in cyber attacks is lateral movement, which plays an important role in the spread of an infection. However, existing dynamic analysis systems are often constructed with only a single machine, making it difficult to observe behaviors that affect multiple machines such as lateral movement. Therefore, we propose STARMAP, a dynamic malware analysis system that can observe behaviors for affecting multiple machines. STARMAP launches multiple machines and observes not only the machine running the malware, but also the communication to the lateral movement target and the post-lateral movement behavior. In addition, since the post lateral movement process is running on a different machine than the machine that executed the malware, it is not directly linked to the malware process. Therefore, STARMAP identifies post lateral movement behavior by extracting processes that are not normally used on the post lateral movement machine. In this paper, we present the design of STARMAP and its implementation method based on the CAPE sandbox. We also analyze real malware on the STARMAP prototype and show that the lateral movement and post-lateral movement behavior can be observed.

Keywords: Malware · Dynamic analysis · Lateral movement

1 Introduction

With the number and sophistication of malware increasing every year, automated dynamic analysis of malware has become widely used. One of the functions of malware is lateral movement, which plays an important role in the spread of an infection. However, existing dynamic analysis systems are often constructed with only one analysis environment, making it difficult to observe behavior that affects multiple machines or behavior after lateral movement.

In this paper, we propose STARMAP (System for Tracking and Analyzing Route of MAlware Propagation), a system that can observe the behavior of malware affecting multiple machines, including lateral movement. STARMAP

J. Zhao and W. Meng (Eds.): SciSec 2024, LNCS 15441, pp. 37–55, 2025.
https://doi.org/10.1007/978-981-96-2417-1_3

starts multiple machines on the same network and observes not only the behavior on the machine running the malware, but also the communication to the other machines and the behavior of the process after lateral movement. STARMAP then identifies the behavior after lateral movement by extracting processes that are not in normal use using an allowlist at the machine after infection spread.

We also show an implementation of STARMAP based on CAPE sandbox [7], one of the de facto dynamic analysis systems. In addition, we implemented a prototype of STARMAP and demonstrated its feasibility for lateral movement and post-propagation behavior through preliminary evaluation using real malwares. Furthermore, we measured the processing time and verified the practicality of the system in daily use.

The contributions of this paper are as follows:

– We designed STARMAP, a malware dynamic analysis system that consists of multiple machines and can automatically observe the behavior of malware affecting multiple machines such as lateral movement. We also presented a concrete implementation of STARMAP using ESXi as the Virtual Machine Monitor (VMM) and CAPE sandbox as the dynamic analysis software.
– We implemented a prototype of STARMAP. We also demonstrated that STARMAP can be used to automatically observe the lateral movement of malware and its behavior after lateral movement through evaluation using real malwares.

2   Background

2.1   Lateral Movement

After an attacker compromises a target machine, the attacker may infiltrate another system on the network in order to compromise more machines, which is known as lateral movement. It is also defined as TA0008 Lateral Movement in *MITRE ATT&CK*® [23], which is one of the frameworks for defining attack techniques. For defenders, it is one of the attack methods that should be especially detected and prevented, since allowing lateral movement can cause further damage.

Lateral movement has also been automated and integrated into malware, contributing to efficient attacks. The most representative examples include mirai [3], which spreads infection via telnet, WannaCry, which spreads infection by exploiting vulnerabilities in file-sharing protocols and dumped authentication information, and NotPetya and BadRabbit, which were developed later [2]. Malware that automatically propagates infection and lateral movement remains active. In addition, not only variants, but new types of malwares are constantly being observed, and it is important to analyze and understand their behavior in order to develop countermeasures.

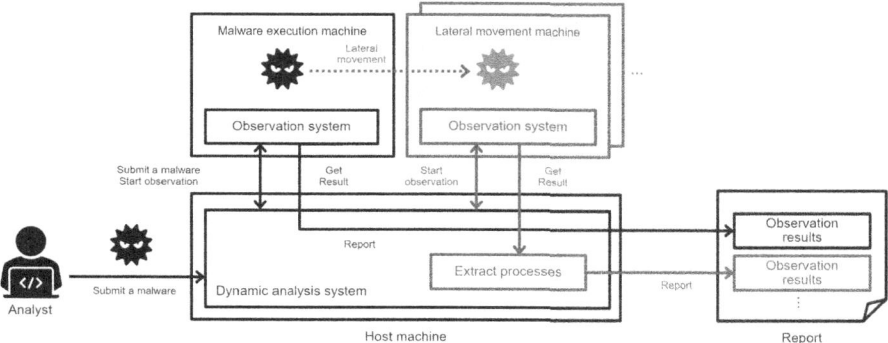

Fig. 1. Overview of our proposed system.

2.2   Malware Dynamic Analysis

Dynamic malware analysis is a widely used method for analyzing the behavior of malware samples by executing them [34]. On the other hand, existing dynamic malware analysis systems often consist of only one analysis environment. As a result, it is difficult to observe behavior that affects multiple machines or behavior after propagation, such as the lateral movement mentioned above.

Therefore, the goal of this paper is to develop an analysis system that can analyze malware affecting multiple machines.

3   Design

3.1   Requirement

As mentioned earlier, the goal of this paper is to develop a dynamic analysis system that can analyze malware that affects multiple machines. The following two points should be considered in order to achieve this system.

Requirement 1. Must be able to handle multiple machines as part of a dynamic analysis system.
Multiple machines must be handled to induce lateral movement behavior and to observe behavior after lateral movement. In addition, it is necessary to be able to observe and record malware-related behavior not only on the machine that launches the malware, but also on other machines.

Requirement 2. Must be able to identify the process after lateral movement.
There are many benign processes and processes unrelated to the behavior of the malware in the analysis machine, and they become noise in the malware analysis. Therefore, in a dynamic analysis system, the behavior of malware is narrowed down by extracting the processes of executed malware samples and their child processes. However, there is no process directly connected to the

malware sample that is the starting point in the machine where the lateral movement is performed. Therefore, it is necessary to use other methods to extract processes related to the behavior of the malware.

3.2   Basic Idea and Overview

First, to satisfy *Requirement 1*, we start multiple machines on the same network and observe not only the malware on the machine running the malware, but also the communication to the propagation target and the behavior of the process after propagation. In addition, we prepare an NTP server to synchronize the time of all the machines so that the behavior across multiple environments can be handled in the same time series. These measures will allow us to handle multiple machines within the dynamic analysis system.

In addition, to satisfy *Requirement 2*, we enumerate the processes that are started during normal use and create an allowlist. The processes that are not on the allowlist in environments other than the one where the malware was executed are then extracted as those that may have been caused by the lateral movement of malware.

Figure 1 shows an overview of STARMAP. The configuration of a general dynamic malware analysis system for a single machine, such as CAPE, is shown in black, and the additional parts for handling multiple machines are shown in red.

3.3   System Components

STARMAP consists of following three machines as shown in Fig. 1.

Host Machine. It operates as a machine that hosts the dynamic analysis system and controls the machines for malware execution and lateral movement described below. It also receives the malware to be executed from the analyst and submits it to the malware execution machine. Furthermore, after the malware has completed its operation, it obtains and parses the observation results from each machine, and summarizes them in a report.

Malware Execution Machine. It operates as a guest machine of the dynamic analysis system and executes malware submitted from the dynamic analysis system. While the malware is being executed, it also observes the behavior of the malware on the analysis machine as a function of the dynamic analysis system. Specifically, network communication and process behavior related to the malware are observed. After the malware is executed, the observation results are delivered to the dynamic analysis system on the host machine.

Table 1. List of techniques for lateral movement (TA0008) and targets of STARMAP (Targeted: ●, Partially targeted: ◐).

ID	Name	Description	Target?
T1210	Exploitation of Remote Services	Exploiting vulnerable remote services (SMB, RDP, etc.)	●
T1534	Internal Spearphishing	Spear phishing (email) within the organization	–
T1570	Lateral Tool Transfer	Transfer of tools for lateral movement	●
T1563	Remote Service Session Hijacking	Hijacking existing sessions on remote services such as RDP	●
T1021	Remote Services	Using valid accounts through remote services such as RDP	●
T1091	Replication Through Removable Media	Using removable media such as USB flash drive	◐
T1072	Software Deployment Tools	Using asset management systems within the corporate network	–
T1080	Taint Shared Content	Using shared storage such as network drives	●
T1550	Use Alternate Authentication Material	Bypassing authentication processes, such as using password hashes	●

Lateral Movement Machine. As with the analysis machine, it operates as a guest machine for the dynamic analysis system. In addition, it is located on the same network to enable lateral movement from the malware execution machine. A network communication and process behavior are observed same as the malware execution machine. After the malware is executed, the observation results are delivered to the dynamic analysis system on the host machine. The malware execution machine that initially executes the malware is a single machine, but the lateral movement machine consists of one or more machines.

3.4   Observation Targets

Table 1 shows a list of techniques for lateral movement defined as TA0008 Lateral Movement in MITRE ATT&CK. MITRE ATT&CK defines sub-techniques for each technique to provide more detail. For example, for T1563 Remote Service Session Hijacking, T1563.001 defines SSH hijacking and T1563.002 defines RDP hijacking as subtechniques. On the other hand, the parent techniques encompass each subtechnique at a high level of granularity and are therefore omitted here.

Since STARMAP analyzes automated malware, we target those of the lateral movement techniques that can be automated. Specifically, as shown in the "Target?" column of Table 1, we target those that can be automated, such as T1210 Exploitation of Remote Services, and those cannot be automated, such as T1534 Internal Spearphishing, which are difficult to fully automate and incor-

Table 2. Implementation environment.

Role	OS/VMM	CPU	Memory	Main software
VMM	ESXi 7.0	Intel Xeon Silver 4316 2.3 GHz (20C/40T)	256 GB	
Host machine	Ubuntu 22.04 LTS 64bit	Same as above (vcpu 2 cores)	2 GB	CAPE sandbox 2.4, chrony 3.2
Malware execution machine	Windows 10 Pro 64bit	Same as above (vcpu 1 core)	1 GB	tshark 3.4.0, Process Monitor 3.87, NtTrace 2335
Lateral movement machine	Windows 10 Pro 64bit	Same as above (vcpu 1 core)	1 GB	Same as above

porate into malware, i.e., manually sending emails, are not included. However, T1091 Replication Through Removable Media was partially included because it can be confirmed up to the infection of the USB flash drive in the preliminary stage, although it cannot be observed after the lateral movement.

In addition, the lateral movement techniques shown as "Target?" in Table 1 can generally be achieved if the endpoints are on the same network. For example, the T1021 Remote Service can achieve lateral movement through a service that is enabled by default, such as Windows Management Instrumentation (WMI)[1], if the credentials of the deployment target are known. On the other hand, some services, such as lateral movement via RDP or exploiting SMBv1 vulnerabilities, rely on services that are not enabled by default. Therefore, we will enable RDP, SMBv1, etc. on the analysis and target machines to induce lateral movement behaviors.

3.5   Identification of Malware-Associated Process

As mentioned above, the processes associated with the malware are not directly connected to the lateral movement machine in the form of a parent-child relationship because the executed malware is located on a different machine. Therefore, it is necessary to identify the processes associated with the executed malware using another method. STARMAP creates an allowlist of processes that are generated within the scope of normal use, and extracts processes observed on the lateral movement machine that are not in the allowlist as processes that may be related to the malware.

A process allowlist is created by observing the processes generated on a lateral movement machine for a certain period of time, which are launched without running malware. After the observation is completed, our system extracts the processes that are not in the list from the processes observed on the lateral movement machine and describes them in a report. Note that the processes running on each machine may differ depending on the installed software and its

---

[1] Installed by default on Windows 2000 and later operating systems.

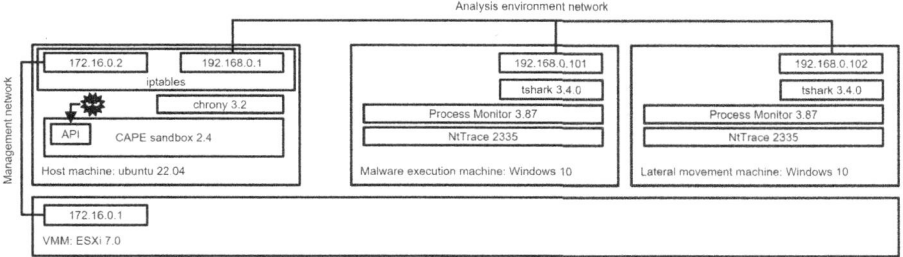

Fig. 2. Overview of implementation environment.

version. Therefore, if there are multiple machines for lateral movement, create a process allowlist for each machine.

4   Implementation

This Section describes the implementation of STARMAP presented in Sect. 3 using CAPE (version 2.4), one of the de facto dynamic malware analysis systems. The software and environments for the implementations discussed in this Section, including the CAPE, are listed in Table 2 and the overview is shown in Fig. 2. The following description assumes the environments and software listed in the table. The CAPE installation directory (/opt/CAPEv2) will be referred to as $CAPE in the following Sections.

In the STARMAP implementation in this Section, each machine is built on a virtual environment and ESXi is used as the VMM. Therefore, the VMM specification ([cuckoo] machinery) in the configuration file ($CAPE/conf/cuckoo.conf) is changed from the default value virtualbox to esx. The host machine, malware execution machine, and lateral movement machine are connected by a local network (analysis environment network). The host machine installs chrony [9] as an NTP server to synchronize the time between the terminals. In addition, the host machine connects to the ESXi to operate the machines for malware execution and lateral movement as described below.

When observing with STARMAP, the first step is to create a process allowlist to extract the processes associated with the malware. STARMAP then activates the lateral movement machine, observes malware behavior on that machine, and reports the observation result, in addition to the standard CAPE operations.

In the following Sections, we describe the methodology for adding function for CAPE. We also describe the additional functions and their implementations.

Listing 1.1. Auxiliary modules notation (host side).

```
1  from lib.cuckoo.common.abstracts import Auxiliary
2
3  class MyAuxiliary(Auxiliary):
4
5      def start(self):
6          # Do something.
7
8      def stop(self):
9          # Stop the execution.
```

4.1 To Customize CAPE Sandbox

CAPE provides several customization modules that can be used to extend and add functions. This section describes the *auxiliary modules* and *processing modules* that are used to implement STARMAP.

Auxiliary Modules. Auxiliary modules can define the processes to be run before and after malware analysis is completed. There are two versions of this module, one for the host and one for the guest. The processes to be executed on the host are written as Python programs and stored in the host's $CAPE/modules/auxiliary/, which are executed on the host before and after the analysis is started. The notation of the host-side auxiliary modules is defined in the form shown in the list 1.1. By inheriting the Auxiliary class and specifying the process to start with *start()* and the process to stop with *stop()*, CAPE executes each process at each point in time.

The process to be run on the guest side is sent from the host to the guest by storing it as a Python program in $CAPE/analyzer/windows/modules/auxiliary/ on the host, and is executed before the malware is run on the guest. However, since CAPE does not have a definition for termination processing in the auxiliary modules on the guest side, it is necessary to start termination processing using a different method. To meet this requirement, CAPE waits without terminating the program after the startup process is executed on guest. The host sends a waitfor signal at the end of the termination process of the host auxiliary modules. The guest resumes processing after receiving the signal, allowing the guest side to execute any process at the time of the end of analysis.

Processing Modules. Processing modules analyze the data obtained from malware execution and store it in a data structure called a global container. The analysis results stored in the global container are finally converted to a reporting format, such as JSON, in reporting modules. Processing can be added to the processing modules. Specifically, by writing a Python program and storing it in $CAPE/modules/processing/ on the host, the program is executed on the host by the processing modules after the analysis is complete.

Each of the mechanisms described below is implemented using the above modules, with the exception of the construction of the analysis-independent process allowlist (Sect. 4.2).

4.2   Construction of Process Allowlist

Prior to analysis, we create a process allowlist on the target machine as described in Sect. 3.5. Specifically, after booting Windows 10 of the lateral movement machine, run Process Monitor [21] for 24 h, acquire the processes that were started during that time, and make them into an allowlist. Since the processes obtained by the above method are only the processes after Process Monitor is started, the processes before that time are added to the allowlist by separately checking the process list immediately after Windows is started. In this environment, winlogon.exe, services.exe, sihost.exe, explorer.exe, and MsMpEng.exe were identified, so we used the allowlist consisting of a total of 108 processes including these processes.

4.3   Operation of Lateral Movement Machine

For the malware execution machine, the default CAPE starts and restores snapshots before analysis and stops the machine after analysis. However, the default CAPE does not operate the lateral movement machine; therefore, the operating function for that machine, such as starting and stopping a machine, must be implemented separately.

CAPE uses libvirt [18] to perform the above operations for the guest machines when the VMM is ESXi. STARMAP also uses libvirt to stop, start, and restore snapshots of the lateral movement machine. Specifically, the IP address and snapshot name of the lateral movement machine are obtained from $CAPE/conf/esx.conf, and the machine is operated by ESXi functions via libvirt over the management network. The startup process at the beginning of analysis and the restoration of snapshots are implemented in the startup process of the host-side auxiliary modules, and the shutdown process at the end of analysis is implemented in the shutdown process of the same modules.

4.4   Observation

STARMAP observes network communications, processes, and Windows APIs, following a general dynamic analysis system such as CAPE. This section describes the implementation method for observing each element.

The default CAPE performs the same observations of the above three elements for the malware execution machine. However, to ensure consistency of observation results across machines, the observation mechanism described in this section is implemented not only on the lateral movement machine, but also on the malware execution machine.

In addition, as described below, the observation results are sent from the malware execution/lateral movement machines to the host machine after the observation is complete. Observation results are sent to the host machine using INET domain socket communication, similar to the function used by CAPE to send files to the host machine (`lib.common.results.upload_to_host()`).

Network Observation. To observe network communications, tshark [31], one of the packet capture tools, is used.

First, tshark is started before analysis begins on the malware execution and lateral movement machines. Specifically, tshark is started by executing commands on the guest using guest-side auxiliary modules and observing network communications that occur during the analysis.

After the analysis is complete, tshark is stopped using the auxiliary modules on the guest side, and the observation results file is output as a pcap file using the tshark function. The output pcap file is then sent to the host machine using the method described above.

Finally, the pcap file is parsed in processing modules. The result is added to the global container, and processing is complete.

Process Observation. To observe processes, Process Monitor [21], a tool that can monitor file operations, registry operations, etc., for processes, is used.

First, before the analysis starts at the analysis malware execution/lateral movement machines, the process allowlist constructed in Sect. 4.2 is sent from the host machine using the host's auxiliary modules. To send files from the host machine, use the CAPE implementation of `cuckoo.core.guest.upload_analyzer()`. This function sends files that are stored in a specific directory (if the guest is Windows, the host side's `$CAPE/analyzer/windows/`), so the process allowlist should be placed in that directory. After that, Process Monitor is launched using auxiliary modules on the guest side to observe the behavior of processes that occur during the analysis. During an analysis, by reading the process allowlist, only processes that are not included in the list are observed as possibly associated with the malware.

After the analysis is complete, Process Monitor is stopped using the auxiliary modules on the guest side, and the observation results file is output as a pml file using the Process Monitor function. The pml file is converted into the csv file by Process Monitor function as well. The output csv file is then sent to the host machine.

Finally, the csv file is parsed in processing modules. The result is added to the global container, and processing is complete.

Windows API Observation. To observe Windows API, NtTrace [25], one of the SYSCALL trace tools, is used. NtTrace can trace Windows APIs executed by a target process by specifying the process ID (PID).

First, before starting the analysis on the malware execution/lateral movement machines, a custom program is started to launch NtTrace using auxiliary modules on the guest side. As mentioned above, NtTrace needs a PID. Therefore, the above custom program captures the process trace by detecting the process start and calling NtTrace with the PID of the started process. Process startup monitoring is achieved by periodically retrieving the process list with the `wmic process` command, and then retrieving the differences as newly started processes. In this case, the process allowlist sent in Sect. 4.4 is used, and only processes that are not included in the list are traced as possibly associated with the malware.

After the analysis is complete, NtTrace is stopped using the auxiliary modules on the guest side, and the observation results file is output as a csv file using the NtTrace function. The output csv file is then sent to the host machine.

Finally, the csv file is parsed in processing modules. The result is added to the global container, and processing is complete.

4.5 Flow of STARMAP

The processing flow of STARMAP when using CAPE as base system is shown in Fig. 3. The parts that are not part of the standard CAPE functions but have been added us are shown as **[Added]** in the following explanations and as red in Fig. 3.

1. Submit malware to CAPE sandbox on the host machine using the CAPE API.
2. **[Added]** Start all the lateral movement machines and restore the snapshots.
3. **[Added]** Start observations by tshark, Process Monitor, and NtTrace on the lateral movement machines.
4. Start the malware execution machine and restore the snapshot.
5. Submit malware to the malware execution machine.
6. Perform CAPE sandbox standard observations on the malware execution machine.
7. **[Added]** Start observations by tshark, Process Monitor, and NtTrace on the malware execution machine.
8. Complete the analysis of the malware and upload the standard CAPE Sandbox observation results from the malware execution machine to the host machine.
9. **[Added]** Terminate observations by tshark, Process Monitor, and NtTrace on the malware execution/lateral movement machines.
10. **[Added]** Upload the result files of tshark, Process Monitor, and NtTrace observed on the malware execution/lateral movement machines to the host machine.
11. **[Added]** Stop the lateral movement machines.
12. Stop the malware execution machine.
13. Add standard CAPE sandbox observations to the global container.
14. **[Added]** Parses the uploaded tshark, Process Monitor, and NtTrace results and adds them to the global container. With respect to lateral movement machines, it then adds only malware-related behavior to the global container by excluding normal processes using the allowlist.
15. Create a report based on the global container.

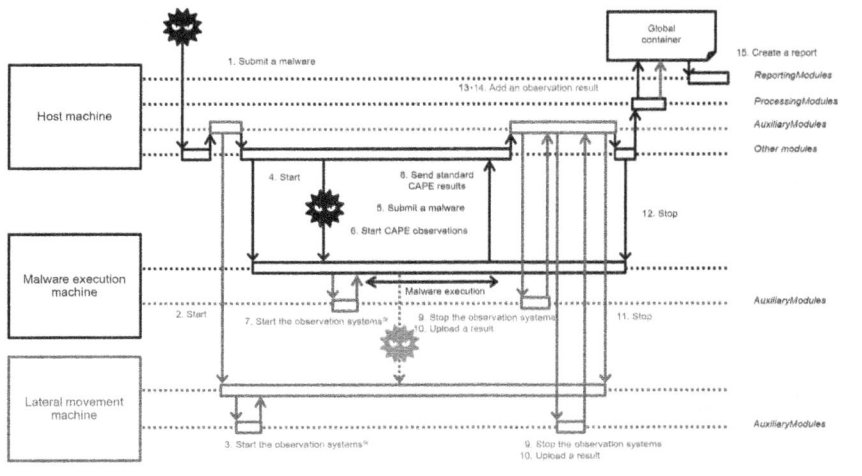

Fig. 3. Flow of STARMAP using CAPE sandbox.

The above flow controls the multi-machine analysis environment using the various features of the CAPE sandbox while observing malware behavior, including behavior that affects multiple machines, such as lateral movement. The observation results uploaded to the host in step 10 are appended to the standard CAPE sandbox report in step 14 in the same format so that they can be used transparently with the existing report.

5   Experiments and Results

In this experiment, we analyzed real malware performing lateral movement on the STARMAP prototype and evaluated its observability. For the evaluation, we constructed the ground truth dataset of malware that automatically performs lateral movement. We referred to analysis articles from security vendors and extracted candidate malware for lateral movement. We then excluded cases, such as RAT (Remote Access Trojan), where attackers attempted to perform lateral movement manually after intrusion, and selected malware that included a lateral movement function in the malware itself. The malwares finally selected for evaluation are shown in Table 3, along with the method of lateral movement and links to analysis articles.

Table 3. Samples for evaluation.

#	Family	MD5	Lateral movement methods	ATT&CK	Link
1	BadRabbit	fbbdc39af1139aebba4da004475e8839	Via SMB sharing	T1080, T1570	[10]
2	Conti	8fe7bfef6ebc53e9047561d35555cd24	Via SMB sharing	T1080	[20]
3	Golang	a37759e4dd1be906b1d9c75da95d31a2	Via vulnerable machines on the Internet	T1210	[6]
4	PlugX	86950b81df2003d08ae4a7869ecf88fe	Via USB flash device	T1091	[28]

5.1   Experimental Setup

To avoid interfering with the operation of the malware, the following settings were made on each machine in the experiment.

- Disable User Account Control (UAC)
- Disable Microsoft Defender Firewall
- Disable Microsoft Defender Antivirus
- Disable Windows Update
- Disable Auto Sleep

In addition, each machine was analyzed without an Internet connection. To observe the effects of lateral movement and other inter-machine communication, we did not restrict inter-machine communication. All communication other than the minimum necessary, such as CAPE sandbox communication, was not allowed.

Furthermore, we implemented settings to induce actions related to lateral movement. Specifically, SMBv1 was enabled as described in Sect. 3.4. In addition, if the ransomware encrypts CAPE sandbox agents or observation systems such as Process Monitor, the analysis may fail. To prevent this, we changed the permissions of files under `C:\Program Files (x86)\Python38-32`, where the CAPE sandbox agent is stored, and `C:\Program Files\Process Monitor`, etc., where the observation systems are stored, to read-only. We also prepared a virtual hard disk (VHD) that mimicked a USB flash drive and mounted it on the E drive to induce lateral movement via the USB flash drive.

Additionally, a survey by Wong et al. found that many malware analysts set the default execution time for dynamic analysis to five minutes [34]. Therefore, in this experiment, we also set the execution time of the malware to 5 min.

5.2   Results

This section shows the results of the STARMAP analysis of the four malware samples listed in Table 3.

Case 1: BadRabbit. BadRabbit is a ransomware that encrypts files on the victim's computer and demands a ransom to decrypt them. It also includes an automatic lateral movement feature [10].

Running this sample on STARMAP confirmed that one of the components of BadRabbit, `infpub.dat`, was copied and executed from the malware execution machine to the lateral movement machine. Specifically, the lateral movement was achieved by being copied by `NtCreateFile()` and `NtWriteFile()` as `UNC\192.168.0.102\admin$\infpub.dat` for the administrative share (`admin$`), which is a shared folder. Infection activity was also performed by executing the copied `infpub.dat` on the lateral movement machine using the svcctl remote procedure call (RPC) from the malware execution machine. In addition to the file and process behavior described above, traces of these lateral movement behaviors were also observed as SMB2 protocol network communications such as `Create Request File: infpub.dat` and `Create Request File: svcctl`.

Furthermore, using the allowlist, we were able to extract the process remotely launched by BadRabbit and record the API calls related to the malware from the lateral movement machine observation results.

In summary, we were able to observe the lateral movement behavior and successfully identify and observe processes after lateral movement.

Case 2: Conti. Conti is a ransomware with the ability to encrypt files in shared folders and other machines on the same network [20].

When this malware was executed on STARMAP, the executed Conti performed a search for a propagation target machine by scanning its own network, 192.168.0.0/24, using the SMB protocol. It then encrypted over the network to 192.168.0.102:445 of the lateral movement machine on the same network. Specifically, each file was overwritten with encrypted data using `NtWriteFile()`. After that, the encrypted files were renamed with the .KCWTT extension using `MoveFileWithProgressTransactedW()`.

As described above, it was confirmed that it is possible to observe the behavior related to the lateral movement as well as the behavior on the machine at the target of the lateral movement.

Case 3: Golang. Golang is a crypto-miner that scans vulnerable machines on the Internet for lateral movement [6].

When run on STARMAP, this malware scanned specific TCP ports (80, 443, 1433, 6379, 6380, 7001, 7002, 8080, 8088, 9200) of 96 random public IP addresses, looking for vulnerable machines that could be compromised. This experiment was performed without connecting to the Internet as described, so there were no machines associated with public IP addresses. Therefore, no post-scan lateral movement behavior was observed, but we were able to confirm the scanning activity prior to the lateral movement.

Case 4: PlugX. This variant of PlugX has the functionality of a USB worm that performs lateral movement via a USB flash drive [30].

When this malware was executed with STARMAP, the `CreateFile()` created the PlugX payloads `CEFHelper.exe`, `wsc.dll` and `AvastAuth.dat` in the E directory of the USB flash drive. Since this malware is a USB worm, infection does not occur until the infected USB flash drive is replaced on another machine, but we were able to confirm the preparation stage.

6   Discussion

6.1   Lateral Movement Observability

The malware used in the evaluation may not have activated behavior in a single machine analysis environment, and even if it does, it is impossible to observe its behavior on another machine after lateral movement. Through evaluation, we have shown that it is possible to activate the lateral movement behavior by preparing the lateral movement machines in the same network in addition to the malware execution machine.

Each malware used in the evaluation performed lateral movement using techniques such as T1080 Taint Shared Content, T1210 Exploitation of Remote Services, T1570 Lateral Tool Transfer, and T1091 Replication Through Removable Media. Although the above techniques do not cover all the target techniques of this study shown in the table, the remaining techniques can be automatically activated when another machine is also present in the network. Therefore, it is concluded that qualitatively, any of the techniques can be activated and observed.

We showed that by observing network communication and process behavior on the lateral movement machine, it is possible to observe behavior after lateral movement, which is difficult with the existing single machine dynamic analysis system. We also showed that it is possible to extract malware processes after lateral movement by excluding normal processes using an allow list, through the observation results of Case 1.

6.2   Limitations

In the implementation of STARMAP in this paper, the environment in which the malware runs consists of only two machines. Therefore, we may not be able to observe behavior that propagates to more than two machines, or that affects non-user machines such as network routers and Active Directory. For example, the behavior of malware, which uses authentication information from Active Directory for lateral movement [26], may not be fully observed. This issue can be mitigated by adding another target system, such as another user machines or authentication server, to the analysis environment.

In addition, this paper describes the design and implementation of STARMAP, which focuses on inducing and observing behavior related to lateral movement, but does not consider analysis evasion by detecting the analysis environment of the malware. Therefore, there is a possibility that the analysis environment will be detected by the analysis evasion function of the malware, and the behavior related to lateral movement may not be activated. There are several known methods to overcome the analysis evasion feature, such as accumulating file and browser usage histories to simulate user usage [22], obfuscating registry entries and responses to various commands that are referenced to detect the analysis environment [29,35], using a bare-metal environment as the analysis environment [16,35], and a method to avoid virtual environment detection by implementing an out-of-the-box monitoring mechanism [16,17]. These methods can be used to mitigate this problem.

As shown in Golang in Case 3, it is impossible to observe the behavior of malware that attempts to spread to random public IP addresses with the current implementation, which has a locally installed machine with a static IP address. In addition to Golang, there are some malware that scan public IP addresses for lateral movement such as WannaCry [19] and Phorpiex [8]. For this reason, it is also desirable to be able to observe a lateral movement for the public IP addresses. Regarding this matter, by preparing a default gateway and redirecting communications to a public IP to a lateral movement machine, it is possible to prevent a situation where the target machine does not exist and improve the observability of lateral movement.

6.3   Research Ethics

Since STARMAP deals with malware, there is a possibility that communication of attacks and spread of infection may occur outside the system. Therefore, as described in the evaluation section, this experiment was conducted in a closed environment with no access to the Internet. In addition, the malwares used in the experiment were managed in an area accessible only to a limited number of members.

7   Related Work

Similar to STARMAP, there are many studies and systems that automatically analyze malware, including CAPE sandbox [7]. One example is CAPE Sandbox, which was used as the base system for STARMAP. There are also online services that provide dynamic analysis systems, such as ANY.RUN [4], Hybrid Analysis [13], Joe Sandbox [14], and Tria.ge [27]. These services implement a number of useful features that make analysis more sophisticated and efficient, such as identifying C2 servers from analysis results, mapping behavior to MITRE ATT&CK, and scoring maliciousness. Moreover, as discussed in Sect. 6.2, some analysis systems are equipped with mechanisms that bypass the analysis evasion function of the malware and manifest malicious behavior [16,17,22,29,35]. Furthermore, some research attempts to improve analysis efficiency by visualizing the results of dynamic analysis of malware [33]. For example, MalView [24] is a study that records malware behavior using Process Monitor in a dynamic analysis system and visualizes process, network, and library calls based on the recorded logs. However, as mentioned earlier, these studies and systems are based on the single machine as the analysis environment, and it is difficult to observe behavior that affects multiple environments, such as lateral movement.

There are also more advanced analysis environments that mimic the user's environment [1, 32]. In addition to multiple machines, these analysis environments often include systems that exist in common organizational networks, such as DNS and Active Directory, making it possible to observe behavior that affects multiple environments. However, these analysis environments are different from the scope of this study, which is to automatically analyze a large number of malware, because they are based on manual analysis and require a large amount of computing resources to build an advanced analysis environment. For example, it is conceivable to collaborate with STARMAP by selecting malwares that have the potential for lateral movement and analyzing them using these advanced analyses.

Several studies have also been conducted to detect lateral movement on the enterprise network. Bai et al. [5] propose a method using machine learning to detect lateral movement using RDP from long-term and multi-user authentication logs in an enterprise network. He et al. [11] propose a method to detect lateral movement through SMBs with high accuracy by installing SMB honeypots on the intranet and combining traffic to the honeypots with the malware API sequence. Jbeil [15] is a framework that uses Graph Neural Network (GNN) and link prediction to detect lateral movement in enterprise networks with high accuracy and high generalization performance. There is also research that uses authentication logs and endpoint logs to detect lateral movement [12] and research that maps malware behavior to *MITRE ATT&CK techniques*, including lateral movement [29]. By applying these techniques to the logs observed by STARMAP, it may be possible to detect behavior related to lateral movement at a higher level.

8   Conclusion

In this paper, we propose STARMAP, a system for observing malware that affects multiple machines, such as lateral movement. STARMAP activates multiple machines in addition to the machine used for analysis, and observes communication to other machines and behavior after propagation. STARMAP also identifies behavior after lateral movement by extracting processes that are not normally in use on the propagated machines. We also implemented a prototype of STARMAP and demonstrated the feasibility of observing lateral movement through preliminary evaluation. Future work includes evaluation using more cases of malware.

References

1. Ahmad, M.A., Woodhead, S., Gan, D.: The V-network testbed for malware analysis. In: 2016 International Conference on Advanced Communication Control and Computing Technologies, ICACCCT 2016, pp. 629–635 (2016)
2. Akbanov, M., Vassilakis, G.V., Michael, D.L.: Wannacry ransomware: analysis of infection, persistence, recovery prevention and propagation mechanisms. J. Telecommun. Inf. Technol. **1**(1), 113–124 (2019)
3. Antonakakis, M., et al.: Understanding the Mirai botnet. In: 26th USENIX Conference on Security Symposium, SEC 2017, pp. 1093–1110 (2017)
4. ANY.RUN: Interactive Online Malware Sandbox. https://any.run/
5. Bai, T., Bian, H., Daya, A.A., Salahuddin, M.A., Limam, N., Boutaba, R.: A machine learning approach for RDP-based lateral movement detection. In: 2019 IEEE 44th Conference on Local Computer Networks, LCN 2015, pp. 242–245 (2019)
6. Barracuda: Threat Spotlight: New cryptominer malware variant. https://blog.barracuda.com/2020/06/25/threat-spotlight-new-cryptominer-malware-variant
7. CAPE Sandbox: CAPE: Malware Configuration And Payload Extraction. https://github.com/kevoreilly/CAPEv2
8. Check Point Research: Phorpiex Arsenal: Part I. https://research.checkpoint.com/2020/phorpiex-arsenal-part-i/
9. chrony: Introduction. https://chrony.tuxfamily.org/
10. Cisco Talos: Threat Spotlight: Follow the Bad Rabbit. https://blog.talosintelligence.com/bad-rabbit/
11. He, D., Gu, H., Zhu, S., Chan, S., Guizani, M.: A comprehensive detection method for the lateral movement stage of apt attacks. IEEE Internet Things J. **11**(5), 8440–8447 (2024)
12. Ho, G., et al.: Hopper: modeling and detecting lateral movement. In: 30th USENIX Conference on Security Symposium, SEC 2021, pp. 3093–3110 (2021)
13. Hybrid Analysis: Free Automated Malware Analysis Service - powered by Falcon Sandbox. https://www.hybrid-analysis.com/
14. Joe Security: Deep Malware Analysis - Joe Sandbox. https://www.joesecurity.org/service-offline
15. Khoury, J., Klisura, D., Zanddizari, H., Parra, G.D.L.T., Najafirad, P., Bou-Harb, E.: Jbeil: Temporal graph-based inductive learning to infer lateral movement in evolving enterprise networks. In: 2024 IEEE Symposium on Security and Privacy, S&P 2024, p. 13 (2024)
16. Kirat, D., Vigna, G., Kruegel, C.: BareCloud: bare-metal analysis-based evasive malware detection. In: 23rd USENIX Conference on Security Symposium, SEC 2014, pp. 287–301 (2014)
17. Lengyel, T.K., Maresca, S., Payne, B.D., Webster, G.D., Vogl, S., Kiayias, A.: Scalability, fidelity and stealth in the DRAKVUF dynamic malware analysis system. In: 30th Annual Computer Security Applications Conference, ACSAC 2014, pp. 386–395 (2014)
18. libvirt: The virtualization API. https://libvirt.org/index.html
19. Mandiant: WannaCry Ransomware Campaign: Threat Details and Risk Management. https://cloud.google.com/blog/topics/threat-intelligence/wannacry-ransomware-campaign/?hl=en
20. MBSD: Unraveling the inner workings of Conti ransomware (in Japanese). https://www.mbsd.jp/research/20210413/conti-ransomware/

21. Microsoft Learn: Process Monitor - Sysinternals. https://learn.microsoft.com/en-us/sysinternals/downloads/procmon
22. Miramirkhani, N., Appini, M.P., Nikiforakis, N., Polychronakis, M.: Spotless sandboxes: evading malware analysis systems using wear-and-tear artifacts. In: 2017 IEEE Symposium on Security and Privacy, S&P 2017, pp. 1009–1024 (2017)
23. MITRE: ATT&CK. https://attack.mitre.org/
24. Nguyen, H.N., Abri, F., Pham, V., Chatterjee, M., Namin, A.S., Dang, T.: MalView: interactive visual analytics for comprehending malware behavior. IEEE Access **10**, 99909–99930 (2022)
25. NtTrace: Native API tracing for Windows. https://rogerorr.github.io/NtTrace/
26. Recorded Future: Targeting of Olympic Games IT Infrastructure Remains Unattributed. https://www.recordedfuture.com/olympic-destroyer-malware
27. Recorded Future: Triage. https://tria.ge/
28. rewterz: Rewterz Threat Alert - APT MustangPanda - Active IOCs. https://www.rewterz.com/rewterz-news/rewterz-threat-alert-apt-mustangpanda-active-iocs-19
29. Sajid, M.S.I., et al.: SODA: a system for cyber deception orchestration and automation. In: Annual Computer Security Applications Conference, ACSAC 2021, pp. 675–689 (2021)
30. Sophos: A border-hopping PlugX USB worm takes its act on the road. https://news.sophos.com/en-us/2023/03/09/border-hopping-plugx-usb-worm/
31. tshark: tshark(1) Manual Page. https://www.wireshark.org/docs/man-pages/tshark.html
32. Tsuda, Y., et al.: STARDUST: large-scale infrastructure for luring cyber adversaries (in japanese). In: Computer Security Symposium 2017, CSS 2017, pp. 472–479 (2017)
33. Wagner, M., et al.: A survey of visualization systems for malware analysis. In: Eurographics Conference on Visualization 2015, EuroVis 2015, pp. 105–125 (2015)
34. Yong Wong, M., Landen, M., Antonakakis, M., Blough, D.M., Redmiles, E.M., Ahamad, M.: An inside look into the practice of malware analysis. In: Proceedings of the 2021 ACM SIGSAC Conference on Computer and Communications Security, CCS 2021, pp. 3053–3069 (2021)
35. Zhang, J., et al.: Scarecrow: deactivating evasive malware via its own evasive logic. In: 2020 50th Annual IEEE/IFIP International Conference on Dependable Systems and Networks, DSN 2020, pp. 76–87 (2020)

LogSHIELD: A Graph-Based Real-Time Anomaly Detection Framework Using Frequency Analysis

Krishna Chandra Roy[1]($\boxtimes$) and Qian Chen[2]

[1] New Mexico Institute of Mining and Technology, Socorro, NM 87801, USA
krishna.roy@nmt.edu
[2] University of Texas at San Antonio, San Antonio, TX 78249, USA
guenevereqian.chen@utsa.edu

Abstract. Anomaly-based cyber threat detection using deep learning is on a constant growth in popularity for novel cyber-attack detection and forensics. A robust, efficient, and real-time threat detector in a large-scale operational enterprise network requires high accuracy, high fidelity, and a high throughput model to detect malicious activities. Traditional anomaly-based detection models, however, suffer from high computational overhead and low detection accuracy, making them unsuitable for real-time threat detection. In this work, we propose **LogSHIELD**, a highly effective graph-based anomaly detection model in host data. We present a real-time threat detection approach using frequency-domain analysis of provenance graphs. To demonstrate the significance of graph-based frequency analysis we proposed two approaches. Approach-I uses a Graph Neural Network (GNN) LogGNN and approach-II performs frequency domain analysis on graph node samples for graph embedding. Both approaches use a statistical clustering algorithm for anomaly detection. The proposed models are evaluated using a large host log dataset consisting of 774M benign logs and 375K malware logs. LogSHIELD explores the provenance graph to extract contextual and causal relationships among logs, exposing abnormal activities. It can detect stealthy and sophisticated attacks with over 98% average AUC and F1 scores. It significantly improves throughput, achieves an average detection latency of 0.13 s, and outperforms state-of-the-art models in detection time.

Keywords: Anomaly Detection · Real-time Detection · Frequency Analysis · Machine Learning · Graph Neural Network

1 Introduction

Anomaly-based intrusion detection approaches are indispensable and widely adopted for zero-day attack detection, but they suffer from a high false-alarm rate compared to signature and rule-based detectors [25]. Although traditional signature and rule-based approaches achieve higher detection accuracy, they fail to detect sophisticated and novel attack campaigns as these detectors depend

J. Zhao and W. Meng (Eds.): SciSec 2024, LNCS 15441, pp. 56–75, 2025.
https://doi.org/10.1007/978-981-96-2417-1_4

on malware signatures. Anomaly-based detectors analyze flow-based data such as NetFlow, leveraging a set of flow fields to model anomaly behavior patterns [13]. However, the flow sequence analysis models suffer from modeling the long-term dependency of the network activity which limits the detection capability. To this end, the provenance graph of system logs provides a better representation of systems and network activities describing the information flow between system processes for anomaly detection [8]. Graph Neural Networks (GNNs) can capture the rich node semantics of provenance graphs by leveraging graph neighborhoods. Therefore, Machine learning/Deep learning detectors show promising anomaly detection performance utilizing GNN-based provenance graph semantics [16, 26].

As the size of provenance graphs grows, graph-based detectors become computationally expensive. The high computational overhead and increased processing time of deep learning algorithms cause the models to struggle with handling a high volume of audit data. Therefore, deep learning-based computation-intensive models can be deployed for offline detection but are challenging to use for real-time detection [3, 27].

Frequency domain analysis (FDA) of audit data is an emerging technique for reducing the computational overload of graph-based detection models. FDA leverages the transformation of graphs into the frequency domain to identify anomalous behaviors [4]. To leverage a graph representation of audit data and design a real-time anomaly detection model we propose LogSHIELD, a host data-driven provenance graph analysis framework exploiting FDA. LogSHIELD utilizes the causal, sequential, and logical relationship of system logs to construct a provenance graph. To demonstrate the trade-off between detection accuracy and detection time, we propose two graph analysis approaches. Approach-I is LogSHIELD with LogGNN, a GNN-based approach for embedding graph nodes and identifying abnormal sequences. Approach-II is LogSHIELD with FDA [4], a Frequency Domain Analysis (FDA) approach for embedding graph nodes. Upon graph embedding both approaches apply a statistical clustering algorithm using the cosine similarity of the node embedding vectors for anomaly detection.

Approach-I, LogSHIELD with LogGNN analyzes the log provenance graph and generates effective log embedding. However, one major drawback is embedding time and detection latency due to the complex embedding network. To handle the higher detection latency of the deep embedding model, approach-II performs frequency domain analysis on the graph. A bidirectional Long Short-Term Memory (BiLSTM) model in LogGNN is replaced with a frequency analysis FDA module. FDA extracts high-level features from the log provenance graph without utilizing computation-intensive deep learning models which is the key advantage of approach-II. LogSHIELD with FDA takes significantly less detection time and achieves higher throughput compromising a small fraction of detection accuracy.

The design and implementation of LogSHIELD analyzing host logs pose the following challenges. The first challenge is constructing a provenance graph using redundant and concurrent host logs. Many redundant and system-level noisy logs are generated which causes the graph to have thousands of redundant nodes and edges. To solve this challenge we pre-process the raw logs and statistically filter the redundant and noisy logs. The log messages contain rich information as event fields such as `EventID`, `ProcessName`, `ProcessID`, `Timestamp`, etc. Event fields are not uniform among all event logs and differ with system environments. For example, the benign host dataset contains 1200 unique fields. Therefore, the second challenge is to down-sample the event fields and select the right fields for content encoding. To solve this challenge we analyzed the 2TB of host data and found the unique and relevant log fields representing log semantics. The third challenge is a needle in a haystack. Hundreds of system logs are generated within seconds depending on the running processes, which makes detection challenging for real-time applications. Moreover, malicious logs are a very small fraction of the benign system logs which makes detection even harder. We solve this challenge by designing a robust and scalable analysis and detection framework. LogSHIELD constructs a provenance graph utilizing the activity logs of the host computer creating causal, sequential, and logical connections. LogSHIELD approach-I utilizes the LogGNN embedding algorithm to learn the underlying representation of the log graph. Random walk-based node sampling in LogGNN provides high throughput in analyzing the logs in large enterprise networks with thousands of users.

The main contributions of this paper are as follows:

- We propose a real-time anomaly detection framework LogSHIELD with FDA to achieve high throughput detection for real-time applications.
- We also propose LogSHIELD with LogGNN, a robust and high-fidelity anomaly detection framework using Graph Neural Network LogGNN to demonstrate the trade-off between detection accuracy and detection time.
- A major challenge in model evaluation is the insufficient real-world host log dataset. To this end, we present a real-world benign and malware dataset. The benign dataset contains more than 774M of host logs for 35 Windows machines and the malware dataset contains logs for 140 malware samples.
- LogSHIELD evaluated with the two real-world log datasets. The results validate the effectiveness of LogSHIELD for real-time and high-accuracy anomaly detection.
- An in-depth ablation study is performed for both the proposed approaches to evaluate functional modules and demonstrate the trade-off between detection accuracy and detection time.

The remainder of the paper is organized as follows. Section 2 presents the related works in real-time and graph-based anomaly detection. In Sect. 3 we introduce the LogSHIELD framework and describe the design approaches. Section 4 presents the experimental design, results, and result evaluations. Finally, Sect. 5 concludes the study.

2   Related Work

This section presents existing research on graph-based anomaly detection, focusing on approaches that utilize deep learning, and machine learning algorithms.

2.1   Real-Time Anomaly Detection

Early detection and prevention of threat actors can mitigate the profound losses incurred by cyber threats. However, sophisticated systems and stealthy attacks require real-time detection. Many researchers working toward reducing the detection time and achieving real-time anomaly detection [4,7–9,14,21–23].

In [22], Wu et al. proposed "Paradise", a real-time, generalized, and distributed provenance-based intrusion detection system. Paradise extracts process feature vectors in a separate environment at the system log level and stores them in high-efficiency memory databases. During the detection phase, it calculates provenance-based dependencies for intrusion detection. Another graph-based real-time outlier detection technique is proposed in [14], which learns all possible process relation semantics. The authors claim that a generalized process relation graph (GPRG) enables the detection of any outlier program with 96% accuracy at any time instance. In [9], Irshad et al. proposed TRACE, a scalable, real-time, enterprise-wide provenance tracking system for APT detection. It uses static analysis techniques to identify unit dependencies and strategies to generate concise provenance graphs, evaluated during five red-team engagements. In [21], Wang et al. proposed LightLog, a lightweight temporal convolutional network (TCN) for log anomaly detection in edge devices. LightLog uses word2vec to generate low-dimensional semantic vectors from logs and TCN for detection, achieving real-time processing and detection. Static log stream feature analysis approaches are also adopted by many researchers for real-time anomaly detection. In [4], Fu et al. proposed "Whisper", a real-time robust malicious network traffic detection method using frequency analysis. The authors experimented with 42 types of attacks and demonstrated that Whisper achieves high throughput with 99% AUC accuracy. While our proposed framework LogSHIELD with FDA is inspired by Whisper, there are notable differences. Unlike Whisper, which uses network traffic features and performs frequency analysis for detection, LogSHIELD constructs a provenance graph from host logs, performs RWR node sampling, and then applies frequency analysis to the sampled node contents.

2.2   Graph-Based Anomaly Detection

Graph representations of multidimensional and heterogeneous data are used in modeling system behavior, thereby detecting abnormalities and threats within a system. Many research works have applied graph-based analysis techniques to threat detection. In [12], Liu et al. proposed a graph-based user behavior modeling approach for insider threat detection using heterogeneous graph embedding. Log2vec uses a set of rules to construct the graph and represents each log entry as a low-dimensional vector using a random walk. In [19], Wang et al. introduced an approach for automatic insider threat detection that utilizes not only self-anomalous behaviors of an employee but also anomalies relative to other employees with similar job roles. It infers the correlation graph among the organization's employees and identifies potential threats using graph signal processing. A graphical analysis and anomaly detection-based approach was proposed in [6], introducing a hybrid framework consisting of a graphical processing unit and an anomaly detection unit to isolate anomalous users from heterogeneous data (logon/logoff, email logs, HTTP logs, etc.). It also incorporated psychometric data. In another paper, they used an attributed graph clustering technique and an outlier ranking mechanism for threat detection [5]. In [11], graph-based semi-supervised machine learning methods, such as label propagation and label spreading, were used for insider threat detection in the CERT insider threat test dataset. Many researchers utilize provenance graphs for anomaly detection. In [8], Han et al. proposed UNICORN, a provenance graph-based advanced persistent threat detection model that uses a histogram and hashing algorithm for graph feature extraction.

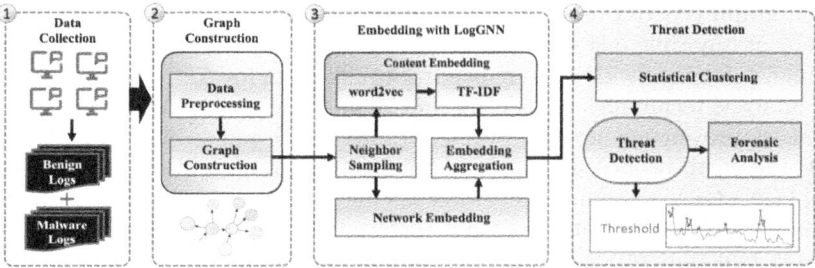

Fig. 1. Approach-I: System architecture of LogSHIELD with LogGNN graph embedding.

3 Proposed LogSHIELD Framework

We introduce two approaches for graph high-level feature extraction and a statistical clustering algorithm for anomaly detection.

3.1 Approach-I: LogSHIELD with LogGNN

Approach-I consists of four main steps: log data collection, graph construction, graph embedding with LogGNN, and anomaly detection. A constructional diagram of LogSHIELD with LogGNN is presented in Fig. 1. LogGNN learns the representation and dependency of the nodes in the provenance graph and obtains a semantic embedding vector for each of the nodes in the graph.

Log Preprocessing. The logs in the benign and malware datasets are collected using Windows Logging Service (WLS) in standard JSON format. For the benign dataset, we recruit 35 legitimate users from a large enterprise network and collect their working PCs' host logs over 90 days upon approval from the Institutional

Review Board (IRB) as human subjects are involved. For the malware dataset, we collected logs for 140 latest malware binaries deployed manually in a testbed with two target machines running Win10 OS. These two physical machines' configuration is the same as the configuration of benign participants' computers.

A total of around 774 million in event logs were collected where each event has its relevant field set with corresponding field values. We perform a data cleaning step before constructing a graph to filter the noisy and redundant system logs. We analyze the dataset and down-sample to 5 event fields `EventID`, `ProcessName`, `BaseFileName`, `LogonType`, `ParentProcessName`, out of 1200 unique fields as the content of the logs. These two steps of pre-processing reduce the volume of the host data to one-fifth of the 2TB raw data size which significantly reduces the log embedding time in LogGNN. Figure 2 presents an example of a host log with EventID 4634, a logoff event.

All events in the processed host logging dataset contain the "EventID" field, which represents the processes triggered by users or computer programs. Hundreds of unique "EventID" values have been collected, and the number of unique "EventID" values varies by user's activity. For example, a sequence of five events has EventID values $4624 \rightarrow 4672 \rightarrow 4798 \rightarrow 4798 \rightarrow 4634$, meaning that (1) the user successfully logon ($EventID = 4624$); (2) special privileges are assigned to new logon ($EventID = 4672$); (3) a user's local group membership is enumerated by "chrome.exe", the Google Chrome application ($EventID = 4798$); (4) the user's local group membership is enumerated by "svchost.exe", a Windows system process to host from one to many services ($EventID = 4798$); and (5) the user is logged off ($EventID = 4634$). A detailed description of host log collection can be found in [1].

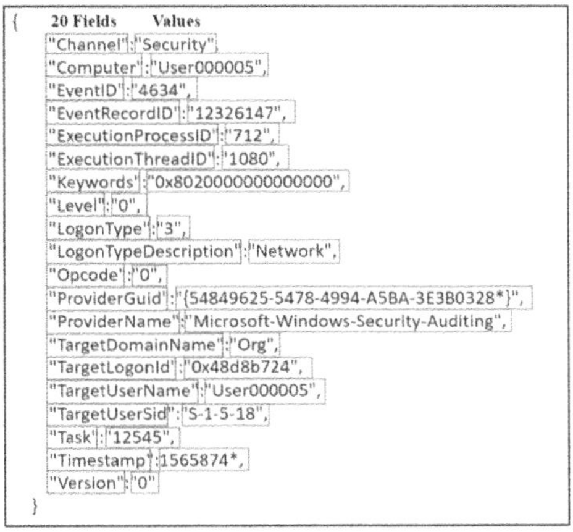

Fig. 2. A host log example for `EventID` **4634**. Red boxes are event fields and green boxes are values of the corresponding fields. (Color figure online)

Graph Construction. To address system anomaly detection as a graph analysis problem, we construct a provenance graph from daily host logs within an enterprise network, capturing the interrelationship of log entries. This is step 2 in Approach-I as shown in Fig. 1. The log provenance graph (LPG) focuses on three types of relationships: causal, sequential, and logical, to learn the contextual representation of event logs. For sequential relationships, we consider the timestamps of log entries to resolve out-of-order issues caused by log arrival delays. Causal relationships among processes (EventID 4688, 4689) are established by creating directional connections between parent and child processes using the fields ProcessName, BaseFileName, and ParentProcessName. Additionally, LogSHIELD builds logical relationships by connecting log entries with the same timestamp. LogSHIELD constructs the graph using five key event fields: EventID, ProcessName, BaseFileName, LogonType, and ParentProcessName. This approach ensures the graph construction follows these specific rules.

- Rule 1: All log entries from the same user are connected sequentially based on their timestamps, forming the main chain of triggered logs.
- Rule 2: Log entries from the same user that are related by a parent-child process are connected with a directional link.
- Rule 3: Log entries from the same user with the same timestamp (concurrent logs) are all connected to the last log of the chain with a distinct timestamp.
- Rule 4: Log entries from the same user with the same timestamp maintain an internal connection between them.

Figure 3 presents an example of a provenance graph constructed using a small fraction of daily benign logs of 3 randomly selected users P1, P2, and P3.

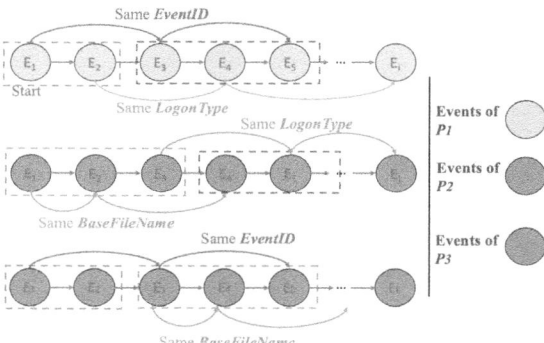

Fig. 3. A provenance graph constructed using a small fraction of daily benign logs of User P1, P2, and P3.

Log Neighbor Sampling. This is the beginning of step 3 in approach-I. The neighbors of a node significantly contribute to its semantics in representation learning. Therefore, in the first step of graph embedding, it samples representative neighbor nodes, performs encoding, and aggregates them to find the embedding of the target node. Many node neighbor sampling algorithms have been adopted recently including random walk sampling [24], direct neighbor sampling with different order [20]. Direct neighbors can differ in number leading to insufficient representation of a node and an inability to capture diverse node types in the graph. Random walk helps to capture the heterogeneous node by sampling the n-hop neighbors.

Random Walk is one of the most popular neighbor sampling algorithms. In this work, we used a modified random walk with a restart strategy. It generates paths traversed by walkers starting from node $v \in V$. The walker travels through n-hop neighbors of the current node and returns back until a fixed number of nodes are sampled which is walk length, the walker traverses through the neighbors. We sample $\mathcal{S}(v)$ neighbor nodes of the node v.

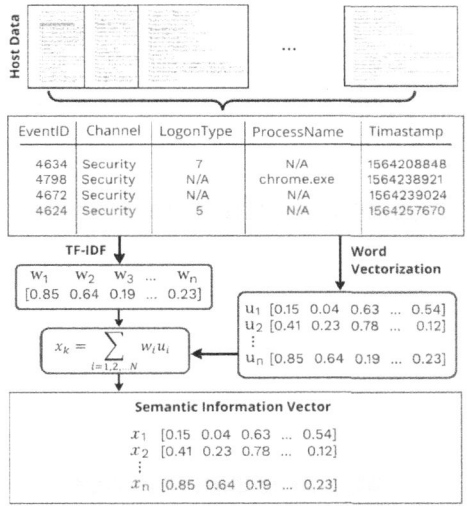

Fig. 4. Log parsing workflow for approach-I. A random set of log event fields is shown in the parsing workflow.

Log and Network Encoding. The next step of graph embedding is encoding heterogeneous contents C_v, which are attributes or texts of the nodes from $v \in V$. LogGNN uses word2vec [2] model for encoding log fields. Node content embedding is performed using a log parsing algorithm in [17]. The details of the encoding text content of event nodes are presented in this section. In a similar fashion to Natural Language Processing (NPL) that converts a group of sentences into tokens, the events in the host log are represented to tokens

$T = [t_1, t_2, \cdots, t_n]$ as presented in Fig. 2. In a host log file, the fields of the dictionary structure (e.g., BaseFileName, EventID) are the tokens, and the number of unique tokens selected from the event fields is n. The log file is converted to a sequence of event vectors $\mathcal{S}(v) = [V_1, V_2, \cdots, V_K]$. Where K represents the $K - th$ node. After that, event vectors are represented as $V_i = [v_1, v_2, \cdots, v_n]$. The next step of the parsing algorithm is word vectorization. We use FastText [10], an extension to the Word2Vec model for word vectorization. Each token parameter value v_j ($j \in [1, n]$) of host events converted to a e-dimension vector u_j (i.e., $v_j \rightarrow u_j = [a_1, a_2, \cdots, a_e]$). Vector V_i is transposed as $V_i^{(T)}$, where $i \in [1, K]$ then $V_i^{(T)}$ can be converted as U_i, which is a $n \times e$ matrix.

Every field information in an event log presents its individual significance. To incorporate the relative importance of a word in a particular document we use the term frequency-inverse document frequency (TF-IDF) weighting method [18]. After applying TF-IDF, the weight vector of an event word vector $V_i = [v_1, v_2, \cdots, v_n]$ ($i \in [1, K]$) is represented as $W_i = [w_1, w_2, \cdots, w_n]$, where w_i is the TF-IDF score of v_i. The content embedding vector $X_k = [X_1, X_2, \cdots, X_K]$ ($k \in [1, K]$) can be obtained using the following equation.

$$X_k = WU = [x_1 \ x_2 \ \cdots \ x_e], \quad x_j = \frac{1}{n} \sum_{i \in [1, n]} W_i U_{ij} \tag{1}$$

The workflow of node content embedding is presented in Fig. 4. The feature vector of the contents is represented as $x \in \mathbb{R}^{e \times 1}$ where e is the embedding dimension. This solves the problem of multidimensional host log embedding. In the next step, LogGNN encodes the network topology between the sampled nodes. To obtain the network embedding for each of the nodes of the input graph we use well known word2vec embedding technique. Word2vec takes the sampled neighbor set of the node $v \in V$ and transforms it to a $e-$dimensional embedding vector $E_{net}(v)$.

Embedding Aggregation. As mentioned in Sect. 3.1, the n-hop neighbor encoding vectors x of each node in the graph are aggregated using the BiLSTM model. The neighbor sampling and aggregation are performed for each node of the graph to obtain embedding for all the nodes.

We use a module to aggregate content embeddings using a sequence-based deep learning model BiLSTM [16]. In Sect. 3.1, we sampled K neighboring nodes for each graph node $v \in K(v)$ using the RWR strategy. Final aggregation is achieved in two steps. In the first step, we concatenate the network embedding $E_{net}(v)$ and content embedding $E_{cont}(v)$. In the second step, the BiLSTM aggregation model takes the concatenated node embedding $E_v = E_{net} + E_{cont}$ of all the K sampled nodes and combines them to capture deep feature interactions. We implement the aggregation model using BiLSTM. The overall aggregation function can be formulated as follows:

$$E_v^{ag} = \frac{\sum_{v \in K(v)} [\overrightarrow{\text{LSTM}}\{E_v\} \oplus \overleftarrow{\text{LSTM}}\{E_v\}]}{K}, \tag{2}$$

where the ⊕ operator denotes the concatenation of left and right directional LSTM output states. The node embedding generated by LogGNN is the input to the statistical clustering and anomaly detection algorithm presented in Sect. 3.3.

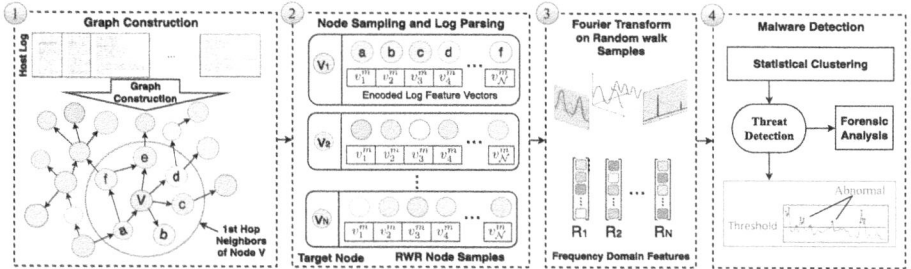

Fig. 5. Approach-II: System Architecture of LogSHIELD with frequency domain analysis (FDA).

3.2 Approach-II: LogSHIELD with FDA

LogSHIELD with FDA uses the same modules of approach-I for graph construction, neighbor sampling, and log parsing as presented in Sect. 3.1 and shown in steps 1 and 2 in Fig. 5.

In this approach, step 3 performs a frequency domain analysis on the sampled nodes to extract the semantic features of the sampled graph nodes. Let's consider there are N nodes in the graph and we sample $\mathcal{N}(v)$ neighbor nodes of the node v. The sampled neighbor set V can be represented as $V = [v_1^m, v_2^m, \cdots, v_{\mathcal{N}}^m]$, where m is the number of fields in each log.

We perform Discrete Fourier Transformation (DFT) on the N nodes of the graph and $\mathcal{N}$ neighbor samples for each node in N. Considering the $\mathcal{N}$ samples as the window length, we obtain the frequency domain features using the following equation.

$$F_{ik} = \sum_{n=1}^{\mathcal{N}} x_{kn} e^{-\frac{2\pi(n-1)(k-1)}{\mathcal{N}}} \quad (1 < k < \mathcal{N}), \tag{3}$$

where $k \in [1 : N]$ which is k^{th} window frequency component of $\frac{2\pi(k-1)}{\mathcal{N}}$. The frequency domain features $F_{ik} = a_{ik} + jb_{ik}$, are complex numbers and need to transform into real numbers for further clustering.

$$\begin{cases} a_{ik} = \sum_{n=1}^{\mathcal{N}} x_{kn} \, cos\frac{2\pi(n-1)(k-1)}{\mathcal{N}} \\[2mm] b_{ik} = \sum_{n=1}^{\mathcal{N}} -x_{kn} \, sin\frac{2\pi(n-1)(k-1)}{\mathcal{N}} \end{cases} \tag{4}$$

We obtain the modulus of the complex features. As the complex feature vectors from DFT are conjugate and symmetric, we further reduce the feature dimension by taking half of the modulus feature vector. Now taking the first half of the modulus,

$$r_{ik} = a_{ik}^2 + b_{ik}^2 \tag{5}$$

$$R_{ik} = \frac{ln(r_{ik} + 1)}{C} \quad (1 < i < N) \tag{6}$$

$$R_{ik} = R_{ik}[1 : L_f] \quad (L_f = *\frac{N}{2} + 1) \tag{7}$$

Now we perform a logarithmic transformation on r_{ik} to obtain a numerically stable feature vector. Finally, we obtain the node feature vector R_{L_f1} with dimension $(L \times f1)$.

$$R_{ki} = \begin{bmatrix} R_{11} & R_{12} & \cdots & R_{1N} \\ R_{21} & R_{22} & \cdots & R_{2N} \\ \vdots & \vdots & \ddots & \vdots \\ R_{L_f1} & R_{L_f2} & \cdots & R_{L_f \times N} \end{bmatrix}, \tag{8}$$

where L is the number of log features.

3.3   Anomaly Detection: Statistical Clustering

We design an unsupervised statistical clustering algorithm to learn the pattern of log neighbor features in both Approach-I and Approach-II. This is step 4 in both Fig. 1 and Fig. 5. Conventional clustering algorithms like k-means clustering and other similar methods are not suitable because of the initialization of cluster centers and the number of clusters k. Therefore, in the detection phase, we cluster the node embeddings $\{R_1, R_2, R_3, ...R_N\}$ using a custom clustering algorithm. We used a cosine-similarity-based log clustering algorithm to cluster system logs and identify malicious activity. We train the clustering algorithm with the benign dataset. Each node embedding x_i from approach-I or feature vector R_i from approach-II where $i \in N$, is assigned to a cluster using their cosine similarity. C_{n_c} clusters are achieved where n_c is the number of clusters.

If R_N is a set of all log feature vectors and δ is a similarity threshold clustering. It calculates the cosine similarity between two embedding vectors to identify cluster members. The clustering process satisfies the following conditions to obtain clusters $\{C_1, C_2, ...C_{n_c}\}$.

$$\begin{cases} \forall e_x \in C_1, \exists e_y \in C_1, sim(e_x, e_y) \geq \delta \\ \forall e_z \in \mathcal{S} \setminus C_1, sim(e_x, e_z) < \delta \end{cases} \tag{9}$$

where $\mathcal{S} \setminus C_1$ is the difference between the two sets. We find the closest cluster center for each embedding or feature vector and calculate the averaged L2-norm representing the training loss.

$$loss_{train} = \frac{1}{N}\sum_{i=1}^{N}\|R_i - \tilde{C}\|^2 \tag{10}$$

In the detection phase, the clustering module calculates the distance between the FDA feature of test data sample R_i^t and the benign cluster centers. The closest cluster center is an estimated cluster of R_i^t and calculates the L2-norm as the prediction loss. If the estimation loss is higher than the training loss the log embedding and feature vector are considered malicious.

$$loss_{test} = min(\|R_i - \tilde{C}\|^2), \tag{11}$$

where $\tilde{C}$ is the assigned cluster for each vector.

4   Experimental Evaluation

In this section, we evaluate the effectiveness of the LogSHIELD framework and its components in anomaly detection. In addition, we measure the performance of different design parameters. We also perform a thorough ablation study to compare the performance of LogSHIELD with its variants.

4.1   Experimental Setup

LogSHIELD consists of ~8.1k lines of python code and is deployed on a server with Intel(R) Xeon(R) CPU E5-2640 v4 @ 2.40 GHz CPU (20 processors) and 48GB memory.

4.2   Dataset

To evaluate the proposed anomaly detection framework, we use two real-world host datasets. The first dataset is a benign host dataset and the second dataset is a malware dataset with 140 malware sample logs we collected in this work.

Threat Model. LogSHIELD endeavors to develop a module for detecting abnormal activities based on host data. We consider only host logs from individual user spaces. As the log resources can be manipulated, the integrity of the host log is out of the scope of this paper and assumes the host logs can be trusted. Due to data limitations, the system environment is limited to Windows OS for benign and malware data samples. The detection system of LogSHIELD is not exposed to any malware data for training and is trained using completely benign data samples. This means that LogSHIELD should be able to detect zero-day attacks. For model evaluation and testing, we collected logs for 140 malware samples of 6 malware families (Ransomware, Rootkit, Trojan, Adware, Backdoor, and Browser Hijack).

Table 1. Benign and Malware Data Description

Dataset	Malware Family Count	#Log	#Node	Data Volume
Benign	x	774M	774M	1.6 TB
Malware	Adware (3)	9K	9K	576 KB
	Backdoor/Trojan (45)	110K	110K	72 MB
	Browser Hijacker (7)	14K	14K	4 MB
	Crypto Miner (5)	12.5K	12.5K	2.5 MB
	Ransomware (72)	210K	210K	86 MB
	Rootkit (4)	8K	8K	2 MB
	Others (6)	11K	11K	670 KB

Benign and Malware Dataset. The benign host log dataset includes system logs from 35 participants collected over 90 days. After preprocessing the dataset contains 774M activity logs with a volume of 1.6 TB.

The malware dataset contains $374.5K$ logs with a volume of 167 MB collected deploying 140 binaries from mostly 6 malware families. Each malware sample was run for 15 min, and the logs from the infected machines were stored on a log server. After each run, the infected machines were re-imaged using the FOG Server before proceeding with the next malware sample. The selected and deployed malware samples belong to some of the most disruptive malware families, as shown in Table 1.

4.3   Train and Test Datasets

LogSHIELD is trained in an unsupervised manner with subsets of the benign dataset. The LogGNN embedding model of LogSHIELD approach-I is trained with 35 users' 7 days of benign data (60M of Benign Logs). The detection models of both approach-I and approach-II are trained with 90 days of benign data with 774M of logs.

To represent the real scenario where malware activities coincide with benign host activities, both approaches of LogSHIELD are evaluated on a malware dataset having around 50% Benign logs (370K). Besides the 50% benign data, each malware log with (a 15-min execution) also contains some benign logs.

Table 2. Datasets for Training and Testing.

Models	Dataset	#Log
LogGNN	Train: Benign	60M
Detection Model	Train: Benign	774M
LogSHIELD	Test: Malware	370K + 372K
(Approach - I & II)	(Benign + Malware Logs)	

4.4   LogSHIELD Implementation Details

The components of the LogSHIELD framework both approach-I and approach-II, including log pre-processing and parsing, graph embedding model LogGNN, FDA feature extraction, and downstream clustering models are implemented in Pytorch [15] and the source code will be available soon.

Hop Count and Sample Size: Node sampling in the LogGNN graph embedding model is implemented with $3 - hop$ Random Walk with Restart(RWR) and sampled neighbor size is selected K = 40 based on parameter tuning experiments presented in Sect. 4.9 to obtain the optimal semantics of the neighboring nodes.

Token Numbers: In the word2vec log encoding model we set a token size of 5 from the most frequently seen fields out of a total of 1,200 unique fields of the training data to obtain the parameter value vectors.

Word Embedding Dimension: The embedding dimension is set to $e = 100$ which is determined based on a separate experiment performed with different embedding dimensions presented in Sect. 4.9. The value of e is passed to the FastText model to encode word parameter value vectors to semantic information vectors.

BiLSTM Model Deployment: We use the open-source machine learning library Pytorch to build the BiLSTM model. It consists of 2 hidden layers each layer having 64 and 32 units. The input dimension is 100. The ReLU activation function and the mean absolute error (MAE) loss function is used to train the embedding model for 100 epochs. All the model training and testing were performed using an NVIDIA A100 GPU server with 40GB of memory. We set the minibatch size to 128, and the learning rate to 0.0015.

Similarity Threshold: The similarity threshold δ between two node embedding or FDA feature vectors is set to 0.72 for optimal clustering performance. A value of δ between 0 to 0.5 outputs a large number of clusters whereas δ between 0.8 to 1 gives a small number of clusters missing many suspicious clusters.

DFT Window Size: In the FDA module the sliding window size of the DFT is set to the number of RWR sampled neighbors $\mathcal{N} = 40$.

4.5   LogSHIELD Detection Performance

We perform an anomaly detection experiment to detect malware logs that are never exposed during training. The detection models are trained with benign data only as summarized in Table 2. We evaluate the detection performance in terms of Accuracy, Precision, Recall, and F-Score. We also use TPR, FPR, and the area under the ROC curve (AUC) to measure LogSHIELD performance.

LogSHIELD anomaly detection performance is evaluated across 6 malware families, including Ransomware, Rootkit, Trojan, Adware, Backdoor, and Browser Hijacker. LogSHIELD demonstrated outstanding detection performance for most malware families. Approach-I using LogGNN embedding, achieved an average AUC of 98%, a TPR of 0.98, and an FPR of 0.038, while Approach-II showed a 97% AUC, a 0.97 TPR, and a 0.079 FPR, as detailed in Table 3. However, LogSHIELD's performance with LogGNN marginally degraded for Rootkit and Adware, with AUCs of 96% and 97%, respectively. In terms of FPR Approach-I shows 2 times better performance than approach-II with FDA. It confirms that LogGNN graph representation learning is more effective than FDA graph feature learning for detection. This is because LogSHIELD with LogGNN better captures the semantic information of the neighboring nodes compared to LogSHIELD with FDA.

Table 3. Malware Detection Results of LogSHIELD [Approach-I: LogSHIELD with LogGNN; Approach-II: LogSHIELD with FDA].

Malware Families	Methods					
	Approach-I			Approach-II		
	TPR	FPR	AUC	TPR	FPR	AUC
Ransomware	0.99	0.02	0.99	0.97	0.06	0.98
Rootkit	0.97	0.08	0.96	0.95	0.12	0.95
Trojan	0.99	0.03	0.99	0.97	0.1	0.95
Adware	0.97	0.07	0.96	0.98	0.085	0.96
Backdoor	0.98	0.02	0.98	0.98	0.07	0.97
Browser Hijack	0.98	0.012	0.99	0.96	0.04	0.98
Average	0.98	0.038	0.98	0.97	0.079	0.97

4.6 LogSHIELD Performance Comparison with Baselines

We performed another study to evaluate LogSHIELD detection AUC, detection time, and throughput in comparison with the other five baseline anomaly detection models (Log2vec, DeepLog, Unicorn, LogRobust, and DeepRan). According to results presented in Table 4, LogSHIELD with LogGNN embedding outperforms the baseline models in terms of detection performance and achieves 98% AUC but the detection time of LogSHIELD with LogGNN is similar to baseline models. Whereas LogSHIELD with FDA achieves an AUC of 96% which is higher than most of the baseline models. It achieves an anomaly detection time of 0.13 s which is significantly less than LogSHIELD with LogGNN and other state-of-the-art models. Frequency domain analysis on the provenance graph of host data in approach-II effectively extracts graph representation with significantly less computation than approach-I. Therefore, approach-II, LogSHIELD with FDA significantly outperforms most of the baseline models and takes a small fraction of the detection time of the other models. As FDA uses DFT frequency analysis for log feature extraction replacing deep graph neural network-based (LogGNN), it has less computational overhead. Therefore, LogSHIELD with FDA can provide real-time detection in an enterprise network.

Table 4. Malware Detection Results of LogSHIELD in Comparison with Five Baseline Methods.

Method	AUC	Detection Time (s)	Throughput (Gbps)
Log2Vec	0.89	1.61	0.35
DeepLog	0.95	1.65	0.5
Unicorn	0.92	0.57	3.2
LogRobust	0.95	1.52	0.4
DeepRan	0.98	1.36	0.72
LogSHIELD with LogGNN	0.98	1.67	0.45
LogSHIELD with FDA	0.96	0.13	5.1

4.7   Detection Latency and Throughput

We conduct experiments on a fused dataset of malware data and 25% of the large benign dataset. The model is trained with the remaining 75% benign data. We measure the anomaly detection time and throughput of the detection model using both approach-I and approach-II. The results are shown in Table 4. LogSHIELD with FDA performs significantly better than LogSHIELD with LogGNN. Approach-I with LogGNN can detect malware events with an average detection time of 1.67 s whereas approach-II with FDA can detect with an average detection time of 0.13 s. The higher computation overhead of Log-GNN is one of the main reasons for the higher detection time of approach-I. In terms of throughput, Approach-II achieves 11.3 times higher throughput than approach-I with LogGNN. We find that LogSHIELD with FDA achieves an average throughput of 5.1 Gbps whereas with the LogGNN graph learning model, it achieves 0.45 Gbps. In summary, the results show that the frequency analysis model slightly suffers from detection AUC but significantly outperforms machine learning-based models in terms of detection time and throughput. The throughput of the baseline models is not calculated and compared.

4.8   Ablation Analysis

We perform an ablation study using benign and malware datasets to evaluate different functional modules of LogSHIELD in anomaly detection. In this study, we present the significance of RWR node sampling, log graph representation learning method LogGNN, and FDA in the overall detection framework as shown in Table 5. In different combinations, we obtain seven ablation models. All the ablation models are trained with 35 users' benign data only and evaluated with the malware dataset. We use the same provenance graph constructed in previous experiments for all the models. According to the results, $LogSHIELD^1$ and $LogSHIELD^2$ differ on the neighbor sampling and it shows that $LogSHIELD^1$ significantly outperforms $LogSHIELD^2$ with AUC of 0.98 and 0.89 respectively as it has the 3-hop neighbor information contributing to

Table 5. Performance Comparison of LogSHIELD with Different Embedding and Clustering Algorithms.

Methods	AUC	Time
$LogSHIELD^1$ (RWR+LogGNN+Statistical Clustering)	0.98	1.67
$LogSHIELD^2$ (LogGNN + Statistical Clustering)	0.89	1.5
$LogSHIELD^3$ (LogGNN + k-means clustering)	0.89	1.51
$LogSHIELD^4$ (RWR + FDA + Statistical Clustering)	0.97	0.13
$LogSHIELD^5$ (RWR + FDA + k-means clustering)	0.95	0.15
$LogSHIELD^6$ (FDA + Statistical Clustering)	0.81	0.11
$LogSHIELD^7$ (FDA + k-means clustering)	0.79	0.11

semantic node feature extraction. The latter model has less computation time. $LogSHIELD^1$ and $LogSHIELD^4$ differ in graph representation learning algorithm and both use the same neighbor sampling and clustering algorithm. It appears that $LogSHIELD^1$ has a higher computation time which is 1.67 s as it uses LogGNN a BiLSTM-based sequence learning model compared to the $LogSHIELD^4$ with a computation time of 0.13 s that uses FDA for high-level features of the neighbor samples. $LogSHIELD^2$ and $LogSHIELD^3$ differ in the clustering algorithm keeping other modules unchanged and the results show that $LogSHIELD^2$ with custom statistical clustering algorithm performs similarly the $LogSHIELD^3$ with k-means clustering in terms of AUC of 0.89. Without RWR sampling the clustering algorithms are not contributing to better results. Therefore, in $LogSHIELD^4$ and $LogSHIELD^5$, RWR node sampling and FDA are included and the models differ in the clustering algorithm. Finally, the other two variants $LogSHIELD^6$ and $LogSHIELD^7$ achieve the lowest detection time but without RWR sampling AUC also drops significantly. Therefore $LogSHIELD^4$ presents a better tradeoff between AUC and detection time.

4.9 LogSHIELD Parameter Performance

LogSHIELD performance depends on some key design parameters such as RWR walk length, number of hops for node sampling, LogGNN embedding dimension, etc. To evaluate the significance of the design parameters and find the optimal value we measure the preset performance metrics in different settings. We experiment with random-walk lengths ranging from 10 to 60 with a hop count of 3 as shown in Fig. 6a. We observe that initially at walk length 10, AUC, Precision, Recall, and F-Score all are very low and gradually increase with higher walk length as more neighbors contribute to the semantic feature of the target node. At a walk length of 40, all the scores reach higher than 0.96 and start gradually decaying at later values. Similar behavior is observed for hop count. We measure performance metrics with hop counts ranging from 1 to 6. At a hop count of 3, LogSHIELD performs with the highest AUC of 0.98 and starts decaying at higher hop counts. The reason for decaying at a higher hop count could be the over-aggregation of unrelated neighbors to target node semantics.

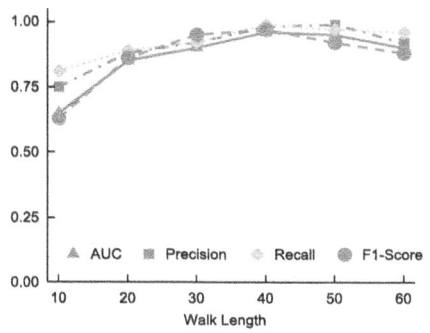

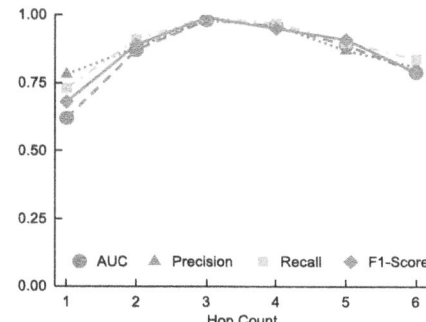

(a) RWR walk length performance measure. Experimented with $\mathcal{N}$ of 10 to 60 nodes in the provenance graph.

(b) Hop count (n-hop) performance measures keeping walk length fixed. Experimented with hop count n of 1 to 6.

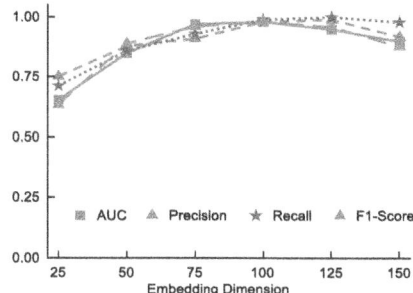

(c) LogGNN embedding dimension d over performance measures. Experimented with embedding dimensions ranging from 25 to 150.

Fig. 6. LogSHIELD performance in terms of AUC, Precision, Recall, and F-Score.

We also measure the model performance with LogGNN embedding dimensions ranging from 25 to 150. The results in Fig. 6c show that the model starts to achieve higher results at an embedding dimension of $e = 50$ and gradually increases at higher embedding dimensions at the same time it increases the computation overhead in LogGNN. We chose an embedding dimension of 100 as optimal for LogSHIELD.

5   Conclusion

In this paper, we propose a provenance graph-based real-time anomaly detection framework LogSHIELD. We present two graph representation learning approaches using the LogGNN model and FDA frequency analysis for real-time anomaly detection to reduce the computational overhead and improve the efficiency of the detection model. LogSHIELD with LogGNN achieves 98% average detection AUC whereas with FDA it achieves 96% AUC. On the other hand,

LogSHIELD with FDA takes 0.13 s for detection time whereas LogGNN takes 1.67 s. It also outperforms baseline models with higher detection AUC and lower detection time. Frequency analysis of host logs with FDA significantly benefits the design of a real-time anomaly detection framework. According to results in Table 5, the novel approach LogSHIELD with RWR and FDA significantly improves anomaly detection efficiency and facilitates real-time detection. A key challenge for LogSHIELD is the preprocessing of high-volume host data and the construction of graphs for detecting advanced persistent threats. In the future, we plan to further analyze the host data to reduce detection time and incorporate more real-host cyber activity. This will allow us to experiment with additional hosts and evaluate the scalability of the framework.

Acknowledgments. This work was supported in part by the U.S. Department of Energy/National Nuclear Security Administration (DOE/NNSA) #DE-NA0003985, NSF Grants #1812599, and Sandia National Laboratories (Department of Energy #DENA000352- 5/PO2048463). Any opinions, findings, conclusions, or recommendations expressed in this material are those of the authors and do not necessarily reflect the views of any of these funding agencies.

References

1. Acquesta, E., et al.: Detailed statistical models of host-based data for detection of malicious activity. Technical report, Sandia National Lab. (SNL-NM), Albuquerque, NM, USA (2019)
2. Church, K.W.: Word2vec. Nat. Lang. Eng. **23**(1), 155–162 (2017)
3. Du, M., Chen, Z., Liu, C., Oak, R., Song, D.: Lifelong anomaly detection through unlearning. In: Proceedings of the 2019 ACM SIGSAC Conference on Computer and Communications Security, pp. 1283–1297 (2019)
4. Fu, C., Li, Q., Shen, M., Xu, K.: Realtime robust malicious traffic detection via frequency domain analysis. In: Proceedings of the 2021 ACM SIGSAC Conference on Computer and Communications Security, pp. 3431–3446 (2021)
5. Gamachchi, A., Boztas, S.: Insider threat detection through attributed graph clustering. In: 2017 IEEE Trustcom/BigDataSE/ICESS, pp. 112–119. IEEE (2017)
6. Gamachchi, A., Sun, L., Boztas, S.: A graph based framework for malicious insider threat detection. arXiv preprint arXiv:1809.00141 (2018)
7. Han, S.H., Lee, D.: Kernel-based real-time file access monitoring structure for detecting malware activity. Electronics **11**(12), 1871 (2022)
8. Han, X., Pasquier, T., Bates, A., Mickens, J., Seltzer, M.: Unicorn: runtime provenance-based detector for advanced persistent threats. arXiv preprint arXiv:2001.01525 (2020)
9. Irshad, H., et al.: TRACE: enterprise-wide provenance tracking for real-time APT detection. IEEE Trans. Inf. Forensics Secur. **16**, 4363–4376 (2021)
10. Joulin, A., Grave, E., Bojanowski, P., Douze, M., Jégou, H., Mikolov, T.: FastText. zip: compressing text classification models. arXiv preprint arXiv:1612.03651 (2016)
11. Le, D.C., Zincir-Heywood, N., Heywood, M.: Training regime influences to semi-supervised learning for insider threat detection. In: 2021 IEEE Security and Privacy Workshops (SPW), pp. 13–18. IEEE (2021)

12. Liu, F., Wen, Y., Zhang, D., Jiang, X., Xing, X., Meng, D.: Log2vec: a heterogeneous graph embedding based approach for detecting cyber threats within enterprise. In: Proceedings of the 2019 ACM SIGSAC Conference on Computer and Communications Security, pp. 1777–1794 (2019)

13. Lo, W.W., Layeghy, S., Sarhan, M., Gallagher, M., Portmann, M.: E-GraphSAGE: a graph neural network based intrusion detection system. arXiv preprint arXiv:2103.16329 (2021)

14. Panda, B., Tripathy, S.N.: Host-specific outlier detection using process relation semantics with graph mining. In: Advances in Data Science and Management, pp. 449–462. Springer (2022)

15. PRODUCTIO, P.F.R.T.: (2022). https://pytorch.org/

16. Roy, K.C., Chen, G.: GraphCH: a deep framework for assessing cyber-human aspects in insider threat detection. IEEE Trans. Dependable Secure Comput. (2024)

17. Roy, K.C., Chen, Q.: DeepRan: attention-based biLSTM and CRF for ransomware early detection and classification. Inf. Syst. Front. 1–17 (2020)

18. Salton, G., Buckley, C.: Term-weighting approaches in automatic text retrieval. Inf. Process. Manage. **24**(5), 513–523 (1988)

19. Wang, J., Aggarwal, S., Ji, F., Tay, W.P., et al.: Learning correlation graph and anomalous employee behavior for insider threat detection. In: 2018 21st International Conference on Information Fusion (FUSION), pp. 1–7. IEEE (2018)

20. Wang, X., et al.: Heterogeneous graph attention network. In: The World Wide Web Conference, pp. 2022–2032 (2019)

21. Wang, Z., Tian, J., Fang, H., Chen, L., Qin, J.: LightLog: a lightweight temporal convolutional network for log anomaly detection on the edge. Comput. Netw. **203**, 108616 (2022)

22. Wu, Y., et al.: Paradise: real-time, generalized, and distributed provenance-based intrusion detection. IEEE Trans. Dependable Secure Comput. (2022)

23. Yang, X., Peng, G., Zhang, D., Lv, Y.: An enhanced intrusion detection system for IoT networks based on deep learning and knowledge graph. Secur. Commun. Netw. **2022** (2022)

24. Zhang, C., Song, D., Huang, C., Swami, A., Chawla, N.V.: Heterogeneous graph neural network. In: Proceedings of the 25th ACM SIGKDD International Conference on Knowledge Discovery & Data Mining, pp. 793–803 (2019)

25. Zhang, X., et al.: Robust log-based anomaly detection on unstable log data. In: Proceedings of the 2019 27th ACM Joint Meeting on European Software Engineering Conference and Symposium on the Foundations of Software Engineering, pp. 807–817 (2019)

26. Zheng, J., Li, Q., Gu, G., Cao, J., Yau, D.K., Wu, J.: Realtime DDoS defense using COTS SDN switches via adaptive correlation analysis. IEEE Trans. Inf. Forensics Secur. **13**(7), 1838–1853 (2018)

27. Zhu, S., et al.: You do (not) belong here: detecting dpi evasion attacks with context learning. In: Proceedings of the 16th International Conference on emerging Networking EXperiments and Technologies, pp. 183–197 (2020)

Completeness Analysis of Mobile Apps' Privacy Policies by Using Deep Learning

Khalid Alkhattabi[1,2]($\boxtimes$) and Chuan Yue[1]($\boxtimes$)

[1] Colorado School of Mines, 1500 Illinois St, Golden, CO 80401, USA
{kalkhattabi,chuanyue}@mines.edu
[2] Tabuk University, King Faisal Road, Tabuk 47512, Saudi Arabia
kalkhattabi@ut.edu.sa

Abstract. Mobile applications (apps) are an essential part of today's connected digital world, and their privacy policies are the legal source that describes how apps collect and use user data. The absence of clear guidelines and outlines to assist app developers with the writing of privacy policies can lead to user confusion and diminished trust in mobile apps. In this work, we study the completeness of privacy policies by developing a novel framework to assess them and applying it to analyze 72,342 privacy policies from the Google Play Store. The Google Play Store platform mandates its app developers to publish a privacy policy but does not enforce a universal standard for their quality. We have observed that 56.2% of these policies are comprehensive, containing the essential information about the privacy policy that should be included. This highlights the necessity for more thorough privacy policies in mobile apps. Our results highlight that it is essential to develop tools that assist mobile app developers to clearly and completely describe how they collect, use, store, and share user data, thereby increasing privacy transparency and accountability.

Keywords: Privacy Policy · Mobile App · Question Answering · Natural Language Processing

1 Introduction

A mobile app's privacy policy is a legal document that explains how the app collects, uses, stores, and shares users' information. Ideally, users read mobile apps' privacy policies to understand how their data is handled by the mobile apps, such as the types of information collected, stored, and shared. Additionally, a mobile app's privacy policy provides information about the services the app offers and when the privacy policy was updated. This information is important for new and current users who use the apps. This information helps users understand the new services, what changes have been made to the app, and what information the updated app will collect.

However, some mobile apps' privacy policies do not provide all the necessary information to their users. We refer to them as incomplete privacy policies. Such

J. Zhao and W. Meng (Eds.): SciSec 2024, LNCS 15441, pp. 76–96, 2025.
https://doi.org/10.1007/978-981-96-2417-1_5

omissions can mislead users into downloading an app and may increase the risk of breaching users' privacy. Also, these incomplete privacy policies can negatively impact the apps' trustworthiness. Therefore, mobile apps' privacy policies should provide all the necessary information to users, as we refer to these privacy policies as completed privacy policies.

There is no universal guideline for the quality of mobile apps' privacy policies; consequently, some app developers provide poorly written policies with insufficient or irrelevant information. Some apps provide links to privacy policies that contain only one or two sentences. The absence of a universal standard can lead to incomplete privacy policies.

Previous research analyzed privacy policies from different angles, including discrepancies between apps' stated policies in privacy policy documents and their actual practices [14,25,26,30,33], techniques to help users understand sometimes complex privacy write-ups [10,20], and internal inconsistencies [13]. However, no studies have yet provided comprehensive information or highlighted how comprehensive mobile app privacy policies are in terms of providing the essential information that users need. In this study, we aim to fill this research gap.

Similar to our work, some existing studies have analyzed the completeness of privacy policies to determine whether they provide specific information [12,19, 22,27,29]. These studies used two different approaches: rule-based and sentence-level analyses. Nevertheless, these approaches have limitations due to the nature of rule-based and sentence-level approaches; privacy policies documents are often unstructured, and relying solely on tokens in rules or sentences can sometimes fail because identifying and locating them is difficult. In addition to this limitation, these studies measure completeness based on specific laws, such as the General Data Protection Regulation (GDPR) [7]. To overcome these limitations, our work employs a paragraph-based approach to capture the text's meaning from the entire paragraph; meanwhile, we analyze the completeness based comprehensive information including global privacy laws (Sect. 3.7).

We define the problem as a binary problem: determining whether a mobile app's privacy policy is complete or incomplete. This can be done by analyzing the content of the policy and evaluating it against a set of questions. The answers to these questions can tell the presence of certain topics, such as the types of data collected, how the data is used, and how data is stored and protected. The answers are aggregated in the final evaluation. We calculated a final score for each mobile app's privacy policy. If any of topics are missing, our system flags it as an incomplete privacy policy.

Through this study, we aim to answer our main research question: How well do mobile apps' privacy policies provide complete information to users? To do this, we conduct a large-scale measurement study to evaluate the completeness of mobile apps' privacy policies using deep learning techniques.

To conduct this study, we built a novel pipeline framework that can automatically assess the completeness of mobile app privacy policies. The framework consists of three main components: a Question Answering (QA) Model, an Answer Selection Module, and a Re-ranking Model. Initially, the QA Model takes a question and its context as inputs to produce preliminary answers. Afterward, the Answer Selection Module groups and adjusts these answers and then passes them into the Re-ranking Model to further process and find the final answer. If the framework finds an accurate answer with a high confidence score, it flags that the answer exists.

The results of our framework showed that out of the 72,342 mobile apps' privacy policies, only 56.2% had complete privacy policies. This result highlights the need for more comprehensive privacy policies for mobile apps. This result indicates that app developers must work harder to protect digital privacy and transparency. The main contributions of this paper are:

- We built a large dataset for mobile apps' privacy policies, including app names, privacy policy URLs, developers' information, and other meta data, for 130,700 mobile apps. This dataset can be used as a basis for our analysis and also for future research in mobile apps' privacy policies analysis.
- We developed a novel pipeline framework that can automatically detect 21 privacy aspects in mobile apps' privacy policies. It has three main components: a QA Model, an Answer Selection Module, and a Re-ranking Model. The Re-ranking Model is crucial as it improves the relevance and accuracy of the selected answers provided by the QA Model.
- We conducted a large-scale measurement and evaluation of privacy policy completeness. This assessment aimed to determine how many of the mobile apps' privacy policies we collected provide comprehensive information to users about their personal data and other essential information.
- Based on our findings, we provided several recommendations.

The rest of the paper is organized as follows: Sect. 2 reviews background and related work. Section 3 describes our research methodology and datasets. Section 4 details the experimental design and results. Section 5 presents a discussion of our study. Section 6 concludes the paper.

2   Background and Related Work

2.1   Background

Complete Privacy Policy. The privacy of app users has recently become one of the top concerns in the digital world due to the growing use of mobile apps. Mobile apps' privacy policies serve as the basis for protecting app users' privacy. Also, mobile apps' privacy policies inform users how data is being collected, used, and protected. They should be clear, comprehensive, and accurate. Furthermore, they should provide the necessary information to ensure its trustworthiness, such as the date of update, services or features provided, and more.

The main issue with existing mobile privacy policies is the lack of standardized guidelines for their development and composition. As a result, some mobile apps present privacy policies that omit crucial details necessary to enhance trust between users and the apps. Therefore, it is essential to establish a standard for drafting privacy policies that contain all required information. In this study, we have determined the essential elements that should be included in any privacy policy, based on our review of current online user privacy laws across various countries and an extensive literature review. All the critical topics that must be incorporated have been collected in Table 2 in the Appendix.

Natural Language Processing (NLP) by Using Deep Learning. NLP uses algorithms from Machine Learning and Deep Learning to process and understand human language very accurately. NLP models learn complex textual data in any language and provide required outcomes.

Using deep learning for NLP requires several steps: 1) Tokenization: it is the initial step in text processing; it involves splitting the text into words or symbols to facilitate subsequent steps. 2) Embedding: When words are embedded, their syntax and semantics are captured by converting them into numerical vectors. Models like GloVe [23] can be used to represent words. 3) Highway Network: This network feeds neural networks with embeddings at both the word and character levels. 4) Contextual Embedding: mechanisms that we use to capture the text's meaning within its context. Models like BiDAF [24] and BERT [17] use attention and other mechanisms to create embeddings that take into account the interaction of each word in a sequence. 5) The last layer of a natural language processing model is the output layer, which finds the section of the text that answers the question.

2.2 Related Work

Recently, significant research has concentrated on automated analysis of privacy policies. Yet, these studies have been mainly concentrated on privacy policies of websites [10, 12, 19, 20, 22, 27], which leaves a gap in the study of mobile apps' privacy policies. Focusing on web privacy policies misses some unique challenges in the mobile environment, such as location data, contact information, and other sensitive information stored on a user's device.

Furthermore, existing studies have used different approaches to automate the analysis of privacy policies, including text-mining and machine-learning techniques [18], topic modeling [16], and deep learning [10, 20]. While traditional machine learning and text-mining techniques can be effective for structured data analysis, privacy policies present unique challenges due to their unstructured and complex legal language. Deep learning demonstrated the powerfulness [21] of these two approaches in understanding language and extracting information. Therefore, we use deep learning as the foundational approach for our framework.

Some research utilizes deep learning for privacy policy analysis by investigating various issues. The authors of [14, 25, 26, 30, 33] investigated discrepancies between apps' behaviors and their stated privacy policies. Other studies, such as those referenced in [20] and [10], aim to help users understand privacy policies using deep learning techniques. Furthermore, several studies [15, 31, 32] have proposed methods that assist app developers in automatically creating legally compliant privacy policies, by asking questions or analyzing the apps' behavior.

In addition to these studies, a few have focused on assessing the completeness of privacy policies using deep learning [12, 19, 22, 27, 29]. These assessments inspect specific legal standards and investigate how closely the documents comply with individual pieces of legislation, such as the GDPR. These studies employed a deep learning approach and used sentence-level-annotated datasets to classify privacy policy content into predefined categories relevant to GDPR compliance. For example, Liu et al. [22] collected a dataset of 304 privacy policies, comprising 36,610 sentences, and annotated into 10 categories.

In contrast, our study takes a broader approach, not limited to a single regulation to ensure general applicability across different legal domains. We look at universal and comprehensive information, such as policy updates. Moreover, our strategy is based on a paragraph-based approach, which allows us to overcome the limitations of the sentence-based approach that may encounter grammatically incorrect sentences, potentially hindering the accuracy of the assessment, as faced by previous studies. Additionally, we offer a broader view of the completeness of privacy policies, beyond compliance with a single regulatory framework.

3 Methodology and Datasets

3.1 Framework Overview

Figure 1 illustrates our pipeline framework, which is applied to analyze the collection of 130k privacy policies gathered for this study.

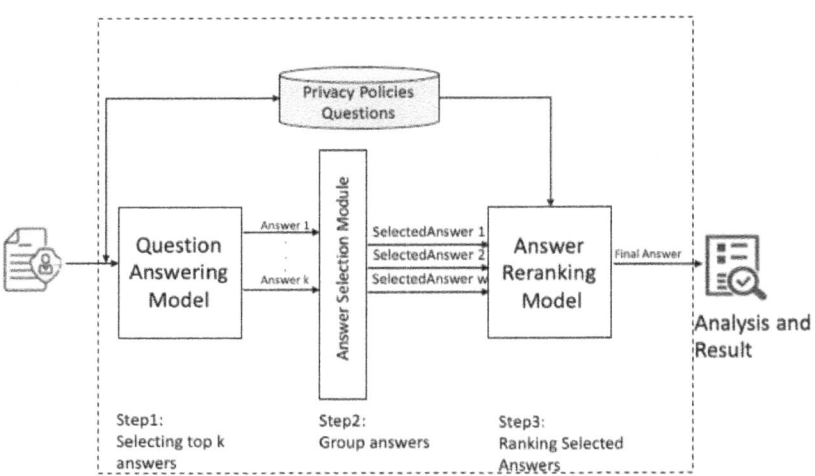

Fig. 1. High-Level Overview.

Our framework initially receives a privacy policy document as input for analysis. Upon receiving this document, our framework divides it into segments using text segmentation techniques. These segments are passed into the QA Model, along with a set of questions.

The QA Model suggests answers for each question in each segment. Then, we arranged the suggested answers into groups using the answer selection module. These groups are based on overlapping and related answers. Next, for each group, we selected the longer answer that provided more contextual information. Afterward, we passed each selected answer into the re-ranking model, which was trained on our Re-ranking dataset to identify the most likely correct answer for the questions. This Re-ranking model ranks the answers based on confidence scores. The answers with high confidence scores indicate that they are more likely to be the correct answers to the question.

3.2 Preprocessing and Segmenting Privacy Policies

Previous studies have shown that preprocessing and segmenting text can enhance NLP analysis [20]. Therefore, before we pass the text of the mobile app privacy policy into our framework, we undertake initial steps. We have developed a Segmentor that is responsible for dividing the mobile app privacy policy into segments and performing further optimizations.

To divide a mobile app's privacy policy into segments, the approach uses heuristic approaches such as stopword removal and the identification of distinct sections and topics within the text. However, privacy policies are often unstructured, which can pose a problem for some privacy policies. Therefore, for long paragraphs, we divide them into smaller chunks (maximum of 512 tokens) that can be processed by QA models.

Furthermore, we also decided to replace pronouns with proper nouns in order to help the NLP models correctly identify what was meant. We first replaced "yourself," "your," and "you" with "the user," and "we" and "us" with the file or company name of an app. Then we used the Stanford CoreNLP library core functionality to handle other pronouns such as "it." We also extracted the main keyword from each section during this process. Once the segments were properly parsed, they were ready for the next phase, which is filtering the segments.

To filter the segments, we trained a Privacy Policy Relevance Filter (PPRF) that checks the relevance of the segments to the privacy policy. The model assigns a score to each segment based on its relevance. Segments with a score higher than a certain threshold (0.5) are considered to be the privacy policy segment and are included into a final privacy policy document. Overall, the PPRF Model accuracy is 96.8. Once the mobile app's privacy policy is segmented and optimized, it is ready to be passed into the first main component, which is the QA Model.

3.3 Question Answer Model

Our QA Model, denoted as QA_{BERT}, employs a BERT-based architecture. We have fine-tuned this QA model using our QA dataset $\mathcal{D}$ to suggest appropriate answers from provided passages.

This dataset exclusively features closed-ended questions and has been optimized to respond to a predefined set of questions. It comprises over 3k instances across 21 key questions per topic regarding mobile apps' privacy policies, which are detailed in Table 2 in the Appendix. Each data entry in our QA dataset is structured as a triplet, consisting of a Passage P, a Question Q, and an Answer A. These triplets cover essential queries pertaining to a range of subjects within mobile apps' privacy policies. Conversely, when the passage P provides the necessary information, the answer A contains the response taken directly from the text with high confidence. However, in testing, we do not rely solely on this confidence score to determine the answer with the highest score, because we deal with it as one of the possible k answers.

The output of the QA_{BERT} model suggests k possible answers based on confidence scores, denoted by $\{A_1, A_2, \ldots, A_k\}$ with corresponding confidence scores $\{C_1, C_2, \ldots, C_k\}$. Then, we passed the answers $\{A_1, A_2, \ldots, A_k\}$ into the next main component, Answer Selection Module (ASM).

3.4   Answer Selection Module (ASM)

The next main component is the Answer Selection Module (ASM), which serves as an intermediary between the QA model (QA) and the re-ranking processes. This component collects all possible answers (A_{all}) provided by the QA model for a specific question (Q) and organizes them into groups (G).

In each group, we select the longer answers (A_{long}) because they include more contextual information. This approach helps to add more semantic detail to the answers, allowing them to be ranked higher based on the contextual information we have. Sometimes, the QA model suggests only a portion of an answer, not a complete one. Therefore, by looking for longer answers with a higher possibility of containing more context, we increase the likelihood that a complete and relevant answer exists if one is available.

Occasionally, to enhance the effectiveness of the re-ranking model, we adjust the QA model to output more suggested answers. The default setting for the QA model is to generate $k = 3$ suggested answers. This adjustment ensures that multiple groups of answers are available for consideration, which significantly improves the Re-ranking model's ability to assess and increases the likelihood of verifying the existence of the correct answer.

3.5   Re-ranking Model (RM)

Our RM model operates through a two-step process to evaluate and order the top-k candidate answers provided by the QA model. Each candidate answer is denoted as A_i where i ranges from 1 to k. The first step involves feature extraction, wherein a set of features is gathered for every answer candidate from the sentence in which the answer is found, as well as from the question itself. These features are standard across QA systems and are extracted directly from the modules without additional overhead. In the second step, we calculate similarities and train the model to identify relationships between the question and answer, along with their contextual features from the corresponding sentence.

For example, we trained the model to recognize that for a question such as "How does an app collect personal information?" an answer stating "from cookies, forms, and by creating an account" is relevant if the contextual features from the sentence support this procedure. Alternatively, if the answer suggests "information stored for 30 days," but the contextual features from the sentence do not support this as a method for collecting information, we train the model to recognize this as irrelevant.

We have tested two methods to train and evaluate ranking models. The first method involves taking questions and identifying the sentences that contain the answers. During training, we focus on the similarity between these sentences and their corresponding questions and answers, considering the context of the answer, the sentence, and the questions.

$$s_{ij} = \text{FeatureCombination}(\text{BERT}(q_i),$$
$$\text{BERT}(a_{ij}), \text{ContextFeatures}(a_{ij})) \tag{1}$$

Here, $\text{BERT}(q_i)$ and $\text{BERT}(a_{ij})$ are the pooled output representations of the question and answer, respectively, from the BERT model, and $\text{ContextFeatures}(a_{ij})$ represents the contextual features extracted from the sentence containing the answer. The final output for a given question q_i is the answer a_{ij*} with the highest similarity score:

$$a_{ij*} = \text{argmax}_j(s_{ij}) \tag{2}$$

Based on these similarity scores, our reranking model determines the final output for a given question q_i to be the answer a_{ij*} that has the highest similarity score. Our findings show a slight difference between using the entire sentence or just the answer when dealing with an extended context. During testing, it can sometimes be difficult to extract sentences because mobile apps' privacy policies have unstructured paragraphs and sentences. Therefore, we only consider the features of the question and the longer answer. This approach ensures that the RM considers the relevance of the answer in the context of its surrounding text.

Generated Re-ranking Dataset: Algorithm 1 outlines our method to build the Re-ranking dataset, which is designed to train the RM. First, the algorithm takes as input the original QA dataset. This dataset is composed of triples, including context, question, and answer. The algorithm starts with an empty dataset named RerankDataset, intended to store the re-ranking training data. For each entry in OriginalData, the algorithm creates the Context, Question, and CorrectAnswer. It then invokes the GenerateIncorrectAnswers function to produce a set of incorrect answers that are randomly generated from the context and do not correspond to the correct answer. The LocateAnswerSentences function is used to identify sentences within the context that contain both the correct and incorrect answers. The ScoreAnswers function assigns relevance scores to all answers based on their accuracy. These scores are vital for training the Re-ranking model, as they help to distinguish between correct and incorrect answers. Using the PrepareEntry function, the algorithm takes each question, its correct answer, and the corresponding scores into a RerankEntry and adds this entry to the RerankDataset, which comprises the context, question, correct answer, and incorrect answers. This process is repeated to all the entries in the original dataset. Lastly, the algorithm returns the complete re-ranking dataset for the training of the Re-ranking model we used.

Algorithm 1. Generating Re-ranking Training Dataset

1: **Input:** Original QA dataset with contexts, questions, and answers.
2: **Output:** Re-ranking training dataset with relevance scores.
3: $OriginalData \leftarrow$ LoadQADataset()
4: $RerankDataset \leftarrow \emptyset$
5: **for** each $item$ in $OriginalData$ **do**
6:    $Context \leftarrow item.context$
7:    $Question \leftarrow item.question$
8:    $CorrectAnswer \leftarrow item.answer$
9:    $IncorrectAnswers \leftarrow$ GenerateIncorrectAnswers($Context, CorrectAnswer$)
10:    $AnswerSentences \leftarrow$ LocateAnswerSentences($Context,$
          $CorrectAnswer, IncorrectAnswers$)
11:    $Labels \leftarrow$ LabelAnswers($Question, CorrectAnswer,$
          $IncorrectAnswers, AnswerSentences$)
12:    $RerankEntry \leftarrow$ PrepareEntry($Question, CorrectAnswer,$
          $IncorrectAnswers, Labels$)
13:    $RerankDataset \leftarrow$ AddToDataset($RerankDataset, RerankEntry$)
14: **end for**
15: **return** $RerankDataset$

3.6   Assessing the Completeness of a Mobile App's Privacy Policy

To formally present the process of deriving and evaluating privacy subjects within mobile apps' privacy policies, we introduce the following notations:

- Let P represent the set of all mobile app privacy policies.
- Let $S = \{s_1, s_2, \ldots, s_{21}\}$ denote the set of 21 specific subjects of interest concerning the privacy laws identified for the mobile apps' privacy policies.
- Define a mapping function $f : P \times S \rightarrow 2^S$ that assigns each mobile app's privacy policy from P to one or more subjects in S, where 2^S denotes the power set of S, containing all possible subsets of S.
- Let Q denote the set of all questions derived from the literature review, which are used to probe the privacy policies for specific information.
- Define a grouping function $g : Q \rightarrow S$ that maps each question in Q to its corresponding subject in S.
- Let A_i represent the privacy policy of a particular mobile app i.
- Define an evaluation function $e : Q \times P \rightarrow \{0,1\}^{|Q|}$, where $e(q, A_i) = 1$ indicates the presence of evidence for the subject related to question q in the privacy policy A_i, and $e(q, A_i) = 0$ indicates its absence.
- Let M be a binary matrix where each element $M_{ij} = e(q_j, A_i)$ for $q_j \in Q$ and $A_i \in P$, representing the presence or absence of subjects in each mobile app's privacy policy.

- The matrix M is then used to measure the completeness of the mobile app privacy policies.

Algorithm 2 is for assessing the completeness of a mobile app's privacy policy. It processes the privacy policy text (A_i) of a specific mobile app and a set of predefined questions (Q), which we used as criteria for evaluation. Line 4 shows the process for text segmentation where the privacy policy text (A_i) is divided into smaller segments. This division enhances the accuracy in the subsequent stages. Lines 8–9 show the Question-answering stage, where the algorithm moves to the QuestionAnswering function. The QuestionAnswering function takes as input each segment with each question to extract relevant answers, which are then gathered in CQA. Line 12 shows GroupingAnswers, where the algorithm groups answer corresponding to the same questions in a function called GroupAnswers (GQA). Line 13 shows the process of selecting the longest answers for each group of answers. The longest answer (SQA) is selected via the SelectLongestAnswers function. Line 14 shows the ReRankAnswers function, where the selected answers are passed to a Re-ranking model, and then re-ranking selected answers are based on their confidence scores. Lines 15–22 show the final stage of the algorithm, which constructs a binary vector (M_i), where each element indicates the presence (1) or absence (0) of the subjects related to each question within the privacy policy. Based on the value of the binary vector (M_i) for each app's privacy policy, we decide its completeness.

3.7   Crafting a Set of 21 Subjects for Mobile App Privacy Policies

We employed a systematic method to identify a set of 21 distinct subjects that encapsulate the essential elements of global privacy laws [1–3,5–7,9], that should be included in mobile apps' privacy policies. These subjects were identified through a comprehensive literature review [10,11,28]. Detailed in Table 2 in the Appendix, the subjects, designated as $S = \{s_1, s_2, \ldots, s_{21}\}$, represent specific elements that should be included within mobile apps' privacy policies.

In the subsequent step, we crafted a set of questions corresponding to these subjects. This process was informed by a detailed literature review of the current state of research on privacy policies [1–3,5–7,9–11,28].

3.8   Experiment Setup

We used the BERT-based model trained for English language on our own QA dataset. The BERT model has 12 layers and a hidden layer of 430,768 neurons. We fine-tuned it for our task by training on our dataset. Our QA model was fine-tuned over 10 epochs using the cross-entropy loss function to compare predicted and correct information. The training process was conducted with a single A-100 GPU on Google Colab.

We used a Re-ranking model based on the BERT-Large architecture. This model was trained on a large text corpus and could capture semantic relationships between questions and their answers. We fine-tuned the model for our specific task, which involved reranking the mobile apps' privacy policy answers.

Algorithm 2. Assessing the Completeness of A Mobile App's Privacy Policy

---

1: **Input:** A_i, privacy policy text of mobile app i to be analyzed
2: **Input:** Q, set of questions to evaluate against the privacy policy
3: **Output:** M_i, binary vector representing the presence or absence of subjects in the privacy policy A_i
4:    $Segments \leftarrow$ TextSegmentation(A_i)
5:    $CQA \leftarrow \emptyset$
6: **for** each $segment \in Segments$ **do**
7:        **for** each $q \in Q$ **do**
8:            $Answers \leftarrow$ QuestionAnswering($segment, q$)
9:            $CQA.add((q, Answers))$
10:        **end for**
11: **end for**
12:    $GQA \leftarrow$ GroupAnswers(CQA)
13:    $SQA \leftarrow$ SelectLongestAnswers(GQA)
14:    $RankedAnswers \leftarrow$ ReRankAnswers(SQA)
15:    $M_i \leftarrow \mathbf{0}$
16: **for** each $(q_j, Answers) \in RankedAnswers$ **do**
17:        **for** each $answer \in Answers$ **do**
18:            **if** ConfidenceScore($answer$) $\geq$ ConfidenceThreshold **then**
19:                $M_{ij} \leftarrow 1$
20:                **break**
21:            **end if**
22:        **end for**
23: **end for**
24: **return** M_i

In our analysis of app developer regions, we extracted the address information for each mobile app and mapped the address with each mobile app and app developer. We assumed that these addresses are where the app developer is located. However, we found the address is not structured and is an incomplete address. To overcome this challenge, we created a custom Named Entity Recognition (NER) model to extract countries, cities, and zip codes within an address. This model helped us to identify the app developer locations.

For filtering the segments, we built the PPRF model. The PPRF checks if the segment is part of the privacy policy. After analyzing the data collection, it became clear to us that some mobile apps do not contain a privacy policy link. Instead, they lead to advertisements for additional products or services. Some are not affiliated with the app server or the companies that own the apps. Filtering these pages is necessary for identifying bad links and web pages as well as preparing a statistic about this issue.

3.9   Privacy Policy (PP) Dataset

To construct a dataset of 130,700 privacy policies for this study, we crawled the Google Play Store to extract the URLs of the apps' privacy policies and their associated metadata. Our tool, developed using unofficial APIs [4], gathered data including privacy policy URLs, app descriptions, and metadata (such as the app's name, developer, developer's address, categories, etc.). In total, we visited over 130,700 Android app webpages.

We integrated Puppeteer [8] into our automation and web crawling tool to retrieve the text of privacy policies from these URLs. The tool accessed each privacy policy URL and waited for the webpage to render fully before extracting its content. Given that some pages require a longer time to load, we established a time limit of 5 minutes per webserver to optimize the response rate.

Consequently, there are 99,179 Android apps with potentially accessible privacy policies. Nonetheless, the dataset excludes privacy policies presented in non-HTML formats, such as PDFs or documents. In addition, we excluded those that have inaccessible privacy policies as we have studied in the results section. After applying these exclusion criteria, the precise number of accessible privacy policies in our dataset is 72,342.

4 Experiments and Analyses

4.1 Performance Analysis

In this section, we compared the performance of our custom-developed framework against two other models used for extracting information from privacy policies, namely the TF-IDF and question answering (QA) model. The TF-IDF model is a statistical measure for identifying commonly used words in a context. The second model is the BERT-based question answering (QA) model, which is widely used to find answers to questions within a context. We chose these two methods (the TF-IDF and BERT-based QA models) for comparison because they are widely used and known for their effectiveness in extracting information from text. These models serve as baselines for our comparison and allow us to measure our custom framework against standard models.

For our training and testing, we used the most common method to partition our dataset, which is an 80/20 split. Our dataset contains 3,205 samples, which we divided into 80% for training and 20% for testing. For TF-IDF, we used the training dataset to create histograms of the words used in each paragraph that contained an answer for our 21 questions. We then utilized these histograms to identify the existence of answers to the questions in the paragraphs of the testing dataset. We utilized the top 10 words in our histograms. For the QA model, we used the training dataset for training and the test dataset for evaluation. For our custom-developed framework, we used the training dataset to train both the QA model and the reranking model with two different structures based on their specific requirements.

For comparing the three models, we used AUC (Area Under the Curve) and accuracy as metrics because they are standard measures that show model performance. AUC reflects the balance between true positives and false positives, while accuracy indicates the proportion of correct predictions.

Figure 2 displays the AUC curves for three different models, illustrating their performance results. With an AUC of 0.91, our model outperforms the other two in determining whether an answer exists or not. The second and third models, with AUC values of 0.82 and 0.63, respectively, perform less effectively but are still functional. Our model achieves an accuracy of 85.3%, compared to the other models, which have accuracies of 74.5% and 65.0%, respectively.

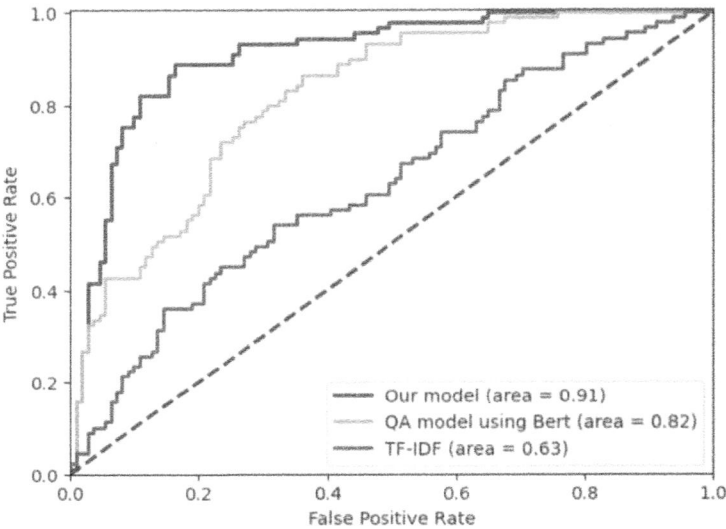

Fig. 2. The AUC curve for three models in determining if an answer exists or not.

4.2   Overall Analysis

Our overall analysis of the completeness of mobile app privacy policies shows that there was a notable imbalance in policy completeness. Only 56.2% (40,676 out of 72,342) of the policies we assessed were complete, providing all the necessary elements of information users need to understand how their data is managed. This leaves a substantial 43.8% (31,666 out of 72,342) of policies that are incomplete, lacking the critical level of details for meeting the expectations and needs of users about data privacy.

Breakdown of Privacy Policy Completeness: Complete Privacy Policies (56.2%): Just over half of the privacy policies we analyzed were complete. These policies typically described data collection, usage, storage, sharing, and user rights. Essentially, they provided users with the information necessary to make meaningful choices about their personal data. In short, these policies aligned with best practices in privacy and data protection.

Incomplete privacy policies (43.8%): The percentage of incomplete policies was alarmingly high, signaling a widespread issue within the mobile app industry. These policies typically omitted critical information, leaving users in the dark as to how data is used and their privacy rights and more.

4.3   Non-existent Privacy Policy URLs

Our analysis unveiled a major issue with respect to non-existent privacy policy URLs. In particular, we found that 5.84% (7,627) URLs - which are supposed to direct users to the privacy policies of the respective mobile apps - were linked to web pages that did not actually exist. This finding can give rise to a false sense of data protection on the part of users, who may think that such a privacy policy is present. Such a finding may also have implications for perceived mobile app trustworthiness. The absence of a functioning link may mislead a user into installing an application under false pretense.

4.4   Fake Privacy Policies

After reviewing the privacy policies, we found that 17.74% (23,194) did not include any privacy practice related information. This high percentage of fake URLs suggests that Google Play may not be checking these policies thoroughly. People trust Google Play to look after their privacy. However, these fake URLs could lead to legal trouble for both Google Play and the app developers, especially if they collect data without permission.

4.5   App Category Differences in Privacy Policy Completeness

Figure 3 presents our results for the top categories identified in our collected dataset. We found the following: for games, 42.4% (5,123 out of 12,081) had complete privacy policies, while 57.6% (6,959 out of 12,081) were incomplete; for dating apps, 32.3% (134 out of 412) were complete and 67.7% (279 out of 412) were incomplete; for social media, 71% (2,126 out of 2,994) were complete and 29% (868 out of 2,994) were incomplete; for productivity apps, 56.1% (19,784 out of 35,283) were complete and 43.9% (15,499 out of 35,283) were incomplete. We also noted that some apps do not fit into specific categories or are unknown. This indicates that certain applications, particularly those related to gaming and dating platforms, often have incomplete privacy policies. These apps, in particular, need to craft privacy policies with greater transparency and provide comprehensive information to ensure users are aware of potential privacy breaches and the security of their personal information.

4.6   Regional Differences in Privacy Policy Completeness

Upon further analysis, we found that privacy policy completeness varied by region. Figure 4 presents our results as follows: 45.6% in Asia (out of 2,120 policies), 72.4% in America (out of 26,340 policies), 68.6% in Europe (out of 19,383

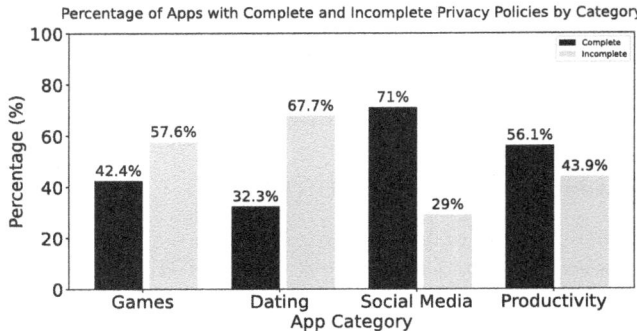

Fig. 3. Privacy policy completeness varies significantly by category.

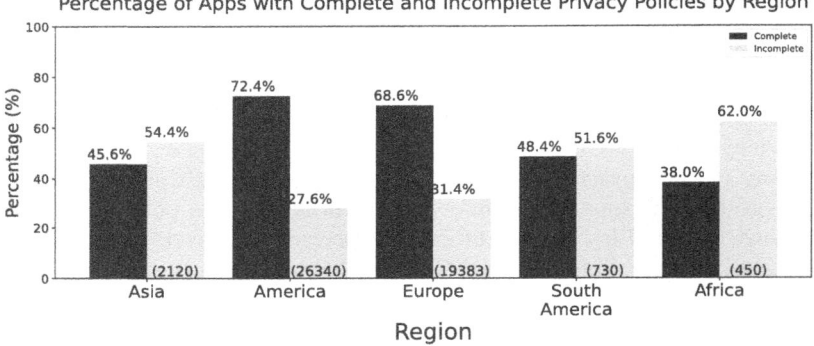

Fig. 4. Privacy policy completeness varies significantly by region.

policies), 48.4% in South America (out of 730 policies), and 38% in Africa (out of 450 policies). Our analysis indicated that apps from Asia, Africa, and South America were more likely to have incomplete privacy policies compared to those from other regions. The disparities may result from differences in legal requirements, awareness, resources, market demands, or cultural attitudes toward privacy.

4.7 Case Study

In this section, we ran our framework to analyze the mobile application with the package name "tw.com.phonebot.fibarulesqatw's," and we identified several key missing pieces of information. Table 1 highlights the main categories of missing privacy policy information. The policy did not provide users with critical information about how the app provides services, protects, retains, or shares their data, which puts users' personal information and data at risk. For example, without mentioning the app's services, users cannot tell what data is being collected or how it is being used. Similarly, the absence of information about the privacy laws or regulations the app follows leaves questions about the app's commitment to safeguarding users' personal data. Moreover, with no information about its data practices, data sharing practices, and data retention policy, the app may not be able to effectively protect users' data. The absence of key details on these areas of the privacy policy signals that the users' essential information and needs might not be a top priority of the app. Finally, The lack of contact information in the privacy policy prevents users from discerning whether the app values transparency or is responsive to user inquiries. In sum, Table 1 shows that tw.com.phonebot.fibarulesqatw's privacy policy was missing several key elements, which may suggest that the app may not adequately protect users' personal information and data. The positive aspects of this privacy policy were that it provides information about the app's name in their privacy policy, which helps users understand that this privacy policy is for this app. In addition, it provided information about what kind of information is used.

Table 1. Categories missing from tw.com.phonebot.fibarulesqatw's privacy policy.

Subjects	Status
Name of the App	Specified
Services Provided by the App	Not specified
Privacy Laws or Regulations Followed by the App	Not specified
Personal Data or Information Collected	Specified
Methods of Collecting Personal Data or Information	Not specified
Use of Collected Personal Data or Information	Not specified
Purpose of Collecting Personal Data or Information	Not specified
Personal Data or Information Collected by Third Parties	Not specified
Personal Data or Information Shared with Third Parties	Not specified
Names of Third Parties Related to Our App	Not specified
Personal Data or Information Sold to Third Parties	Not specified
Purpose of Sharing Personal Information or Data with Third Parties	Not specified
User Rights	Not specified
Notification Process for Changes to This Privacy Policy	Not specified
Effective and Update Date of This Privacy Policy	Not specified
Security Measures for Protecting Personal Data or Information	Not specified
Information Collected from Children	Not specified
Data Retention Policy	Not specified
International Data Transfer Policy	Not specified
Contact Information for Queries	Not specified

5   Discussion

5.1   Implications and Recommendations

Verification of Privacy Policy URLs by Google Play: The number of fake URLs detected makes it clear that Google Play is not checking for fake URLs provided by the app's developers that their store endorses. That can cause legal trouble, as users trust the Google Play store to protect them. Furthermore, developers of the apps themselves could end up getting sued for taking information that users did not consent to giving them.

Risks Associated with Downloadable Privacy Policies: Downloadable privacy policies also serve as a legal liability due to the possibility of malware being downloaded to a user's device. This can cause damage to users' devices or lead to phishing. It is better to have a viewable privacy policy without downloading to protect users from these threats.

Hosting of Privacy Policies: There are many issues with Google Play hosting URLs for a privacy policy. Apps that point to malicious URLs and redirect the users to a malicious website. Therefore, we suggest that Google host static websites for each app developer to post their privacy policy instead of hosting a URL that can potentially redirect the users to a malicious website.

Downloadable Files as Privacy Policies: We also found that some URLs point to a downloadable file with a privacy policy. However, a downloadable file may contain malware or malicious content.

Need for Better Privacy Policy Guidelines and Tools: We also found that there is a need for a set of guidelines or tools to help app developers write better privacy policies.

5.2   Limitations and Future

Our work has several limitations. First, we utilized fixed questions to evaluate the effectiveness of mobile apps' privacy policies. Our framework cannot answer every type of question because it is trained to respond to only 21 major questions concerning mobile apps' privacy policies.

Second, when filtering the segments related to the mobile app privacy policy, sometimes our model incorrectly predicts a small segment and leads to false negatives.

In the future, we will attempt to build and leverage more powerful components to increase the accuracy of evaluating mobile privacy policies. We also aim to develop an AI service to verify all necessary information and ensure that mobile apps' privacy policies provide all essential details. Additionally, we aim to examine the relationship between app updates and privacy policy revisions in terms of app practices and the types of services provided.

6   Conclusion

This paper presents the results of a framework examining 72,342 privacy policies, revealing that only 56.2% of them were complete, emphasizing the need for more comprehensive privacy policies for mobile apps. We developed a pipeline framework featuring an Answer Selection Module, which categorizes answers generated by a QA model into groups based on similarity, and then selects the most representative answers from each group.

Our analysis also revealed several security and privacy issues with Google Play's hosting URLs for privacy policies, including the presence of fake URLs and the potential for users to be redirected to malicious websites. We suggested that Google hosts static websites for each app developer's privacy policy to mitigate this risk. Furthermore, we discovered that some URLs point to downloadable files containing privacy policies, which could potentially harbor malware or malicious content. The study concludes that there is a clear need for guidelines or tools to assist app developers in crafting better privacy policies.

Acknowledgment. We thank the reviewers for their comments. Khalid was supported by the Tabuk University. All authors were supported by the Colorado School of Mines.

A   Appendix

Table 2. A set of 21 subjects for the major aspects of mobile app privacy policies.

Privacy Subject	Description
Name of the App	The unique identifier of the app as it appears in the app store and on the device
Services Provided by the App	Detailed functionalities and services offered to the user by the app
Privacy Laws or Regulations Followed by the App	Specific local or international privacy laws that the app complies with
Personal Data or Information Collected	Types of user data collected, such as name, age, location, etc.
Methods of Collecting Personal Data or Information	How the app collects data, e.g., through user input, sensors, etc.
Use of Collected Personal Data or Information	The intended use of the data, such as personalization, advertising, etc.
Purpose of Collecting Personal Data or Information	The purpose for which the app uses the collected data
Personal Data or Information Collected by Third Parties	If and what data is collected by third parties through the app

(*continued*)

Table 2. (*continued*)

Privacy Subject	Description
Personal Data or Information Shared with Third Parties	What user data, if any, is shared with third parties and under what conditions
Names of Third Parties Related to Our App	Specific third parties that are affiliated with or integrated into the app
Personal Data or Information Sold to Third Parties	Whether the app sells user data to third parties and the nature of that data
Purpose of Sharing Personal Information or Data with Third Parties	Why the app shares data with third parties and the expected benefits
User Rights	What rights the user has regarding their data, such as the right to access, correct, or delete their data
Notification Process for Changes to This Privacy Policy	How users are informed about changes to the privacy policy
Effective Date of This Privacy Policy	The date when the privacy policy comes into effect and when it was last updated
Update Date of This Privacy Policy	The specific dates when the policy was updated
Security Measures for Protecting Personal Data or Information	The technical and organizational measures in place to protect user data
Information Collected from Children	Special considerations and practices for collecting data from children, if applicable
Data Retention Policy	How long the app retains user data and the criteria for data deletion
International Data Transfer Policy	How the app handles data transfer across international borders
Contact Information	Contact details for privacy concerns or questions

References

1. Brazilian general data protection law (LGPD, English translation). https://iapp.org/resources/article/brazilian-data-protection-law-lgpd-english-translation/
2. California Consumer Privacy Act (CCPA)—state of California - department of justice - office of the attorney general. https://oag.ca.gov/privacy/ccpa
3. GAPP privacy: 10 generally accepted privacy principles. https://linfordco.com/blog/the-10-generally-accepted-privacy-principles/
4. Github - nomore201/googleplay-api: Google play unofficial python API. https://github.com/NoMore201/googleplay-api
5. How to read a privacy policy—state of California - department of justice - office of the attorney general. https://oag.ca.gov/privacy/facts/online-privacy/privacy-policy

6. Pipeda fair information principles - office of the privacy commissioner of Canada. https://www.priv.gc.ca/en/privacy-topics/privacy-laws-in-canada/the-personal-information-protection-and-electronic-documents-act-pipeda/
7. The principles—ico. https://ico.org.uk/for-organisations/guide-to-data-protection/guide-to-the-general-data-protection-regulation-gdpr/principles/
8. Puppeteer—puppeteer. https://pptr.dev/
9. Read the Australian privacy principles - home. https://www.oaic.gov.au/privacy/australian-privacy-principles/read-the-australian-privacy-principles
10. Ahmad, W., Chi, J., Tian, Y., Chang, K.W.: PolicyQA: a reading comprehension dataset for privacy policies. In: Findings of the Association for Computational Linguistics: EMNLP (2020)
11. Alkhattabi, K., Bird, D., Miller, K., Yue, C.: Question answering models for privacy policies of mobile apps: are we there yet? In: Science of Cyber Security: 4th International Conference. SciSec (2022)
12. Amaral, O., Abualhaija, S., Torre, D., Sabetzadeh, M., Briand, L.C.: AI-enabled automation for completeness checking of privacy policies. IEEE (2021)
13. Andow, B., et al.: Policylint: investigating internal privacy policy contradictions on google play. In: USENIX Security Symposium (2019)
14. Andow, B., et al.: Actions speak louder than words: entity-sensitive privacy policy and data flow analysis with policheck. In: Proceedings of the USENIX Security Symposium (2020)
15. Bateni, N., Dara, R.: Automated generation of privacy policy using deep models. In: Proceedings of the IEEE International Symposium on Technology and Society (ISTAS) (2021)
16. Costante, E., Sun, Y., Petković, M., Den Hartog, J.: A machine learning solution to assess privacy policy completeness: (short paper). In: Proceedings of the 2012 ACM Workshop on Privacy in the Electronic Society (2012)
17. Devlin, J., Chang, M.W., Lee, K., Toutanova, K.: BERT: pre-training of deep bidirectional transformers for language understanding. In: Conference of the North American Chapter of the Association for Computational Linguistics: Human Language Technologies (2019)
18. Guntamukkala, N., Dara, R., Grewal, G.: A machine-learning based approach for measuring the completeness of online privacy policies. In: 2015 IEEE 14th International Conference on Machine Learning and Applications (ICMLA). IEEE (2015)
19. Hamdani, R.E., Mustapha, M., Amariles, D.R., Troussel, A., Meeùs, S., Krasnashchok, K.: A combined rule-based and machine learning approach for automated GDPR compliance checking. In: Proceedings of the Eighteenth International Conference on Artificial Intelligence and Law (2021)
20. Harkous, H., Fawaz, K., Lebret, R., Schaub, F., Shin, K.G., Aberer, K.: Polisis: automated analysis and presentation of privacy policies using deep learning. In: Proceedings of the USENIX Security Symposium (2018)
21. Liu, L., Luo, J.: A question answering system based on deep learning. In: Intelligent Computing Methodologies: 14th International Conference, ICIC 2018, Wuhan, China, 15–18 August 2018, Proceedings, Part III 14 (2018)
22. Liu, S., Zhao, B., Guo, R., Meng, G., Zhang, F., Zhang, M.: Have you been properly notified? Automatic compliance analysis of privacy policy text with GDPR article 13. In: Proceedings of the Web Conference (2021)
23. Pennington, J., Socher, R., Manning, C.D.: Glove: global vectors for word representation. In: Proceedings of the 2014 Conference on Empirical Methods in Natural Language Processing (EMNLP) (2014)

24. Seo, M.J., Kembhavi, A., Farhadi, A., Hajishirzi, H.: Bidirectional attention flow for machine comprehension. In: Proceedings of the International Conference on Learning Representations (2017)
25. Slavin, R., et al.: Toward a framework for detecting privacy policy violations in android application code. In: Proceedings of the International Conference on Software Engineering (2016)
26. Wang, X., Qin, X., Hosseini, M.B., Slavin, R., Breaux, T.D., Niu, J.: Guileak: tracing privacy policy claims on user input data for android applications. In: Proceedings of the International Conference on Software Engineering (2018)
27. Wilson, S., et al.: The creation and analysis of a website privacy policy corpus. In: Proceedings of the 54th Annual Meeting of the Association for Computational Linguistics (Volume 1: Long Papers) (2016)
28. Wilson, S., et al.: The creation and analysis of a website privacy policy corpus. In: Annual Meeting of the Association for Computational Linguistics (2016)
29. Xiang, A., Pei, W., Yue, C.: Policychecker: analyzing the GDPR completeness of mobile apps' privacy policies. In: Proceedings of the 2023 ACM SIGSAC Conference on Computer and Communications Security (2023)
30. Yu, L., Luo, X., Liu, X., Zhang, T.: Can we trust the privacy policies of android apps? In: Annual IEEE/IFIP International Conference on Dependable Systems and Networks (DSN) (2016)
31. Yu, L., Zhang, T., Luo, X., Xue, L., Chang, H.: Toward automatically generating privacy policy for android apps. IEEE Trans. Inf. Forensics Secur. (2016)
32. Zimmeck, S., Goldstein, R., Baraka, D.: Privacyflash pro: automating privacy policy generation for mobile apps. In: Proceedings of the Network and Distributed System Security Symposium. NDSS (2021)
33. Zimmeck, S., et al.: Automated analysis of privacy requirements for mobile apps. In: Annual Network and Distributed System Security Symposium, NDSS (2017)

Characterizing the Evolution of Psychological Tactics and Techniques Exploited by Malicious Emails

Theodore Longtchi and Shouhuai Xu$^{(\boxtimes)}$

Department of Computer Science, University of Colorado Colorado Springs,
Colorado Springs, CO, USA
sxu@uccs.edu

Abstract. The landscape of malicious emails and cyber social engineering attacks in general are constantly evolving. In order to design effective defenses against these attacks, we must deeply understand the Psychological Tactics (PTacs) and Psychological Techniques (PTechs) that are exploited by these attacks. In this paper we present a methodology for characterizing the evolution of PTacs and PTechs exploited by malicious emails. As a case study, we apply the methodology to a real-world dataset. This leads to a number insights, such as which PTacs (PTechs) are more often exploited than others. These insights shed light on directions for future research towards designing psychologically-principled solutions to effectively counter malicious emails.

Keywords: Malicious Email · Cyber Social Engineering Attack · Psychological Tactic (PTac) · Psychological Technique (PTech) · Psychological Factor (PF)

1 Introduction

Malicious emails are one major form of cyber social engineering attacks. For example, the 2023 Anti-Phishing Working Group (APWG) report [6] shows that the number of phishing attacks had tripled since May 2020 and that more than one million domain names were used for phishing purposes in the year of 2023. These alarming facts highlight that malicious emails and cyber social engineering attacks in general remain to be a rampart cyber threat. This is somewhat ironic because there have been many studies on designing defenses against these attacks, suggesting that existing defenses have achieved very limited success. Thus, it is imperative to understand the threat landscape. In particular, this has motivated studies to understand the psychological aspects of cyber social engineering attacks, including malicious emails (e.g., [30,50,54]).

In term of understanding psychological aspects of cyber social engineering attacks, three concepts have been investigated. One concept is known as Psychological Tactic (PTac) [32,38,50], which describes the overall deliberate thoughtfulness (and objective in a sense) of attackers in framing, for example, the malicious content of an email to influence an email recipient. Another concept is

J. Zhao and W. Meng (Eds.): SciSec 2024, LNCS 15441, pp. 97–117, 2025.
https://doi.org/10.1007/978-981-96-2417-1_6

Psychological Technique (PTech) [32,38,50], which describes the psychologically relevant textual and imagery elements (e.g.) in email contents. The final of the three concepts is Psychological Factor (PF) [32], which describes an individual's psychological attributes or characteristics that may be exploited by cyber social engineering attacks. Studies, such as [9,32,33,59], show that these psychological concepts have not been adequately considered in existing defenses, explaining their ineffectiveness. This motivates the present study to deepen our understanding of the exploitation PTacs and PTechs by malicious emails.

Our Contributions. In this paper we make three contributions. First, we propose a methodology for characterizing the evolution of PTacs and PTechs exploited by malicious emails. The methodology includes the task of identifying the PTacs and PTechs exploited by malicious emails. The methodology can be applied to any dataset of malicious emails to characterize the evolution of PTacs and PTechs exploited by these emails. Second, we conduct a case study by applying the methodology to a real-world dataset of 1,260 malicious emails over two decades (2004–2024), with 60 malicious emails per year. The case study leads to a number of insights, including: (i) the exploitation of PTacs and PTechs is potentially greatly affected by the emergence of major events such as the COVID pandemic; (ii) future defenses should pay more attention on countering the PTacs and PTechs that are often exploited by attackers, as we have yet to observe that attackers are forced to change their PTacs and/or PTechs; (iii) future defenses should strive to deal with the PTacs and PTechs that are often exploited together. These insights shed light on promising research directions.

Ethical Issue. After consulting with our institution's Internal Review Board (IRB), it is determined that no IRB approval is needed because the emails we analyze are from third parties (except for 108 malicious emails that are collected from our own email box).

Related Work. There are many studies on cyber social engineering attacks, such as [2–4,7,10,14,19,20,22,25,29,32,51,56,58,60,61,69] and the references therein. The most closely related prior studies are: [32], which studied the evolution of PFs exploited by malicious emails; [33], which presented the first systematization on the 16 PTechs and 46 PFs that have been exploited by cyber social engineering attacks; and [50], which analyzed 7 PTacs and 8 PTechs exploited by 1,036 malicious emails. However, these studies did not consider the evolution of PTacs or PTechs exploited by malicious emails. By contrast, we investigate the evolution of PTacs or PTechs in the span of 2004–2024.

Loosely related prior studies include: De Bona and Paci [15] showed, based on 191 participants, that urgency is more likely to make employees susceptible to phishing attacks than authority; Gallo et al. [24] studied how cognitive vulnerabilities are exploited by phishing emails based on a dataset of 2-year span; Wang et al. [60] investigated susceptibility to phishing emails via visceral triggers; Gallo et al. [23] designed a system to detect persuasive elements in phishing email; [19,20,22,58] investigated how PTechs were exploited by malicious emails; and [14] leveraged PTechs to detect malicious emails. Each of these studies focused on one or very few PTechs. By contrast, we consider the evolution of 7 PTacs and 9 PTechs over two decades.

Paper Outline. Section 2 reviews and refines PTacs and PTechs. Section 3 presents our methodology. Section 4 describes a case study based on 1,260 malicious emails. Section 5 discusses our limitations. Section 6 concludes the paper.

2 Reviewing and Refining PTacs and PTechs

The concepts of PTac and PTech were inspired by the concepts of Tactic and Technique in the MITRE ATT&CK framework [35], but in the setting of cyber social engineering attacks [32,38,50]. At a high level, a PTac can exploit one or multiple PTechs, and a PTech can exploit one or multiple PFs; PFs represent root causes of humans' susceptibility to cyber social engineering attacks including malicious emails. The investigation of the evolution of PFs is an independent work [32], where 20 PFs are reconciled from the 46 PFs proposed in [33].

2.1 Reviewing the Concept of PTac

Definition 1 (PTac [38,50]). *In the context of malicious emails, a PTac describes the overall deliberate thoughtfulness (and objective) of the attacker in framing the malicious email content in order to victimize an email recipient.*

PTacs can be seen as a measure of the attacker's effort to craft a malicious email with respect to an objective. There are 7 PTacs [50], all of which will be considered in this study. (i) `Familiarity`, which refers to how an attacker attempts to establish a positive/trust relationship or association with a recipient (of a malicious email) [1,36]. (ii) `Immediacy`, which refers to the amplification of a time constraint to reduce a recipient's skepticism or scrutiny [37,40]. (iii) `Reward`, which refers to an exchange of something (physical or social) that is valuable to a recipient [26,31]. (iv) `Threat of Loss`, which refers to an appeal to the recipient's desire to maintain a certain status and/or prevent losing an opportunity [26,53]. (v) `Threat to Identity`, which refers to an attacker's efforts at manipulating a recipient's desire to maintain a positive and/or socially valuable reputation [36,53]. (vi) `Claim to Legitimate Authority`, which refers to the exploitation of a legitimate power to obscure or deter a recipient's scrutiny [20,53]. (vii) `Fit & Form`, which refers to an attacker's effort at mimicking the expected style of authentic emails in a malicious email [26,49].

2.2 Reviewing and Refining the Concept of PTech

Definition 2 (PTech [32,38,50]). *In the context of malicious emails, a PTech describes the psychologically relevant visual (textual and imagery) elements that an attacker employs in an email message to lure a recipient of the email.*

PTechs can be seen as the visual elements exploited by a malicious email, such as highlighted texts or images that would have a psychological significance to a recipient to act as intended by the attacker. There are the 16 PTechs [32]. For the purpose of the present study, we propose focusing on the PTechs that meet the following criteria: (i) A PTech must be distinct from the other PTechs as we observe that some of the 16 existing PTechs may overlap with each other. (ii) A PTech should be exploited in a single email interaction to lure a victim rather than multiple interactions. This is a practical consideration when one cannot obtain emails of multiple interactions between an attacker and a recipient.

The preceding criteria guide us to focus on the following 9 PTechs (out of the 16 PTechs presented in [33]), while noting that 8 (out of the 16) PTechs are considered in [50]. (i) **Persuasion**, which is the use of arguments pertaining to Cialdini's 6 Principles of Persuasion (i.e., authority, reciprocity, liking, scarcity, social proof, consistency/commitment) to convince a recipient [20,22]. (ii) **Pretexting**, which is the use of a made-up story (or pretext) to justify contacting a recipient to gain the recipient's trust [1,26]. (iii) **Impersonation**, which is the use of a false (and sometimes known) persona/entity to build trust with a recipient [5,20]. (iv) **Visual Deception**, which is the use of visual but deceptive elements to gain the trust of a recipient [37,39]. (v) **Incentive & Motivator**, which is the use of textual elements to show financial benefit or gain to a recipient [8,38]. (vi) **Urgency**, which is the use of textual elements to show a time constraint to urge a recipient to act quickly [13,59]. (vii) **Attention Grabbing**, which is the use of graphics or text to draw a recipient's attention [21,40]. (viii) **Personalization**, which is about addressing a recipient by name or some personal identifiable information [27,28]. (iv) **Contextualization**, which is the use of current or ongoing events (e.g., Covid-19) to engage a recipient [26,44].

3   Methodology

The methodology has three steps: preparing a dataset; identifying PTacs and PTechs exploited by malicious emails while noting the identification of PFs is described in [32]; analysis.

3.1   Preparing Dataset

We use the following guidance to help prepare datasets. (i) Researchers should determine the scope of their studies, including specifying the kinds of attacks they plan to study. This is relevant because there are different kinds of malicious emails, such as spear phishing vs. general phishing. (ii) Researchers should determine the time span for collecting malicious emails. In principle, the longer the time span, the better. Moreover, researchers should determine the time granularity unit, such as year vs. month vs. week vs. day. These factors determine the effort made at analyzing a dataset, meaning that a feasible trade-off may need to be made. (iii) Researchers should assure the quality of malicious emails. This means that the sources of malicious emails should be trusted, and the malicious emails should be re-examined to confirm their maliciousness. Moreover, each malicious email should contain all of its original information/content. This is relevant because an email may contain some missing information, such as logos.

3.2 Identifying PTacs and PTechs Exploited by Malicious Emails

Inspired by (but different from) [50], we propose identifying or grading PTacs and PTechs from a given malicious email as follows. We define the exploitation of a PTac or PTech based on the degree of application of the PTac or PTech in a malicious email: 0 for no application, 1 for implicit application, and 2 for explicit application. This allows us to analyze not only the absence (score 0) vs. presence (scores 1 and 2) of PFs, but also implicit vs. explicit exploitation (i.e., score 1 vs. score 2). The latter is important because implicit exploitation is stealthier than explicit exploitation and thus harder to defend against because recognizing implicit exploitation of PTacs and PTechs would require understanding semantics of email contents, but recognizing explicit exploitation would only require recognition of their names (and variants) via keyword search.

We grade an email with respect to a PTac as follows: (i) We assign a score 0 to an email if the PTac is neither implicitly nor explicitly exploited by the email. For instance, an email containing *"Your account is scheduled to be suspended in 10 days"* would receive a score 0 with respect to the Reward PTac. (ii) We assign a score 1 to an email if the PTac is implicitly exploited by the email. For instance, an email containing *"Your subscription has expired, update now"* would receive a score 1 with respect to the Immediacy PTac because the word "now" implies an implicit, but not explicit, exploitation of Immediacy. (iii) We assign a score 2 to an email if the PTac is explicitly exploited by the email. For instance, an email containing *"Your reward is ready"* would receive a score 2 with respect to the Reward PTac because the word "reward" is mentioned in the email.

Similarly, we grade an email with respect to a PTech as follows. (i) We assign a score 0 to an email if the PTech is neither implicitly nor explicitly exploited in the email. For instance, an email containing *"I've sent you the form. Please fill it out and send to me whenever you are ready."* would receive a score 0 with respect to the Urgency PTech because the recipient has the control over the time of response. (ii) We assign a score 1 to an email if elements of the PTech are implicitly, but not explicitly, exploited in the email; for instance, an email containing *"Download the file now"* would receive a score 1 with respect to the Urgency PTech because the word "now" signifies a time constraint on a recipient of the email. (iii) We assign a score 2 to an email if the PTech is explicitly exploited by the email. For instance, an email containing *"Treat this as a matter of urgency"* would receive a score 2 with respect to the Urgency PTech because of the word "urgency" is mentioned in the email.

3.3 Analysis

The analysis is centered at characterizing the evolution of PTacs and PTechs exploited by malicious emails. An analysis can be centered at answering many Research Questions (RQs), such as the following.

- RQ1: How frequently are PTacs and PTechs exploited by malicious emails? Answering this question would allow us to draw insights into if PTacs (PTechs) are equally frequently exploited by malicious emails or not.
- RQ2: Which PTacs and PTechs have been increasingly, decreasingly, or constantly exploited? Answering this question would allow us to draw insights into if attackers are forced to change their strategies in exploiting PTacs (PTechs) because the previously exploited PTacs (PTechs) are no longer effective (owing to effective defense).
- RQ3: Which PTacs and/or PTechs are often exploited together? Answering this question would allow us to draw insights into thwarting the collective exploitation of PTacs and/or PTechs because it is neither wise nor feasible to design one defense against each PTac and/or PTech.
- RQ4: What PTacs often exploit which PTechs, and what PTechs often exploit which PFs? Answering this question would allow us to draw insights into how defenses should be designed to thwart the collective exploitation of PTacs and PTechs (PTechs and PFs) because we cannot afford to design one defense against an individual pair of (PTac, PTech) or (PTech, PF).

4   Case Study

4.1   Preparing Dataset

The dataset is the same as the one analyzed in [32]. That is, it is a dataset of malicious emails over the last 21 years (2004–2024). For each year, we select 60 emails, or 1,260 emails in total. The 1,260 emails are collected from various sources, including: the Anti-Phishing Working Group (APWG) (250), researchers (317), organizations (298), malicious emails reaching our own email box (108), and the Internet (287). The 1,260 emails are separated into 3 different types: phishing 74%, scam 24%, and spam 2%. We manually checked the 1,260 emails and find that the spam emails are incorrectly labelled as phishing emails by their respective data sources. We accomplish this by verifying the senders, while noting that many companies and institutions send out spam emails.

4.2   Identifying PTacs and PTechs from the 1,260 Malicious Emails

We identify or grade PTacs and PTechs from the 1,260 malicious emails as described in the methodology, while noting that the identification of PFs is described in [32]. Owing to the amount of work and demand of expertise required for the study, we can only afford to have one PhD student grade the 1,260 emails. To mitigate the potential inconsistency in between emails by the same grader (i.e., the PhD student), we use tables with examples of psychological elements or cues pertaining to the PTacs and PTechs.

Table 1 is an excerpt of the real table that was used as assistant during the grading process. Examples of psychological elements corresponding to each PF is presented in the second column, while noting that these elements provide evidence to the PTechs listed in column 3 and the PTacs listed in column 4. The reason for ordering the table as "PF → Example elements or cues → PTech → PTac" is that PTechs exploit PFs through elements or cues and PTacs exploit PTechs with the objective to lure recipients of an email.

Table 1. Examples of PTechs exploiting PFs, where "Cog. miser stands for Cognitive miser, "Ind. diff." stands for Individual difference, "Soc. proof stands for Social proof, "Incentive & Mot." stands for Incentive & Motivator, and "Th. of Loss" stands for Threat of Loss, "Le. Authority" stands for Claim to Legitimate Authority.

PF	Example elements or cues	PTech	PTac
Impulsivity	"Claim your price now"	Urgency	Immediacy
Trust	University or Company logo	Visual Deception	Familiarity
Curiosity	"Congratulation! You won!"	Incentive & Mot.	Reward
Cog. Miser	"Verify your account"	Attention Grabbing	Th. of Loss
Ind. Diff.	"Someone logged in your account"	Impersonation	Familiarity
Greed	"You have won £552,000.00..."	Incentive & Mot.	Reward
Liking	"We're committed to serving you"	Impersonation	Le. Authority
Laziness	"John, we're here for you."	Personalization	Fit & Form
Sympathy	"She lost both parents in.."	Incentive & Mot.	Familiarity
Vigilance	"G00GLE.COM" for Google.com	Visual Deception	Fit & Form
Authority	"I'm Colonel...."/(Court logo)	Impersonation	Le. Authority
Commitment	Brand name (e.g., Walmart)	Persuasion	Familiarity
Soc. Proof	"As a member of our party"/	Contextualization	Familiarity
Expertise	(Technical of subject matter)	Contextualization	Le. Authority
Defenselessness	"Your account has been blocked"	Urgency	Immediacy
Workload	"Download and review it ASAP"	Urgency	Immediacy
Negligence	(Hyperlink photo in an email)	Attention Grabbing	Fit & Form
Scarcity	"10 bottles for the first 10 callers"	Incentive & Mot.	Reward
Loneliness	"Jasmine wants to meet you"	Pretexting	Familiarity
Reciprocity	"...give back to the university..."	Persuasion	Le. Authority

Figure 1 presents an end-to-end example showing the PTacs, PTechs, and PFs that are exploited by a phishing email. In terms of grading PTacs, the email receives a score 1 with respect to the Familiarity PTac because it is implicitly exploited via the following elements: (i) the impersonation of a known company to make the email appear familiar to the recipient; and (ii) addressing the recipient by name to engender trust and familiarity with the attacker. The email receives a score 0 with respect to the `Fit & Form` PTac because its composition failed in mimicking an email from McCarthy staffing company, namely the attacker's mistake in including the signature twice, which is a red flag alarm.

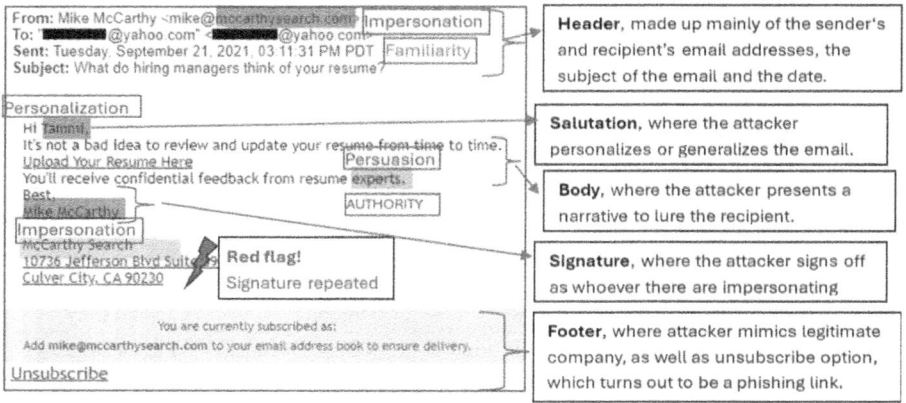

Fig. 1. A real-world phishing email showing which PTacs, PTechs, and PFs are exploited, where the exploited `Familiarity` PTac, Personalization and Persuasion PTechs, and AUTHORITY PF are highlighted. The email receives a score 0 with respect to the `Fit & Form` PTac because the email has a mistake, which is the repeated Signature (i.e., an email from the McCarthy staffing company will not have this mistake, highlighted with the red flag). (Color figure online)

In terms of PTechs, the email receives a score 1 with respect to the Impersonation PTech because it is implicitly exploited in the header of the email (highlighted in yellow), namely that the attacker impersonates the McCarthy Search Staffing company by spoofing its emails address. The email receives a score 1 with respect to Personalization because it is implicitly exploited in the salutation, namely that the attacker addresses the recipient by name (i.e., "Hi Tammi"). The email receives a score 1 with respect to the Persuasion PTech because it is implicitly exploited via the "expert" element that reflects the AUTHORITY PF [32], according to Cialdini's Principles of Persuasion [19].

In terms of PFs, the email receives a score 1 with respect to the AUTHORITY PF because the word "experts" is used in the email to describe the email sender's expertise. Note that the email receives a score 0 with respect to the EXPERTISE PF because the word "experts" describes the expertise of the sender, rather than the expertise of the recipient.

Table 2. Counts of PTacs, PTechs, and PFs exploited by the 1,260 emails per year, as well as the Total of PTac (TPTac), the Total of PTech (TPTech), and the Total of PF (TPF) from 2004 to 2024.

PTac	2004	2005	2006	2007	2008	2009	2010	2011	2012	2013	2014	2015	2016	2017	2018	2019	2020	2021	2022	2023	2024	TPTac
Familiarity	35	21	31	19	30	23	18	20	24	31	52	46	49	47	32	35	40	33	17	41	25	669
Immediacy	25	15	27	20	22	12	9	16	15	20	30	26	31	23	28	32	37	22	22	32	23	487
Reward	18	39	15	30	31	30	28	16	17	4	7	5	7	12	10	11	21	13	13	4	16	353
Threat of Loss	19	4	19	13	14	13	7	13	24	23	51	39	47	35	21	25	16	20	14	28	18	463
Threat to Identity	2	1	0	1	0	0	0	0	3	0	0	2	0	0	1	2	1	1	1	1	1	16
Claim to Legit Authority	27	23	28	27	35	26	28	25	34	36	55	47	52	52	36	39	50	42	31	42	38	772
Fit & Form	33	44	42	45	50	43	50	51	46	52	52	52	54	44	51	57	53	53	42	47	44	1004
Total	159	147	162	155	182	147	140	144	162	179	244	219	237	216	174	194	212	192	140	195	164	3764
PTech	2004	2005	2006	2007	2008	2009	2010	2011	2012	2013	2014	2015	2016	2017	2018	2019	2020	2021	2022	2023	2024	TPTech
Urgency	24	31	32	22	15	19	17	21	13	19	22	29	32	24	28	32	32	25	18	26	24	565
Visual Deception	14	5	19	11	22	3	7	13	10	21	15	6	3	8	16	11	16	20	19	25	18	289
Incentive & Motivator	17	43	14	32	31	33	28	28	18	21	5	7	5	7	16	11	13	22	21	4	20	390
Persuasion	11	22	17	30	25	16	8	13	17	11	18	13	18	26	9	6	22	8	9	10	6	315
Impersonation	41	42	41	39	50	43	41	28	35	42	56	47	56	52	39	45	56	50	38	47	43	931
Contextualization	8	33	16	17	25	20	26	18	10	13	5	3	6	6	11	6	32	24	12	5	11	302
Pretexting	32	40	36	39	37	27	15	27	24	37	49	52	46	49	30	36	36	41	32	46	41	772
Personalization	5	5	1	4	3	0	4	12	12	9	2	3	3	7	6	5	4	15	5	10	5	117
Attention Grabbing	26	37	53	52	55	42	57	55	55	57	51	57	48	46	52	31	53	48	53	53	52	1030
Total	178	258	229	246	263	203	203	209	194	230	218	210	223	227	201	211	242	258	202	226	220	4651
PF	2004	2005	2006	2007	2008	2009	2010	2011	2012	2013	2014	2015	2016	2017	2018	2019	2020	2021	2022	2023	2024	TPF
Individual Difference	15	13	11	15	9	10	13	15	16	16	10	17	16	28	18	23	16	22	24	14	15	336
Trust	25	26	16	14	19	17	22	26	31	26	50	50	40	44	26	35	38	41	27	41	32	646
Impulsivity	32	36	32	45	47	40	41	28	46	50	55	54	50	52	43	49	48	54	47	57	53	959
Vigilance	9	11	7	0	0	0	0	2	0	0	1	0	1	1	0	0	0	0	0	0	0	34
Greed	17	26	5	17	21	26	12	10	3	9	1	1	2	1	8	2	8	8	5	7	7	187
Sympathy	1	1	0	10	10	5	3	4	2	2	1	1	1	1	0	1	0	5	5	1	0	45
Liking	3	2	0	11	7	5	3	5	5	2	3	2	4	4	11	8	13	8	3	7	6	113
Curiosity	17	33	20	39	36	40	35	30	32	31	6	13	17	17	33	28	34	33	36	31	36	597
Laziness	3	10	4	1	1	2	8	1	2	2	2	0	6	4	5	4	4	1	0	3	0	54
Cognitive Miser	15	14	12	6	12	8	11	10	10	18	42	37	43	42	25	31	35	23	19	24	21	469
Social Proof	4	2	2	2	2	3	5	5	1	0	2	3	1	1	2	3	0	4	1	0	0	43
Authority	17	5	13	17	25	13	5	14	20	21	49	46	50	43	24	28	37	20	14	25	22	508
Expertise	0	0	1	1	2	1	0	2	1	0	2	0	0	0	0	0	0	0	0	0	0	11
Scarcity	0	6	8	5	2	4	3	4	3	5	17	16	1	0	2	1	0	0	4	0	0	81
Commitment	6	0	10	13	12	11	7	11	3	8	16	9	7	14	22	20	18	16	13	18	17	251
Negligence	7	1	0	0	1	4	0	0	5	6	3	6	7	10	11	6	4	6	7	11	6	101
Defenselessness	2	3	5	2	3	1	9	6	14	11	7	7	10	9	30	20	26	28	28	30	23	278
Loneliness	4	4	1	4	3	8	10	8	2	6	6	0	1	0	3	0	0	1	0	0	0	59
Workload	6	1	1	2	2	2	2	0	9	5	16	15	27	29	8	13	10	13	18	16	17	212
Reciprocity	1	0	0	0	0	0	1	1	0	1	0	0	0	1	1	0	2	0	0	0	0	5
Total	184	195	148	204	213	201	187	180	215	221	281	283	279	298	273	272	291	279	248	282	255	4989

4.3   Analysis

Our analysis first considers the absence vs. presence of a PTac or PTech in an email, namely an email receiving a score of 0 vs. a score of 1 or 2 with respect to a PTac or PTech. We may further consider the distinction of score 1 vs. 2 (i.e., implicit vs. explicit exploitation) when the need arises, especially for the purpose of determining how evasive the attacks are.

Addressing RQ1: How frequently are PTacs and PTechs exploited by malicious emails? To answer this RQ, we only need to differentiate the absence vs. presence of a PTac or PTech, namely a score of 0 according to the grading method described in the methodology vs. a non-zero score (i.e., score 1 or 2 with respect to a PTac or PTech according to the grading method described in the methodology). Table 2 presents the detailed results.

Figure 2(a) plots the occurrence of all the PTacs over the 21 years, namely the aggregated frequency (i.e., the sum of the instances) of PTacs exploited by malicious emails, where occurrence in each year is upped bounded by 420 (recalling that we have 60 emails per year and 7 PTacs). We make the following observations. (i) The exploitation of PTacs has, by and large, somewhat increased over the 21 years. There is a surge in 2014, which is mainly caused by the surge in the exploitation of `Familiarity`, `Threat of Loss`, and `Claim to Legitimate Authority`. By looking into the emails, we find that this is caused by surges in the impersonation of authorities, such as a bank or the U.S. Internal Revenue Service (IRS), in the impersonation of familiar entities, such as PayPal or university IT desk demanding an account update, and in the use of attacks containing messages like "*update your password or your account will be suspended.*" (ii) Almost all the PTacs have been exploited by malicious emails on a yearly basis. (iii) The `threat to identity` PTac has been rarely exploited over the 21 year.

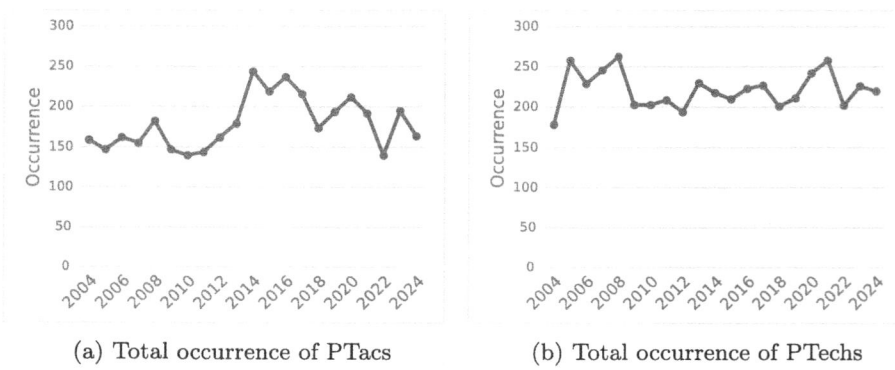

(a) Total occurrence of PTacs   (b) Total occurrence of PTechs

Fig. 2. Evolution of exploited PTacs and PTechs in malicious emails from 2004 to 2024

Figure 2(b) plots the occurrence of all the PTechs over the 21 years, where the occurrence is upper bounded by 540 (as we have 60 emails per year and 9

PTechs). We make the following observations. (i) There is, by and large, a general trend that PTechs are increasingly exploited. There is an explosive increase in 2005 perhaps because the Nigerian Prince (or 419) scams became popular as 26 out of the 60 emails are of this kind. The drop in the exploitation of PTechs in 2009 to 2012 is caused by the drop in the exploitation of the Incentive & Motivator, Persuasion, and Pretexting PTechs. The increase in the exploitation of PTechs in 2021 may be attributed to the emergence of the Covid-19 pandemic whereby malicious emails attempted to scam people with Covid-19 related messages. This may be justified by the fact that 22 out of the 60 emails are in this category, while 8 emails explicitly use Covid-19 as a pretext and 14 emails use work from home offers during the Covid-19 lockdown. However, these increases and decrease could well be caused by the small sample size (i.e., 60 emails per year). (ii) Almost all PTechs have been exploited on a yearly basis.

Insight 1. *The exploitation of PTacs and PTechs over time is potentially largely affected by the emergence of major events that provide attacks with opportunities to wage new kinds of malicious emails.*

Insight 1 resonates findings by other researchers in closely related contexts such as malicious websites (e.g., [44, 45, 48]). Moreover, it stresses the importance in designing proactive defenses against cyber social engineering attacks, including but not limited to, malicious emails that exploit emerging events to wage new kinds of attacks.

Addressing RQ2: Which PTacs and PTechs have been increasingly, decreasingly, or constantly exploited? Figure 3 plots the evolution of PTacs exploited by the 1,260 malicious emails over the last 21 years; each year, the number of occurrences (the y-axis) is the number of instances that a PTac is exploited by the 60 emails (i.e., bounded from above by 60).

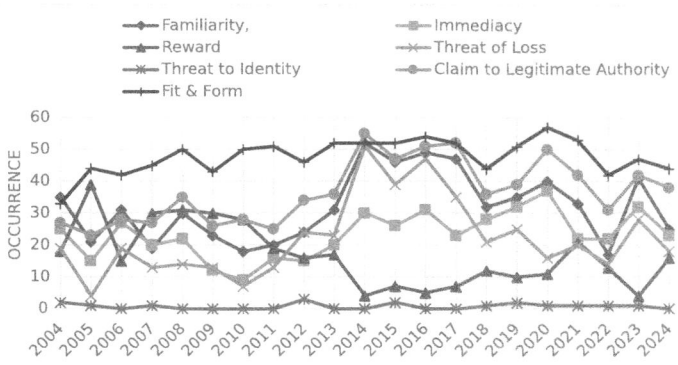

Fig. 3. Evolution of the exploitation of individual PTacs

We make the following observations. (i) `Fit & Form` is the most exploited PTac and `Familiarity` is a frequently exploited PTac. This is different from the finding presented in [50] that `Familiarity` was the most exploited PTac. The discrepancy can be attributed to the fact that the two studies used two different datasets. (ii) `Threat to identity` is the least exploited PTac over the 21 years, with a total of 16 instances in the 1,260 emails (i.e., 0 exploitations in some years). The second least exploited PTac is `Reward`, with a total of 353 instances in the 1,260 emails. (iii) `Reward` is the only PTac that shows a relatively decreasing exploitation from its peak in 2005 (39 instances in 60 emails) to its lowest exploitation in 2014 (4 instances in 60 emails) and remaining relatively low before rising again in 2021, then dropping again in 2022 and 2023. Although the sample (i.e., 60 emails) is too small to draw reliable conclusions, we speculate that this trend may be due to the gradual decline of scams from the early days of the Internet boom and the increase of Covid-19 relative scams that peaked in 2021 with stimulus check scams. (iv) The `Immediacy` and *Threat to Identity* PTacs have a relatively constant exploitation from 2004 to 2024. This suggests that the attackers see no need to change tactics that have been working for them.

Similarly, Fig. 4 plots the evolution of PTechs exploited by the 1,260 malicious emails over the 21 years, where the number of occurrence each year is also bounded from above by 60 (i.e., 60 emails per year). We make the following observations. (i) Attention Grabbing PTech is the most exploited PTech with a total of 1,030 instances in 1,260 emails, which is followed by Impersonation with 931 instances and Pretexting with 772 instances. (ii) Personalization is the least exploited PTech with 117 instances in 1,260 emails. This may be attributed to some of the following reasons: it is difficult for attackers to get personal details of recipients in order to send personalized emails, attackers can be too lazy to carry out meaningful reconnaissance, some attackers may not be equipped with tools and skills to carry out reconnaissance, and the personalization of emails does not give attackers the leverage to send bulk of emails. (iii) Visual Deception, Pretexting, Impersonation, and Attention Grabbing have all increased from 2004 to 2024. This may be attributed

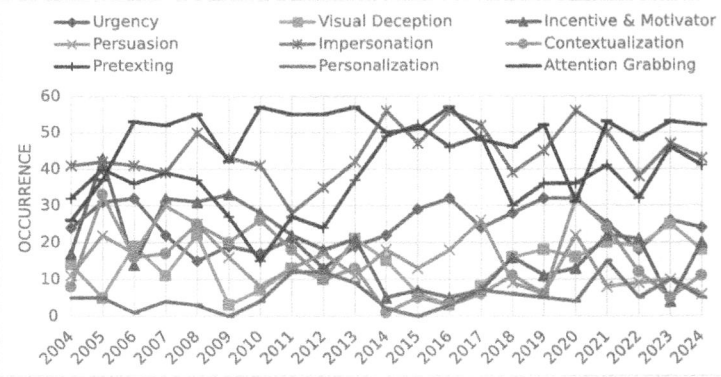

Fig. 4. Evolution of the exploitation of individual PTechs

to attackers applying more visuals in malicious emails to grab recipients' attention and using false narratives, such as pretexts, to disarm recipients' suspicion. (iv) Incentive & Motivator has seen a relative decrease in exploitation from its peak in 2005 (43 instances in 60 emails) to 2014 (5 instances in 60 emails), then levels up to in 2017 before a bumpy rise in 2024 with 20 instances in 60 emails. This decrease may be attributed to the decrease of scam emails offering incentives, such as the Nigerian Prince scams, since they became known to the public and thus decreasingly successful. (v) Urgency and Personalization have seen a relatively constant exploitation by malicious emails from 2004 to 2024. This may be attributed to the fact that attackers often employ time-constrained elements of Urgency to force recipients to act without thoughtfulness.

Insight 2. *Future defenses should pay more attention on coping with the PTacs (e.g.,* Fit & Form*) and PTechs (e.g.,* Attention Grabbing*) that are more exploited than others, as we have yet to witness that attackers are forced to change their PTacs and/or PTechs.*

Addressing RQ3: Which PTacs and/or PTechs are often exploited together? To address this question, we use the Pearson correlation coefficient [52], denoted by r, to characterize the correlations between PTacs (i.e., PTac-PTac correlation), PTechs (i.e., PTech-PTech correlation), and PTacs and PTechs (i.e., PTac-PTech correlation). These coefficients are computed by treating the 1,260 emails as a sample. Figure 5 presents the correlation coefficients, where intuitive abbreviations are used (e.g., "Rew" for Reward). Elaborations follow.

For PTac-PTac correlations, we make the following observations from Fig. 5(a). (i) The strongest positive correlation is between Familiarity and Claim to Legitimate Authority with coefficient $r = 0.878$, perhaps because attackers usually impersonate entities of authority that are known to a recipient. Another strong positive correlation occurs between Familiarity and Immediacy with $r = 0.826$, perhaps because attackers often impersonate an entity familiar to a recipient while urging the recipient to take quick actions. We use one example email from the dataset to illustrate these positive correlation: An attacker impersonates a university IT desk and sends an email stating, "You've used 97% of your email storage. You have to verify your email within 24 h, or you will lose your email and your free storage." In this attack, the claimed university IT is familiar to the recipient (Familiarity); the attacker claims to be the university IT desk (Claim to Legitimate Authority); the recipient (e.g., a student) will lose their email storage if they do not act within 24 h (Immediacy); and the claimed sender has the power to take the recipient's email away (Threat of Loss). (ii) The strongest negative correlation is between Reward and Threat of Loss with $r = -0.80$. This is natural because attackers do not offer reward and incur loss to recipients in the same email. Another strong negative correlation is between Claim to Legitimacy and Reward with $r = -0.76$. This also natural because a legitimate power would not offer reward to a recipient.

For PTech-PTech correlations, we make the following observations from Fig. 5(b). (i) The strongest positive correlation is between Contextualization and Incentive & Motivator with $r = 0.756$ because attackers often leverage current events to create emails that offer incentives to the email recipients. For example, one email from the dataset states, "$2400 has been allocated to you as part of the Covid-19 aid stimulus bill", where Covid-19 pandemic is the context (Contextualization) and the stimulus check is the incentive to motivate the recipient to act quickly (Incentive & Motivator). Another positive correlation is between Pretexting and Impersonation with $r = 0.642$, perhaps because attackers often need to impersonate a trusted entity in order to present the pretext for contacting a recipient. (ii) The strongest negative correlation is between Pretexting

	Fam	Imm	Rew	Loss	Iden	Legit	F&F
Familiarity	1	0.726	-0.74	0.877	-0.01	0.878	0.439
Immediacy	0.726	1	-0.73	0.552	0.124	0.673	0.261
Reward	-0.74	-0.73	1	-0.8	-0.12	-0.76	-0.3
Threat of Loss	0.877	0.552	-0.8	1	0.006	0.825	0.39
Threat to Identity	-0.01	0.124	-0.12	0.006	1	-0.05	-0.23
Claim to Legitimate Authority	0.878	0.673	-0.76	0.825	-0.05	1	0.662
Fit & Form	0.439	0.261	-0.3	0.39	-0.23	0.662	1

(a) Correlation between PTacs and PTacs.

	Urg	Vis D	I&M	Persu	Imp	Contx	Pretx	Perso	Att G
Urgency	1	-0.06	-0.36	-0.02	0.373	-0.06	0.476	-0.25	-0.3
Visual Deception	-0.06	1	-0.19	-0.32	-0	-0.05	0.085	0.355	0.146
Incentive & Motivator	-0.36	-0.19	1	0.207	-0.44	0.756	-0.49	-0.01	-0.13
Persuasion	-0.02	-0.32	0.207	1	0.265	0.24	0.265	-0.26	-0.14
Impersonation	0.373	-0	-0.44	0.265	1	-0.11	0.642	-0.3	-0.11
Contextualization	-0.06	-0.05	0.756	0.24	-0.11	1	-0.43	0.054	-0.26
Pretexting	0.476	0.085	-0.49	0.265	0.642	-0.43	1	-0.15	0.029
Personalization	-0.25	0.355	-0.01	-0.26	-0.3	0.054	-0.15	1	0.228
Attention Grabbing	-0.3	0.146	-0.13	-0.14	-0.11	-0.26	0.029	0.228	1

(b) Correlation between PTechs and PTechcs.

	Urg	Vis D	I&M	Persu	Imp	Contx	Pretx	Perso	Att G
Familiarity	0.464	0.008	-0.81	0.115	0.783	-0.55	0.736	-0.17	-0.06
Immediacy	0.654	0.411	-0.74	-0.02	0.592	-0.39	0.587	-0.15	-0.19
Reward	-0.33	-0.22	0.979	0.266	-0.38	0.764	-0.47	-0.04	-0.13
Threat of Loss	0.24	-0.09	-0.83	0.032	0.582	-0.79	0.67	-0.13	0.237
Threat to Identity	0.059	-0.02	-0.12	-0.15	-0.21	-0.17	-0.07	0.222	-0.28
Claim to Legitimate Authority	0.302	0.062	-0.77	0.084	0.815	-0.46	0.68	-0.04	0.111
Fit & Form	0.145	-0.03	-0.32	0.153	0.527	0.071	0.31	0.151	0.411

(c) Correlation between PTacs and PTechs.

Fig. 5. Correlations in color coding: positive correction increases from white to a depth of blue, and negative correction (absolute value) increase from white to a depth of red. (Color figure online)

and Incentive & Motivator with $r = -0.49$ because these PTechs should not be exploited in a single email.

For PTac-PTech correlations, we make the following observations from Fig. 5(c). (i) The strongest positive correlation is between the Reward PTac and the Incentive & Motivator PTech with $r = 0.979$, perhaps because attackers usually use monetary reward as incentives to lure victims. Another strong positive correlation is between the Claim to Legitimate Authority PTac and the Impersonation PTech with $r = 0.815$, because attackers have to impersonate some authority. Yet another strong positive correlation is between the Familiarity PTac and the Impersonation PTech with $r = 0.783$, because attackers always try to impersonate a personality or entity that is familiar to the recipient. (ii) The strongest negative PTac-PTech correlation is between the Threat of Loss PTac and Incentive & Motivator PTech with $r = -0.83$, because the former uses loss as a threat and the latter uses gain as incentive (i.e., they are not compatible).

Insight 3. *There are strong positive PTac-PTac and PTac-PTech correlations, but also not-so strong positive PTech-PTech correlations. Future defenses should strive to deal with the PTacs and PTechs that are often exploited together.*

Addressing RQ4: What PTacs often exploit which PTechs, and what PTechs often exploit which PFs? Figure 6 shows the relationships between the PTacs and PTechs in the malicious emails. We determine which PTac exploits which PTech based on their definitions. We say PTac A exploits PTech B if PTech B is semantically related to PTac A. Thus, if a malicious email exploits PTech B, then the email also exploits PTac A, when is semantically related to PTech B. See concrete examples below. We observe that all 7 PTacs exploit multiple PTechs. For example, the Familiarity PTac exploits the following 7 PTechs: (i) Persuasion, by using the Principles of Persuasion (e.g., when the Commitment Principle is employed in an email together with a brand name that is familiar to a recipient); (ii) Impersonation, by impersonating a familiar authority (e.g., one's boss); (iii) Visual Deception, by using visuals, such as logos, of popular brands to deceive a recipient; (iv) Incentive & Motivator by using well-known or brand goods as an incentive (e.g., Walmart gift card); (v) Urgency, by leveraging familiar situations, such as the Covid-19 infection, to encourage quick action; (vi) Attention Grabbing, by using symbols that are familiar to a recipient to draw the recipient's attention; and (vii) Contextualization, by using common societal events that are familiar to a recipient (e.g., using the War in Ukraine to ask for donation). As another example, the Threat to Identity PTac exploits the following 2 PTechs: (i) Impersonation, by assuming an entity that is tied to one's identity (e.g., "Your memberships expires in 48 h, please pay your dues to keep your membership"; and (ii) Personalization, by addressing the recipient by name (for example) threatening a recipient about pictures that the attacker will send to the wife if the recipient does not give in to the demands of the attacker.

Insight 4. *While all PTacs have exploited multiple PTechs, the Fit & Form PTac has exploited all 9 PTechs.*

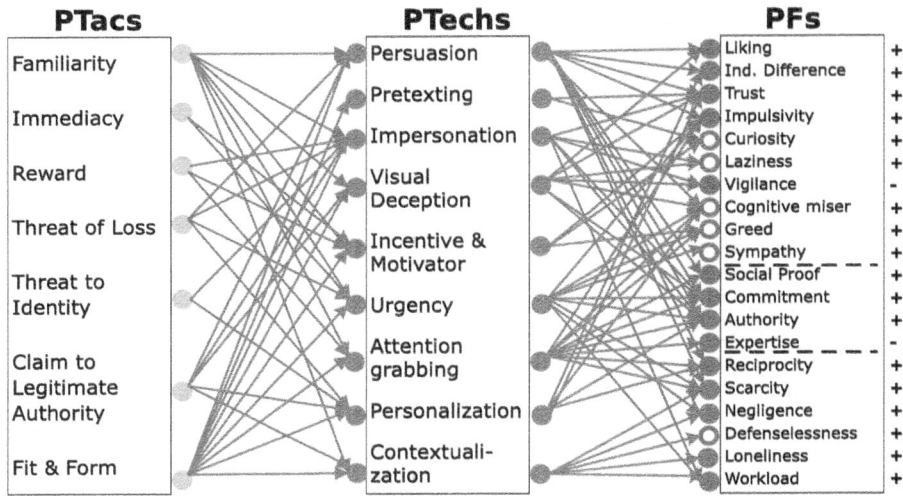

Fig. 6. Mappings of relationships between PTacs, PTechs, and PFs, where "$A \rightarrow B$" indicates "A exploits B," a dashed line inside a box indicates different categories, a filled circle indicates PFs with empirical quantitative studies and an empty circle otherwise, the "+" ("-") sign indicates that a factors increases (decreases) human susceptibility to attacks [33]. "Ind." is short for Individual.

Figure 6 also highlights the PTechs exploit PFs. We observe that all 9 PTechs exploit multiple PFs (out of the 20 PFs specified in [32]). For example, the Urgency PTech exploits the following 9 PFs: (i) the IMPULSIVITY PF, by using time constraint elements in a malicious email (e.g. "Click here to get your free copy"); (ii) the COGNITIVE MISER PF, by leveraging elements that give the recipient no time for any thoughtful decision (e.g., "Free membership ends today, claim your free membership now"); (iii) the GREED PF, by using elevated rewards and gains to bait greediness or the desire to get more (e.g., "...6 million US dollars, and you will get 30%"); (iv) the SYMPATHY PF, by using a pathetic narrative that needs immediate action (e.g., "Thieves stole all my belongings in an Airbnb, please send me $1000 through Western Union. I will pay you back when I return"); (v) the COMMITMENT PF, by using what the recipients may be dedicated to, or doing with consistency (e.g., "Please, donate now to support your party..."); (vi) the AUTHORITY PF, by impersonating someone with some power over the recipient (e.g., emails purported to come from one's boss, saying "Fill the attached form and send it to me urgently"); (vii) the SCARCITY PF, by saying a good or service is limited (e.g., "Only few remaining, get yours now:); (viii) the NEGLIGENCE PF, by leveraging a situation that the recipient must be negligent; and (ix) the WORKLOAD PF, by leveraging situations where the recipient is overwhelmed with work. Note that the last two PFs are opportunistic, because the attacker can only hope that the email will find the recipient in such situations (e.g., "Review the attached file and send it to me asap").

Similarly, the Impersonation PTech exploits 6 PFs, namely TRUST, LAZINESS, VIGILANCE, COMMITMENT, AUTHORITY, and WORKLOAD. For example, the TRUST PF is exploited by assuming the personality of a trusted entity via the logo of the World Health Organization where the message reads, *"Download the attached Covid-19 treatment update"*. The Contextualization PTech exploits 5 PFs, namely SCARCITY, NEGLIGENCE, DEFENSELESSNESS, LONELINESS, and WORKLOAD. For example, it exploits the SCARCITY PF by using elements that indicate a fewer than normal in the availability of goods and services, such as in the email message, *"Sign up here to be among the first people to get a free Covid-19 home test. Supply is limited."*

Insight 5. *The* Attention Grabbing *PTech exploits most PFs, while the* IMPULSIVITY *PF is most exploited by PTechs.*

5 Limitations

This study has two limitations that need to be addressed in the future. First, the empirical study is based on one investigator because it demands a substantial amount of time and expertise in identifying PTacs and PTechs. This means that the results may be biased. However, our methodology is equally applicable when there are multiple "graders" for identifying PTacs and PTechs from malicious emails, but would need to be extended with a calibration process, as shown in [50]. Second, our study is based on 1,260 malicious emails over 21 years, or 60 emails per year. Thus, the results may not be as representative as desired. It would be ideal to analyze a much larger dataset (e.g., 60 emails per month).

6 Conclusion

We have presented a methodology and applied it to a case study on characterizing the evolution of PTacs (Psychological Tactics) and PTechs (Psychological Techniques) exploited by malicious emails. We have drawn a number of insights, such as which PTacs and PTechs are exploited more often than others and which PTacs and/or PTechs are often exploited together. However, we have not observed the desired phenomenon that defenses have forced attackers to change their PTacs and/or PTechs because they become ineffective.

The insights and limitations of the present study shed light on future research directions toward designing effective defenses against malicious emails. Moreover, it is important to characterize the evolution of PTechs and PTacs that are exploited by the other kinds of attacks, such as malicious websites (e.g., [46–48,62,63]) and online social engineering attacks and the attacks that are geared towards specific sectors (e.g., healthcare and finance). It is interesting to model and forecast the evolution of PTechs and PTacs, perhaps together with the evolution of PFs, in the same fashion as in [17,18,42,43,55,57,64,65,67,68] so as to enable proactive defense. It is also interesting to systematically define metrics to quantify the susceptibility of humans to cyber social engineering attacks in principled fashion [11,12,16,34,41,66].

Acknowledgement. We thank the reviewers for their comments. This research was supported in part by NSF Grant #2115134 and Colorado State Bill 18-086. This research work is also a contribution to the International Alliance for Strengthening Cybersecurity and Privacy in Healthcare (CybAlliance, Project no. 337316).

References

1. Al-Hamar, M., Dawson, R., Guan, L.: A culture of trust threatens security and privacy in Qatar. In: 2010 10th IEEE International Conference on Computer and Information Technology, pp. 991–995. IEEE (2010)
2. Aleroud, A., Zhou, L.: Phishing environments, techniques, and countermeasures: a survey. Comput. Secur. **68**, 160–196 (2017)
3. Algarni, A., Xu, Y., Chan, T.: An empirical study on the susceptibility to social engineering in social networking sites: the case of Facebook. Eur. J. Inf. Syst. **26**(6), 661–687 (2017)
4. Alharbi, A., Dong, H., Yi, X., Tari, Z., Khalil, I.: Social media identity deception detection: a survey. ACM Comput. Surv. (CSUR) **54**(3), 1–35 (2021)
5. Allodi, L., Chotza, T., Panina, E., Zannone, N.: The need for new antiphishing measures against spear-phishing attacks. IEEE Secur. Priv. **18**(2) (2019)
6. APWG: Phishing activity trends report - unifying the global response to cybercrime. Technical report, Anti-Phishing Working Group, APWG (2023)
7. Asiri, S., Xiao, Y., Alzahrani, S., Li, S., Li, T.: A survey of intelligent detection designs of html URL phishing attacks. IEEE Access (2023)
8. Beckmann, J., Heckhausen, H.: Motivation as a function of expectancy and incentive. In: Motivation and Action, pp. 163–220. Springer (2018)
9. Canham, M., Tuthill, J.: Planting a poison SEAD: using social engineering active defense (SEAD) to counter cybercriminals. In: The 16th International Conference Augmented Cognition (AC 2022), pp. 48–57. Springer (2022)
10. Chanti, S., Chithralekha, T.: Classification of anti-phishing solutions. SN Comput. Sci. **1**(1), 1–18 (2020)
11. Cho, J., Hurley, P., Xu, S.: Metrics and measurement of trustworthy systems. In: Proceedings of IEEE MILCOM (2016)
12. Cho, J.H., Xu, S., Hurley, P.M., Mackay, M., Benjamin, T., Beaumont, M.: Stram: measuring the trustworthiness of computer-based systems. ACM Comput. Surv. **51**(6), 128:1–128:47 (2019)
13. Chowdhury, N.H., Adam, M.T., Skinner, G.: The impact of time pressure on cybersecurity behaviour: a systematic literature review. Behav. Inf. Technol. **38**(12), 1290–1308 (2019)
14. Cidon, A., Gavish, L., Bleier, I., Korshun, N., Schweighauser, M., Tsitkin, A.: High precision detection of business email compromise. In: 28th USENIX Security Symposium, pp. 1291–1307 (2019)
15. De Bona, M., Paci, F.: A real world study on employees' susceptibility to phishing attacks. In: Proceedings of the 15th International Conference on Availability, Reliability and Security, pp. 1–10 (2020)
16. Du, P., Sun, Z., Chen, H., Cho, J.H., Xu, S.: Statistical estimation of malware detection metrics in the absence of ground truth. IEEE T-IFS **13**(12), 2965–2980 (2018)
17. Fang, X., Xu, M., Xu, S., Zhao, P.: A deep learning framework for predicting cyber attacks rates. EURASIP J. Inf. Secur. **2019**, 5 (2019)

18. Fang, Z., Xu, M., Xu, S., Hu, T.: A framework for predicting data breach risk: leveraging dependence to cope with sparsity. IEEE T-IFS **16**, 2186–2201 (2021)
19. Ferreira, A., Coventry, L., Lenzini, G.: Principles of persuasion in social engineering and their use in phishing. In: International Conference on Human Aspects of Information Security, Privacy, and Trust, pp. 36–47. Springer (2015)
20. Ferreira, A., Lenzini, G.: An analysis of social engineering principles in effective phishing. In: Workshop on Socio-Technical Aspects in Security and Trust (2015)
21. Flores, W.R., Holm, H., Nohlberg, M., Ekstedt, M.: Investigating personal determinants of phishing and the effect of national culture. Inf. Comput. Secur. (2015)
22. Frauenstein, E.D., Flowerday, S.: Susceptibility to phishing on social network sites: a personality information processing model. Comput. Secur. (2020)
23. Gallo, L., Gentile, D., Ruggiero, S., Botta, A., Ventre, G.: The human factor in phishing: collecting and analyzing user behavior when reading emails. Comput. Secur. **139**, 103671 (2024)
24. Gallo, L., Maiello, A., Botta, A., Ventre, G.: 2 years in the anti-phishing group of a large company. Comput. Secur. **105**, 102259 (2021)
25. Goel, D., Jain, A.K.: Mobile phishing attacks and defence mechanisms: state of art and open research challenges. Comput. Secur. **73**, 519–544 (2018)
26. Goel, S., Williams, K., Dincelli, E.: Got phished? Internet security and human vulnerability. J. Assoc. Inf. Syst. **18**(1), 2 (2017)
27. Hirsh, J.B., Kang, S.K., Bodenhausen, G.V.: Personalized persuasion: tailoring persuasive appeals to recipients' personality traits. Psychol. Sci. **23**(6), 578–581 (2012)
28. Jagatic, T.N., Johnson, N.A., Jakobsson, M., Menczer, F.: Social phishing. Commun. ACM **50**(10), 94–100 (2007)
29. Kearney, W.D., Kruger, H.A.: Can perceptual differences account for enigmatic information security behaviour in an organisation? Comput. Secur. **61**, 46–58 (2016)
30. Khonji, M., Iraqi, Y., Jones, A.: Phishing detection: a literature survey. IEEE Commun. Surv. Tutor. **15**(4), 2091–2121 (2013)
31. Longtchi, T., Rodriguez, R.M., Al-Shawaf, L., Atyabi, A., Xu, S.: SoK: why have defenses against social engineering attacks achieved limited success? arXiv preprint arXiv:2203.08302 (2022)
32. Longtchi, T., Xu, S.: Characterizing the evolution of psychological factors exploited by malicious emails. In: Proceedings of International Conference on Science of Cyber Security (SciSec 2024) (2024)
33. Longtchi, T.T., Rodriguez, R.M., Al-Shawaf, L., Atyabi, A., Xu, S.: Internet-based social engineering psychology, attacks, and defenses: a survey. Proc. IEEE **112**(3), 210–246 (2024)
34. Mireles, J., Ficke, E., Cho, J., Hurley, P., Xu, S.: Metrics towards measuring cyber agility. IEEE Trans. Inf. Forensics Secur. **14**(12), 3217–3232 (2019)
35. MITRE: MITRE ATT&CK (2023). https://attack.mitre.org/
36. Montañez, R., Atyabi, A., Xu, S.: Social engineering attacks and defenses in the physical world vs. cyberspace: a contrast study. In: Cybersecurity and Cognitive Science, pp. 3–41. Elsevier (2022)
37. Montañez, R., Golob, E., Xu, S.: Human cognition through the lens of social engineering cyberattacks. Front. Psychol. **11**, 1755 (2020)
38. Montañez Rodriguez, R., Xu, S.: Cyber social engineering kill chain. In: Science of Cyber Security: 4th International Conference, SciSec 2022, Matsue, Japan, 10–12 August 2022, Revised Selected Papers, pp. 487–504. Springer (2022)

39. Moreno-Fernández, M.M., Blanco, F., Garaizar, P., Matute, H.: Fishing for phishers. improving internet users' sensitivity to visual deception cues to prevent electronic fraud. Comput. Hum. Behav. **69**, 421–436 (2017)

40. Nelms, T., Perdisci, R., Antonakakis, M., Ahamad, M.: Towards measuring and mitigating social engineering software download attacks. In: 25th USENIX Security Symposium, pp. 773–789. USENIX Association, Austin, TX (2016)

41. Pendleton, M., Garcia-Lebron, R., Cho, J.H., Xu, S.: A survey on systems security metrics. ACM Comput. Surv. **49**(4), 62:1–62:35 (2016)

42. Peng, C., Xu, M., Xu, S., Hu, T.: Modeling and predicting extreme cyber attack rates via marked point processes. J. Appl. Stat. **44**(14), 2534–2563 (2017)

43. Peng, C., Xu, M., Xu, S., Hu, T.: Modeling multivariate cybersecurity risks. J. Appl. Stat. 1–23 (2018)

44. Pritom, M., Schweitzer, K., Bateman, R., Xu, M., Xu, S.: Characterizing the landscape of COVID-19 themed cyberattacks and defenses. In: IEEE International Conference on Intelligence and Security Informatics, pp. 1–6 (2020)

45. Pritom, M., Schweitzer, K., Bateman, R., Xu, M., Xu, S.: Data-driven characterization and detection of COVID-19 themed malicious websites. In: IEEE International Conference on Intelligence and Security Informatics, pp. 1–6 (2020)

46. Pritom, M., Schweitzer, K., Bateman, R., Xu, M., Xu, S.: Characterizing the landscape of covid-19 themed cyberattacks and defenses. In: IEEE ISI 2020 (2020)

47. Pritom, M., Schweitzer, K., Bateman, R., Xu, M., Xu, S.: Data-driven characterization and detection of covid-19 themed malicious websites. In: IEEE ISI 2020 (2020)

48. Pritom, M., Xu, S.: Supporting law-enforcement to cope with blacklisted websites: framework and case study. In: IEEE CNS 2022 (2022)

49. Rajivan, P., Gonzalez, C.: Creative persuasion: a study on adversarial behaviors and strategies in phishing attacks. Front. Psychol. **9**, 135 (2018)

50. Rodriguez, R.M., et al.: Quantifying psychological sophistication of malicious emails. In: The 5th International Conference Science of Cyber Security (SciSec 2023). Lecture Notes in Computer Science, vol. 14299, pp. 319–331. Springer (2023)

51. Schaab, P., Beckers, K., Pape, S.: Social engineering defence mechanisms and counteracting training strategies. Info. Comput. Secur. **25**(2), 206–222 (2017)

52. Schober, P., Boer, C., Schwarte, L.A.: Correlation coefficients: appropriate use and interpretation. Anesth. Analg. **126**(5), 1763–1768 (2018)

53. Stajano, F., Wilson, P.: Understanding scam victims: seven principles for systems security. Commun. ACM **54**(3), 70–75 (2011)

54. Steves, M.P., Greene, K.K., Theofanos, M.F., et al.: A phish scale: rating human phishing message detection difficulty. In: Workshop on Usable Security (USEC) (2019)

55. Sun, Z., Xu, M., Schweitzer, K., Bateman, R., Kott, A., Xu, S.: Cyber attacks against enterprise networks: characterization, modeling and forecasting. In: Proceedings of SciSec 2023 (2023)

56. Syafitri, W., Shukur, Z., Asma'Mokhtar, U., Sulaiman, R., Ibrahim, M.A.: Social engineering attacks prevention: a systematic literature review. IEEE Access **10**, 39325–39343 (2022)

57. Trieu-Do, V., Garcia-Lebron, R., Xu, M., Xu, S., Feng, Y.: Characterizing and leveraging granger causality in cybersecurity: framework and case study. EAI Endorsed Trans. Secur. Saf. **7**(25), e4 (2020)

58. Van Der Heijden, A., Allodi, L.: Cognitive triaging of phishing attacks. In: 28th USENIX Security Symposium 2019), pp. 1309–1326 (2019)

59. Vishwanath, A., Herath, T., Chen, R., Wang, J., Rao, H.R.: Why do people get phished? Decis. Support Syst. **51**(3), 576–586 (2011)
60. Wang, Z., Zhu, H., Sun, L.: Social engineering in cybersecurity: effect mechanisms, human vulnerabilities and attack methods. IEEE Access **9**, 11895–11910 (2021)
61. Williams, E.J., Beardmore, A., Joinson, A.N.: Individual differences in susceptibility to online influence: a theoretical review. Comput. Hum. Behav. **72**, 412–421 (2017)
62. Xu, L., Zhan, Z., Xu, S., Ye, K.: An evasion and counter-evasion study in malicious websites detection. In: IEEE CNS, pp. 265–273 (2014)
63. Xu, L., Zhan, Z., Xu, S., Ye, K.: Cross-layer detection of malicious websites. In: ACM CODASPY 2013, pp. 141–152 (2013)
64. Xu, M., Hua, L., Xu, S.: A vine copula model for predicting the effectiveness of cyber defense early-warning. Technometrics **59**(4), 508–520 (2017)
65. Xu, M., Schweitzer, K.M., Bateman, R.M., Xu, S.: Modeling and predicting cyber hacking breaches. IEEE T-IFS **13**(11), 2856–2871 (2018)
66. Xu, S.: SARR: a cybersecurity metrics and quantification framework. In: Third International Conference on Science of Cyber Security (SciSec 2021), pp. 3–17 (2021)
67. Zhan, Z., Xu, M., Xu, S.: Characterizing honeypot-captured cyber attacks: statistical framework and case study. IEEE T-IFS **8**(11), 1775–1789 (2013)
68. Zhan, Z., Xu, M., Xu, S.: Predicting cyber attack rates with extreme values. IEEE Trans. Inf. Forensics Secur. **10**(8), 1666–1677 (2015)
69. Zieni, R., Massari, L., Calzarossa, M.C.: Phishing or not phishing? A survey on the detection of phishing websites. IEEE Access **11**, 18499–18519 (2023)

Matching Knowledge Graphs for Cybersecurity Countermeasures Selection

Kéren A. Saint-Hilaire[1,2](✉) ⓘ, Christopher Neal[1,2] ⓘ, Frédéric Cuppens[1] ⓘ,
Nora Boulahia-Cuppens[1] ⓘ, and Makhlouf Hadji[2] ⓘ

[1] Polytechnique Montréal, Montréal, Canada
keren-a.saint-hilaire@polymtl.ca
[2] IRT SystemX, Palaiseau, France

Abstract. As cyberattacks continue to increase, detecting and performing remediation actions m is essential. This paper presents an approach to automate the countermeasures selection process to deal with a vulnerability exploitation performed by a cyberattack. We propose an approach to match two knowledge graphs, one from a vulnerability ontology, Vulnerability Description Ontology (VDO), and the other is the countermeasures knowledge graph, D3FEND, to mitigate cyberattack impacts. Our approach uses machine learning and an inference system to match entities from VDO and D3FEND to select candidate countermeasures to an attack. Our contribution aims to automatically select countermeasures intended to be part of an incident response playbook for a vulnerability. We show our approach application to a WannaCry use-case scenario. We validate our countermeasures selection approach by comparing the countermeasures automatically selected with those proposed in the literature for a WannaCry attack.

Keywords: Graph Matching · Machine Learning for Cybersecurity · Incident Response

1 Introduction

As cyberattacks increase, advancing our capabilities to thwart them is crucial. In this paper, we consider cyberattacks that are performed by exploiting a vulnerability. Detection is the first step in stopping a cyberattack. An Intrusion Detection System (IDS) generates an alert when it detects a cyberattack. A Security Information and Event Management (SIEM) system correlates the alerts an IDS generates. The next step consists of selecting appropriate countermeasures to manage the attack. As there are many alerts, countermeasures must be automatically activated. Security Orchestration, Automation, and Response (SOAR) systems represent a step towards automation; a SOAR allows one to write playbooks and scripts to execute to cope with an attack. However, presently, experts manually write playbooks, and a SOAR requires extensive configuration.

Integrating an organization's security tools within the SOAR is a substantial task; it is even more complex when an organization has several instances of tools from different vendors. The SOAR configuration requires a significant investment in time and financial resources, and there is no standard to ease SOAR interoperability. To reduce the manual work experts should do regarding SOAR configuration, we propose to automatically select countermeasures aimed to generate automated playbooks that will be integrated into a SOAR.

Resolving this gap in cyberattack response automation consists of automatically identifying which countermeasures are effective in dealing with a cyberattack. We address this problem in this paper. As a system is monitored in real-time, metadata of generated alerts of detected intrusions is available. By semantically mapping this metadata with system knowledge, such as existing vulnerabilities, it is possible to determine the exploited vulnerabilities. Multiple vulnerability databases exist that contain the required conditions for the exploitation of vulnerabilities as well as the post-conditions of their exploitation.

In order to mitigate adversary actions and block any possible future action, it is necessary to know which countermeasures are related to the consequences of vulnerability. We propose to generate the individuals for each vulnerability description information to create a Knowledge Graph (KG) for the Vulnerability Description Ontology (VDO)[1], proposed by the National Institute of Standards and Technology (NIST). Then, the data can be available and machine-readable for automation tasks. Additionally, a popular KG of countermeasures exists, D3FEND[2]. We propose matching the VDO KG with D3FEND to identify the countermeasures to mitigate an ongoing vulnerability exploitation.

To perform the proposed KG matching we utilize a knowledge base of vulnerabilities, the targets of cyberattacks, and defensive countermeasures.

- **VDO** To represent vulnerabilities, we use VDO, a standardized vulnerability ontology proposed by NIST. VDO represents various attributes for characterizing software vulnerabilities.
- **Digital Artifact Ontology (DAO)**. DAO is used to represent the target of an attack. DAO is an ontology that specifies the concepts needed to classify and represent digital objects of interest for cybersecurity analysis. The use of DAO makes it possible to associate the offensive techniques offered by ATT&CK[3] with the defensive techniques of D3FEND.
- **D3FEND** To select countermeasures, we use D3FEND, a KG created by MITRE that describes specific technical functions within cyber technologies in a common language of defensive techniques. The D3FEND taxonomy inherits artifacts from DAO. ATT&CK is incorporated into the D3FEND KG by mapping its concepts directly to D3FEND's defensive techniques and artifacts model. Throughout this paper, when mentioning D3FEND, we also reference the inherited concepts from DAO and ATT&CK.

---

[1] https://github.com/usnistgov/vulntology.
[2] https://d3fend.mitre.org/.
[3] https://attack.mitre.org/.

Our contributions consist of creating corpora based on entities from VDO and D3FEND, matching each impact and exploitation method of a CVE with an offensive technique of D3FEND, and selecting the defensive techniques related to the offensive technique for the countermeasure plan construction. We show the application of the proposed model in a real-world situation. We validate the approach by comparing the countermeasures automatically selected for a WannaCry use-case scenario with those proposed for this kind of attack.

The paper is organized as follows: Sect. 2 covers techniques for matching KGs. Section 3 reviews related works and compares them to our approach. Section 4 details our methodology for matching VDO and D3FEND to select countermeasures. Section 5 evaluates the approach's efficiency and demonstrates its application in a use case. Section 6 discusses the advantages and challenges of our approach. Finally, Sect. 7 concludes the paper.

2   Background

This section introduces the related concepts used in our approach.

Attack Graph. An Attack Graph (AG) represents all the paths an adversary can take to release a detrimental event on an Information System.

Attack-Defense Graph. An attack defense graph (ADG) is a directed acyclic graph in which nodes represent threats arising from existing vulnerabilities and countermeasures to mitigate these threats.

Ontologies. An ontology is the concrete and formal representation of a domain. An ontology is a set of terms and the links between them. The ontology ensures that no contradictions exist between these terms. Description Logic (DL) makes it possible to represent an ontology. Ontologies allow automating inference and enabling operability between applications [12]. The open-world assumption governs ontologies and states that what is not known is assumed to be true. It is common to confuse the concepts of ontology and KG, however they have nuanced differences [11]. In the next paragraph, we explain the concept of KG.

Knowledge Graph (KG). A KG is a data graph aimed at accumulating and disseminating knowledge of the real world. A KG uses an ontology as a framework to describe a given instance of a domain. A concrete example would be creating an ontology to describe a countermeasure. The ontology comprises all the characteristics that all countermeasures share. This ontology would, therefore, have classes for the concepts of countermeasure, asset, adversarial action, and properties like mitigates. These terminologies are represented in the TBox. Definition 1 represents the TBox in the DL language. However, a knowledge graph must be created to represent a particular countermeasure, such as **Update software**. The asset mitigated by this countermeasure is a software. Nodes represent entities of interest, such as Software, and edges of potentially different relationships between nodes, such as impacts, mitigates. The constructed KG makes it possible to represent the following reality: *A countermeasure **Update software** mitigates an asset that is a software. This software is impacted by an adversarial action **Run virtual instance**. This KG does not make it possible to know if **Update software** and **Run virtual instance** are two different entities, namely that the countermeasure and adversarial action classes are disjoint. It is the ontology that makes it possible to define such restrictions.

Definition 1. *Terminologies domain in DL language*

$$Countermeasure \sqcap AdversarialAction \equiv \bot$$
$$Countermeasure \equiv mitigates.Asset$$

SPARQL Protocol and RDF Query Language (SPARQL). KGs are often modeled using the Resource Description Framework (RDF) in the N-triple format. This is a line-based, plain text serialization format for RDF graphs. For applications to interact with data, queries must be made on KGs. SPARQL is used to query an RDF graph. SPARQL is a query language that allows searching, adding, modifying, or deleting available RDF data.

b-matching of a Graph. Based on a, possibly weighted, graph with a positive integer b_v for every vertex v of the graph, a b-matching of a graph is a multiset M of its edges such that, for every vertex v, the number of edges of M incident to v does not exceed b_v [9].

Graph Matching. A matching of a graph is a case of b-matching in which $b_v = 1$ for every vertex v. We use a word embedding model to proceed to the b-matching of the two KGs.

3   Related Work

In recent years, researchers have proposed several countermeasure selection approaches. However, they come with limitations that are presented in this section. To fill this gap, we base our approach on graph matching. The research into graph matching has followed several approaches. In this section, we also present the different graph-matching approaches and their limitations and explain how we address them with our work to propose an automated countermeasures selection approach.

In their survey of countermeasures selection approaches [20], Nespoli et al. define a countermeasure strategy. A countermeasure strategy comprises methodologies, procedures, and processes that aim to react to and eradicate security incidents. They present a list of necessary components to define a countermeasure strategy.

Some necessary components [20] are a monitored system, detection tools, and countermeasure knowledge. Others are a system model that synthesizes information gathered from the monitored system, atomic countermeasure options, and a list of possible countermeasures. A threat model representing attack patterns in AGs or ATs is essential. Finally, countermeasures were selected while looking for a trade-off between security level and cost of reaction. A crucial component is a prediction reward that can lead to a model attacker decision where the threat model is updated based on all previously cited components. The system operator's decision is also important.

In our proposed approach, we consider these components. From a vulnerability report and a system model, we automatically generate an AG that is updated in real time based on system monitoring. Countermeasure knowledge is available in the literature. In this paper, we propose to select a list of possible countermeasures by correlating countermeasures knowledge with vulnerability reports. Other components, such as cost of reaction, are part of our automatically optimal security playbook generation approach [26] that can lead to the update of the threat model.

Nespoli et al. [20] survey the approaches based on 7 comparison features such as attack modeling, countermeasures provision techniques, and used standards. The surveyed approaches [8,15,29] highlight the lack of countermeasure standards. This issue is linked to the lack of countermeasure knowledge. The authors only present a limited set of countermeasures used to counteract specific attacks reported to the monitored system. These countermeasure selection processes lose relevance and effectiveness because they do not apply to another type of attack. The solutions also rely on the administrator's knowledge of each threat [10,29], implying a limitation in the correlation between atomic mitigation steps and attacks.

In [8,15], the authors use the Common Remediation Enumeration (CRE) standard to develop a countermeasure model. Nespoli et al. argue that CRE can not contribute significantly to the automation of countermeasures selection. Kotenko et al. [15] also assume that the system already has a pool of countermeasures that can be selected for an ad hoc algorithm. This assumption requires much effort from the expert filling the knowledge database of countermeasures. Our countermeasure selection approach is based on countermeasure and vulnerability standards. Using D3FEND, security experts do not have to fill any countermeasures knowledge base.

Our countermeasure selection approach is part of an ADG generation process. Compared with the approach in [28], our approach selects countermeasures for a system monitored in real-time instead of a formal system model. The AG generated in an initial state is updated in real-time based on newly deducted knowledge from VDO [5]. We choose this ontology because it is a standardized ontology proposed by NIST. It allows the representation of vulnerabilities based on the natural language description of a vulnerability in the National Vulnerability Database (NVD).

VDO is an ontology that describes vulnerabilities, the required conditions for exploitation, the consequences, and the product concerned. In our approach, we implement VDO in DL and automatically generate individuals for specific vulnerabilities to create the KG. In order to select appropriate countermeasures to a vulnerability, we need other knowledge bases focusing on countermeasures. We investigate several countermeasure knowledge bases such as the Security Control Catalogue ITSG-33 (from the Canadian Centre for Cyber Security) [6], D3FEND, and CIS Security Controls [7].

In [6], the Canadian Centre for Cyber Security provides security control definitions suitable for Departmental Security Officers (DSOs), IT security coordinators, and security practitioners. Each security control provides guidance on the best practices the security department should have in developing departmental and domain security control profiles, as well as, defining and implementing departmental IT security functions. In our approach, we focus on alleviating the work of security practitioners by automatically selecting countermeasures aimed to create automatic playbooks that can be integrated into a SOAR. Therefore, we discard these security controls.

In [14], Kaloroumakis et al. present a countermeasure knowledge graph, D3FEND. The Digital Artifact Ontology (DAO) is an ontology that specifies the concepts needed to classify and represent digital objects of interest for cybersecurity analysis. The use of DAO makes it possible to associate the offensive techniques proposed by ATT&CK with the defensive techniques of D3FEND based on the relationship of each technique with digital objects.

The CIS controls aim to share insights about attacks and attackers to identify root causes and translate them into defensive action classes. Each protective measure is general advice to enhance an organization's security. In comparison, the D3FEND defensive techniques provide more granularity by proposing specific actions applicable to a specific artifact. Additionally, D3FEND is standardized and machine-readable, which implies that integration for automated countermeasures selection is more accessible than the other knowledge-based methods.

In [22], Pershina et al. present an algorithm for aligning instances in large knowledge bases using Holistic Entity Matching (HolisticEM). Their approach involves generating entity pairs from two KGs and matching them based on attribute similarity. However, this approach's heuristic optimization phase results in varying matches between executions. Our approach aims to obtain a stable and equivalent output from a graph-matching procedure.

In [4], Azmy et al. propose an approach to match entities from two different KGs focusing on ambiguous entities from DBpedia and Wikidata. The approach consists of matching an entity from DBpedia with an entity in Wikidata corresponding to the same real-world entity and vice versa. The datasets are created thanks to existing cross-ontology links (i.e. OWL:sameAs predicate) between DBpedia and Wikidata. They focus on entities from the KGs that share the same name, for example, the foaf:name predicate in DBpedia and the rdfs:label predicate in Wikidata. This approach is not adaptable to our needs because there are no existing cross-ontology links between VDO and D3FEND.

In [23], Portish et al. propose an approach to align graphs through a graph embedding algorithm. The ontologies are embedded, and an approach known as absolute orientation aligns the two embedding spaces. To match the two KGs, they use the Euclidian distance to assign each node of a graph to its closest node in the other graph. However, the approach only works well on similarly structured graphs because they only match entity with entity. In our approach, we are matching a sub-graph from VDO with a sub-graph from D3FEND, and the sub-graphs are structured differently using cosine distance defined in Appendix A.

Currently, there is no contribution in the cybersecurity domain that proposes to match KGs. However, some existing contributions propose the matching of behavioral graphs in cybersecurity and the matching of AGs. In [21], Park et al. propose a malware classification method based on maximal joint sub-graph detection. In [16], Li et al. propose an approach to automatically extract behavior-based AGs from CTI reports and identify the associated attack techniques. They identify the associated techniques by matching the AGs with technique templates.

Hung et al., in [13], also propose an approach to address the radicalization detection problem. They propose graph pattern matching to be used to track individual-level indicators using data merged from available public and government/law enforcement databases. The approach provides a quantification method that allows for checking the occurrence of the indicators that are beneficial in prioritizing investigative efforts and resources for planned attack prevention.

In [22], the resulting matching varies from one case to another. Our approach aims to obtain a stable output from the graph-matching process. The methods proposed in [4] and [23] are promising; however, they are not adaptable to our goal as the approaches are limited to either KGs that share exiting cross-ontology links or KGs structured similarly. In the approaches in [13] and [21], the researchers propose matching behavioral graphs to prevent adversarial actions against a system or a government. In this paper, we propose a solution that can be adapted to KGs representing attacker behavior and countermeasures actions to block attackers' actions.

4   Methodology

4.1   Overview

We propose an approach for selecting countermeasures based on KG matching. This approach allows selecting countermeasure actions for an attack scenario in our ADG generation process, represented in Fig. 1. The components in blue in Fig. 1 represent the ones involved in our AG enrichment process, part of another contribution [25]. In Step 1, we scan the system. In Step 2, the scanning output serves as input to a logical reasoner, allowing the AG generation to occur in Step 3. When an adversary exploits a vulnerability, as shown in Fig. 2, the procedural reasoner matches the alert information with the AG information in Step 4. In Step 5, the vulnerability ontology infers new information. When the procedural reasoner receives the inferred information in Step 6, it releases new rules to the logical reasoner in Step 7 and generates an enriched AG with a new attack path in Step 3.

After, as shown in Fig. 2, the ADG generation process can start. The components in gray in Fig. 1 are involved in the ADG generation process. Even if the procedural reasoner does not release the AG enrichment process, the ADG generation process starts as shown in the flow chart presented in Fig. 2. An ADG is generated in Step 10 by mapping the AG with an Incident Response (IR) playbook for the exploited vulnerability in Step 9. An IR playbook consists of the steps and procedures an organization should follow when addressing and mitigating occurred incidents. If there is no existing playbook for this vulnerability, our solution automatically generates it from a list of countermeasures as shown in Step 8 of Fig. 1.

If the playbook is not generated because of the countermeasures absence for the exploited vulnerability, the automated countermeasures selection process is executed (see Fig. 2). The components involved in the countermeasures selection are in white, as well as the output of the countermeasures selected. The dashed box represents the graph-matching process. The graph matching involves VDO and D3FEND KGs. Humans can periodically execute the graph matching process, as well as the scanning of the system. However, it is automatically released for a specific vulnerability exploited when there are no countermeasures available for this vulnerability. It outputs the countermeasures in Step 7 that are required for generating the playbook necessary for the ADG generation. In this paper, we focus on the selection of countermeasures through graph matching. In the next section, we present the dataset involved in the graph-matching process.

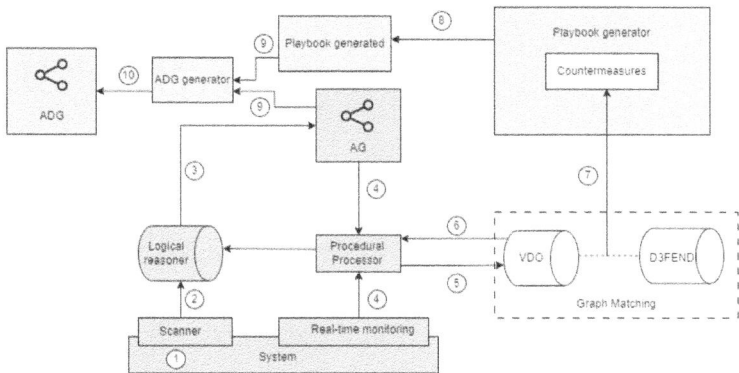

Fig. 1. Our ADG generation process (Color figure online)

4.2 Experimental Dataset

The dataset in our approach consists of a sub-graph of VDO KG and a sub-graph of D3FEND KG. VDO describes the pre and post-conditions of a vulnerability exploitation. The countermeasures KG, derived from D3FEND [14], provides

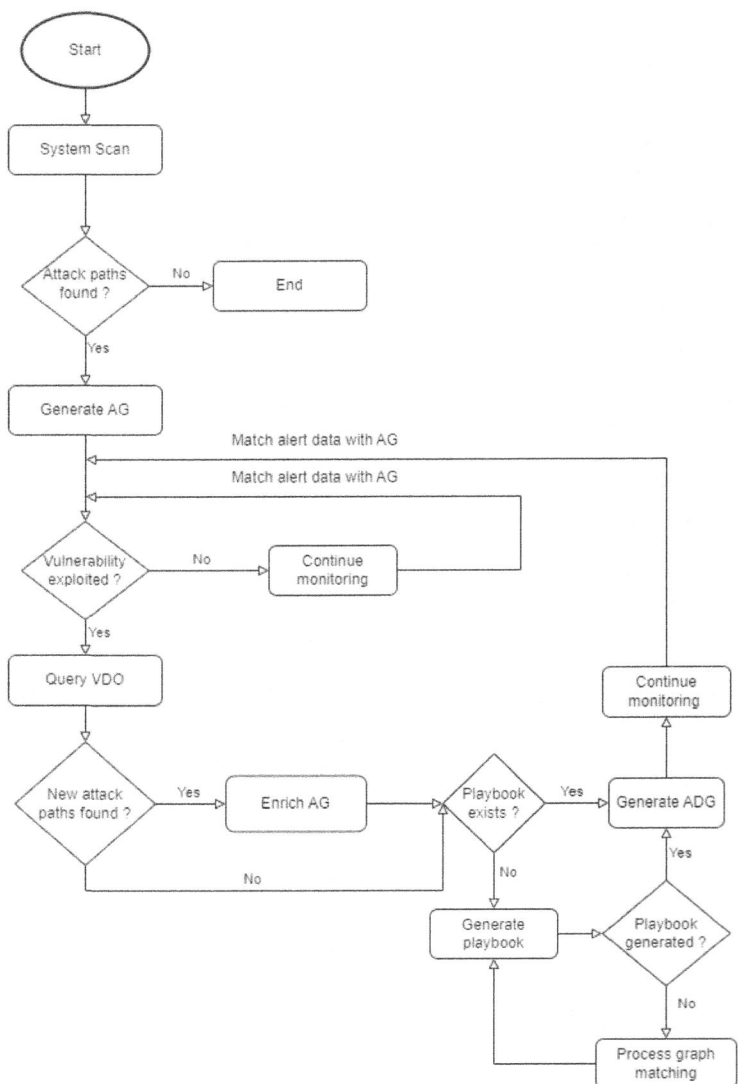

Fig. 2. A flow chart of the ADG generation process

corrective actions to perform in the face of exploitations of particular systems. We analyze VDO and D3FEND to choose the classes of interest for the countermeasures selection approach. Figure 3 represents the selected sub-graphs for D3FEND and VDO, respectively.

In D3FEND, these classes are *OffensiveTechnique*, *Artifact*, and *DefenseTechnique*. We select these classes to simplify the graph embedding and avoid noise in the matching process. The class *OffensiveTechnique* refers to the methods an adversary can employ to attack a system and the impacts of the attack. The class *Artifact* refers to the components affected by an offensive technique, and the class *DefensiveTechnique* refers to the countermeasures that should be applied to an artifact. We, therefore, select the sub-graph composed of all the entities whose type is either *OffensiveTechnique*, *DefensiveTechnique* and *Artifact* and the properties allowing the relation between those entities.

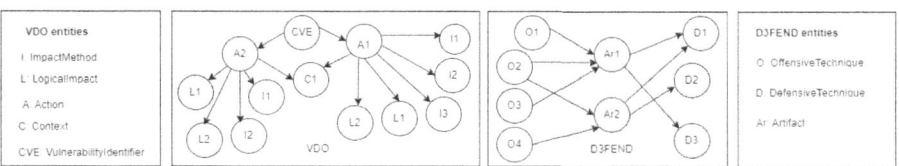

Fig. 3. VDO and D3FEND sub-graphs

In VDO, we choose the classes *LogicalImpact, ImpactMethod, Action, Context*, and *VulnerabilityIdentifier*. We do not choose the other classes to simplify the graph embedding and avoid noise in the matching process. These classes are essential to understanding the assets of a system affected by a scenario, how an adversary can exploit a vulnerability and the consequences of this vulnerability. *LogicalImpact* refers to the consequences of exploiting a vulnerability. *ImpactMethod* relates to the methods applied by an adversary to exploit a vulnerability. *Context* refers to the software and hardware concerned by a vulnerability. *VulnerabilityIdentifier* represents the CVE ID. However, some of the classes are not directly linked. Therefore, we consider the class *Action* to make the complete meaningful sub-graph required for the matching. The class *Action* allows linking an entity of the class *VulnerabilityIdentifier* with entities of the classes *ImpactMethod* and *LogicalImpact*.

4.3   Graph Matching

Our graph matching approach takes as input two KGs, O and O'. O contains the set of classes $O = C_O^0, C_O^1, .., C_O^i$ and each class of O contains the set of entities $C_O^i = e_0^i, e_1^i, ..., e_n^i$. Similarly, O' contains the set of classes $O' = C_{O'}^0, C_{O'}^1, .., C_{O'}^k$ and each class of O contains the set of entities $C_{O'}^k = e_0^k, e_1^k, ..., e_n^k$. We base our matching approach on graph embedding, word embedding, and SPARQL queries.

Figure 4 gives an overview of our solution architecture. The input is the two subgraphs of D3FEND and VDO KGs. The automated pre-processing phase of the KGs involves parsing the KGs, text processing, and corpora creation. This phase is necessary before starting the training of Word2Vec (see Appendix A) models and the embedding of KGs. Afterward, the matching process starts automatically.

In the matching phase, the cosine similarity is calculated automatically between the offensive techniques from D3FEND and the impact and method from VDO. A SPARQL query then allows the automated retrieval of artifact entities linked to the offensive techniques. If no artifact entity is retrieved, the cosine similarity is automatically calculated between the artifact entities of D3FEND and the context entities of VDO to find the artifact entities necessary to query candidate countermeasures for each CVE. The output from the matching process consists of a list of candidate countermeasures for each CVE.

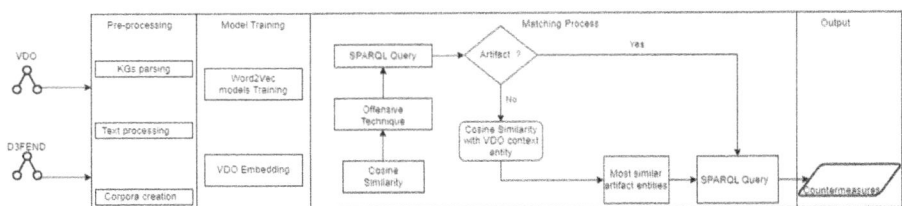

Fig. 4. Architecture of the proposed solution

4.4   Matching Models Validation

The value of the parameters of a Word2Vec model impacts its performance. Window size is the size of the context window used to predict surrounding words. A smaller window size captures a more specific, local context, which leads to more precise word associations. We fix the window size to 5. The value of min_count is a threshold, where words that appear fewer times than the value are ignored during training; we fix its value to 1. The number of workers refers to the number of CPU cores used for parallel processing. A higher workers number leads to faster training. We fix the number of workers to 20. The vector size is the dimensionality of the word vector. Higher dimensions capture more semantic nuances and context but can lead to overfitting. The number of iterations over the training corpus is the number of epochs. Fewer epochs reduce training time but can lead to underfitting. More epochs allow the model to learn better and converge more fully.

We perform different Word2Vec model training for the method and impact corpora by modifying the vector size and the number of epochs. To choose the best Word2Vec model, we consider the number of VDO entities for which we get the correct matches and the similarity score for the matches with the most similar D3FEND entity. Table 1 compares different models based on the number of correct matching for impacts and methods in VDO and the average cosine similarity score for the correct prediction.

The blue rows in Table 1 indicate the best models for *LogicalImpact* and *ImpactMethod*, respectively. For 250 vector size and 150 epochs, 100% of the

Table 1. Word2Vec model prediction evaluation

Entity class	vector size	epochs	Percentage of correct predictions	Average Similarity Score
LogicalImpact	250	150	100%	0.46
ImpactMethod	250	150	50%	0.5
LogicalImpact	250	50	100%	0.45
ImpactMethod	250	50	100%	0.69
LogicalImpact	250	200	75%	0.38
ImpactMethod	250	200	25%	0.25
LogicalImpact	400	50	100%	0.44
ImpactMethod	400	50	50%	0.33
LogicalImpact	400	150	100%	0.45
ImpactMethod	400	150	75%	0.43
LogicalImpact	400	200	75%	0.38
ImpactMethod	400	200	50%	0.5
LogicalImpact	200	150	100%	0.44
ImpactMethod	200	150	50%	0.5
LogicalImpact	200	50	100%	0.44
ImpactMethod	200	50	50%	0.38
LogicalImpact	200	200	75%	0.5
ImpactMethod	200	200	25%	0.25

predictions for the entities of *LogicalImpact* are correct. Most importantly, for this model, the average similarity score is higher. The average similarity score is less than 0.5 because the higher similarity score for the match of privilege escalation is 0.33. This value low similarity score for this entity is due to the impact of corpus size. Since the dataset is small, the model lacks variety, leading to poor generalization. For more epochs, the model learns specific patterns and noise from the training data rather than the underlying trends, which results in overfitting. We, therefore, fix the threshold value for the similarity score to 0.33.

4.5   Countermeasure Selection

To extract the entities required for the model's training and the matching process, we parse the KGs using techniques such as SPARQL queries and splitting. We create an impact corpus composed of the entities of *ImpactTechnique* and *PrivilegeEscalationTechnique*, both subclasses of *OffensiveTechnique*, and *LogicalImpact*. We also create a method corpus composed of the entities of the other subclasses of *OffensiveTechnique* and the entities of the classes *ImpactMethod*.

We create an artifact corpus with the entities of *Artifact* and *Context*. A general corpus is created with the entities of all the classes.

We train 2 Word2Vec models with the impact and method corpora. We fix the parameters for the best model in Table 1 in Sect. 4.4 for Word2Vec impact and method models. A Word2Vec model receives a corpus text and produces the word vectors as output. It first constructs a vocabulary from the training text data and then learns the vector representation of words. To get the embeddings of the KGs with RDF2Vec, we use the general corpus for the Word2Vec embedder.

We get the literals of the VDO graph embedding for each CVE. Then, for each entity classified as an impact from VDO, we calculate the cosine similarity between this entity and each offensive technique entity of the impact corpus using the trained model for the impact corpus. As output, we get the most similar offensive technique for the impact. We also calculate the cosine similarity between each entity categorized as a method from VDO and each offensive technique entity of the method corpus using the trained model for the exploitation method corpus. The output consists of the most similar offensive technique for the exploitation method.

The similarity score varies from 0 to 1. When the score is equal to 1 for an entity, this means that this entity is similar to the input entity. However, it is common for an entry entity to have several matches. We choose the one whose score is higher than the given similarity threshold from those entities. Therefore, if all entity scores exceed this threshold, we keep them all. Then, we execute a SPARQL query on D3FEND KG to get the artifact entities related to the offensive techniques. However, it is possible not to get a specific artifact entity from the SPARQL query. In this case, we calculate the cosine similarity for each artifact's entities of D3FEND for the context entity of each CVE using a trained Word2Vec model with the artifact corpus. Then, we get the artifact with the highest score as output. Afterward, we proceed to a SPARQL query to get all the countermeasures related to the artifact entities.

5   Implementation Results

5.1   Countermeasures Selection Evaluation

We use the trained models highlighted in blue from Table 1 for the countermeasures selection process[4]. Our solution automatically executes a SPARQL query on D3FEND for each selected offensive technique to retrieve related artifacts and their defensive techniques. If no artifact is linked to an offensive technique, cosine similarity is used to find the top 5 matching artifacts that match the context entity in VDO most. A SPARQL query follows to obtain the corresponding defensive techniques for each match. We evaluate the efficiency of the countermeasures selected using the precision, recall, and F1 score metrics, defined in Appendix A.

We count the FN, TP, and FP for the defensive techniques matched with each method and impact. Then, we calculate the precision, recall, and F1 score.

---

[4] https://github.com/phDimplKS/graph-matching.

Table 2. Evaluation of the prediction for the methods

Impact Method	Metric		
	Precision	Recall	F1 Score
Code Execution	1	0.75	0.86
Man-in-the-Middle	1	1	1
Authentication Bypass	1	1	1
Trust Failure	1	1	1
Average	1	0.94	0.97

Table 3. Evaluation of the prediction for the impacts

Logical Impact	Metric		
	Precision	Recall	F1 Score
Manipulation	1	1	1
Discovery	1	0.88	0.93
Shutdown	1	1	1
Privilege Escalation	1	1	1
Average	1	0.97	0.98

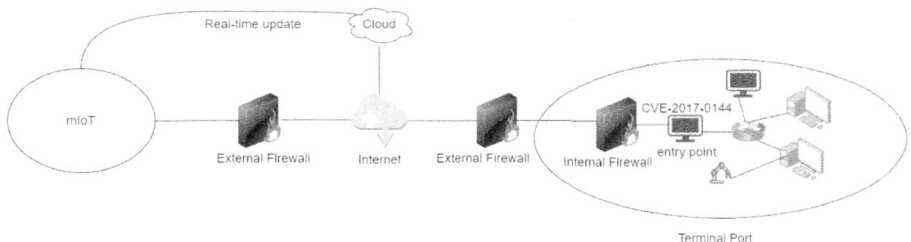

Fig. 5. A Maritime Transportation System use case

Finally, we calculate the macro-value for these metrics by calculating the average recall, precision, and F1 score value for the method and impact, respectively. Table 2 and 3 represents the value for these metrics for the methods and impacts respectively.

The average F1 score for the countermeasures selected for both the impacts and the methods is higher than 0.9. Thus, we conclude that our approach for automatically selecting countermeasures is efficient.

5.2   Illustrative Use Case

This section introduces an illustrative use case represented in Fig. 5 to validate our approach. We choose a WannaCry use case because ransomware is the attack

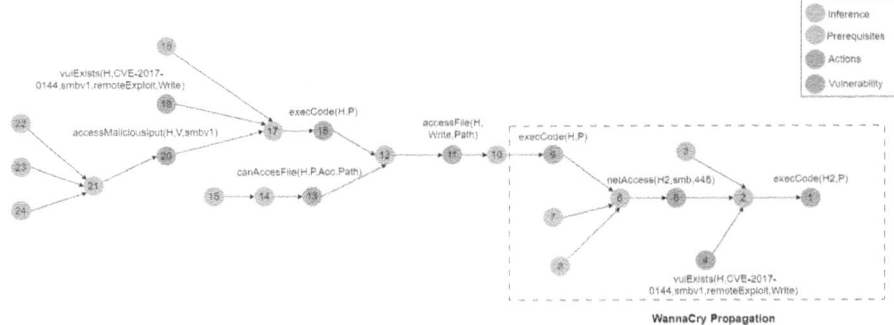

Fig. 6. An AG generated for the use case scenario

that has impacted most organizations recently [18,27]. The use case concerns a Maritime Transportation System (MTS). We choose an MTS because this is a critical sector; in the past, attacks such as the one impacting Maersk [19] have shown how a ransomware attack against an MTS organization can have a severe financial impact.

There is a terminal port with several workstations and a robot hoist that unloads merchandise from ships. The terminal port receives data from the cloud through an internet connection. External and internal firewalls exist between the port network and the internet. The Maritime Internet of Things (IoT) system communicates in real time with the cloud. A user on the entry point machine vulnerable to EternalBlue (CVE-2017-0144) downloads a malicious input with the WannaCry ransomware. The WannaCry propagates to all vulnerable local hosts via port 445, encrypts files, and ransom requests follow. The terminal port must disconnect from the internet and work in degraded mode. There is no communication between the terminal port and the mIoT system.

Let us explain how our architecture integration allows us to deal with this scenario. For this use case scenario, an AG is generated first, as shown in Fig. 6. The node in the dashed box represents the WannaCry propagation on the vulnerable local hosts. Thanks to detection rules created on an IDS installed over the SIEM, the SIEM allows the detection of EternalBlue exploit attempts: alert tcp any − > any 445 (msg: "SURICATA **SMB** Trans2 Request"; flow:to_server,established; content: "—FF 53 4D 42 32—"; depth:5; content: "—00 00—"; distance:30; within:2; content: "—00 00 00 00 00 00 00 00 00 00 00 00 00 00 00 00—"; distance:10; within:16; reference:cve,2017-0143; classtype:attempted-admin; sid:1000003; rev:1;).

Our solution maps the alert information with the AG generated for the scenario. The alert is mapped with the Node 19 representing the EternalBlue vulnerability on the entry point. Then, it launches the countermeasures selection process to get the related countermeasures that can block the attacker from launching the ransomware. This is the list of countermeasures proposed by our solution for CVE-2017-0144:

- There are countermeasures about the smbv1: Asset Vulnerability Enumeration, Software Update, and Restore Software.
- Some countermeasures aim to control files allowed to be downloaded or executed: File Encryption, Local File Permissions, Decoy File, File Analysis, File Removal, File Integrity Monitoring, Restore File, Dynamic Analysis, Emulated File Analysis, Executable Allowlisting and Executable Denylisting.
- Other countermeasures concern the network traffic: Client-server Payload Profiling, Network Traffic Community Deviation, Per Host Download-Upload Ratio Analysis, Protocol Metadata Anomaly Detection, Remote Terminal Session Detection, User Geolocation Logon Pattern Analysis, and Network Traffic Filtering.
- Several countermeasures selected concern a process: Process Spawn Analysis, Hardware-based Process Isolation, Mandatory Access Control, System Call Analysis, and System Call Filtering.

6 Discussion

We compare the countermeasures selected by our approach with the countermeasures proposed in the literature for ransomware [1,3,18] and, more precisely, WannaCry [2] use cases. In the literature, most of the proposed countermeasures concern prevention to avoid infection. The proposed actions consist of setting up spam filters to quarantine suspicious emails and attachments [1] and using predictive models to detect malicious behavior [3]. Other countermeasures proposed frequently are patching the vulnerability [3], excluding kill-switch domains from firewall rules, and blocking the SMB ports [2]. Our approach selects countermeasures that are part of the prevention phase but also from the containment, eradication, and recovery phases. File removal is an action from the eradication phase, which consists of removing the ransomware from the system. The recovery of encrypted files may follow.

All the countermeasures mentioned in Sect. 5.2 are the candidate countermeasures to the mitigation plan. This paper only aims to select countermeasures, a basis for the playbook generation that is part of another contribution [26]. Thanks to the real-time generation, the AG will help determine which countermeasure actions should be applied quickly based on the attacker's position knowledge. So, based on how far the attacker has gone, only some countermeasures will be instantiated to the AG, leading to the ADG generation, which is part of future work. In the future, we will evaluate our approach performance to select countermeasures in real-time for different system complexity, particularly for large-scale infrastructure for which many security incidents may be generated. We will consider the impact of real-time countermeasure selection performance on the ADG generation process.

7 Conclusion

As part of our ADG generation, we propose a countermeasures selection approach based on graph matching. The approach matches a cybersecurity vulnerability KG, VDO, with a countermeasures KG, D3FEND, thanks to 4 corpora creation for the Word2Vec models training. Graph and word embedding follow to calculate the cosine distance between the entities of the two KGs. The similarity between the entities from the two KGs allows us to select countermeasures. However, some candidate countermeasures only apply to specific assets. The countermeasures are also different in terms of complexity and impact. The selected countermeasures are used for automated playbook generation based on system and organization constraints. Using the AG will help us to determine at which point of the network the playbook actions should be applied to block the adversary from advancing in the system.

Acknowledgements. This work was supported by the Mitacs Accelerate International program, IRT SystemX, and the Cyber Resilience of Transport Infrastructure and Supply Chains (CRITiCAL) Chair.

Appendix A   Additional Background Definitions

Word2Vec. Word2Vec [17] is a popular word embeddings model used to address the limitations of the bag-of-words model [30], which is a type of vector space model that simplifies text data representation in Natural Language Processing (NLP) and Information Retrieval (IR). A bag-of-words vector represents text describing the occurrence of words within a document. In Word2Vec, each token becomes a vector with the length of a determined number.

RDF2Vec. RDF2Vec, created by Ristoski et al. [24], is an unsupervised technique built on Word2Vec. RDF2Vec first creates sentences that can be fed to Word2Vec by extracting walks of a certain depth from a KG to make the embedding. A vector of latent numerical features represents each entity in the KG. We calculate the similarity of the vectors to match their embeddings using a distance metric.

Cosine Similarity. Cosine similarity is a metric for measuring distance when the magnitude of the vectors does not matter. Mathematically, cosine similarity calculates the cosine of the angle between two vectors projected in a multidimensional space. Considering two vectors A and B; we can measure their cosine similarity using Formula 1.

$$\cos(AB) = \frac{A.B}{||A||||B||} \tag{1}$$

where, $A.B$ is the dot product of the vectors A and B, $||A||$ and $||B||$ are the length (magnitude) of the two vectors A and B, and $||A||||B||$ is the regular product of the vectors A and B.

If $A = B$, $\cos(AB) = 1$; in this case A and B are fully similar. If $A.B = 0$, then A and B are in opposite directions, so $A.B$ is negative, and one or both vectors are zero vectors, $\cos(AB) = 0$; in this case A and B are opposite.

Precision. Precision measures the number of positive prediction correctly predicted. So, it is calculated by dividing the number of true positive prediction (TP) by all positive prediction i.e. True Positive (TP) + False Positive (FP).

$$Precision = \frac{TP}{TP + FP}$$

Recall. Recall gives a percentage of true positives instances by a model. It is the number of well predicted positives divided by the total number of positives (True Positive + False Negative (FN)).

$$Recall = \frac{TP}{TP + FN}$$

F1 Score. Either precision and recall can not evaluate a machine learning model separately. F1 score allows combining precision and recall. So, it can provide a good evaluation of a model performance. It is calculating as follow:

$$F1Score = 2 \cdot \frac{Recall \cdot Precision}{Recall + Precision}$$

References

1. Adamov, A., Carlsson, A.: The state of ransomware. Trends and mitigation techniques. In: 2017 IEEE East-West Design Test Symposium (EWDTS), pp. 1–8 (2017)
2. Aljaidi, M., et al.: NHS WannaCry ransomware attack: technical explanation of the vulnerability, exploitation, and countermeasures. In: 2022 International Engineering Conference on Electrical, Energy, and Artificial Intelligence (EICEEAI), pp. 1–6. IEEE (2022)
3. Alshaikh, H., Ramadan, N., Ahmed, H.: Ransomware prevention and mitigation techniques. Int. J. Comput. Appl. **177**(40), 31–39 (2020)
4. Azmy, M., Shi, P., Lin, J., Ilyas, I.F.: Matching entities across different knowledge graphs with graph embeddings. arXiv preprint arXiv:1903.06607 (2019)
5. Booth, H., Turner, C.: Vulnerability description ontology (VDO): a framework for characterizing vulnerabilities. Technical report, National Institute of Standards and Technology (2016)
6. CCCS: Security control catalogue (2012). https://www.cyber.gc.ca/en/guidance/annex-3a-security-control-catalogue-itsg-33
7. CIS: CIS critical security controls (2024). https://www.cisecurity.org/controls
8. Doynikova, E., Kotenko, I.: Countermeasure selection based on the attack and service dependency graphs for security incident management. In: Risks and Security of Internet and Systems: 10th International Conference, CRiSIS 2015, Mytilene, Lesbos Island, Greece, 20–22 July 2015, Revised Selected Papers 10, pp. 107–124. Springer (2016)
9. Emek, Y., Kutten, S., Shalom, M., Zaks, S.: Hierarchical b-matching. In: Bureš, T., et al. (eds.) SOFSEM 2021. LNCS, vol. 12607, pp. 189–202. Springer, Cham (2021). https://doi.org/10.1007/978-3-030-67731-2_14

10. Gupta, M., Rees, J., Chaturvedi, A., Chi, J.: Matching information security vulnerabilities to organizational security profiles: a genetic algorithm approach. Decis. Support Syst. **41**(3), 592–603 (2006)
11. Hogan, A., et al.: Knowledge graphs. ACM Comput. Surv. **54**(4) (2021). https://doi.org/10.1145/3447772
12. Horridge, M., Knublauch, H., Rector, A., Stevens, R., Wroe, C.: A practical guide to building owl ontologies using the protégé-owl plugin and co-ode tools edition 1.0. University of Manchester (2004)
13. Hung, B.W.K., Jayasumana, A.P., Bandara, V.W.: Detecting radicalization trajectories using graph pattern matching algorithms. In: 2016 IEEE Conference on Intelligence and Security Informatics (ISI), pp. 313–315 (2016)
14. Kaloroumakis, P.E., Smith, M.J.: Toward a knowledge graph of cybersecurity countermeasures. Technical report (2021)
15. Kotenko, I., Doynikova, E.: Countermeasure selection in SIEM systems based on the integrated complex of security metrics. In: 2015 23rd Euromicro International Conference on Parallel, Distributed, and Network-Based Processing, pp. 567–574. IEEE (2015)
16. Li, Z., Zeng, J., Chen, Y., Liang, Z.: Attackg: constructing technique knowledge graph from cyber threat intelligence reports. In: European Symposium on Research in Computer Security, pp. 589–609. Springer (2022)
17. Ma, L., Zhang, Y.: Using word2vec to process big text data. In: 2015 IEEE International Conference on Big Data (Big Data), pp. 2895–2897. IEEE (2015)
18. Malik, A.W., Anwar, Z., Rahman, A.U.: A novel framework for studying the business impact of ransomware on connected vehicles. IEEE Internet Things J. **10**(10), 8348–8356 (2023). https://doi.org/10.1109/JIOT.2022.3209687
19. Mos, M.A., Chowdhury, M.M.: The growing influence of ransomware. In: 2020 IEEE International Conference on Electro Information Technology (EIT), pp. 643–647 (2020). https://doi.org/10.1109/EIT48999.2020.9208254
20. Nespoli, P., Papamartzivanos, D., Gómez Mármol, F., Kambourakis, G.: Optimal countermeasures selection against cyber attacks: a comprehensive survey on reaction frameworks. IEEE Commun. Surv. Tutor. **20**(2), 1361–1396 (2018). https://doi.org/10.1109/COMST.2017.2781126
21. Park, Y., Reeves, D., Mulukutla, V., Sundaravel, B.: Fast malware classification by automated behavioral graph matching. In: Proceedings of the Sixth Annual Workshop on Cyber Security and Information Intelligence Research. CSIIRW 2010. Association for Computing Machinery, New York (2010)
22. Pershina, M., Yakout, M., Chakrabarti, K.: Holistic entity matching across knowledge graphs. In: 2015 IEEE International Conference on Big Data (Big Data), pp. 1585–1590. IEEE (2015)
23. Portisch, J., Costa, G., Stefani, K., Kreplin, K., Hladik, M., Paulheim, H.: Ontology matching through absolute orientation of embedding spaces. In: The Semantic Web: ESWC 2022 Satellite Events: Hersonissos, Crete, Greece, 29 May–2 June 2022, Proceedings, pp. 153–157. Springer (2022)
24. Ristoski, P., Paulheim, H.: RDF2Vec: RDF graph embeddings for data mining. In: Groth, P., et al. (eds.) ISWC 2016. LNCS, vol. 9981, pp. 498–514. Springer, Cham (2016). https://doi.org/10.1007/978-3-319-46523-4_30
25. Saint-Hilaire, K., Cuppens, F., Cuppens, N., Garcia-Alfaro, J.: Automated enrichment of logical attack graphs via formal ontologies. In: Meyer, N., Grocholewska-Czuryło, A. (eds.) ICT Systems Security and Privacy Protection, pp. 59–72. Springer, Cham (2024). https://doi.org/10.1007/978-3-031-56326-3_5

26. Saint-Hilaire, K., Cuppens, F., Cuppens-Boulahia, N., Hadji, M.: Optimal automated generation of playbooks. 38th Annual IFIP WG 11.3 Conference on Data and Applications Security and Privacy (DBSec 2024) (2024)
27. Šulc, V.: Current ransomware trends. In: International Days of Science, vol. 31 (2021)
28. Swarup, V.: Remediation graphs for security patch management. In: IFIP International Information Security Conference, pp. 17–28. Springer (2004)
29. Viduto, V., Maple, C., Huang, W., López-Peréz, D.: A novel risk assessment and optimisation model for a multi-objective network security countermeasure selection problem. Decis. Support Syst. **53**(3), 599–610 (2012)
30. Zhang, Y., Jin, R., Zhou, Z.H.: Understanding bag-of-words model: a statistical framework. Int. J. Mach. Learn. Cybern. **1**, 43–52 (2010)

Graph-Based Profiling of Dependency Vulnerability Remediation

Fernando Vera Buschmann[1]([✉]), Palina Pauliuchenka[1], Ethan Oh[1],
Bai Chien Kao[2], Louis DiValentin[2], and David A. Bader[1]

[1] Department of Data Science, New Jersey Institute of Technology, Newark, NJ, USA
{fv54,pp272,eo238,bader}@njit.edu
[2] Accenture PLC, Arlington, VA, USA
{bai.chien.kao,louis.divalentin}@accenture.com

Abstract. This research presents an enhanced Graph Attention Convolutional Neural Network (GAT) tailored for the analysis of open-source package vulnerability remediation. By meticulously examining control flow graphs and implementing node centrality metrics-specifically, degree, norm, and closeness centrality-our methodology identifies and evaluates changes resulting from vulnerability fixes in nodes, thereby predicting the ramifications of dependency upgrades on application workflows. Empirical testing on diverse datasets reveals that our model challenges established paradigms in software security, showcasing its efficacy in delivering comprehensive insights into code vulnerabilities and contributing to advancements in cybersecurity practices. This study delineates a strategic framework for the development of sustainable monitoring systems and the effective remediation of vulnerabilities in open-source software.

Keywords: Graph Attention Convolutional Neural Network (GAT) ·
Package Vulnerability Analysis · open source · package upgrade ·
Knowledge Graph · Node Centrality Metrics · Cybersecurity · Deep
Learning Applications · Network Analysis · Code Vulnerability
Mitigation

1 Introduction

The increasing complexity of software systems and the reliance on third-party libraries have significantly heightened the importance of vulnerability analysis in open-source packages. Despite progress in identifying and mitigating vulnerabilities, significant gaps remain in understanding the interdependencies and impact of remediation efforts. This study focuses on three real-world applications: a Python-based data processing package, a Java-based web application, and a mixed-language analytics tool. By analyzing these diverse codebases, we aim to demonstrate the versatility and effectiveness of our graph-based approach.

Preventing and analyzing vulnerabilities is crucial for averting cyber attacks and protecting programs and components [31]. The growing complexity and

J. Zhao and W. Meng (Eds.): SciSec 2024, LNCS 15441, pp. 138–157, 2025.
https://doi.org/10.1007/978-981-96-2417-1_8

abstraction of dependencies and version management increase the likelihood of vulnerabilities within application code in this dynamic, continuously updating world [15]. During code development, it is essential to identify and address the weakest link or the most apparent vulnerability and to pursue continuous solutions with each code update [14]. To effectively protect against cyber attacks, three steps are often followed: 1) know the vulnerabilities, 2) understand their impact on the application, and 3) act on them in depth.

Vulnerability remediation in open source packages is vital for the open-source community and user security. This process involves identifying vulnerabilities, reporting them, and analyzing their impact [2]. Open source maintainers then triage these vulnerabilities and bugs, prioritize updates, and release new package versions that fix the affected functions and remove bugs.

Various tools facilitate the discovery of vulnerabilities in dependencies, including CVEfixes, which replicates the comprehensive database from the U.S. National Vulnerability Database (NVD). Static and dynamic code analysis capabilities, such as CodeQL used by the GitHub community, can identify and modify functions to discover new vulnerabilities [25,26]. Software Composition Analysis (SCA) techniques create a Software Bill of Materials (SBOM) listing the specific versions of dependencies used. This SBOM is correlated with existing vulnerability repositories to identify vulnerabilities in each dependency, targeting them for upgrades to newer versions where issues are fixed. However, upgrading dependencies can sometimes cause compilation errors or functional issues in the application [12,29].

Understanding the impact of dependency version changes, along with functional interconnections and network structures, is crucial for estimating the likelihood of specific package upgrades causing breaking changes in application code. Given the distinct and adaptable nature of applications, tools that generalize across different applications are needed to provide insights into upgrade complexity. Thus, we turn to knowledge graphs [13,24]. Constructing a graph allows us to comprehend interactions between functions within the code. In this research, we represent functions as entities in a graph space, constructing an Inter Procedural Control Flow graph of function interactions. This enables examination of the application code and dependencies, with the caller-callee relationship as the connecting link [32]. Observing and analyzing these interactions assists in identifying the impact of dependency changes on application workflow and downstream effects. This relational understanding supports the automation of package upgrades with minimal functionality impact, directly translating to improved security by enabling automated processes to upgrade vulnerable package versions with minimal disruptions.

Understanding the impact of changed function nodes between two application versions using a graph necessitates a profound comprehension of the network's topology and connectivity attributes [33]. This knowledge is indispensable for identifying the significance of affected nodes to the network's overall operation and making informed decisions on upgrade impacts [4]. This process involves analyzing centrality, density, connected components, degree assortativity coefficient, and pathways. Such comprehensive evaluations facilitate strategic decision-making for optimally mitigating vulnerabilities, ensuring a robust and secure network architecture. This is particularly critical in software, where functions exhibit complex nested dependencies and feedback loops.

The remainder of this paper is organized as follows: Sect. 2 outlines our proposed approach. Section 3 reviews the background and related work. Section 4 elaborates on the methodology, including the problem statement and workflow. Section 5 describes the experimental setup and presents the results and analysis. Finally, Sect. 6 discusses the findings and provides the conclusion.

2   Proposed Approach

The central issue addressed in this paper is the challenge of accurately identifying and mitigating vulnerabilities in open-source packages without causing functional disruptions. Our methodology leverages a Graph Attention Network (GAT) [28] to analyze the interdependencies of software components and predict the impact of remediation efforts on software workflows. This approach is formalized by examining control flow graphs and applying node centrality metrics [5].

2.1   Problem Statement

The primary problem is to develop a model that can pinpoint changed nodes and forecast the impact of dependency updates. This involves analyzing the structural and functional attributes of the code to ensure that upgrades do not introduce new vulnerabilities or disrupt existing functionalities [6, 26].

2.2   Workflow

Our workflow involves four main stages: data collection, graph construction, vulnerability analysis, and impact assessment. Each stage is designed to incrementally build a comprehensive understanding of the software's vulnerability landscape.

2.3   Domain-Specific Example

To illustrate our methodology, consider an open-source Python-based data processing package. In this context: - **Nodes**: Represent functions and classes within the package. For example, a node might represent a function responsible for data encryption. - **Features**: Include degree centrality, which indicates the number of direct connections a function has, and closeness centrality, which measures how quickly a function can interact with other functions in the network.

When a vulnerability is identified in a critical function (e.g., the encryption function), our approach maps the control flow graph for the application and analyzes each function for differences between the current and remediated version of the open source package. The GAT model is then applied to analyze how the changed functions impact related functions and the overall system performance. By doing so, we can predict potential disruptions and ensure that remediation efforts do not affect existing functionalities.

This concrete scenario bridges the gap between the high-level overview of our methodology and its technical implementation, enhancing the clarity and applicability of our approach.

3   Background and Related Work

Traditional methods of package function analysis primarily rely on static and dynamic analysis techniques. While static analysis provides a broad overview, it often lacks the context-specific insights gained from dynamic runtime analysis [19]. Additionally, these techniques typically focus on the implementations of functions within the application, rather than vulnerabilities in the application's dependencies. Some scanning techniques and vendors address these dependency vulnerabilities by upgrading to a version where the vulnerability is resolved and recompiling the code. If the remediated version causes compilation failures or unit test failures, further code updates are required to ensure compatibility. Understanding the effects of upgrading to a remediated package version remains a complex and opaque process.

Inter Procedural Control flow graph analysis is a useful tool for understanding software behavior, allowing developers to easily visualize the flow of execution within a program. By reducing abstract and complex code into manageable graphs, it provides information about the structure of the program and its possible vulnerabilities and affected functions. This analysis is crucial to optimize code performance and ensure strong cybersecurity measures. In our case, we enhance the inter procedural control flow graph with a knowledge graph related to the use of algorithms and measures focused on large-scale analysis.

Graph Attention Networks (GAT) have emerged as a promising tool in similar analytical contexts. Originally conceptualized by Veličković et al. [28], GATs leverage the attention mechanism to provide node-specific contextual insights, enhancing the accuracy of feature representation in graph-structured data. This has been effectively utilized in various domains, including bioinformatics, social network analysis, and natural language processing [30]. In package function analysis, GATs offer an innovative opportunity to address the limitations of traditional methods. They provide a scalable approach to analyze the intricate interrelationships and dependencies among package functions represented as nodes in a graph. This method aligns with recent trends in utilizing deep learning techniques for software engineering challenges, as documented in several contemporary studies [8].

4   Methodology

This section is outlining key definitions and delving into the critical elements that underpin our methodological approach. A portion of our research is the deployment of a modified Graph Attention Network (GAT), an advanced model that is being specifically designed to analyze feature code interactions and study the importance of changed functions represented as nodes in the graph.

4.1 Modified Graph Attention Neural Network

The adaptation of the GAT model is motivated by the lack of current visibility into the effects of upgrading package versions, which directly affects developers ability to stay on top of package vulnerabilities originating from dependencies. Unlike traditional GAT models based solely on norm [28], our modified version incorporates essential node centrality metrics (degree centrality, norm, and closeness) to evaluate the importance and interrelationships of various functions within the code, with a particular focus on identifying and evaluating vulnerable nodes. This enhancement increases the ability of GATs to provide a more refined analysis of the relational dynamics within the code, offering a comprehensive perspective on the impact of changes in dependency versions and the potential risk of breakage resulting from changed code.

Rationale for Modification

Utilizing the Knowledge Graph for network analysis, particularly emphasizing on graph attention Neural Network (GAT) as defined in the work of Veličković et al. [28], is enabling us to train extensive datasets for the purpose of evaluating and quantifying the significance of nodes within a network. This base model is facilitating accurate detection of nodes based on both transductive methods (such as Cora, Citeseer, Pubmed) and inductive methodologies. Presented as a convolutional-like neural network operating on knowledge graph-structured data, this model assigns an importance metric to nodes within their neighborhoods.

The input to our layer consists of a set of node features, $h = \{\mathbf{h}_1, \mathbf{h}_2, \ldots, \mathbf{h}_N\}$, with $\mathbf{h}_i \in \mathbb{R}^F$, where N denotes the number of nodes and F the number of features per node. This layer is producing a new set of node features, $h' = \{\mathbf{h}'_1, \mathbf{h}'_2, \ldots, \mathbf{h}'_N\}$, with a potentially different cardinality F'. A shared linear transformation, parameterized by a weight matrix $\mathbf{W} \in \mathbb{R}^{F' \times F}$, is applied to each node. The GAT then performs a self-attention mechanism a, yielding attention coefficients e_{ij} as: $e_{ij} = a(\mathbf{W}\mathbf{h}_i, \mathbf{W}\mathbf{h}_j)$

These coefficients indicate the importance of node j's features to node i. The GAT model is assessing the first-order neighbors of i, necessitating a normalization of coefficients across all node features j using the softmax function:

$\alpha_{ij} = \text{softmax}_j(e_{ij}) = \exp(e_{ij}) / \sum_{k \in \mathcal{N}_i} \exp(e_{ik})$

it is calculated the coefficient most relevant of each feture of node using the norm. It identified as α_{ij}. Replacing equation e_{ij} in equation α_{ij}

$$\alpha_{ij} = \frac{\exp\left(\text{LeakyReLU}\left(a^\top [\mathbf{W}h_i | \mathbf{W}hj]\right)\right)}{\sum k \in \mathcal{N}(i) \exp\left(\text{LeakyReLU}\left(a^\top [\mathbf{W}h_i | \mathbf{W}h_k]\right)\right)} \tag{1}$$

In line with the original GAT model [28], with the attention mechanism a as a single-layer feedforward neural network, we define a parameterized weight matrix

$\mathbf{a} \in \mathbb{R}^{2F'}$ and apply the LeakyReLU nonlinearity (negative input slope of 0.2). $a^{\top}$ is a learnable weight vector in the single-layer feedforward neural network which constitutes the attention mechanism. $\mathcal{N}(i)$ denotes the neighborhood of node i in the graph. $\|\|$ represents the concatenation operation.$e_{ij} = a(\mathbf{Wh}_i, \mathbf{Wh}_j)$ This process calculates the most relevant coefficient for each feature of a node using the norm, identified as α_{ij}. To enhance the focus and applicability of these coefficients, particularly in code analysis, we are introducing a modification β_{ij} to highlight the robustness and criticality of all nodes. This modification encompasses functions such as degree centrality, the norm, and closeness centrality metrics, relevant to nodes i and j. Consequently, the revised attention coefficient, α'_{ij}, is defined as: $\alpha'_{ij} = \alpha_{ij} \cdot \beta_{ij}$

This novel approach is proving instrumental in evaluating the importance of specific functions within their neighborhoods, offering insights into the criticality of an affected function in a code structure. The modified model enables the generation of a normalized score ranging from 0 to 1, which reflects the importance of nodes based on metrics such as degree, norm, and centrality. This scoring system facilitates a more nuanced understanding of node significance within the graph structure, particularly in the context of package code analysis.

4.2   Mapping and Mitigating Vulnerability in a Code Development

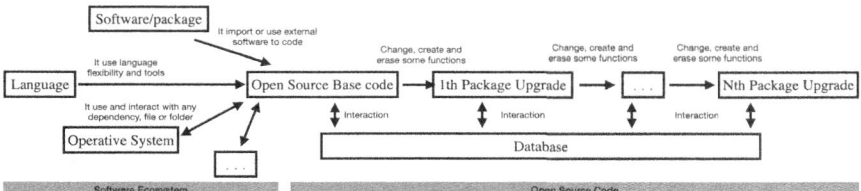

Fig. 1. In this figure, we are providing a conceptual description of the actors considered behind the open source. We are taking into account a generic software ecosystem of code from key elements such as the operating system, language, and software, packages or others that enable the code to function and connect with the real world. A code under development typically begins with the Open Source Base code (primary functions) that is interacting with the repository, the operating system, and other components of the ecosystem. This base code is undergoing updates Nth times to enhance its functionality. It is crucial to emphasize that within each component, there could exist certain flaws which might potentially manifest as vulnerabilities within the code.

Vulnerabilities are recorded in the NVD repositories from various sources during the development process. These can be inherent from the base of the code (Fig. 1) to the programming language used, integrated through specific packages or software, or due to operating system flaws [7,9]. These vulnerabilities are remediated by standardizing community-driven coding practices [7], like SAST and

DAST scanning, and bug reporting. Nevertheless, vulnerability mapping is continuously evolving, with new vulnerabilities emerging or being discovered over time, as well as new tools and methods for finding them [3,27]. Thus, it is proving beneficial to maintain an updated, dynamic, and comprehensive connection map, which serves as a guide to pinpoint the source and address these vulnerabilities swiftly and precisely.

From the foundational layer of code, the language, environment, configurations [2], and built-in packages [1] emerge. The core code develops from this foundation, forming the program's heart with its principal functions. An initial version of the code is committed, and subsequent updates add functionality and features, resolving errors and introducing new functions and packages. Each update may integrate previous functions into the core or leave them as branches. Vulnerabilities can appear in any update or trace back to the base code, and a vulnerable fragment can extend its impact across updates to critical branches, potentially affecting the entire application's core structure. Understanding the network of functions within the code is crucial to mitigate risks and ensure updates do not compromise the application's integrity.

4.3   Knowledge Graph in Cybersecurity

According to Hogan [11], a debate is ongoing regarding the precise definition of a Knowledge Graph, yet consensus exists about its remarkably high adaptability. In the context of this discussion, the knowledge graph G is defined as $G = (E, R, S)$, where E represents the entity (node), R symbolizes the relationship (edge), and S denotes relationship facts (node-node relationship). A triplet constitutes a typical form of knowledge representation within this framework. Entities, serving as foundational elements of the Knowledge Graph, encompass a wide range of classifications, such as collections, categories, object types, and thing categories (e.g., domain, host, etc.). Relationships interconnect these entities to formulate the graph's structure, while attributes encapsulate features and parameters, exemplified by entities like google.com, windows, and similar.

To construct a dataset, it is necessary to study the relationships existing between functions through their callee or caller interactions. A database is generated considering the node entities E_i, their relations R_i, and their relation S_i, which can be a call or a caller. This approach enables the generation of a knowledge graph containing the identified entities. The data structure provides critical information, encompassing the function's path, its name, and whether it has been modified in the latest update. Additionally, it indicates whether the function is vulnerable and specifies its role as either a callee or a caller.

4.4   Building the Dataset

To build a comprehensive dataset, we compare subsequent versions of code where upgraded package versions are intended to remediate vulnerabilities in open source packages [1]. We use the open source software Syft to generate the software bill of materials (SBOM) for the current version of the target application source code. Curl fetches vulnerability data for individual package versions in the SBOM from the Open Source Vulnerabilities (OSV) database, which provides an accessible query interface for all known dependency vulnerabilities. This information is mapped to the SBOM to identify existing vulnerabilities in the dependencies. If a package with a vulnerability is detected, we search for the updated version of the package and clone it. Using CodeQL, a semantic code analysis engine, we model different versions of the code and construct control flow graphs of the execution paths for the repository using both the vulnerable and fixed packages. Tree-sitter performs incremental analysis, constructs, and maintains a syntax tree, and builds the dataset. The impact of a single package upgrade is measured by comparing the control flow for matching functions in the inter procedural graph of the application before and after the upgrade. The updated version control flow graph serves as the base for the knowledge graph, with changed or affected nodes marked, along with nodes causing compilation errors. Multiple upgrades are performed in two sets for each repository: one set with graphs introducing errors resulting in code breakage, and another set with graphs not impacting code functionality.

In this research, a preliminary approach is being used, initially focusing on three application source repositories; The first case in a code base comprising 9621 features with 27 vulnerabilities in its broken update; the second case with 19,569 functions and 3 vulnerabilities. Subsequently, in Cases 3 (15908 functions and 6 vulnerabilities) and 4 (16095 functions and no vulnerabilities), an application is observed with a similar order of magnitude but different dependencies regarding the base functions.

5   Results and Analysis

Our experimental setup involves analyzing three diverse codebases in Python and Java, employing Graph Attention Networks (GAT) for vulnerability assessment [28]. The results, underscore the efficacy of our approach in accurately identifying critical vulnerabilities and predicting the impact of remediation efforts. For instance, in one case study, our model effectively pinpointed a vulnerable encryption function and forecasted its implications on the data processing workflow, thereby demonstrating its practical utility in real-world scenarios.

In this section, we discuss the methods of data analysis and describe the metrics we will use. Once the graph structure is built, we can utilize tools like NetworkX in Python for small-scale analysis [20] or Arachne for large-scale projects [21]. In this case, NetworkX is used exclusively to facilitate calculation and analysis, given the volume of data (less than 20,000 nodes). This analysis adopts a two-pronged approach. Firstly, it involves an exploratory examination of graph connectivity, focusing on the distribution and other metrics that provide insights into the differences between package upgrades that break dataflow and those that do not. This includes evaluating the structural design and robustness of the network, ensuring that each functional node and its connections contribute to the overall integrity and efficiency of the system. Secondly, we analyze the modified

GAT scores to obtain a normalized measure of the interaction of vulnerabilities as nodes within the graph. The model provides an advanced mechanism to measure the importance of nodes within a network, allowing for a targeted strategy for classifying and assessing vulnerabilities and simplifying the assessment of remediation effectiveness. The practical benefits of this approach include prioritizing development efforts, improving software integrity, formulating strategic planning initiatives, and deepening knowledge about software architecture. This comprehensive strategy not only improves the immediate security posture of software systems but also lays the foundation for long-term sustainable software development and maintenance practices.

The results highlight the model's ability to maintain system stability while addressing vulnerabilities, underscoring its potential for broader application in cybersecurity.

5.1   Graph Analysis

It is crucial when analyzing the knowledge graph created with the data set to measure the network and understand the meaning and relevance of the entities and relationships within it [5]. It is necessary to quantify the connectivity, the paths and their strength through the centrality analysis of each objective element.

Preliminary Graph Analysis

We are constructing a knowledge graph, denoted as $G = (\hat{E}, \hat{R}, \hat{S})$, where $\hat{E} = V$ signifies the entities or nodes. Our primary focus is on the callee-caller relationship, considered as directional edges. This simplifies the triplet $(\hat{R}, \hat{S}) = E$ into the graph structure $G = (V, E)$ [5], aligning with the GAT model [28].

This knowledge graph features a dynamic representation of functions, with vertices n and directed edges m. Each edge $e \in E$ is assigned a specific weight $w(e)$, reflecting the significance and connectivity strength between functions. Paths within the graph are sequences of edges $\langle u_i, u_{i+1} \rangle$, where $u_0 = s$ and $u_l = t$ denote the start and end vertices. The distance between two vertices s and t, represented as $d(s,t)$, is indicated by the shortest paths σ_{st}. We quantify the number of shortest paths traversing a specific vertex v using $\sigma_{st}(v)$, consistent with Bader et al. [5].

Degree Centrality: We are measuring the degree centrality of a vertex v, denoted as $\deg(v)$, to quantify the extent of interactions a node has within its neighborhood. This metric reflects the node's importance based on the number of caller-callee connections it maintains [5].

The Norm: By definition of the Euclidean norm of a vector used by velivckovic et al. [28] in GATs model, $\mathbf{x} \in \mathbb{R}^n$ as: $\|\mathbf{x}\|_2 = \sqrt{\sum_{i=1}^n x_i^2}$ where $\mathbf{x} = (x_1, x_2, \ldots, x_n)$ represents a vector in an n-dimensional real space, and x_i corresponds to the i-th element of the vector.

Closeness Centrality: Closeness centrality measures the degree of proximity of a node to all other nodes in the graph, based on distance. For any node n, its closeness centrality is calculated as the average length of the shortest path from n to every other node. A node with higher closeness centrality is more centrally located in the network, indicating greater importance or influence within the network's structure. This metric is determined by the inverse of the sum of the shortest distances from the node to all other nodes [5,18]:

$$CC(v) = \frac{1}{\sum_{u \in V} d(v, u)}$$

In our implementation, this translates to assessing how interconnected a specific function is relative to the rest of the functions within the code.

Betweenness Centrality: Betweenness centrality quantifies how often a node appears on the shortest paths between other nodes, acting as a critical bridge within the network [5]. In software systems, a function with high betweenness centrality is pivotal in the flow of information or processes, significantly influencing other functions. The pairwise dependency $\delta_{st}(v)$, representing the fraction of shortest paths between nodes s and t passing through node v, is defined as: $\delta_{st}(v) = \sigma_{st}(v)/\sigma_{st}$ This leads to the formulation of betweenness centrality for a node v, where $s, v, t \in V$:

$$BC(v) = \sum_{s \neq v \neq t} \delta_{st}(v)$$

Betweenness centrality reflects a node's ability to control information or resource flow by bridging the shortest paths between nodes. Nodes with high betweenness centrality are essential for network connectivity, facilitating communication and interactions by being part of numerous shortest paths connecting various node pairs.

Connected Components: Connected components are defined as subgraphs where any two vertices are connected by paths, without external connections [10]. Our research defines a connected component as a set of nodes where each node can access all other nodes within the same set. This concept is crucial for analyzing network structures, detecting isolated clusters, and examining connectivity. The study distinguishes between undirected graphs, where a connected component comprises the largest set of interconnected nodes, and directed graphs, which include strongly connected components defined by bidirectional paths between all pairs of nodes. Our approach leverages the concept of connected components to gain insights into the network's architecture. This analysis is pivotal in identifying potential vulnerabilities or areas of improvement within the network, particularly in cybersecurity and software engineering. By understanding the formation and interaction of these components, we can devise more effective strategies for network optimization and vulnerability mitigation.

Clustering Coefficient: The clustering coefficient for a node v, denoted as C_i, assesses the likelihood of connectivity between two randomly chosen neighbors of this node. This measure indicates the number of triangles in which the i-th node participates, normalized by the maximum possible number of such triangles. We compute the average clustering coefficient by averaging these individual values across all nodes in the graph:

$$C_i = \frac{2t_i}{k_i(k_i - 1)},$$

where t_i is the number of triangles around node i, and k_i is the degree of node i. As this coefficient approaches 1, it suggests increasing completeness of the graph with a predominant cohesive component. Higher coefficients indicate triadic closure, observed in denser graphs with prevalent triangular formations [22]. The average clustering for a graph is calculated as:

$$\bar{C}_i = \frac{1}{n} \sum_{i=1}^{n} C_i.$$

This metric provides an overall indication of the degree of clustering within the network, reflecting how closely nodes tend to cluster together, thus offering insights into the network's structural density and connectivity patterns.

Degree Assortativity Coefficient: We measure the degree assortativity coefficient to quantify the tendency of nodes in a network to connect with other nodes of similar degree, providing insights into assortative or disassortative mixing [17]. This metric determines whether high-degree nodes are more likely to connect with other high-degree nodes or with low-degree nodes. The coefficient ranges from -1 to 1, where values close to 1 indicate assortative mixing and values close to -1 indicate disassortative mixing. A value around 0 indicates no particular connectivity preference. The degree assortativity coefficient r is defined as:

$$r = \frac{\sum_{jk} jk(e_{jk} - q_j q_k)}{\sigma_q^2},$$

where e_{jk} is the fraction of edges connecting nodes of degree j and k, q_j is the distribution of the remaining degrees of nodes, and σ_q^2 is the variance of q. This coefficient is crucial for understanding the structural tendencies of the network and how nodes preferentially form connections based on their degrees.

Cyclomatic Complexity. Cyclomatic complexity serves as a quantitative measure for evaluating the number of linearly independent paths within a code, estimating the program's complexity [6]. This metric is crucial for understanding the intricacy and structural complexity of a program [23]. Adapting McCabe's definition [16], the cyclomatic complexity $V(G)$ of a control flow graph G is defined as: $V(G) = E - N + 2P$ where E represents the number of edges, N signifies the number of nodes, and P stands for the number of connected components. This measure provides insight into potential paths and decision points in a program's structure, helping to understand software maintainability and areas for refactoring. Cyclomatic complexity is thus a practical tool for guiding the development and maintenance of robust and efficient software systems.

5.2   Dataset Results and Interpretation

This preliminary study analyzes three different Python and Java applications, leveraging open source packages with control flow graphs of base versions, broken post-upgrade, and non-broken post-upgrade. CodeQL analysis flags critical features, identifying changes that cause build failures. The goal is to understand how the structural and functional attributes of the functions in the graph affect the probability of code breaking, especially due to modifications and new features or patches to fix vulnerabilities.

CASE 1: (Table 1) We analyzed an application code base with 9626 functions, identifying 27 as critical, where remediation upgrades caused code breakage. Post-update, we observed an increase in the number of nodes and average degree, while network density remained constant. The number of connected components increased, and the degree assortativity coefficient remained negative and constant. Notably, the degree assortativity coefficient for the Non-Broken Upgrade was much lower than that for the Broken Upgrade, and closeness centrality was higher for Non-Broken Upgrades.

Table 1. (Case1) Comparative analysis of code metrics from code base, non-broken update and broken upgrade.

	Base	Non Broken Upgrade		Broken Upgrade	
Number of unique functions	9621	9621		9621	
nodes kind		**Unchanged**	**Changed**	**Unchanged**	**Changed**
Number of nodes	9621	9614	19	9613	23
Number of edges	15186	15178	19	15176	24
Average degree	3.1568	3.1575	2.0	3.1574	2.087
Density	0.0003281	0.0003285	0.1111	0.0003285	0.09486
Num. of connected components	327	327	2	327	2
Average clustering	0.06809	0.06815	0.0	0.06815	0.0
Degree assortativity coefficient	−0.08515	−0.08519	−0.3996	−0.08521	−0.3464
Avg. betweenness centrality	0.00036	0.000361	1.63973	0.00036	2.13699
Avg. closeness centrality	0.15857	0.15864	0.11057	0.15871	0.0909
Cyclomatic Complexity	6219	6218	4	6217	5

CASE 2: (Table 2) We analyzed an application code base with 19,569 functions and 37,615 caller-recipient interactions, identifying 3 functions as critical in their failed update, causing code breakage. Both upgrades significantly affected the number of functions, although the total number of functions remained the same. Cyclomatic complexity decreased after both upgrades, while the number of connected components increased. The degree assortativity coefficient for the Non-Broken Upgrade was lower than that for the Broken Upgrade, as was the

Table 2. (CASE 2) Comparative analysis of code metrics from code base, non-broken upgrade and broken upgrade.

	Base	Non Broken Upgrade		Broken Upgrade	
Number of unique functions	19569	19569		19569	
nodes kind		**Unchanged**	**Changed**	**Unchanged**	**Changed**
Number of nodes	19569	17635	2629	17255	3338
Number of edges	37615	33180	4586	32234	5850
Average degree	3.8443	3.7630	3.4888	3.7362	3.5051
Density	0.00019646	0.00021339	0.00132754	0.00021654	0.00105037
Number of connected components	440	426	80	429	97
Average clustering	0.05552	0.05505	0.04415	0.05467	0.04482
Degree assortativity coefficient	−0.07772	−0.08047	−0.12601	−0.07853	−0.12377
Avg. betweenness centrality	0.00018	0.00019	0.000124	0.00019	0.00012
Avg. closeness centrality	0.17790	0.17686	0.18617	0.18716	0.17640
Cyclomatic Complexity	18926	16397	2117	15837	2706

density of the changed nodes, while closeness centrality was higher for Non-Broken Upgrades.

CASE 3: (Table 3) We analyzed an application code base with 15,908 functions and 29,449 caller-recipient interactions, identifying 6 critical functions in the upgrade. Post-upgrade, cyclomatic complexity decreased, and the number of connected components increased. The degree assortativity coefficient for the Non-Broken Upgrade was higher than that for the Broken Upgrade, while the density of changed nodes remained lower and closeness centrality was higher for Non-Broken Upgrades.

Table 3. (CASE 3) Comparative analysis of code metrics from code base, non-broken update and broken upgrade

	Base	Non Broken Upgrade		Broken Upgrade	
Number of unique functions	15908	15908		15908	
nodes kind		**Unchanged**	**Changed**	**Unchanged**	**Changed**
Number of nodes	15908	15553	603	15127	1015
Number of edges	29449	28616	912	27695	1816
Average degree	3.7024	3.6798	3.0249	3.6617	3.5783
Density	0.00023275	0.00023661	0.00502471	0.00024208	0.00352892
Number of connected components	337	341	19	335	28
Average clustering	0.05004	0.04834	0.06402	0.04688	0.05890
Degree assortativity coefficient	−0.10958	−0.11123	−0.10286	−0.11203	−0.13705
Avg. betweenness centrality	0.00022	0.00022	0.00018	0.00022	0.00022
Avg. closeness centrality	0.18028	0.18032	0.17876	0.18049	0.17684
Cyclomatic Complexity	14215	13745	347	13238	857

Statistical Differences Between Base and Subgraphs: Given the consistent differences in closeness centrality across the three cases, we further explored these variations. Conducting a package upgrade generates two subgraphs of the control flow graph: functions affected by the upgrade and those not affected. We hypothesize that the subgraphs resulting from broken package upgrades are statistically different from randomly drawn subgraphs due to higher concentrations of vulnerabilities. This hypothesis is based on the premise that vulnerabilities are likely to disrupt the normal connectivity patterns, which can be captured through centrality metrics.

To test this hypothesis, we used the closeness centrality values for each node and applied T-tests and Kolmogorov-Smirnov (K-S) tests to evaluate the differences in means and distributions between the subgraphs. The T-test assesses whether the average closeness centrality of nodes in the affected subgraph differs significantly from the baseline, while the K-S test compares the overall distributions to identify broader variations in connectivity patterns. These statistical methods are appropriate because they provide a robust framework for detecting changes in node centrality, which is crucial for understanding the impact of upgrades on the network structure.

For Broken Case 1, we observed a T-statistic of -4.881 (p-value $= 0.0012$) and a K-S statistic of 0.6916 (p-value $= 8.16e-05$), indicating significant deviations in closeness centrality between changed nodes and all nodes. In contrast, Broken Case 2 had a T-statistic of 8.399 (p-value $= 6.47e-17$) and a K-S statistic of 0.1161 (p-value $= 1.38e-27$), suggesting less variation but still significant differences. Broken Case 3 showed significant differences with a K-S statistic of 0.2176 (p-value $= 9.81e-34$) and a T-statistic of -2.349 (p-value $= 0.019$). These results validate the hypothesis by confirming that the mean values and distributions of closeness centrality are statistically different between the subgraphs resulting from upgrades that break functionality and the base repositories.

The results from the three cases analyzed here exhibit a trend where cyclomatic complexity tends to keep or decrease as upgrades are made to package. This observation suggests a relationship between the resolution of vulnerabilities and the simplification of code structure that may justify further exploration. This ongoing analysis is crucial for understanding the dynamic nature of package vulnerabilities and their impact on overall code complexity.

Modified GAT Results. The next step of the analysis attempts to analyze the interconnectedness of the critical functions causing the package upgrades to fail. When we apply the modified GAT model Fig. 2, the scores obtained. It is necessary to see each case as a specific case. For this, an average GAT score was obtained for each case. Having a high GAT score above average indicates that the vulnerabilities are more critical and necessary since the network depends more on them.

Principal Component Analysis (PCA) and t-Distributed Stochastic Neighbor Embedding (t-SNE) [28] are dimensionality reduction techniques applied alongside GATs to enhance visualization and interpretability as we do in the Case 2 Fig. 5. While PCA projects data into a lower-dimensional space preserving vari-

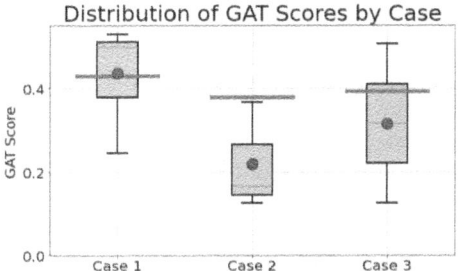

	Case 1	Case 2	Case 3
NC	27	3	6
MSC	0.5295	0.3662	0.5045
mSC	0.2461	0.1254	0.1260
ASC	0.4359	0.2185	0.3228
AGS	0.4287	0.3785	0.3153

Fig. 2. NC: number of Critical Function, MSFC:Max Score Critical, mSFC:Min Score Critical, ASC:Avr Score Critical, and AG:Avr Gat Score for all nodes(Red line) (Color figure online)

ance, t-SNE focuses on maintaining local relationships, making it particularly useful for visualizing high-dimensional data generated by GATs in a way that highlights patterns and relationships within the graph structure.

The enhanced GAT Score significantly improves our ability to discern the connectivity and importance of critical functions within a system's context. By employing a normalized approach, where scores closer to 1 denote higher importance, we gain nuanced insights into specific vulnerabilities. This is exemplified in CASE 1, where the average GAT score for the entire graph is 0.4287, with most of the 27 critical functions scoring above this value (see Fig. 2). In CASE 2, the GAT scores for all critical functions are below the average, with a maximum critical function score of 0.3662, while the average score for the entire graph is 0.3785, indicating lower criticality. These scores integrate a weighted combination of degree, norm, and centrality metrics, rather than relying solely on connectivity. This holistic approach allows us to identify functions that, despite having lower connectivity, hold substantial significance within the network's overall architecture. Such insights underscore the complexity of network dynamics and the crucial role of advanced analytical tools in unveiling the intricate interplay of functions within a software system.

6 Discussion

The results of our extensive analysis of several code bases reveal an intricate dynamic between package vulnerabilities and the complexity of the code. This study contributes to the current discourse in package development by highlighting the nuanced relationship between the vulnerabilities after the publication and evolution of the code.This requires not only immediate code updates, but also a deeper understanding of the underlying dynamics of these vulnerabilities.

Our analysis of CASES 1, 2, and 3 reveals several key insights into the impact of vulnerability resolution on code structure. We observed a trend of decreased or consistent cyclomatic complexity when comparing base cases with broken and

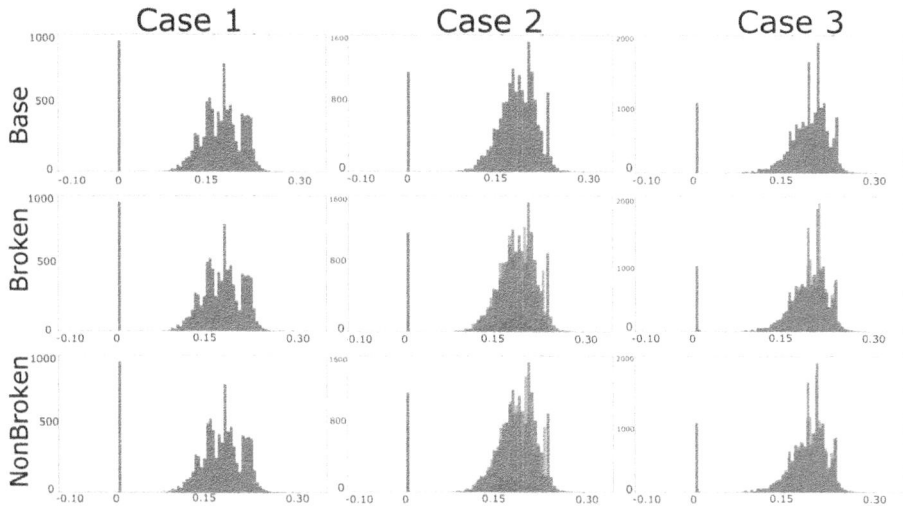

Fig. 3. Closeness Centrality Histogram, y-axes counts and x-axes Closeness Centrality. Cases 1, 2, 3, and 4 with their respective base code, broken upgrade, and non-broken upgrade as applicable. Light blue indicates the centrality of all nodes, light red represents unchanged nodes, and green denotes changed nodes. (Color figure online)

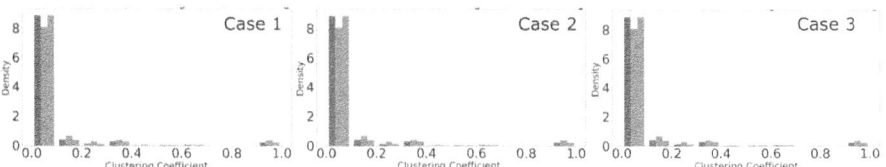

Fig. 4. Normalized cases of clustering coefficient histogram. Blue: All nodes, Red: Changed nodes, Green: Non changed nodes. (Color figure online)

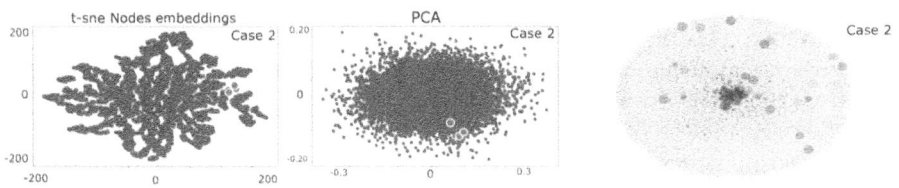

Fig. 5. Visualization of communities for Case 2, using t-SNE and PCA applied to a modified Graph Attention Network (GAT) as usually worked in GAT data analysis, The red dot is the representation of the vulnerability in their respective spaces. The third figure is a graphical representation of the communities within the graph, each distinguished by a unique color. (Color figure online)

intact vulnerability resolutions, indicating that addressing vulnerabilities simplifies the code structure and aligns with best practices for maintainable and less error-prone packages. However, managing the complexity of package vulnerabilities remains challenging. Notably, the number of connected components increased from the base case to the non-broken update, suggesting that remediation efforts improve connectivity and robustness. Despite this, even minor vulnerabilities can affect the entire application, highlighting the need for a comprehensive approach to understanding and mitigating the impacts of package updates. This complexity is further underscored by cross-centrality graphs (3, 4), which show significant variations in node centrality, emphasizing the critical importance of detailed graph-based analyses for managing software vulnerabilities.

In addition, our analysis sheds light on the connectivity patterns of the function within the code bases. The coefficient of negative degree supply consistently observed in multiple studies indicates that the code functions tend to connect with a diverse variety of other functions, instead of predominantly linked to similar functionalities. This diversity in connectivity patterns has deep implications to understand how vulnerabilities could spread through a code base and affect their general integrity.

This ambiguity highlights a critical gap in current understanding and requires more research. It emphasizes the importance of a strategic approach to code updates, where essential functionalities such as configuration are refined and maintained, instead of being completely eliminated.

The significance of the modified GAT score is underscored by our observations of the wide dispersion of nodes and low density within the analyzed graphs. This dispersion requires a nuanced approach to understanding the role of each node, focusing on both its connections and the paths traversing through it. The modified GAT model provides a normalized view of how interconnected a function is concerning its degree, norm, and closeness centrality metrics. This comprehensive perspective is crucial for effective vulnerability management and software maintenance, highlighting critical nodes that warrant prioritized attention. By integrating GAT score insights with node propagation and density observations, stakeholders are equipped with a powerful tool to identify and address significant vulnerabilities, optimizing resource allocation and enhancing software security and reliability. This blend of quantitative measures and attention-based analysis represents a significant advancement in strategically mitigating vulnerabilities and strengthening software infrastructure.

Our study underscores the dynamic nature of package vulnerabilities and their impact on code complexity, emphasizing the need for continuous vigilance and strategic intervention throughout the package development lifecycle. By hypothesizing that broken package upgrade subgraphs differ statistically from randomly drawn subgraphs due to higher vulnerability concentrations, we validated our approach using closeness centrality values for each node and comparative T-tests and Kolmogorov-Smirnov tests, confirming the reliability of our vulnerability assessments. To illustrate the versatility and effectiveness of our graph-based methodology, we analyzed three real-world applications: a Python-based data processing package, a Java-based web application, and a mixed-language analytics tool. This comprehensive analysis demonstrates the applicability of our methods across different programming environments, highlighting critical functions within each application that contribute to increased vulnerability risks. By integrating insights from our modified Graph Attention Network (GAT) with node propagation and density observations, we offer valuable strategic planning for vulnerability mitigation, ultimately enhancing software security and reliability across diverse software ecosystems.

7 Conclusion

This paper embarks on a comprehensive exploration, leveraging the intricate capabilities of knowledge graphs to delve into the dynamics of opensource package function networks. Central to our investigation is the identification and analysis of vulnerable functions, where we scrutinize their interactions and assess the impact of their mitigation on the codebase. Our research reveals a notable insight: targeted remediation of specific vulnerabilities tends to preserve the overall network's connectivity, underscoring the package structure's inherent resilience.

During our analysis, we noted a pattern of decreasing vulnerabilities following successive package updates, prompting a pivotal inquiry: are these vulnerabilities conclusively resolved, or do they subtly embed themselves into subsequent features? This ambiguity signals a compelling need for further work into the lifecycle of vulnerabilities within package development, promising to enrich the discourse in this field significantly.

At the heart of our methodology is the odified Graph Attention Network (GAT), especially its attention mechanism. Through the integration of node-centric metrics-such as degree centrality, norm, and closeness centrality-our approach refines the network's ability to discern detailed aspects of the graph's architecture and the nuances of node characteristics. This methodological advancement facilitates a nuanced portrayal of the network, yielding a comprehensive understanding of node interrelations and their significance.

Furthermore, our investigation brings to light the existence of latent vulnerabilities within the most critical segments of the code, initially perceived as flawless. These covert vulnerabilities represent significant security risks, with the potential to compromise vital components, including databases and core functionalities. To address these issues, we advocate for an in-depth and ongoing code analysis from its inception. Utilizing knowledge graphs as both historical and dynamic monitoring tools enables proactive surveillance of vulnerabilities.

For future works, our research direction will focus on methodological improvements, particularly in dissecting the interconnectivity between functions. The number of case analyzed was small with significant complexity, necessitating additional application repositories for comprehensive analysis to draw more robust conclusions. By evaluating various aspects such as variable types, execution times, and functional dependencies, we aim to unravel the importance of specific functions within the network. This comprehensive strategy is designed to offer deeper insights into the structural integrity and vulnerabilities of software systems, thereby making a substantial contribution to enhancing package security and dependability. Our endeavors are geared towards the development of robust software systems capable of navigating the complexities of contemporary cyber threats, marking a significant stride forward in the realm of package vulnerability analysis and cybersecurity.

Acknowledgment. This research was supported in part by NSF grant CCF-2109988.

References

1. Alanazi, M., Mahmood, A., Chowdhury, M.J.M.: SCADA vulnerabilities and attacks: a review of the state-of-the-art and open issues. Comput. Secur. **125**, 103028 (2023)
2. Alfadel, M., Costa, D.E., Shihab, E.: Empirical analysis of security vulnerabilities in python packages. Empirical Softw. Eng. **28**(3), 59 (2023)
3. Althar, R.R., Samanta, D., Kaur, M., Singh, D., Lee, H.N.: Automated risk management based software security vulnerabilities management. IEEE Access **10**, 90597–90608 (2022)
4. Arulselvan, A., Commander, C.W., Elefteriadou, L., Pardalos, P.M.: Detecting critical nodes in sparse graphs. Comput. Oper. Res. **36**(7), 2193–2200 (2009)
5. Bader, D.A., Madduri, K.: Parallel algorithms for evaluating centrality indices in real-world networks. In: 2006 International Conference on Parallel Processing (ICPP 2006), pp. 539–550 (2006)
6. Ebert, C., Cain, J., Antoniol, G., Counsell, S., Laplante, P.: Cyclomatic complexity. IEEE Softw. **33**(6), 27–29 (2016)
7. Fan, J., Li, Y., Wang, S., Nguyen, T.N.: A C/C++ code vulnerability dataset with code changes and CVE summaries. In: Proceedings of the 17th International Conference on Mining Software Repositories, MSR 2020, pp. 508–512. Association for Computing Machinery, New York (2020). https://doi.org/10.1145/3379597.3387501. ISBN 9781450375177
8. Giray, G.: A software engineering perspective on engineering machine learning systems: State of the art and challenges. J. Syst. Softw. **180**, 111031 (2021)
9. Goseva-Popstojanova, K., Perhinschi, A.: On the capability of static code analysis to detect security vulnerabilities. Inf. Softw. Technol. **68**, 18–33 (2015)
10. He, L., Ren, X., Gao, Q., Zhao, X., Yao, B., Chao, Y.: The connected-component labeling problem: a review of state-of-the-art algorithms. Pattern Recogn. **70**, 25–43 (2017)
11. Hogan, A., et al.: Knowledge graphs. ACM Comput. Surv. **54**(4), 1–37 (2021)
12. Imtiaz, N., Thorn, S., Williams, L.: A comparative study of vulnerability reporting by software composition analysis tools. In: Proceedings of the 15th ACM/IEEE International Symposium on Empirical Software Engineering and Measurement (ESEM), pp. 1–11 (2021)
13. Jia, Y., Qi, Y., Shang, H., Jiang, R., Li, A.: A practical approach to constructing a Knowledge Graph for cybersecurity. Engineering **4**(1), 53–60 (2018)
14. Liu, K., Wang, F., Ding, Z., Liang, S., Zhengfei, Yu., Zhou, Y.: Recent progress of using knowledge graph for cybersecurity. Electronics **11**(15), 2287 (2022)

15. Liu, K., Wang, F., Ding, Z., Liang, S., Yu, Z., Zhou, Y.: A review of knowledge graph application scenarios in cybersecurity. *arXiv preprint*arXiv:2204.04769 (2022)

16. McCabe, T.J.: A complexity measure. IEEE Trans. Softw. Eng. **4**, 308–320 (1976)

17. Newman, M.E.J.: Mixing patterns in networks. Phys. Rev. E **67**(2), 026126 (2003)

18. Nieminen, U.J.: On the centrality in a directed graph. Soc. Sci. Res. **2**(4), 371–378 (1973)

19. Park, N., Kan, A., Dong, X.L., Zhao, T., Faloutsos, C.: Estimating node importance in knowledge graphs using graph neural networks. In: Proceedings of the 25th ACM SIGKDD International Conference on Knowledge Discovery Data Mining, pp. 596–606 (2019)

20. Platt, E.L.: Network Science with Python and NetworkX Quick Start Guide: Explore and Visualize Network Data Effectively. Packt Publishing Ltd (2019)

21. Rodriguez, O.A., Du, Z., Patchett, J., Li, F., Bader, D.A.: Arachne: an arkouda package for large-scale graph analytics. In: 2022 IEEE High Performance Extreme Computing Conference (HPEC), pp. 1–7. IEEE (2022)

22. Saramäki, J., Kivelä, M., Onnela, J.-P., Kaski, K., Kertesz, J.: Generalizations of the clustering coefficient to weighted complex networks. Phys. Rev. E **75**(2), 027105 (2007)

23. Sarwar, M.M.S., Shahzad, S., Ahmad, I.: Cyclomatic complexity: the nesting problem. In: Eighth International Conference on Digital Information Management (ICDIM 2013), pp. 274–279. IEEE (2013)

24. Sikos, L.F.: Cybersecurity knowledge graphs. Knowl. Inf. Syst. 1–21 (2023)

25. Sparks, S., Embleton, S., Cunningham, R., Zou, C.: Automated vulnerability analysis: Leveraging control flow for evolutionary input crafting. In: Twenty-Third Annual Computer Security Applications Conference (ACSAC 2007), pp. 477–486. IEEE (2007)

26. Szabó, T.: Incrementalizing production CodeQL analyses. arXiv preprint-arXiv:2308.09660 (2023)

27. Varela-Vaca, Á.J., Borrego, D., Gómez-López, M.T., Gasca, R.M., Márquez, A.G.: Feature models to boost the vulnerability management process. J. Syst. Softw **195**, 111541 (2023)

28. Veličković, P., Cucurull, G., Casanova, A., Romero, A., Lio, P., Bengio, Y.: Graph attention networks. *arXiv preprint*arXiv:1710.10903 (2017)

29. Xia, B., Bi, T., Xing, Z., Lu, Q., Zhu, L.: An empirical study on software bill of materials: where we stand and the road ahead. *arXiv preprint*arXiv:2301.05362 (2023)

30. Xia, F., Chen, X., Yu, S., Hou, M., Liu, M., You, L.: Coupled attention networks for multivariate time series anomaly detection. IEEE Trans. Emerg. Top. Comput. (2023)

31. Yan, Z., Liu, J.: A review on application of Knowledge Graph in cybersecurity. In: 2020 International Signal Processing, Communications and Engineering Management Conference (ISPCEM), pp. 240–243. IEEE (2020)

32. Zhu, K., Yuliang, L., Hui Huang, L.Y., Zhao, J.: Constructing more complete control flow graphs utilizing directed gray-box fuzzing. Appl. Sci. **11**(3), 1351 (2021)

33. Zomorodian, A.: Computational topology. Algorithms and Theory of Computation Handbook, vol. 2, no. 3 (2009)

Characterizing the Evolution
of Psychological Factors Exploited
by Malicious Emails

Theodore Longtchi and Shouhuai Xu[✉]

Department of Computer Science, University of Colorado Colorado Springs,
Colorado Springs, CO, USA
sxu@uccs.edu

Abstract. Cyber attacks, including cyber social engineering attacks, such as malicious emails, are always evolving with time. Thus, it is important to understand their evolution. In this paper we characterize the evolution of malicious emails through the lens of Psychological Factors (PFs), which are humans' psychological attributes that can be exploited by malicious emails (i.e., attackers who send them). For this purpose, we propose a methodology and apply it to conduct a case study on 1,260 malicious emails over a span of 21 years (2004–2024). Our findings include: (i) attackers have been constantly seeking to exploit many PFs, especially the ones that reflect human traits; (ii) attackers have been increasingly exploiting 9 PFs and mostly in an implicit or stealthy fashion; (iii) some PFs are often exploited together. These insights shed light on how to design future defenses against malicious emails.

Keywords: Psychological Factors · Malicious emails · Cyber social engineering attack · Cybersecurity · Impulsivity · Trust · Curiosity · Defenselessness

1 Introduction

Cyber social engineering attacks keep increasing despite efforts at defending against them. For example, phishing is perhaps the most prolific cyber social engineering attack [2,34,43,83]. The 2023 Anti-Phishing Working Group (APWG) report [3] states that 2023 has been the worst on record so far, with more than 5 million attacks, while 2022 and 2021 were respectively the record holder. The 2022 APWG report [2] states that the number of phishing attacks reported to APWG has quadrupled since early 2020. That is, the situation continues to worsen despite the endeavors at defending against phishing emails.

A recent study [34] shows that attackers employ psychologically charged elements in crafting malicious emails in order to exploit human Psychological Factors (PFs), which are humans' psychological characteristics or attributes that can be exploited by cyber social engineering attackers including malicious emails (i.e., the attackers who send them). However, existing defenses rarely consider

J. Zhao and W. Meng (Eds.): SciSec 2024, LNCS 15441, pp. 158–178, 2025.
https://doi.org/10.1007/978-981-96-2417-1_9

PFs [34,62], highlighting a mismatch between cyber social engineering attacks and defenses against them. This motivates us to investigate how PFs have been exploited by malicious emails over time, so as to shed light on designing more effective defenses to counter these attacks.

Our Contributions. We make three contributions. First, we reconcile the 46 PFs presented in [34] into 20 PFs. This is important because the reconciliation would ease many tasks, such as categorizing and prioritizing PFs for defense purposes, and because the 46 PFs contain some overlaps. The resulting 20 PFs can serve as a foundation or new baseline for future studies. Second, we propose a methodology for characterizing the evolution of PFs exploited by malicious emails over any time span and at any granularity (e.g., monthly if not weekly basis), including how to identify PFs exploited by malicious emails. The methodology can be adopted/adapted to identify the PFs that are exploited by other cyber social engineering attacks, or to identify other PFs (e.g., another reconciliation/refinement of the 46 PFs [34] than what we propose) used by malicious emails and other cyber social engineering attacks. Third, we apply the methodology to study the evolution of the PFs that are exploited by malicious emails over the span of 21 years (2004–2024), by analyzing 1,260 malicious emails, or 60 per year, so as to draw insights into the evolution of adversary's exploitation of PFs. Our findings include: (i) all 20 PFs have been exploited by malicious emails during the last 21 years; (ii) attackers have been constantly seeking to exploit multiple PFs, especially the ones that reflect human traits and are dubbed Inherent PFs in this paper; (iii) attackers have been increasingly exploiting 9 PFs in a mostly implicit or stealthy fashion rather than an explicit fashion perhaps because they do not feel wise or needing to do so; (iv) some PFs are often exploited together. These insights shed light on future studies, especially the need of designing defenses to counter the exploitation of certain PFs (especially the ones that are increasingly exploited) and their collective exploitation (rather than designing defenses to counter the exploitation of individual PFs).

Ethical Issue. We consulted with our institution's Internal Review Board (IRB), and were advised that we do not need the IRB's approval to conduct this study, as the dataset analyzed in this paper is provided by third parties (in addition to the 108 malicious emails from our own email box).

Related Work. Cyber social engineering attacks have been extensively studied (cf., e.g., [1,12,22–24,31,34,37] and the references therein). However, the *root cause* of these attacks is not adequately understood despite that the notion of PF has been implicitly and occasionally explicitly mentioned as humans' weaknesses (e.g., [45]). The literature mainly discusses Cialdini's 6 Principles of Persuasion as PFs [16], often with a very small dataset (e.g., 52 emails in [15]). Few studies discuss PFs beyond these 6 Principles: IMPULSIVITY reflecting how fast an individual reacts to a malicious email [19], LONELINESS in the context of Covid-19 pandemic lockdown [10], and CURIOSITY in phishing attacks [5]. Gallo et al. [18] investigate how phishing emails exploit cognitive vulnerabilities based on a dataset of 2-year span. Wang et al. [63] study the impact of visceral triggers on phishing susceptibility, without tying to PFs. By contrast, we consider 20 (reconciled from 46) PFs for over 21 years. In [33] we study the evolution of Psychological Techniques and Tactics exploited by malicious emails.

Not until very recently was the first systematization of PFs presented in [34], including 46 PFs in 5 categories (i.e., social psychology, personality and individual differences, cognition, emotion, and workplace), which serve as a baseline for the community to scrutinize. Some of these PFs have been leveraged to quantify the degree of sophistication of malicious emails [42]. The present study is the first to scrutinize the 46 PFs presented in [34] by reconciling them into 20 PFs. We further propose a methodology to guide the identification of PFs exploited by malicious emails, and apply it to analyze 1,260 malicious emails over a span of 21 years to characterize the evolution of PFs exploited by malicious emails.

Paper Outline. Section 2 presents our reconciliation of the 46 PFs proposed in [34] into 20 PFs. Section 3 presents our methodology for characterizing the evolution of the PFs exploited by malicious emails. Section 4 applies the methodology to analyze malicious emails over any time span and at any granularity. Section 5 discusses the limitations of the present study. Section 6 concludes the paper.

2   Reconciling the 46 PFs Presented in [34] Into 20 PFs

We reconcile the 46 PFs to ease our analysis and provide more effective guidance on designing future defenses. This is important because it is neither wise nor practical to design defenses against the exploitation of each individual PF.

2.1   Designing Rules to Guide the Reconciliation of PFs

For this purpose, we propose using the following rules to guide the PF reconciliation process, with no particular precedence.

Rule 1: When reconciling a set of PFs into a single one, it is intuitive to choose the one that is *representative*. A representative PF is one that has been studied in a quantitative/empirical fashion, or has been studied extensively in a qualitative fashion. If there is still a tie, we prefer to preserving the PF that has been most exploited by attackers. Since this is a subjective matter, we use our domain knowledge to break a tie.

Rule 2: In the ideal case, each PF is orthogonal or complementary to the other PFs so that there is no overlapping or redundancy among the PFs. Even though the 46 PFs introduced in [34] are carefully defined such that no two PFs have the same meaning, we observe that there are *overlaps* and *congruence*, as elaborated below, meaning that we need to reconcile them. (i) Some PFs may be overlapping, meaning that one PF may be replaced by multiple PFs that can collectively accommodate what is described by the particular PF. For instance, the FEAR PF could be replaced by AUTHORITY and SCARCITY because they each accommodates one sense of FEAR. (ii) Some PFs may be congruent, meaning that they deal with human characteristics or attributes that are inherently related to each other; as a result, one defense can, or should, cope with the exploitation of these PFs. For instance, one defense could simultaneously deal with the PFs known as AUTHORITY, RESPECT, and SUBMISSIVENESS, suggesting that we can reconcile them into a single PF, namely AUTHORITY (per Rule 1).

Rule 3: Assure that the input PFs (i.e., the 46 PFs in this study) are accommodated by the resulting smaller set of PFs, namely that the latter can adequately describe what the former can. This means that any PF that is not reconciled into, or replaced by, another PF, should be preserved.

Note that these rules are not specific to the 46 PFs we reconcile, but can be adopted/adapted to reconcile other PFs of interest. For example, it is possible that these rules can be applied to produce another reconciliation of the 46 PFs than what is presented in this paper.

2.2 Reconciling the 46 PFs [34] Into 20 PFs

Our reconciliation proceeds in two steps: leveraging the preceding rules to reconcile the input PFs (i.e., the 46 PFs in this study); and categorizing the resulting PFs into a small number of families. Both steps require substantial domain expertise, meaning that the result is inevitably subjective.

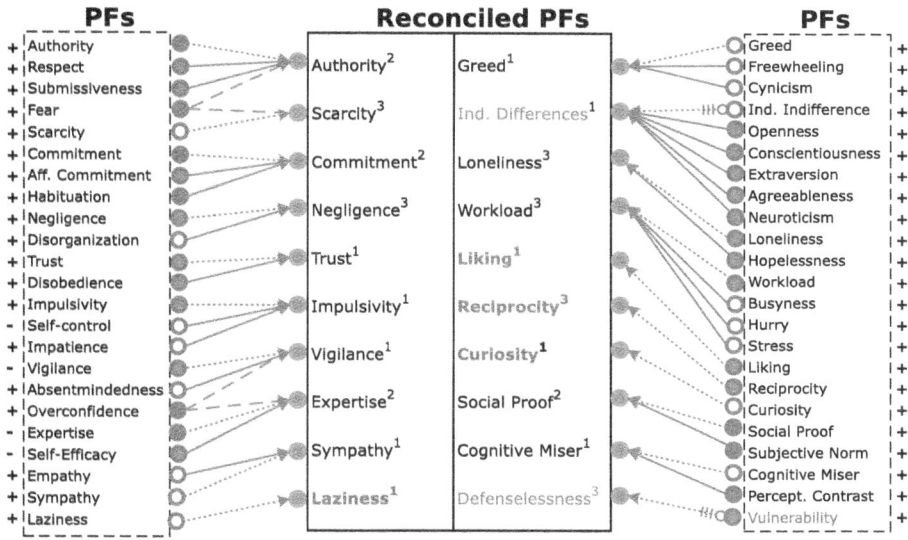

Fig. 1. Summary of the reconciliation of the 46 PFs (columns 1 and 4) into the 20 PFs, including two being renamed (in green) and 4 being preserved (in blue). A filled circle in the original PFs (i.e., columns 1 and 4) indicates that a PF has been studied quantitatively and an empty circle indicates otherwise; the "+" ("-") sign indicates that a higher PF value means a higher (lower) susceptibility to cyber social engineering attacks (e.g., malicious emails) [34]. "Ind." is short for Individual, and "Percept." is short for perceptual. Superscripts of the reconciled PFs indicate the PF family to which it belongs (1 for Inherent PFs, 2 for Social PFs, and 3 for Situational PFs). (Color figure online)

Step 1: Reconciling Overlapping/Congruent PFs. Figure 1 summarizes how the 46 PFs are reconciled into the 20 PFs, where a solid arrow indicates that the source PF is fully incorporated into the destination PF, a dashed arrow indicates that the source PF is partially incorporated into the destination PF (e.g., FEAR is partly incorporated into AUTHORITY and partly incorporated into SCARCITY), a dotted arrow indicates that the source PF is preserved while accommodating other PFs (e.g., GREED, FREEWHEELING, and CYNICISM are reconciled into GREED) or not (i.e., LIKING, RECIPROCITY, CURIOSITY, LAZINESS, and DEFENSELESSNESS which is renamed from VULNERABILITY though), a dotted arrow with a tail indicates that the source PF is renamed to a destination PF (e.g., VULNERABILITY renamed to DEFENSELESSNESS). Details follow.

We reconcile the following four PFs AUTHORITY, RESPECT, SUBMISSIVENESS, and FEAR [34] into a resulting AUTHORITY PF. This can be justified by the observation that an effective defense against the exploitation of AUTHORITY could be effective against the exploitation of the three other PFs because AUTHORITY could demand or accommodate the three other PFs [8], while noting that AUTHORITY may imply FEAR in some cultures [64] and that another sense of FEAR will be accommodated by the SCARCITY PF below. The preceding Rule 1 prompts us to preserve the PF name AUTHORITY among the 4 PFs because it has been more extensively studied in a quantitative fashion than the three other PFs and because it is one of the most exploited principle of persuasion [15].

We reconcile SCARCITY and FEAR [34] into SCARCITY. Although FEAR of authority has been incorporated into AUTHORITY as discussed above, it is appropriate to incorporate FEAR of missing out [26,80] into SCARCITY, which has been highly exploited by cyber social engineering attackers to make the recipient of an email to act quickly according to an attacker's proposition (e.g., "Bitcoin price is soaring, click here to buy yours"). Rule 3 prompts us to preserve SCARCITY because FEAR has been partially incorporated into AUTHORITY.

We reconcile COMMITMENT, CONSISTENCY, AFFECTIVE COMMITMENT, and HABITUATION [34] into COMMITMENT, because these 4 PFs are congruent in the sense that they are all about one's commitment to something or one's habit of doing something [17,61]. Rule 1 prompts us to preserve COMMITMENT because it is the most studied PF among these four and it is one of the most exploited principle of persuasion [17,64].

We reconcile SOCIAL PROOF, CONFORMITY, and SUBJECTIVE NORM [34] into SOCIAL PROOF, because they are congruent in the sense that they define people's behaviour according to social norms with respect to a group of people (e.g., workplace, a sport team, school). Rule 1 prompts us to preserve SOCIAL PROOF because it is the most extensively studied among the three PFs (and is synonymous to CONFORMITY per Cialdini's Principles of Persuasion [17]).

We reconcile NEGLIGENCE and DISORGANIZATION [34] into NEGLIGENCE because they are congruent in the sense that DISORGANIZATION can lead to NEGLIGENCE and vice versa. Therefore, one defense that can effectively cope with the exploitation of NEGLIGENCE should also be able to deal with the exploitation of DISORGANIZATION. Rule 1 prompts us to preserve NEGLIGENCE because it has been studied more extensively and in a quantitative fashion [44].

We reconcile INDIVIDUAL INDIFFERENCE and the Big 5 personality traits which are treated as PFs—OPENNESS, CONSCIENTIOUSNESS, EXTRAVERSION, AGREEABLENESS, and NEUROTICISM [34]—into INDIVIDUAL INDIFFERENCE because these 6 PFs are congruent in the sense that they all deal with human traits and characteristics. Thus, one defense should effectively cope with the exploitation of them. Moreover, we propose renaming INDIVIDUAL INDIFFER-ENCE to INDIVIDUAL DIFFERENCE because the latter encompasses the former and the Big 5 [36] and because the latter is most investigated among these 6 PFs [4,53].

We reconcile TRUST and DISOBEDIENCE [34] into TRUST because a breach of trust (e.g., at work) would cause DISOBEDIENCE [32]. Studies show individuals who are more trusting and obedient to authority are more susceptible to cyber social engineering attacks [24] and willful disobedience of employees has been exploited by cyber social engineering attacks [27]. Thus, an effective defense against the exploitation of TRUST would also be able to deal with the exploitation of DISOBEDIENCE. Rule 1 prompts us to preserve TRUST because it has been more extensively studied in a quantitative fashion than DISOBEDIENCE [25].

We reconcile IMPULSIVITY, IMPATIENCE, and SELF-CONTROL [34] into IMPUL-SIVITY. Note that IMPULSIVITY and IMPATIENCE are congruent as both indicate the lack of SELF-CONTROL, which makes one less susceptible to cyber social engineering attacks [60]. Thus, one defense that can effectively cope with the exploitation of IMPULSIVITY could also deal with the exploitation of the other two. Rule 1 prompts us to preserve IMPULSIVITY because it has been more exten-sively studied than IMPATIENCE and SELF-CONTROL [9].

We reconcile VIGILANCE, OVERCONFIDENCE, and ABSENTMINDEDNESS [34] into VIGILANCE because OVERCONFIDENCE may lead to lack of VIGILANCE and a lack of VIGILANCE is ABSENTMINDEDNESS (i.e., VIGILANCE is the opposite of ABSENTMINDEDNESS) while noting that a high VIGILANCE makes an individual less susceptible to cyber social engineering attacks. Thus, one defense that can effectively enhance VIGILANCE would address the exploitation of OVERCONFI-DENCE and ABSENTMINDEDNESS at the same time. Rule 1 prompts us to pre-serve VIGILANCE because it has been more extensively studied in a quantitative fashion than OVERCONFIDENCE and ABSENTMINDEDNESS PF [59].

We reconcile EXPERTISE, OVERCONFIDENCE, and SELF-EFFICACY [34] into EXPERTISE because it usually implies SELF-EFFICACY (i.e., they are congruent) and because EXPERTISE and SELF-EFFICACY may lead to OVERCONFIDENCE. Rule 1 prompts us to preserve EXPERTISE because it has been more extensively studied than the other two [28].

We reconcile GREED, FREEWHEELING, and CYNICISM [34] into GREED because they are congruent, namely that GREED and CYNICISM involve self-interest and GREED and FREEWHEELING are about getting something effortlessly. This means that one effective defense against the exploitation of GREED would deal with the exploitation of the other two. Rule 1 prompts us to preserve GREED because it is more studied than the other two [56].

We reconcile SYMPATHY and EMPATHY [34] into SYMPATHY; even though they have a semantic difference, both deal with feeling others' pain. Thus, they are congruent and an effective defense against the exploitation of one PF would also be effective in coping with the exploitation of the other. Rule 1 prompts us to preserve SYMPATHY because it is more studied [64].

We reconcile LONELINESS and HOPELESSNESS [34] into LONELINESS because an effective defense against the exploitation of LONELINESS would also be able to deal with the exploitation of HOPELESSNESS. Rule 1 prompts us to preserve LONELINESS because it is more extensively studied than HOPELESSNESS [10].

We reconcile WORKLOAD, BUSYNESS, HURRY, and STRESS [34] into WORK-LOAD because the other three are often congruent to, or incurred by, workload. Rule 1 prompts us to preserve WORKLOAD because it is the most extensively studied among the four PFs [41].

We reconcile COGNITIVE MISER and PERCEPTUAL CONTRAST into COGNITIVE MISER because it leads to PERCEPTUAL CONTRAST, where a person quickly compares items without thoughtfulness. Rule 1 prompts us to preserve COGNITIVE MISER because it is more extensively studied [38].

At this point, there are five PFs that have not been reconciled with others: (i) LIKING, which means that people lean favorably to, or have the desire to become, the people they like [15]; (ii) RECIPROCITY, which means paying back for an favor that one received earlier [64]; (iii) CURIOSITY, which is the degree of desire to know/learn something; (iv) LAZINESS, which is the unwillingness to push pass one's comfort zone [77]; and (v) VULNERABILITY, which occurs under conditions such as failing memory or incapability of taking an informed decision. We propose renaming VULNERABILITY to DEFENSELESSNESS to avoid potential confusions because some PFs (e.g., OVERCONFIDENCE) are human vulnerabilities. The term DEFENSELESSNESS has been used in [39] to describe senior citizens or mentally declining people. Rule 3 prompts us to preserve these 5 PFs.

Step 2: Classifying the Resulting PFs into a Small Number of Families.
Having reconciled the 46 PFs into the 20 PFs mentioned above, now we classify the 20 PFs into a small number of families. This is important not only because the classification deepens our understanding but also because it sheds light on designing future defenses (i.e., different defenses may be required when dealing with the exploitation of different families of PFs).

Table 1. The three families of the resulting 20 PFs

Family	PFs
Inherent PFs	LIKING, INDIVIDUAL DIFFERENCES, TRUST, IMPULSIVITY, CURIOSITY, LAZINESS, VIGILANCE, COGNITIVE MISER, GREED, SYMPATHY
Social PFs	SOCIAL PROOF, COMMITMENT, AUTHORITY, EXPERTISE
Situational PFs	RECIPROCITY, SCARCITY, NEGLIGENCE, DEFENSELESSNESS, LONELINESS, WORKLOAD

We propose classifying the resulting 20 PFs into three families highlighted in Table 1, which summarizes the superscripts in Fig. 1. The three families are: Inherent PFs, which are human traits such as IMPATIENCE; Social PFs, which are the characteristics that humans may learn by training, such as EXPERTISE; Situational PFs, which are the behaviors that occur only when an individual is subjected to a situation or condition that represents an external stimulus that may cause one to act in ways to offset the stimulus, such as WORKLOAD.

We stress that there may be other ways to classify the 20 PFs because such a classification is inevitably subjective. The one presented above is what we deem an appropriate classification. Moreover, even for a given classification, it is still subjective to determine to which family a PF should belong. For example, we classify VIGILANCE into the Inherent PFs family even though some researchers may not treat it as a personality trait on its own but is affected by personality traits instead. This is demonstrated in an empirical study with 96 participants, which shows that VIGILANCE is related to (in our terminology) Situational PFs, such as WORKLOAD, but the degree of VIGILANCE is directly associated with the personality of a participant [52]. As another example, we classify AUTHORITY and EXPERTISE into Social PFs because the society creates a hierarchy among people owing to a difference in knowledge or accomplishments; as a consequence, attackers exploit the role of a high persona in the hierarchy to attack victims.

Insight 1. *The 20 PFs (reconciled from the 46 PFs) can serve as a new baseline for future studies on teaching and understanding cyber social engineering attacks, and possibly for guiding the design of future defenses against these attacks.*

3   Methodology for Characterizing the Evolution of PFs

Given a set of PFs (such as the 20 PFs mentioned above), the objective is to use them to characterize their evolution via a set of malicious emails. For this purpose, we propose the following methodology of three steps: (i) preparing a dataset of malicious emails; (ii) identifying the PFs exploited by the malicious emails or grading the malicious emails with respect to the PFs; (iii) characterizing the evolution of PFs exploited by the malicious emails over time.

3.1   Preparing Dataset

To prepare a high-quality dataset, we propose proceeding as follows. First, one should determine the scope of a study in terms of the kinds of malicious emails, such as phishing vs. non-phishing emails and general phishing vs. spear phishing. Second, one should determine the time span and granularity of the malicious emails that are to be studied. It would be ideal to collect malicious emails for the longest time span possible (e.g., from the time when emails became popular) to the present time, and at the finest granularity (e.g., daily rather than monthly). However, a trade-off needs to be made in terms of the feasibility of the study, as identifying which of the 20 PFs are exploited and how by a single email is already time-consuming. Third, one should assure that the malicious emails under study are of high quality. This implies the following: the sources of malicious emails are trusted and credible to have a good baseline of quality data; each email must be legible, which is relevant when an email is provided in the form of image or screenshot (especially for older emails); and each email must have all of its original content in its entirety, including logos if applicable.

3.2   Identifying PFs Exploited by Malicious Emails

For a given malicious email, we propose identifying the PFs exploited by it as follows: we grade the application of a PF as 0 for no application, 1 for implicit application, and 2 for explicit application. That is, each email will receive a score vector, where the value of each element belong to $\{0, 1, 2\}$. This offers us a flexibility in conducting analysis because (i) we can characterize the absence (score 0) vs. presence (scores 1 and 2) of PFs in emails and (ii) we can leverage the distinction between score 1 and score 2 to analyze implicit exploitations of PFs. Note that (ii) is important because implicit exploitations of PFs are stealthy when compared with explicit exploitations of PFs, and are perhaps harder to defend against owing to the fact that explicit exploitations could be detected by recognizing PF names in emails. Note that one email may implicitly and/or explicitly exploit multiple PFs.

To illustrate how we grade an email, which is inevitably subjective, let us look at some examples. If an email contains the statement *"We have updated our login system, please click here to login"*, from which we observe that PFs such as GREED, SYMPATHY and RECIPROCITY are neither implicitly nor explicitly exploited by the email, then we assign a score 0 to the email with respect to each of these three PFs. If an email contains the statement *"Send it now!"*, from which we observe that IMPULSIVITY, AUTHORITY, and WORKLOAD are (arguably) implicitly exploited (because of the word "now"), then we assign a score 1 to the email with respect to each of these three PFs. If an email contains the statement *"For your commitment to staying a loyal customer, we're giving you a $50 Walmart gift card. Click here to claim it"*, from which we observe COMMITMENT is explicitly exploited; thus we assign a score 2 to the email with respect to COMMITMENT. If an email contains the statement *"We truly appreciate your trust in our brand, see what is new"*, from which we observe TRUST is explicitly employed; thus, we assign a score 2 to the email with respect to TRUST. The last email is more interesting because it also implicitly exploits other PFs, namely CURIOSITY and IMPULSIVITY because of "see what is new"; thus, we further assign a score of 1 to the email with respect to these 2 PFs, respectively.

3.3   Characterizing Evolution of PFs Exploited by Malicious Emails

Having graded a set of malicious email with respect to the 20 PFs, we can characterize the evolution of the PFs that have been exploited by malicious emails via a range of Research Questions (RQs), such as:

- RQ1: How frequently are PFs exploited by malicious emails? This allows us to understand which PFs are more exploited than others.
- RQ2: Which PFs have been increasingly, constantly, or decreasingly exploited by malicious emails? Answering this would deepen our understanding of the trends in adversary's exploitation of PFs.
- RQ3: Are some PFs often exploited together or all of them are exploited randomly? Answering this would shed light on designing future defenses.

4   Case Study

4.1   Preparing Dataset

Under the guidance of the methodology, we prepare a dataset of malicious emails as follows. First, we decide to focus on 3 kinds of malicious emails: phishing, scam, and spam. A phishing email is one that provides a file for the recipient to download or a link to click on; a scam email is one that provides a pretext to ask for the recipient's login details or that provides a narrative so that the recipient may contact the scammer; and a spam email is one that attempts to market goods or services and is considered malicious at least when the truth is hidden from the recipient or when it uses deceptive marketing (e.g., snake-oil salesman). A key difference between phishing and scam is that phishing would provide a link to click or a document to download, but scam would provide a narrative for the recipient to provide login details, possible via a different email address to get back to the sender (i.e., the scammer).

Second, we decide to collect malicious emails for the span of 2004–2024 (i.e., 21 years) and at the granularity of year. Owing to the amount of work that is required to identify the PFs that are exploited by each malicious email, we collect 60 emails per year, leading to a dataset of 1,260 malicious emails. We collect emails from 2004 because we could not collect 60 quality emails per year for the years prior to 2004.

Third, we collect emails from the following sources: the Anti-Phishing Working Group (APWG) (250), universities or academic researchers (317), websites of tech companies and organizations (298), our own mailboxes (108), and real examples found in the Internet (287). The intent for considering these diversified sources (i.e., the entities that responded to our request for data) is to offer a comprehensive view of the malicious email landscape. However, this implies that the sources may have a spectrum of credibility. To address this issue, we use the same process as [42]: we first use ScamPredictor [54] to determine whether an email is phishing, spam, or scam, and then manually check each email to confirm that it is indeed malicious and correctly labelled in the three categories mentioned above (i.e., phishing, scam, and spam). We indeed observe that some emails are originally mistakenly labelled as phishing emails while they are not.

Among the 1,260 emails, 933 (or 74%) are phishing emails, 305 (or 24%) are scam emails, and 25 (or 2%) are spam emails. This discrepancy is inherent to the set of emails we received from the sources. Note that 27 (out of the 1,260) emails are originally in French (as one co-author is fluent in French).

4.2 Identifying PFs Exploited by the 1,260 Malicious Emails

Even though we only have one grader to identify PFs from malicious emails, there may be an inconsistency in treating emails. To mitigate this inconsistency, we use Table 2 as a reference, which is always shown on one screen of the grader. The process for identifying PFs from a single email takes about 10 min, meaning that grading 1,260 emails takes about 210 h.

Table 2. Examples showing identification of PFs from email components.

PF	Email component from which a PF is identified
IMPULSIVITY	"Click here for more"/"Special offer for you"
TRUST	"Hey Bill, let me know if you're available"/(logos known entities)
CURIOSITY	"Check my hot photos"/"See you credit score"
COGNITIVE MISER	"We detected suspicious behaviour on your PayPal account. Verify"
INDIVIDUAL DIFFERENCES	"Faculty/Personal/Student/Alumni, We received your resume application via..."/"verify your account, and sorry for the inconvenience"
GREED	"You will receive 30% of the money"/"Confirm your 3 million here
LIKING	"Protecting your account is our primary concern"
LAZINESS	"We got you covered..."/"We offer the best price. Look no further"
SYMPATHY	"I'm in vacation, and I lost everything to thieves. I need your help..."
VIGILANCE	"Annazon.com" for Amazon.com. (replacing "m" with double "n")
AUTHORITY	"I'm Police chief..."/"I'm Dr...."/(Use of logos such as FBI, IRS)
COMMITMENT	"As a value Walmart customer, please help us take a survey..."
SOCIAL PROOF	"I've just joined Jhoo, click here to join" (I = your idol)
EXPERTISE	"Download Covid-19 treatment procedure" (impersonating WHO)
DEFENSELESSNESS	"You owe $380 of utility bill. Make payment now"
WORKLOAD	"Here is the updated salary file"/"Here is the final report"
NEGLIGENCE	"Send $4,740 transfer immediately to XYZ for tomorrow's supply"
SCARCITY	"...limited supply"/"Stock is running out, get yours now..."
LONELINESS	"Meet cuties in your area"/"Judith sent you an invite"
RECIPROCITY	"As a former recipient yourself, please consider giving back to..."

4.3 Characterizing the Evolution of the PFs

In addressing these RQs, we deal with the presence vs. absence of a PF with respect to an email (a PF is or is not exploited). For this purpose, we treat a score 1 or 2 described in the methodology as the presence of a PF, and 0 otherwise. When the need arises, we further look at the distinction of implicit vs. explicit exploitation of a PF (i.e., score 1 vs. 2).

Addressing RQ1: How Frequently Exploited are PFs by Malicious Emails? In total, the 20 PFs are exploited for 4,989 times (or instances) by the 1,260 emails, meaning that one emails can exploit multiple PFs.

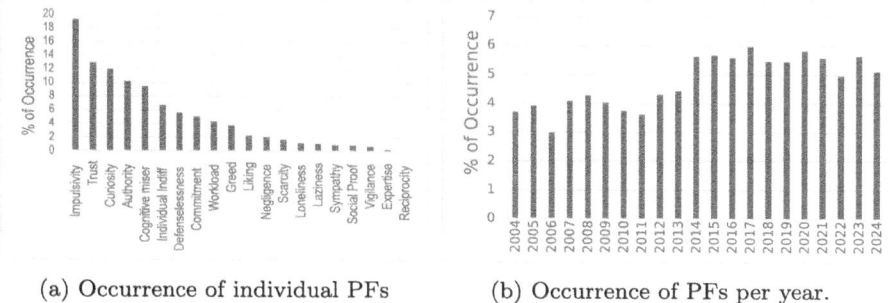

(a) Occurrence of individual PFs   (b) Occurrence of PFs per year.

Fig. 2. Frequency of PFs exploited by malicious email from 2004 to 2024.

Figure 2(a) plots the frequency (i.e., occurrence) of individual PFs that are exploited by the 1,260 malicious emails, namely the ratio between the total number of instances of a PF that is exploited by the 1,260 emails (while noting that this number for each PF is upper bounded by 1,260 because a PF can only be exploited at most once by an email) and 4,989. We make two observations. (i) All the 20 PFs have been exploited, highlighting that attackers are always identifying and exploiting PFs. (ii) The degree of exploitation of the 20 PFs vary: IMPULSIVITY is the most exploited (959 times or 19.22%), followed by TRUST (646 or 12.95%), CURIOSITY (597 or 11.97%), AUTHORITY (508 or 10.18%), and COGNITIVE MISER (469 or 9.40%). Among these 5 most exploited PFs, AUTHORITY is a Social PF and the other 4 are Inherent PFs. The most exploited PF in the Situational PFs family is DEFENSELESSNESS (278 or 5.57%), ranked #7 among the 20 PFs. This hints that attackers have exploited Inherent PFs, much more than Social PFs and Situational PFs. Thus, future research should prioritize on designing defenses to thwart adversary's exploitation of Inherent PFs (e.g., designing training schemes to educate people by focusing on these PFs).

Figure 2(b) plots the frequency (or occurrence) of PFs per year, namely the ratio between the total number of instances of the 20 PFs that are exploited by the 60 emails each year and 4,989. We observe that overall PFs are increasingly exploited by malicious emails. In particular, there are 1,948 instances of PFs on average between 2004 and 2013, and this number increases to 3,041 between 2014 and 2024. This means that attackers have been making effort at exploiting more PFs. By looking into the grades, we find that many PFs are exploited in all years, such as INDIVIDUAL DIFFERENCE, TRUST, IMPULSIVITY, CURIOSITY, COGNITIVE MISER, and AUTHORITY, while noting that all these are Inherent PFs. However, there are a few PFs that have not been constantly exploited, such as VIGILANCE (an Inherent PF), SYMPATHY (an Inherent PF), EXPERTISE (a Social PF), and RECIPROCITY (a Situational PF); in particular, VIGILANCE and RECIPROCITY have not been exploited during the last 6 years (the caveat is that the sample is small, with 60 emails per year). This means that attackers have been constantly seeking to exploit most PFs, but do not appear to exploit the PFs that (i) make individuals less susceptible, such as VIGILANCE and EXPERTISE, or (ii) are perhaps hard to exploit, such as RECIPROCITY.

Insight 2. *Attackers have been constantly seeking to exploit most of the 20 PFs, especially the Inherent PFs for which a higher value makes individuals more susceptible (e.g., IMPULSIVITY and CURIOSITY) but not the PFs for which a higher value makes individuals less susceptible (e.g., VIGILANCE and EXPERTISE).*

Insight 2 suggests that future defenses should focus on dealing with the PFs that are often exploited by attackers, especially the PFs that are easy to exploit.

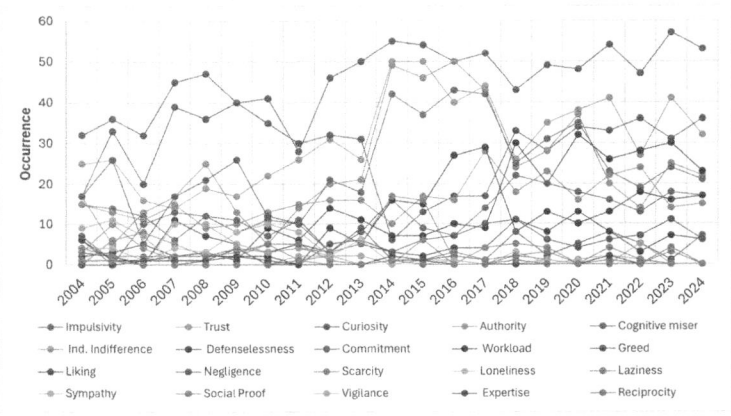

Fig. 3. Plots of occurrences of PFs over the 21 years where Ind. is short for Individual.

Addressing RQ2: Which PFs have Been Increasingly, Constantly, or Decreasingly Exploited? Figure 3 plots the evolution of the exploited PFs over the 21 years, where occurrence (*y*-axis) is the number of instances a PF is exploited among the 60 emails in a specific year (i.e., upped bounded by 60). Owning to the fact that there are 20 PFs, the visual effect is not perfect. However, it does show that some PFs are increasingly, roughly constantly, or decreasingly exploited. This prompts us to zoom into the detail as described below.

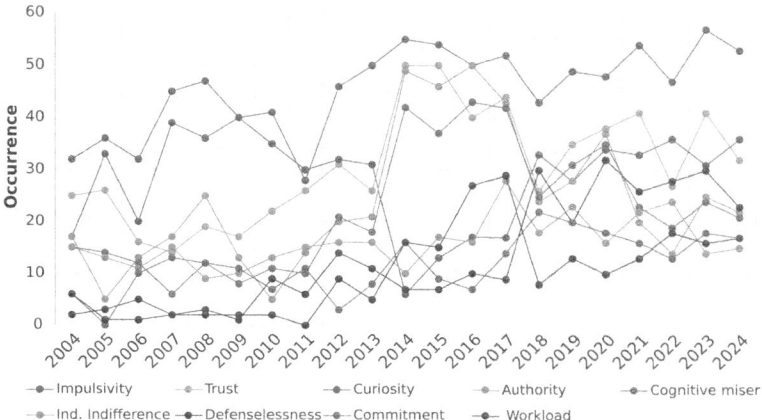

(a) Plots of PFs that were increasingly exploited in the past 21 years.

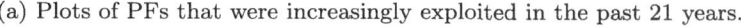

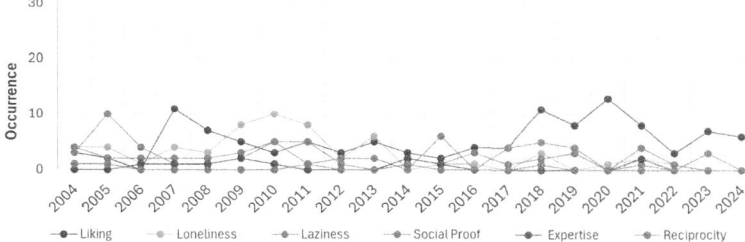

(b) Plots of PFs that were constantly exploited in the past 21 years.

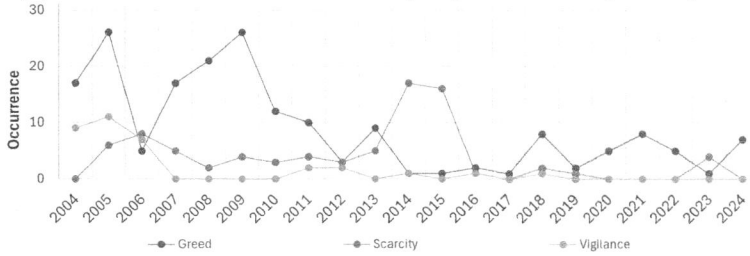

(c) Plots of PFs that were decreasingly exploited in the past 21 years.

Fig. 4. Plots of individual PFs, showing which PFs have been increasingly, decreasingly, and constantly exploited by malicious emails during the 21 years.

Figure 4(a) plots the 9 PFs that have been increasingly exploited over the 21 years. We observe: (i) IMPULSIVITY (an Inherent PF) is not only the most exploited PF, but also increasingly exploited overall, suggesting that existing defenses against the exploitation of IMPULSIVITY has not been successful. (ii)

DEFENSELESSNESS (a Situational PF) exhibits the highest increase over the 21 years, from 2 instances in 2004 to 32 instances in 2020 before leveling to 2024.

Figure 4(b) plots the 6 PFs that are, roughly speaking, constantly exploited over the 21 years. We make three observations. (i) Among these PFs, LIKING and LAZINESS are Inherent PFs, SOCIAL PROOF and EXPERTISE are Social PFs, and LONELINESS and RECIPROCITY are Situational PFs. This means that each PF family has some PFs that are roughly constantly exploited. (ii) These PFs are not frequently exploited (roughly speaking, they are exploited by less than 16% of the emails), while recalling that RECIPROCITY (a Situational PF) is the least exploited. (iii) When taking into account the implicit vs. explicit exploitation of a PF, we find that only 2 (out of the 9) PFs, namely TRUST (an Inherent PF) and COMMITMENT (a Social PF), are explicitly exploited by emails. This means that attackers find a way of achieving their objectives without raising suspicion. For example, an attacker would not outwardly say, "You have to trust me, it is a legitimate deal". Instead, attackers often implicitly instill TRUST in email recipients by leveraging logos of well recognized companies/institutions or by impersonating a known personality.

Figure 4(c) plots the 3 PFs that have been decreasingly exploited over the 21 years, despite occasional increases. The 3 PFs include 2 Inherent PFs, GREED and VIGILANCE, and 1 situational PF, SCARCITY. We make two observations. (i) The exploitation of GREED has significantly decreased from 2004 to 2024. This may be attributed to the employment of training individuals on the consequence of being greedy when encountering malicious emails, especially scam emails such as the Nigerian Price (or 419). As a result, attackers became less interested in exploiting this PF. (ii) VIGILANCE is rarely exploited perhaps because a high VIGILANCE indicates a low susceptibility and it is not clear how an attacker can reduce an individual's vigilance (i.e., this PF may be hard to exploit).

Insight 3. *Attackers have been increasingly exploiting 9 PFs and mostly exploiting PFs in an implicit or stealthy fashion, suggesting that future defenses should adequately address these kinds of exploitations.*

Addressing RQ3: Are some PFs Exploited Together or All of them are Exploited Randomly? We use the Pearson correlation [55], denoted by r, to quantify the relationships between the 20 PFs based on the 1,260 emails (i.e., we treat the dataset as a whole without considering the years).

Figure 5 presents the result, from which we make two observations. First, some PFs are more likely exploited together by malicious emails (i.e., positive correlation). The highest correlation is between COGNITIVE MISER (an Inherent PF) and AUTHORITY (a Social PF), with $r = 0.91$; followed by the correlation between COGNITIVE MISER and TRUST (both are Inherent PFs), with $r = 0.89$, then by the correlation between COGNITIVE MISER and WORKLOAD (both are Situational PFs), with $r = 0.84$. Moreover, the correlation between TRUST and AUTHORITY is $r = 0.79$, the correlation between TRUST and WORKLOAD is $r = 0.76$, and the correlation between AUTHORITY and WORKLOAD is $r = 0.73$. That is, COGNITIVE MISER, AUTHORITY, TRUST, and WORKLOAD are often exploited together by malicious emails. This sheds light on designing effective defense: A defense should cope with the PFs that are often exploited together.

	Impuls	Trust	Curios	Autho	Cognit	Indivi	Defen	Comm	Workl	Greed	Liking	Neglig	Scarci	Loneli	Lazine	Symp	Social	Vigila	Exper	Recip
Impulsivity	1	0.688	-0.13	0.629	0.637	0.308	0.469	0.439	0.688	-0.54	0.259	0.517	0.111	-0.59	-0.28	-0.2	-0.56	-0.66	0.119	-0.48
Trust	0.688	1	-0.61	0.787	0.891	0.382	0.351	0.26	0.76	-0.7	-0.03	0.508	0.293	-0.6	-0.06	-0.62	-0.26	-0.28	-0.01	-0.01
Curiosity	-0.13	-0.61	1	-0.67	-0.66	-0.06	0.275	0.092	-0.46	0.506	0.482	-0.18	-0.55	0.412	-0.08	0.502	-0.04	-0.21	-0.08	-0.28
Authority	0.629	0.787	0.67	1	0.911	0.211	0.116	0.309	0.733	-0.62	0.103	0.41	0.304	-0.61	-0.28	-0.34	-0.32	-0.38	0.019	-0.1
Cognitive miser	0.637	0.891	-0.66	0.911	1	0.419	0.322	0.304	0.836	-0.73	0.036	0.54	0.185	-0.67	-0.1	-0.61	-0.32	0.28	-0.12	-0.06
Individual Diff	0.308	0.382	-0.06	0.211	0.419	1	0.454	0.293	0.606	-0.47	0.151	0.567	-0.42	-0.36	0.012	-0.26	-0.11	-0.27	-0.37	-0.21
Defenselessness	0.469	0.351	0.275	0.116	0.322	0.454	1	0.65	0.388	-0.51	0.522	0.584	-0.39	-0.44	-0.23	-0.4	-0.39	0.4	-0.32	-0.19
Commitment	0.439	0.26	0.092	0.309	0.304	0.293	0.65	1	0.298	-0.39	0.673	0.413	-0.15	-0.41	-0.36	-0.13	-0.21	-0.57	0.065	-0.1
Workload	0.688	0.76	-0.46	0.733	0.836	0.606	0.388	0.298	1	-0.69	-0.04	0.687	-0.11	-0.69	-0.25	-0.47	-0.34	-0.37	-0.21	-0.26
Greed	-0.54	-0.7	0.506	-0.62	-0.73	-0.47	-0.51	-0.39	-0.69	1	0.027	-0.47	-0.14	0.632	0.251	0.562	0.329	0.401	0.232	0.248
Liking	0.259	-0.03	0.482	0.103	0.036	0.151	0.522	0.673	-0.04	0.027	1	0.183	-0.39	-0.08	-0.31	0.183	-0.16	-0.46	-0.08	-0.1
Negligence	0.517	0.508	-0.18	0.41	0.54	0.567	0.584	0.413	0.687	-0.47	0.183	1	-0.28	-0.48	-0.02	0.5	-0.4	-0.26	-0.37	-0.07
Scarcity	0.111	0.293	-0.55	0.304	0.185	-0.42	-0.39	-0.15	-0.11	-0.14	-0.39	-0.28	1	-0.06	0.229	-0.07	-0.09	0.081	0.468	0.232
Loneliness	-0.59	-0.6	0.412	-0.61	-0.67	-0.36	-0.44	-0.41	-0.69	0.632	-0.08	-0.48	-0.06	1	0.161	0.411	0.582	0.081	0.181	0.191
Laziness	-0.28	-0.06	-0.08	-0.28	-0.1	0.012	-0.23	-0.36	-0.25	0.251	-0.31	-0.02	0.229	0.161	1	-0.23	0.093	0.542	-0.14	0.277
Sympathy	-0.2	-0.62	0.502	-0.34	-0.61	0.26	0.4	-0.13	-0.47	0.562	0.183	0.5	-0.07	0.411	-0.23	1	0.154	-0.21	0.225	-0.17
Social Proof	0.56	-0.26	-0.04	-0.32	-0.32	-0.11	-0.39	-0.21	-0.34	0.329	-0.16	-0.4	-0.09	0.582	0.093	0.154	1	0.212	0.319	0.349
Vigilance	-0.66	-0.28	-0.21	-0.38	0.28	-0.27	-0.4	-0.57	-0.37	0.401	-0.46	-0.26	0.081	0.081	0.542	-0.21	0.212	1	-0.18	0.566
Expertise	0.119	-0.01	-0.08	0.019	-0.12	-0.37	-0.32	0.065	-0.21	0.232	-0.08	-0.37	0.468	0.181	-0.14	0.225	0.319	-0.18	1	-0.09

Fig. 5. Correlation between the PFs, where positive corrections are highlighted in green (with a greener color indicating a higher positive correlation) and negative corrections are highlighted in red (with a deeper red indicating a higher negative coefficient). (Color figure online)

Second, some PFs are less likely exploited together by malicious emails (i.e., negative correlation). The highest negative correlation is the correlation between COGNITIVE MISER and GREED, with $r = -0.73$, followed by the correlation between TRUST and GREED, with $r = -0.7$, then by the correlation between GREED and WORKLOAD, with $r = -0.69$, then by the correlation between AUTHORITY and CURIOSITY, with $r = -0.67$. This can be explained as follows: GREED is an action that the recipient has likely thought about the potential gains in scenarios such as the Nigerian Prince scam; in these scenarios, neither of COGNITIVE MISER, TRUST, and WORKLOAD would be relevant. Similarly, an individual with a high CURIOSITY would (for example) likely click on the link presented in a malicious email without being persuaded by an entity of AUTHORITY.

Insight 4. *While* IMPULSIVITY *is the most exploited PF,* COGNITIVE MISER, AUTHORITY, TRUST, *and* WORKLOAD *are most often exploited together. This means that future defense should cope with not only the PFs it targets but also the other PFs that are often exploited together with the target PFs.*

5   Limitations

The present study has three limitations. First, the study is time-consuming in identifying the PFs that are exploited by malicious malicious, forcing the grading work to be done by one PhD student, meaning that the results may be biased. However, the methodology can be equally applied by multiple researchers to reduce such biases. Second, even though the present study is time-consuming, the sample we analyzed (i.e., 1,260 malicious emails over 21 years span) is small compared to the large number of malicious emails on a yearly basis. Future studies need to consider a much larger sample. Third, our refinement of the 46 PFs into the 20 PFs in three categories (i.e., Inherent, Social, and Situational PFs) might not be perfect and may oversimplify nuances between some PFs. It is an open problem to determine what would be the smallest set of PFs that would be necessary and sufficient.

6   Conclusion

Attackers have been exploiting human PFs (Psychological Factors) to wage cyber social engineering attacks including malicious emails. We presented a methodology for understanding the evolution of PFs exploited by malicious emails. We conducted a case study by applying the methodology to 1,260 malicious emails during the span of 21 years (2004–2024). The case study has led to a number of useful insights, which shed light on how future defenses may be designed (e.g., defenses should adequately deal with the exploitation of the PFs that are increasingly exploited by attackers, and one defense should be able to cope with a set of PFs that are often exploited together).

The aforementioned limitations of the present study represents exciting open problems for future research. In addition, we highlight the following research directions. First, it is important to study the evolution of PFs exploited by malicious websites (e.g., [49–51,65,66]), online social networks, and attacks targeting specific sectors (e.g., the healthcare sector). Second, it is important to study mathematical, statistical, and machine learning models to forecast the evolution of PFs, as well as PTechs and PTacs, in a fashion similar to [13,14,47,48,57,58,67,68,78,79]. The resulting forecasting capability would allow us to design adaptive and proactive defense mechanisms (e.g., leveraging the PFs, PTechs, and PTacs that are predicted to be exploited by attackers). Third, it is important to define metrics to quantify the susceptibility of humans to these attacks [6,7,11,40,46,76]. Fourth, the results obtained in the preceding research directions would formulate a body of new knowledge that will become an integral component of the Cybersecurity Dynamics framework [70,74,75], which can already accommodate types of cyber social engineering attacks (e.g., malicious websites-incurred drive-by download or "pull-based" attacks [21,30,71,72,82]). Systematically incorporating cyber social engineering attacks and the associated psychological aspects will lead to the modeling and accommodation of humans in advanced mathematical models of preventive, reactive, adaptive, proactive, and active cyber defenses (see, e.g., [20,29,35,69,73,81]).

Acknowledgement. We thank the reviewers for their comments. This research was supported in part by NSF Grant #2115134 and Colorado State Bill 18-086. This research work is also a contribution to the International Alliance for Strengthening Cybersecurity and Privacy in Healthcare (CybAlliance, Project no. 337316).

References

1. Aleroud, A., Zhou, L.: Phishing environments, techniques, and countermeasures: a survey. Comput. Secur. **68**, 160–196 (2017)
2. APWG: Phishing activity trend report 4th quarter. Technical report, APWG (2022)
3. APWG: Phishing activity trends report - unifying the global response to cybercrime. Technical report, Anti-Phishing Working Group, APWG (2023)
4. Cantillo, V., Amaya, J., Ortúzar, J.D.D.: Thresholds & indifference in stated choice surveys. Transp. Res. Part B: Methodol. **44**(6), 753–763 (2010)
5. Chiew, K.L., Yong, K.S.C., Tan, C.L.: A survey of phishing attacks: their types, vectors and technical approaches. Expert Syst. Appl. **106**, 1–20 (2018)
6. Cho, J., Hurley, P., Xu, S.: Metrics and measurement of trustworthy systems. In: Proceedings of IEEE MILCOM (2016)
7. Cho, J.H., Xu, S., Hurley, P.M., Mackay, M., Benjamin, T., Beaumont, M.: Stram: measuring the trustworthiness of computer-based systems. ACM Comput. Surv. **51**(6), 128:1–128:47 (2019)
8. Dai, Y., Li, H., Xie, W., Deng, T.: Power distance belief and workplace communication: the mediating role of fear of authority. Int. J. Environ. Res. Public Health **19**(5), 2932 (2022)
9. Das, A., Baki, S., El Aassal, A., Verma, R., Dunbar, A.: SoK: a comprehensive reexamination of phishing research from the security perspective. IEEE Commun. Surv. Tutorials **22**(1), 671–708 (2019)
10. Deutrom, J., Katos, V., Ali, R.: Loneliness, life satisfaction, problematic internet use and security behaviours: re-examining the relationships when working from home during covid-19. Behav. Inf. Technol. **1**(1), 1–15 (2021)
11. Du, P., Sun, Z., Chen, H., Cho, J.H., Xu, S.: Statistical estimation of malware detection metrics in the absence of ground truth. IEEE T-IFS **13**(12), 2965–2980 (2018)
12. Du, Y., Xue, F.: Research of the anti-phishing technology based on e-mail extraction and analysis. In: 2013 International Conference on Information Science and Cloud Computing Companion, pp. 60–65. IEEE (2013)
13. Fang, X., Xu, M., Xu, S., Zhao, P.: A deep learning framework for predicting cyber attacks rates. EURASIP J. Inf. Secur. **2019**, 5 (2019)
14. Fang, Z., Xu, M., Xu, S., Hu, T.: A framework for predicting data breach risk: leveraging dependence to cope with sparsity. IEEE T-IFS **16**, 2186–2201 (2021)
15. Ferreira, A., Coventry, L., Lenzini, G.: Principles of persuasion in social engineering and their use in phishing. In: International Conference on Human Aspects of Information Security, Privacy, and Trust, pp. 36–47. Springer (2015)
16. Ferreira, A., Lenzini, G.: An analysis of social engineering principles in effective phishing. In: Workshop on Socio-Technical Aspects in Security and Trust (2015)
17. Frauenstein, E.D., Flowerday, S.: Susceptibility to phishing on social network sites: a personality information processing model. Comput. Secur. (2020)
18. Gallo, L., Maiello, A., Botta, A., Ventre, G.: 2 years in the anti-phishing group of a large company. Comput. Secur. **105**, 102259 (2021)
19. Greitzer, F.L., Li, W., Laskey, K.B., Lee, J., Purl, J.: Experimental investigation of technical and human factors related to phishing susceptibility. ACM Trans. Soc. Comput. **4**(2), 1–48 (2021)
20. Han, Y., Lu, W., Xu, S.: Characterizing the power of moving target defense via cyber epidemic dynamics. In: HotSoS, pp. 1–12 (2014)

21. Han, Y., Lu, W., Xu, S.: Preventive and reactive cyber defense dynamics with ergodic time-dependent parameters is globally attractive. IEEE TNSE **8**(3), 2517–2532 (2021)
22. He, D., Lv, X., Xu, X., Chan, S., Choo, K.K.R.: Double-layer detection of internal threat in enterprise systems based on deep learning. IEEE Trans. Inf. Forensics Secur. (2024)
23. He, D., Lv, X., Zhu, S., Chan, S., Choo, K.K.R.: A method for detecting phishing websites based on tiny-bert stacking. IEEE Internet Things J. (2023)
24. Jampen, D., Gür, G., Sutter, T., Tellenbach, B.: Don't click: towards an effective anti-phishing training. A comparative literature review. Human-centric Comput. Inf. Sci. **10**(1), 1–41 (2020)
25. Kano, Y., Nakajima, T.: Trust factors of social engineering attacks on social networking services. In: 2021 IEEE 3rd Global Conference on Life Sciences and Technologies (LifeTech), pp. 25–28. IEEE, Osaka, Japan (2021)
26. Khetarpal, M., Singh, S.: "limited time offer": impact of time scarcity messages on consumer's impulse purchase. J. Promot. Manage. **30**(2), 282–301 (2024)
27. Kirlappos, I., Parkin, S., Sasse, M.A.: Learning from "shadow security": why understanding non-compliance provides the basis for effective security. In: Workshop on Usable Security, pp. 1–10 (2014)
28. Klimburg-Witjes, N., Wentland, A.: Hacking humans? social engineering and the construction of the "deficient user" in cybersecurity discourses. Sci. Technol. Hum. Values **46**(6), 1316–1339 (2021)
29. Li, X., Parker, P., Xu, S.: A stochastic model for quantitative security analyses of networked systems. IEEE TDSC **8**(1), 28–43 (2011)
30. Lin, Z., Lu, W., Xu, S.: Unified preventive and reactive cyber defense dynamics is still globally convergent. IEEE/ACM ToN **27**(3), 1098–1111 (2019)
31. Livara, A., Hernandez, R.: An empirical analysis of machine learning techniques in phishing e-mail detection. In: 2022 International Conference for Advancement in Technology (ICONAT), pp. 1–6. IEEE (2022)
32. Longtchi, T., Rodriguez, R.M., Al-Shawaf, L., Atyabi, A., Xu, S.: Sok: why have defenses against social engineering attacks achieved limited success? arXiv preprint arXiv:2203.08302 (2022)
33. Longtchi, T., Xu, S.: Characterizing the evolution of psychological tactics and techniques exploited by malicious emails. In: Proceedings of International Conference on Science of Cyber Security (SciSec 2024) (2024)
34. Longtchi, T.T., Rodriguez, R.M., Al-Shawaf, L., Atyabi, A., Xu, S.: Internet-based social engineering psychology, attacks, and defenses: a survey. Proc. IEEE (2024)
35. Lu, W., Xu, S., Yi, X.: Optimizing active cyber defense dynamics. In: Proceedings of GameSec 2013, pp. 206–225 (2013)
36. Marengo, D., Davis, K.L., Gradwohl, G.Ö., Montag, C.: A meta-analysis on individual differences in primary emotional systems and big five personality traits. Sci. Rep. **11**(1), 7453 (2021)
37. Mashtalyar, N., Ntaganzwa, U.N., Santos, T., Hakak, S., Ray, S.: Social engineering attacks: recent advances and challenges. In: International Conference on Human-Computer Interaction, pp. 417–431. Springer (2021)
38. McAlaney, J., Benson, V.: Cybersecurity as a social phenomenon. In: Cyber Influence and Cognitive Threats, pp. 1–8. Elsevier, USA (2020)
39. Mihelič, A., Žvanut, B.: (in) secure smart device use among senior citizens. IEEE Secur. Priv. **20**(1), 62–71 (2021)
40. Mireles, J., Ficke, E., Cho, J., Hurley, P., Xu, S.: Metrics towards measuring cyber agility. IEEE Trans. Inf. Forensics Secur. **14**(12), 3217–3232 (2019)

41. Montañez, R., Golob, E., Xu, S.: Human cognition through the lens of social engineering cyberattacks. Front. Psychol. **11**, 1755 (2020)
42. Montañez Rodriguez, R., et al.: Quantifying psychological sophistication of malicious emails. In: International Conference on Science of Cyber Security, pp. 319–331. Springer (2023)
43. Montañez Rodriguez, R., Xu, S.: Cyber social engineering kill chain. In: Science of Cyber Security: 4th International Conference, SciSec 2022, Matsue, Japan, August 10–12, 2022, Revised Selected Papers, pp. 487–504. Springer (2022)
44. Ndibwile, J.D., Luhanga, E.T., Fall, D., Miyamoto, D., Blanc, G., Kadobayashi, Y.: An empirical approach to phishing countermeasures through smart glasses and validation agents. IEEE Access **7**, 130758–130771 (2019)
45. Parsons, K., Butavicius, M., Delfabbro, P., Lillie, M.: Predicting susceptibility to social influence in phishing emails. Int. J. Hum Comput Stud. **128**, 17–26 (2019)
46. Pendleton, M., Garcia-Lebron, R., Cho, J.H., Xu, S.: A survey on systems security metrics. ACM Comput. Surv. **49**(4), 62:1–62:35 (2016)
47. Peng, C., Xu, M., Xu, S., Hu, T.: Modeling and predicting extreme cyber attack rates via marked point processes. J. Appl. Stat. **44**(14), 2534–2563 (2017)
48. Peng, C., Xu, M., Xu, S., Hu, T.: Modeling multivariate cybersecurity risks. J. Appl. Stat. 1–23 (2018)
49. Pritom, M., Schweitzer, K., Bateman, R., Xu, M., Xu, S.: Characterizing the landscape of COVID-19 themed cyberattacks and defenses. In: IEEE ISI'2020 (2020)
50. Pritom, M., Schweitzer, K., Bateman, R., Xu, M., Xu, S.: Data-driven characterization and detection of covid-19 themed malicious websites. In: IEEE ISI'2020 (2020)
51. Pritom, M., Xu, S.: Supporting law-enforcement to cope with blacklisted websites: framework and case study. In: IEEE CNS'2022 (2022)
52. Rose, C.L., Murphy, L.B., Byard, L., Nikzad, K.: The role of the big five personality factors in vigilance performance and workload. Eur. J. Pers. **16**(3), 185–200 (2002)
53. Rozgonjuk, D., Sindermann, C., Elhai, J.D., Montag, C.: Individual differences in fear of missing out (FoMO): age, gender, and the big five personality trait domains, facets, and items. Pers. Individ. Differ. **171**, 110546 (2021)
54. SAS, H.: Scamdoc.com. https://www.scamdoc.com/. Accessed 04 Nov 2023
55. Schober, P., Boer, C., Schwarte, L.A.: Correlation coefficients: appropriate use and interpretation. Anesth. Analg. **126**(5), 1763–1768 (2018)
56. Stajano, F., Wilson, P.: Understanding scam victims: seven principles for systems security. Commun. ACM **54**(3), 70–75 (2011)
57. Sun, Z., Xu, M., Schweitzer, K., Bateman, R., Kott, A., Xu, S.: Cyber attacks against enterprise networks: characterization, modeling and forecasting. In: Proceedings of SciSec'2023 (2023)
58. Trieu-Do, V., Garcia-Lebron, R., Xu, M., Xu, S., Feng, Y.: Characterizing and leveraging granger causality in cybersecurity: framework and case study. EAI Endorsed Trans. Secur. Saf. **7**(25), e4 (2020)
59. Tu, H., Doupé, A., Zhao, Z., Ahn, G.J.: Users really do answer telephone scams. In: 28th USENIX Security Symposium Security 19), pp. 1327–1340 (2019)
60. Uebelacker, S., Quiel, S.: The social engineering personality framework. In: 2014 Workshop on Socio-Technical Aspects in Security and Trust, pp. 24–30 (2014)
61. Van Der Heijden, A., Allodi, L.: Cognitive triaging of phishing attacks. In: 28th USENIX Security Symposium 2019), pp. 1309–1326 (2019)
62. Vishwanath, A., Herath, T., Chen, R., Wang, J., Rao, H.R.: Why do people get phished? Decis. Support Syst. **51**(3), 576–586 (2011)

63. Wang, J., Herath, T., Chen, R., Vishwanath, A., Rao, H.R.: Research article phishing susceptibility: an investigation into the processing of a targeted spear phishing email. IEEE Trans. Prof. Commun. **55**(4), 345–362 (2012)
64. Wang, Z., Zhu, H., Sun, L.: Social engineering in cybersecurity: effect mechanisms, human vulnerabilities and attack methods. IEEE Access **9**, 11895–11910 (2021)
65. Xu, L., Zhan, Z., Xu, S., Ye, K.: An evasion and counter-evasion study in malicious websites detection. In: IEEE CNS, pp. 265–273 (2014)
66. Xu, L., Zhan, Z., Xu, S., Ye, K.: Cross-layer detection of malicious websites. In: ACM CODASPY 2013, pp. 141–152 (2013)
67. Xu, M., Hua, L., Xu, S.: A vine copula model for predicting the effectiveness of cyber defense early-warning. Technometrics **59**(4), 508–520 (2017)
68. Xu, M., Schweitzer, K.M., Bateman, R.M., Xu, S.: Modeling and predicting cyber hacking breaches. IEEE T-IFS **13**(11), 2856–2871 (2018)
69. Xu, M., Xu, S.: An extended stochastic model for quantitative security analysis of networked systems. Internet Math. **8**(3), 288–320 (2012)
70. Xu, S.: The cybersecurity dynamics way of thinking and landscape (invited paper). In: ACM Workshop on Moving Target Defense (2020)
71. Xu, S., Lu, W., Xu, L.: Push- and pull-based epidemic spreading in networks: thresholds and deeper insights. ACM TAAS **7**(3) (2012)
72. Xu, S., Lu, W., Xu, L., Zhan, Z.: Adaptive epidemic dynamics in networks: thresholds and control. ACM TAAS **8**(4) (2014)
73. Xu, S., Lu, W., Zhan, Z.: A stochastic model of multivirus dynamics. IEEE Trans. Dependable Secure Comput. **9**(1), 30–45 (2012)
74. Xu, S.: Cybersecurity dynamics. In: Proceedings of Symposium on the Science of Security (HotSoS 2014), pp. 14:1–14:2 (2014)
75. Xu, S.: Cybersecurity dynamics: a foundation for the science of cybersecurity. In: Proactive and Dynamic Network Defense, vol. 74, pp. 1–31. Springer (2019)
76. Xu, S.: SARR: a cybersecurity metrics and quantification framework. In: Third International Conference on Science of Cyber Security (SciSec 2021), pp. 3–17 (2021)
77. Zafar, H., Randolph, A., Gupta, S., Hollingsworth, C.: Traditional seta no more: investigating the intersection between cybersecurity and cognitive neuroscience. In: Proceedings of the 52nd Hawaii International Conference on System Sciences. pp. –. ScholarSpace/AIS Electronic Library (AISeL), USA (2019)
78. Zhan, Z., Xu, M., Xu, S.: Characterizing honeypot-captured cyber attacks: statistical framework and case study. IEEE T-IFS **8**(11), 1775–1789 (2013)
79. Zhan, Z., Xu, M., Xu, S.: Predicting cyber attack rates with extreme values. IEEE Trans. Inf. Forensics Secur. **10**(8), 1666–1677 (2015)
80. Zhang, Z., Jiménez, F.R., Cicala, J.E.: Fear of missing out scale: a self-concept perspective. Psychol. Mark. **37**(11), 1619–1634 (2020)
81. Zheng, R., Lu, W., Xu, S.: Active cyber defense dynamics exhibiting rich phenomena. In: Proceedings of HotSoS (2015)
82. Zheng, R., Lu, W., Xu, S.: Preventive and reactive cyber defense dynamics is globally stable. IEEE TNSE **5**(2), 156–170 (2018)
83. Zieni, R., Massari, L., Calzarossa, M.C.: Phishing or not phishing? A survey on the detection of phishing websites. IEEE Access **11**, 18499–18519 (2023)

An Enhanced Firewall for IoT Security

Sai Veerya Mahadevan[1]([⊠]), Yuuki Takano[2], and Atsuko Miyaji[1,3]

[1] Osaka University, Osaka, Japan
saiveerya@cy2sec.comm.eng.osaka-u.ac.jp, saiveeryamahadevan@gmail.com ,
miyaji@comm.eng.osaka-u.ac.jp
[2] Tier IV Inc., Tokyo, Japan
[3] Japan Advanced Institute of Science and Technology, Ishikawa, Japan

Abstract. The Message Queuing Telemetry Transport (MQTT) protocol is a machine-to-machine communication protocol widely used in Internet of Things (IoT) applications due to its lightweight nature. However, ensuring security in such systems remains a challenge due to the resource-constrained environments it operates in. This paper presents the design and implementation of a packet filtering firewall for MQTT traffic by detecting and filtering packets early in the packet processing pipeline using a Linux feature called the extended Berkeley Packet Filter (eBPF) with established performance implications.

Keywords: firewall · MQTT · XDP · eBPF · IoT Security

1 Introduction

The Internet of Things (IoT), is no more an emerging technological term and now holds significant [25] relevance worldwide. With increasing adoption of IoT devices for a range of application domains, there is an increased need to understand the communication protocols that govern these systems as they are adapted to various resource-constrained environments.

Among the various protocols used [14], MQTT (Message Queuing Telemetry Transport) [5] is a lightweight, session layer, publish-subscribe messaging protocol commonly used for machine-to-machine communication in low bandwidth environments. It follows a client-server architecture with a central message broker that handles the distribution of messages. With the expansion of the IoT device landscape, there is a natural consequent requirement for increased IoT security [8] as well, especially in critical applications and services at risk like healthcare and renewable energy industry. Key challenges include authentication and physical threats to IoT devices, confidentiality risks during data transmission, data integrity issues from attacks like DoS/DDoS, and privacy violations from unauthorized access to user information [8]. These multi-layered security concerns, combined with the complexity of integrating diverse IoT technologies and standards, necessitate robust cybersecurity solutions.

© The Author(s), under exclusive license to Springer Nature Singapore Pte Ltd. 2025
J. Zhao and W. Meng (Eds.): SciSec 2024, LNCS 15441, pp. 179–195, 2025.
https://doi.org/10.1007/978-981-96-2417-1_10

IoT firewalls are usually implemented as solutions to these challenges at the industry level, using commercial vendor solutions. These firewalls integrate functions like packet filtering and inspection with Intrusion Detection and Prevention Systems (IDPS) that monitor network traffic, detect threats and protect the confidentiality, integrity, and availability of IoT based smart environments. However, addressing the security challenges requires a lightweight, high-performance IDS tailored for the resource constraints and real-time requirements of IoT systems. Given the resource constraints, security often needs to be evaluated as a trade off. As a consequence, even with the research trends and summaries of Machine-learning based IDS solutions [3,9,10] emerging as better solutions to traditional signature based IDSes and anomaly based IDSes [22], there is a need to assess if these can be deployed from a practical standpoint and assess their efficacy accordingly.

In a bid to prioritize performance benefits while also maintaining a security standpoint, a Linux feature called the extended Berkeley Packet Filter (eBPF) [13] is being widely adopted by companies such as Facebook, Netflix, Google to manage very heavy loads of network traffic in different capacities. eBPF provides powerful capabilities for monitoring and controlling network traffic, allowing enforcement of security policies in real-time. Some key use cases include network filtering, DDoS protection, container network security, load balancing, traffic control [20]. The ability to run eBPF bytecode at various hook points in the kernel's network data path, combined with its efficiency and programmability, make it a powerful tool for implementing a wide range of network security and traffic management solutions without modifying the kernel source code.

In this paper, we propose a packet filtering based firewall system that leverages the eBPF's XDP [24] and TC hooks [24] from eBPF to address the challenges of traditional signature based IDS systems like Snort [21] and Suricata [16] and enhance the security of MQTT-based IoT systems, a space that is yet to be fully explored. The rest of this paper is structured as follows: Sect. 2 presents the preliminaries of the MQTT protocol, eBPF and the traditional IDS systems Snort and Suricata. 3 presents a summary of related work and where this research fits in. Section 4 presents the architecture of our proposed packet filtering system. Section 5 presents the functional evaluation and Sect. 6 discusses the work in progress and future work. Section 7 concludes the paper.

2   Background and Motivation

In order to understand the difference with regular internet traffic processing, we need to highlight the preliminaries in our research.

2.1   The MQTT Protocol

As briefly described earlier, the MQTT (Message Queuing Telemetry Transport) is a lightweight, publish/subscribe messaging protocol designed for constrained devices and low-bandwidth, high-latency or unreliable networks [5]. It is widely used in IoT applications.

Packet Structure of MQTT Protocol: MQTT defines several types of packets for different purposes like connection establishment, publishing messages, subscribing to topics, etc. The basic packet structure as shown in Fig. 1, consists of:

- **Fixed Header (2 bytes):** Contains the packet type, flags, and remaining length.
- **Variable Header:** Contains packet-specific data like topic name, packet identifier, etc.
- **Payload:** Contains the application data being sent (for some packet types).

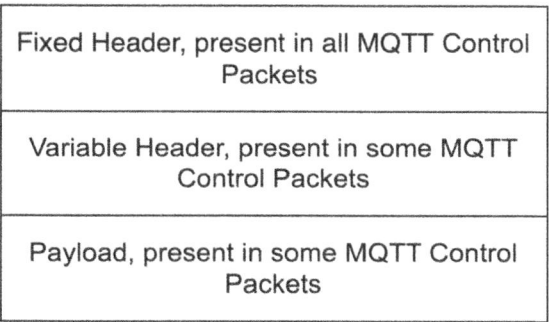

Fig. 1. MQTT Packet structure

Port Values for an MQTT Packet:

- **Unencrypted MQTT:** Uses port 1883 by default for unencrypted communication.
- **Encrypted MQTT:** Uses port 8883 by default when using SSL/TLS encryption.

The latest version is MQTT v5.0 standardized by OASIS in 2019. It introduces new features like reason codes, enhanced authentication, request/response patterns, and shared subscriptions [5]. Key MQTT Packets are CONNECT, PUBLISH, SUBSCRIBE, UNSUBSCRIBE, and PINGREQ/PINGRESP [5]. MQTT assures reliability by three Quality of Service (QoS) levels: 1) Fire and forget, 2) Delivered at least once, and 3) Delivered exactly once. MQTT also supports security; it uses username and password, which can be handled by SSL or independently by MQTT itself. However, the lightweight nature also makes it vulnerable to various security threats such as message spoofing, unauthorized access, and denial-of-service attacks. Traditional IDS solutions are often ineffective in detecting and mitigating these threats due to their reliance on signature-based detection mechanisms and limited support for protocol-specific anomalies. Therefore, there is a need for a more robust and efficient security solution to protect MQTT traffic from such threats.

2.2   Traditional IDS/IPS Tools

Snort [21] is a widely used open-source network intrusion detection and prevention system (NIDS/NIPS) that can perform real-time traffic analysis and packet logging on IP networks. It can be used for packet filtering by leveraging its rule-based detection engine, which allows users to define custom rules to detect and block malicious traffic patterns, known vulnerabilities, and other security threats. Snort supports both signature-based and anomaly-based detection techniques, making it versatile in identifying known and unknown threats.

Suricata [16] is another popular open-source NIDS/NIPS that is designed to be highly scalable and capable of handling high-speed network traffic. Like Snort, Suricata utilizes a rule-based detection engine that can be customized to filter and block malicious network traffic based on predefined rules or signatures. It supports various protocols, including MQTT, and can perform deep packet inspection and pattern matching to identify potential threats. Both Snort and Suricata can be integrated into network security solutions to provide packet filtering capabilities. They can be deployed as standalone intrusion detection systems or as part of a more comprehensive security architecture, such as a firewall or a web application firewall (WAF). In the context of MQTT-based IoT environments, Snort and Suricata can be used to inspect and filter MQTT traffic, detecting and mitigating potential threats like malformed packets, unauthorized access attempts, or malicious payloads. By defining appropriate rules or signatures, these tools can identify and block MQTT-specific attacks, such as topic hijacking, message spoofing, or denial-of-service attacks.

2.3   XDP and TC

XDP (eXpress Data Path) is the lowest layer of the Linux kernel network stack. It enables programs to be run at the earliest possible point in the network stack, before packets are processed by the kernel space. XDP is used as follows:

–   XDP programs run in the driver space, allowing them to inspect and filter packets before they enter the kernel's network stack [24].
–   XDP programs can perform actions such as dropping, forwarding, or redirecting packets based on custom filtering rules [24].
–   XDP programs are limited in functionality due to the constraints of the driver space environment [24].

More on the quantitative implications of this in Sect. 5.

TC (Traffic Control) on the other hand, operates at a higher level in the network stack, specifically in the kernel space in Layer 2 or the Data Link Layer in the OSI model [17]. TC eBPF programs can perform more complex packet processing and manipulation tasks and even handle transmitting packets or egress traffic. TC eBPF is used as follows:

- TC eBPF programs run in the kernel space, allowing them to access more kernel data structures and perform more complex operations [24].
- TC eBPF programs can filter, modify, and redirect packets based on various criteria, such as IP addresses, ports, and protocol headers [24].
- Multiple TC eBPF programs can be attached to a network interface, enabling a more modular and flexible approach to packet processing [24].
- TC eBPF programs have access to more kernel resources and can perform operations that are not possible with XDP, such as accessing socket buffers and performing complex packet modifications [24].

In summary, based on the level of packet processing needed, either XDP or TC or a combination of both with even more kernel hooks [24] can be used for early packet processing.

2.4 XDP and TC in the Context of MQTT Traffic Processing

Some advantages of applying XDP and TC hooks for packet processing include:

- **Flexibility and Granular Control**: XDP and TC provide more granular control and customization options for packet filtering rules, enabling targeted and efficient handling of MQTT traffic.
- **Reduced Overhead**: By handling packet filtering at the kernel level, XDP and TC can reduce the overhead and resource utilization compared to user-space tools that require context-switching between kernel and user space.
- **Specialized MQTT Handling**: XDP and TC can be configured to have a deeper understanding of the MQTT protocol, allowing for more effective filtering of MQTT-specific traffic. This includes:
  - Parsing MQTT packet headers and payloads.
  - Identifying MQTT control packet types (CONNECT, PUBLISH, SUB-SCRIBE, etc.).
  - Detecting MQTT protocol violations or anomalies.
  - Client fingerprinting based on connection patterns and message characteristics.
  - Advanced pattern matching on MQTT message payloads.
  - Anomaly detection based on MQTT protocol behavior.
  - Granular filtering rules targeting MQTT-related parameters (client IDs, topic names, QoS levels, etc.).

3 Related Work

This section outlines the relevant research in eBPF applications, intrusion detection systems, and IoT security, highlighting the gaps our work aims to address:

- **eBPF Applications:** The extended Berkeley Packet Filter (eBPF) is widely considered the equivalent of the Linux OS of the kernel due to its diverse range of applicability. eBPF has been used in various network monitoring and security applications, including DDoS mitigation, load balancing and network function virtualization [20].
- **eBPF-based IDS:** Recent research has explored the use of eBPF in designing and implementing Intrusion Detection Systems (IDS) in the Linux kernel [26]. Wang and Chang proposed an NIDS using eBPF in the Linux kernel alongside Snort for fast-pattern matching, claiming a maximum throughput that outperforms Snort by a factor of 3.
- **Limitations of Existing Solutions:** The work by Wang and Chang [26] does not address encrypted packet payloads and focuses on HTTP rather than MQTT traffic, making it less suitable for IoT environments. It also does not specifically target IoT security challenges.
- **MQTT Security:** Existing research has explored various approaches to securing MQTT traffic, including encryption mechanisms, access control policies and anomaly detection techniques. However, most of these solutions focus on application-layer security and may not provide adequate protection against low-level network attacks.
- **System-level Monitoring:** Research by Deri et al. [7] has focused on tools like *ntopng* that correlate network traffic with system-level metadata. While *ntopng* operates at Layer 7, our work aims to implement a lower-level packet filtering based firewall system.
- **XDP and eBPF for IoT:** The advancement of XDP and eBPF has opened up new possibilities for enhancing network security. Our proposed system leverages these technologies to perform deep packet inspection and filtering at wire speed, specifically tailored for IoT environments using MQTT protocol.

4   Architecture, Design and Implementation

Our proposed packet filtering system consists of three main components: a packet capture module, an analysis engine, and a filtering mechanism. The packet capture module intercepts incoming MQTT traffic at the network interface using XDP, allowing for high-performance packet processing with minimal overhead. The analysis engine inspects the captured packets for suspicious patterns and anomalies using eBPF programs, while the filtering mechanism selectively drops or redirects malicious traffic based on predefined rules and policies. Figure 2 illustrates the overall architecture of our proposed system before the user-space processing of MQTT packets via an MQTT broker and later on, MQTT client.

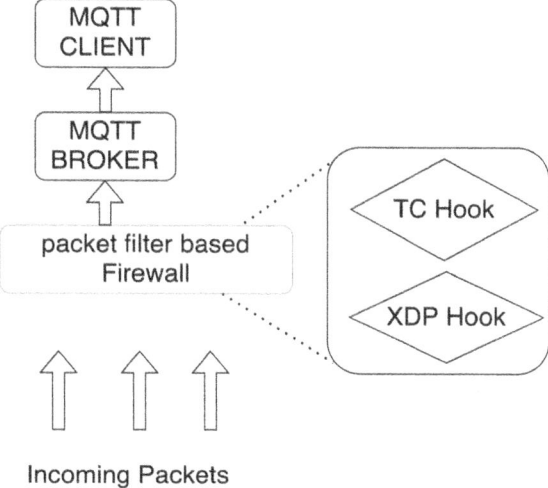

Fig. 2. Overall architecture of proposed system

4.1 Architecture and Design

The proposed system essentially apply the core ideas of fast packet processing from the paper [24], with some MQTT specific customizations. The inner workings of our architecture can be seen in Fig. 3. The incoming MQTT packets are received in the driver space, specifically in the RX queue. These packets are initially filtered by an XDP (eXpress Data Path) program attached to an XDP hook before entering the network stack. The XDP program checks the packet headers for indicators of potential malicious behavior or non-conformance with the MQTT packet format. If a packet violates the expected format, it is dropped; otherwise, it is passed up into the network stack. In the Data Link Layer (Layer 2 of the OSI model), the TC (Traffic Control) hook of eBPF is used for further packet processing. If the IP address of the packet belongs to a blocklist, it is filtered out; otherwise, it is passed up the stack. This step is performed at the TC level because it allows for more complex filtering rules and packet manipulation compared to the XDP level, which is primarily focused on early packet filtering and forwarding decisions. As MQTT is a session layer protocol, the actual MQTT packet payload can be inspected in the application layer. In this layer, Nginx [15] is used as a reverse proxy, and ModSecurity Web Application Firewall (WAF) [1] rules are applied to the captured packets. These rules filter out packets containing restricted topics, payloads with excessive size, and known signatures of malicious payloads. The IP addresses of the packets that match these criteria are stored in an eBPF map in user space, which is accessible by both the TC program hook and the XDP program hook. Packets continue to be filtered out as the XDP and TC programs attached to the network interface constantly check the eBPF maps in user space for updates.

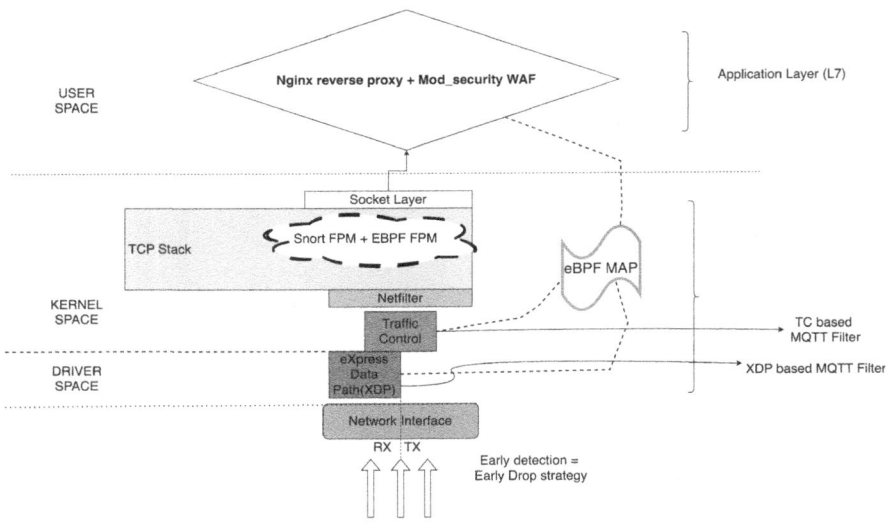

Fig. 3. Inner workings of our system, with previous research highlighted for positioning our research

4.2   Implementation Method and Algorithm

This entire flow is described in the flowchart in Fig. 4.

Level 1:XDP Filtering. The XDP program is compiled, accepted by the eBPF verifier (a static, in-kernel verifier [12]) and then attached to a network interface (e.g., eno3) directly using the 'ip link' command. Although the actual application payload cannot be inspected at the XDP level, the headers of the incoming packet at the interface can be checked to decide if it should be passed up the stack. Assuming the MQTT packets are over the TCP protocol, the following checks are performed:

1. Ethernet header, IP header, and TCP header:
   (a) If the Ethernet header is incomplete, drop the packet using XDP_DROP.
       i. If the Ethernet header protocol is not IPv4, drop the packet.
   (b) If the IP header is incomplete, drop the packet.
       i. If the IP header protocol is not TCP, drop the packet.
   (c) If the TCP header is incomplete, drop the packet.
       i. If the destination port is not 1883 or 8883 (SSL/TLS encrypted), drop the packet.
   (d) Based on the format of the MQTT packet data in the parsed dataset, an optional filter can be added to obtain the payload length offset and check if the MQTT payload length is within a reasonable range (customizable based on preference, as very large payloads are suspicious, and fragmented and sequenced packets are not intended to be handled at this point); otherwise, drop the packet.

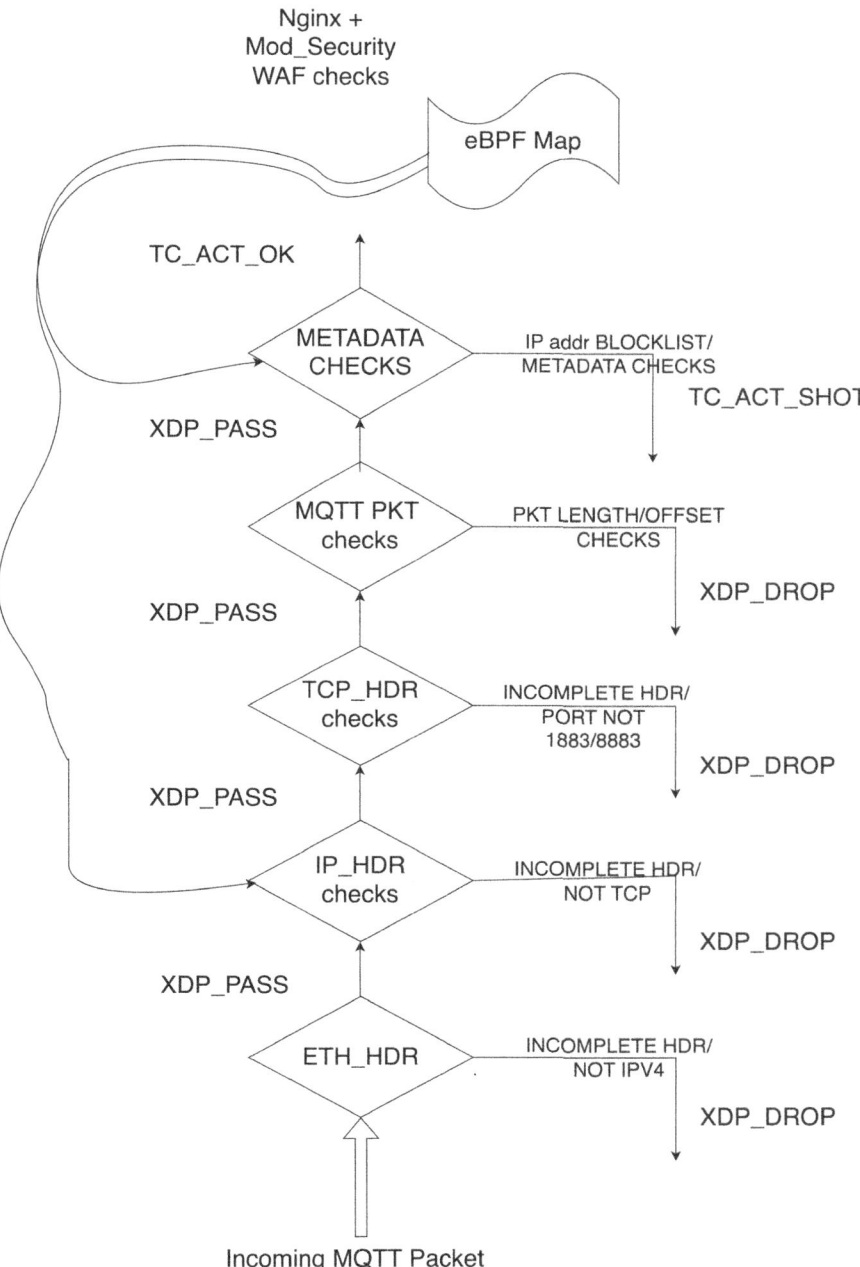

Fig. 4. Flow of MQTT packets inside our system

Level 2 and Level 3: TC Filtering and User-Space Collaboration. After XDP based filtering, at the TC level, the packets that arrive are checked. As multiple TC programs can be used (as opposed to the XDP program, which can only be attached once to an interface [18]), the logic of the TC program can be framed based on the metadata to be manipulated. Packets are filtered at the TC level by accessing the socket buffer (__sk_buff). In the TC program, a simple check is performed by comparing the IP address of the packet against a static array of IP addresses in a blocklist and dropping packets using TC_ACT_SHOT if needed; otherwise, the packet is allowed to pass up the stack using TC_ACT_OK. When the packets finally reach user space, Nginx [15] acts as a reverse proxy, intercepting MQTT traffic on port 1883 and forwarding it to the MQTT broker at 10.0.0.8:1883. The modsecurity_rules_file directive [1] specifies the path to the file containing ModSecurity rules for MQTT filtering. By using Nginx as a reverse proxy, ModSecurity's rule engine can be integrated with the MQTT traffic passed up from the kernel via XDP_PASS, TC (eBPF), or even if its redirected via XDP_REDIRECT to an AF_XDP socket for direct user-space processing, bypassing the kernel stack [24]. Here, the packet payload is inspected to identify malicious payloads based on topics, anomalies detected, and other criteria. The IP addresses of the identified malicious packets are stored in an eBPF map in user space. This eBPF map is continuously looked up via bpf_map_lookup_element() from the lower-layer XDP and TC programs. The detailed explanation and relevance of program structure and functions used, checks needed by the eBPF verifier, command-line commands used to attach programs to an interface, the difference between TC and XDP, available map types, and user-space to kernel-space interaction are described in depth across various resources [24], [6,18,20].

5   Evaluation

There are several kinds of evaluation that must be done to decide the scope of our work. We broadly divide our evaluation into three categories. 1. Qualitative Evaluation 2. Functional Evaluation 3. Quantitative Evaluation .

5.1   Qualitative Evaluation

First, we determine the factors that need to be considered when choosing between traditional IDS systems like Snort, Suricata or earlier research work for processing MQTT traffic, as opposed to an eBPF based system and our proposal in particular. A summary of these points is given in Table 1.

5.2   Functional Evaluation

In order to evaluate our proposal, we test it on publicly available MQTT datasets and analyze the results.

Table 1. Comparison of Snort/Suricata and eBPF-based firewall system

Point of Comparison	Snort/Suricata	eBPF-based firewall system
Protocol Support	Extensive support for well-established protocols like HTTP and SMTP.	Flexible support for a broader range of protocols, including custom or less common ones like MQTT, through custom eBPF programs.
Performance	Can handle high traffic loads but may be resource-intensive, leading to potential latency or resource constraints in IoT or high-throughput environments. Context-switching overhead.	Operates in the kernel and driver spaces, making it efficient and lightweight, well-suited for resource-constrained IoT environments and providing lower latency.
Real-time Response	Operate in user-space, leading to some delay in response time when mitigating threats.	Can provide real-time response mechanisms, such as dropping or redirecting traffic, directly at the kernel level, reducing response times.
Protocol Complexity	May struggle with complex binary protocols like MQTT, requiring advanced protocol decoding capabilities.	Can be adapted to handle complex and custom protocols effectively as it can be programmed in a low-level language such as C.
Encrypted Traffic	May have limited visibility into encrypted traffic without decryption capabilities.	Can analyze encrypted traffic if encryption and decryption is implemented from user-space, using uprobes and uretprobes [18].

MQTT IoT IDS Dataset Overview. The MQTT IoT IDS dataset utilized in this research was sourced from the IEEE DataPort repository [11]. This dataset, named MQTT-IoT-IDS2020, consists of packet capture (pcap) files generated from a simulated MQTT network architecture. This architecture comprises twelve sensors, a broker, a simulated camera, and an attacker node. Five distinct scenarios are recorded within the dataset:

1. Normal operation,
2. Aggressive scan,
3. UDP scan,
4. Sparta SSH brute-force attack,
5. MQTT brute-force attack

Each scenario is captured in a separate pcap file, allowing for focused analysis and evaluation. Additionally, the dataset provides a comprehensive view of network activities during both normal operation and attack scenarios.

How We Used This Dataset: Upon acquiring the raw pcap files, we simulated the network scenario using the following steps:

1. The XDP and TC programs to test were attached to the respective hooks needed in the driver and kernel spaces at specific interfaces available.
2. Using tcpreplay [23], a tool which enables network packets captured to be replayed at the same pace they entered the network, the pcap files were made to run at the assigned network interface in a sandboxed environment. This environment was hosted on a high-performance system running on an AMD EPYC 7601 32-Core Processor with 128 logical CPUs, operating at a base clock speed of 2.2 GHz and a max turbo frequency of 3.2 GHz. The system was equipped with 503 GB of RAM, using Ubuntu 20.04.6 LTS and the GNU/Linux 5.15.0-113-generic kernel.
3. Two internal counters were set up using 1. eBPF maps in the kernel space eBPF programs and 2. by explicitly printing out the count of packets dropped after the entire pcap file is entirely parsed using kernel level debugging function bpf_printk() to print to the kernel's trace_pipe.
4. The count of packets at the XDP and TC level and subsequently using the data retrieved from user-space eBPF maps, was crosschecked by using tcpdump [2] and the appropriate filters to filter non-mqtt packets or malicious packets based on IP address, packet format and other criteria mentioend earlier.
5. Although limited MQTT datasets exist, in addition to this dataset [11], we have obtained access to other MQTT datasets [4] and we are in the process of evaluating our programs against the data available.

5.3 Quantitative Evaluation

To evaluate our current system, we utilized the same MQTT-IoT-IDS2020 dataset from IEEE DataPort as used above in our functional evaluation. This dataset simulates an MQTT network with various scenarios, including normal operations and different attack types. We replayed the pcap files using tcpreplay in a high-performance sandboxed environment, with our XDP programs attached to a specific network interface. Packet counts were tracked using eBPF maps and kernel debugging functions, and cross-checked with tcpdump. This approach allowed us to assess the effectiveness of our filtering mechanisms in a controlled yet realistic MQTT network environment. While our current evaluation focuses on this dataset, we are expanding our analysis to include additional MQTT datasets and performing a more comprehensive evaluation. This expanded analysis aims to ensure a thorough and unbiased assessment of our programs' performance across diverse MQTT network scenarios, comparing it against baseline metrics such as packet processing rate, CPU utilization, memory usage, and latency. Our evaluation methodology, which is currently a work in progress, includes setting up a controlled test environment with various packet replay speeds and filtering rules, and comparing our eBPF-based system against established intrusion detection systems. This approach will allow us to rigorously assess the performance benefits of our XDP and TC programs in the context of MQTT network security. However, currently we use the metrics established in previous research to show the performance implications of using XDP over traditional filtering options [19]. The setup for our evaluation is presented in Fig. 5.

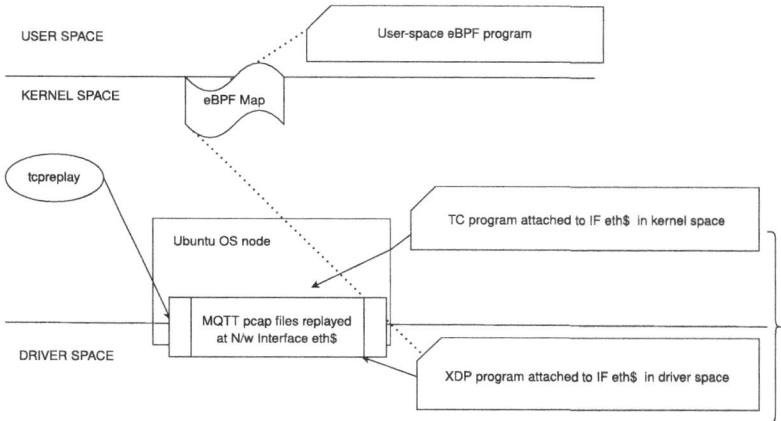

Fig. 5. Current evaluation setup of system

Advantages of XDP for Packet Filtering. This paper highlights the following key advantages of using XDP (eXpress Data Path) for packet filtering:

- **High Packet Processing Rate**: XDP can process over 14 million packets per second on a single CPU core, significantly higher than traditional kernel-based solutions like *iptables* (around 1 Mpps).
- **Low CPU Utilization**: For 10Gbps traffic, XDP utilizes only around 20% of a single CPU core compared to over 80% CPU utilization for kernel-based packet filtering.
- **Low Per-Packet Processing Overhead**: XDP has a per-packet processing overhead of only around 100 cycles, much lower than kernel-based solutions which can take over 1000 cycles per packet.
- **Reduced Tail Latency**: XDP shows lower latency outliers compared to kernel-based filtering, with 99th percentile latency being 2–3 times lower.

- **Efficient Memory Access**: XDP processes packets in a linear data stream, avoiding costly per-packet memory access and cache misses.
- **Flexible Programmability**: eBPF programs used in XDP allow implementing complex filtering logic while maintaining high performance.

In addition to the above points of quantitative evaluation, we must note that evaluation metrics may vary based on some specific conditions as well. To list a few:

1. Message Variability: MQTT messages can vary widely in structure and content, making it challenging to define universal filtering rules.
2. Real-time Processing: Filtering MQTT traffic in real-time requires efficient packet processing mechanisms to minimize latency and performance overhead.
3. Protocol Compliance: Ensuring compliance with the MQTT protocol specifications while filtering traffic is essential to prevent legitimate messages from being discarded. So a false positive analysis is due as well, based on a more diverse group of datasets.

6 Discussion

The core idea we intend to contribute is how we handle MQTT traffic. Our current packet filtering strategies at the lowest levels of the network stack are unbiased towards encrypted and non-encrypted datasets as we operate on packet headers and not payloads and hence, our results with respects to these programs would not change irrespective of packet encryption status. Also, while filtering based on headers, we ensure that packets that are encrypted prior to entering the network stack are handled by analyzing the destination port value. If transport-level encryption has been used (using TLS/SSL), typically the port is 8883. However, due to the lack of availability of encrypted MQTT datasets, in order to explicitly test this ourselves, we need to encrypt the available datasets ourselves in order to explain how we intend to handle encrypted datasets and differentiate our research from existing research that does not prioritize encryption [26]. Further, ModSecurity [1] is made to deal with encrypted payloads using pattern matching with existing malicious signatures. In this regard, it must be noted that although SSL/TLS encryption is recommended from an IoT security standpoint even for MQTT traffic, due to the low resources available, existing IDSes like Snort [21] do not encourage handling encrypted application payloads so as to not compromise on performance. However, as pointed out in the textbook [18], there is a mechanism within eBPF to follow up on tracing out the decrypted content of encrypted packets. We can do this by hooking into calls made to user space libraries such as OpenSSL or BoringSSL. An application sends unencrypted data to an OpenSSL function called SSL_write() for encryption and retrieves plaintext data by making a call to the corresponding function of SSL_read(). Hence, by attaching eBPF programs to 'uprobes' that access SSL_write() and 'uretprobes' that access SSL_read(), we are able to access plaintext data before it is encrypted and after it is decrypted.

6.1 Additional Points of Consideration in Evaluation of MQTT Datasets

We have currently not considered more complex MQTT traffic scenarios such as large or fragmented packets, implementing load balancing scenarios from a security standpoint (to avoid overwhelming any MQTT device) using eBPF programs and integration with existing IoT platforms in userspace and also, we have not considered using eBPF programs to leverage eBPF maps and store the necessary encryption keys (TLS session keys or MQTT client keys) to decrypt application payloads.

7 Conclusion

In this paper, we proposed a novel approach to enhance the security of MQTT-based IoT systems by leveraging the power of the extended Berkeley Packet Filter (eBPF) technology. Our packet filtering firewall system utilizes the XDP and TC hooks provided by eBPF to perform efficient packet inspection and filtering at various stages of the network data path. The proposed solution addresses the limitations of traditional signature-based Intrusion Detection Systems (IDS) like Snort and Suricata, which can be resource-intensive and may not be suitable for resource-constrained IoT environments. By leveraging eBPF's programmability and efficiency, our system can provide real-time security enforcement without significantly impacting system performance. Through our functional evaluation, we demonstrated the effectiveness of our approach in detecting and filtering malicious MQTT traffic. The system's ability to operate at the kernel level and process packets at line-rate speeds makes it a promising solution for securing IoT systems that require low latency and high throughput. While our work represents a significant step towards enhancing the security of MQTT-based IoT systems, there are still areas for further exploration and improvement. Future work may include integrating machine learning techniques for more advanced threat detection, exploring additional eBPF hooks for broader security coverage, and conducting comprehensive performance evaluations under various workload scenarios. Overall, our research highlights the potential of leveraging emerging technologies like eBPF to address the unique security challenges faced by IoT systems. By combining efficient packet filtering with the flexibility of programmable data paths, we can pave the way for more secure and resilient IoT deployments in critical application domains.

Acknowledgements. This work is partially supported by JSPS KAKENHI Grant Number JP21H03443 and SECOM Science and Technology Foundation.

References

1. OWASP ModSecurity (2024). https://owasp.org/www-project-modsecurity/ Accessed 01 June 2024
2. tcpdump website (2024). https://www.tcpdump.org/. Accessed 11 May 2024
3. Alasmari, R., Alhogail, A.: Protecting smart-home IoT devices from MQTT attacks: an empirical study of ML-based IDS. IEEE Access 1–1 (2024). https://doi.org/10.1109/ACCESS.2024.3367113
4. Alatram, A., Sikos, L.F., Johnstone, M., Szewczyk, P., Kang, J.J.: DoS/DDoS-MQTT-IoT: a dataset for evaluating intrusions in IoT networks using the MQTT protocol. Comput. Netw. **231**, 109809 (2023). https://doi.org/10.1016/j.comnet.2023.109809
5. Banks, A., Briggs, E., Borgendale, K., Gupta, R.: MQTT version 5.0 (2019). http://docs.oasis-open.org/mqtt/mqtt/v5.0/os/mqtt-v5.0-os.html, latest version: http://docs.oasis-open.org/mqtt/mqtt/v5.0/mqtt-v5.0.html
6. Calavera, D., Fontana, L.: Linux Observability with BPF. O'Reilly Media, Inc. (2019)
7. Deri, L., Sabella, S., Mainardi, S.: Combining system visibility and security using eBPF. In: Italian Conference on Cybersecurity (2019). https://api.semanticscholar.org/CorpusID:59616648
8. Elrawy, M.F., Awad, A.I., Hamed, H.F.A.: Intrusion detection systems for IoT-based smart environments: a survey. J. Cloud Comput. **7**(1), 21 (2018) https://doi.org/10.1186/s13677-018-0123-6
9. Hayette, Z., Mehdi, B., Chikh, R.: Securing MQTT protocol for IoT environment using IDS based on ensemble learning. Int. J. Inf. Secur. **22**, 1–12 (2023). https://doi.org/10.1007/s10207-023-00681-3
10. Hindy, H., Bayne, E., Bures, M., Atkinson, R., Tachtatzis, C., Bellekens, X.: Machine learning based IoT intrusion detection system: an MQTT case study (MQTT-IoT-IDS2020 dataset). In: International Networking Conference, pp. 104650–104675. Springer (2020)
11. Hindy, H., Tachtatzis, C., Atkinson, R., Bayne, E., Bellekens, X.: MQTT-IoT-IDS2020: MQTT internet of things intrusion detection dataset (2020). https://doi.org/10.21227/bhxy-ep04
12. Linux Kernel Documentation: eBPF verifier (2024). https://www.kernel.org/doc/. Accessed 07 May 2024
13. Linux Kernel Organization: BPF and XDP documentation (2024). https://www.kernel.org/doc/html/latest/bpf/. Accessed 01 June 2024
14. Mansour, M., et al.: Internet of things: a comprehensive overview on protocols, architectures, technologies, simulation tools, and future directions. Energies **16**(8) (2023). https://doi.org/10.3390/en16083465, https://www.mdpi.com/1996-1073/16/8/3465
15. NGINX Documentation: NGINX reverse proxy (2024). https://docs.nginx.com/nginx/admin-guide/web-server/reverse-proxy/. Accessed 01 June 2024
16. Open Information Security Foundation: Home - suricata (2023). https://suricata.io/. Accessed 01 June 2024
17. OSI Model, ISO: OSI model (2024). https://www.iso.org/about-us.html. Accessed 07 May 2024
18. Rice, L.: Learning eBPF: Programming the Linux Kernel for Enhanced Observability, Networking, and Security. O'Reilly Media, 1st edn. (2023)

19. Scholz, D., Raumer, D., Emmerich, P., Kurtz, A., Lesiak, K., Carle, G.: Performance implications of packet filtering with linux eBPF. In: 2018 30th International Teletraffic Congress (ITC 30), vol. 01, pp. 209–217 (2018). https://doi.org/10.1109/ITC30.2018.00039

20. Sharaf, H., Ahmad, I., Dimitriou, T.: Extended Berkeley packet filter: an application perspective. IEEE Access **10**, 126370–126393 (2022). https://doi.org/10.1109/ACCESS.2022.3226269

21. Snort.org: Snort (2023). https://www.snort.org/. Accessed 01 June 2024

22. Spadaccino, P., Cuomo, F.: Intrusion detection systems for IoT: opportunities and challenges offered by edge computing and machine learning. arXiv preprint arXiv:2012.01174 (2022)

23. Tcpreplay Documentation: Tcpreplay (2024). https://tcpreplay.appneta.com/wiki/tcpreplay-man.html. Accessed 01 June 2024

24. Vieira, M.A.M., Castanho, M.S., Pacífico, R.D.G., Santos, E.R.S., Júnior, E.P.M.C., Vieira, L.F.M.: Fast packet processing with eBPF and XDP: concepts, code, challenges, and applications. ACM Comput. Surv. **53**(1) (2020). https://doi.org/10.1145/3371038

25. Vishwakarma, A.K., Chaurasia, S., Kumar, K., Singh, Y.N., Chaurasia, R.: Internet of things technology, research, and challenges: a survey. Multimed. Tools Appl. (2024). https://doi.org/10.1007/s11042-024-19278-6

26. Wang, S.Y., Chang, J.C.: Design and implementation of an intrusion detection system by using extended BPF in the linux kernel. J. Netw. Comput. Appl. **198**, 103283 (2022). https://doi.org/10.1016/j.jnca.2021.103283

Exploring the Effects of Cybersecurity Awareness and Decision-Making Under Risk

Jan Hörnemann[1]([✉])[iD], Oskar Braun[1][iD], Daniel Theis[2][iD],
Norbert Pohlmann[2][iD], Tobias Urban[2][iD], and Matteo Große-Kampmann[1,3][iD]

[1] AWARE7 GmbH, Gelsenkirchen, Germany
{jan,oskar}@aware7.de
[2] Institute for Internet Security, Westphalian University of Applied Sciences,
Gelsenkirchen, Germany
{theis,pohlmann,urban}@internet-sicherheit.de
[3] Rhine-Waal University of Applied Sciences, Kamp-Lintfort, Germany
matteo.grosse-kampmann@hochschule-rhein-waal.de

Abstract. This paper challenges the conventional assumption in cybersecurity that users act as rational actors. Despite numerous technical solutions, awareness campaigns, and organizational strategies aimed at bolstering cybersecurity, these often overlook the prevalence of non-rational user behavior. Our study, involving a survey of 208 participants, empirically demonstrates this aspect. We found that a significant portion of users (55.3%) would accept a substantial risk (35%) to click on a potentially malicious link or attachment. This propensity increases to 61% when users are led to believe there is a 65% chance of facing no adverse consequences. To address this irrationality, we explored the efficacy of nudging mechanisms within email systems. Our qualitative user study revealed that incorporating a simple colored nudge in the email intably enhance the ability of users to discern malicious emails, improving decision-making accuracy by an average of 10%.

Keywords: Economics of Cybersecurity · User Behavior · Behavioral Economics

1 Introduction

Threats to internet-facing users and systems are manifold. Ransomware, spam, fraud, and malware delivery are just a few to be named. The delivery vector email is a threat to users and organizations especially. Human users are often emphasized and framed as *the last line of defense*, yet little is known about the economics of decision-making in cybersecurity. Malicious actors use different delivery vectors for various kinds of illicit activities, ranging from stealing data and compromising single machines to whole networks and compromising the privacy of victims. The victims do not recognize that they are victims of fraud because the attack is deceptive. Thus, building awareness among users is essential when protecting modern information systems. This paper aims to understand how decision-making under risk is done and how awareness measures help users improve these decisions. We do this using an online survey, in which 208 participants took part. Furthermore, we wanted to understand how users perceive the warning marker. Therefore, we conducted a qualitative user study asking 31 participants to determine if a mail is malicious or legit. We used a simple color nudge during our experiment to evaluate the effectiveness of this approach and whether this changed the participant's detection capabilities. Our results show that participants perform better regarding email classification if they are nudged in their inbox and that a misclassified email, whether false negative or false positive, is correctly classified.

In summary, we make the following contributions:

- To the best of our knowledge, this is the first work to empirically analyze behavioral economics in cybersecurity with a focus on decision-making under risk and stress.
- We show that 'stress' affects the chances that users might click on malicious emails and that awareness measures help users perceive themselves or their organization as more secure.
- We empirically show that more than half of our surveyed users would take a risk in clicking a potentially malicious link or attachment, and if they are framed to believe there is a chance that nothing will happen, the share of risk-takers rises.
- We conducted an experiment with a follow-up survey that shows that placing nudges in email inboxes helps users decide if an email is malicious or legitimate. It also seems to raise their confidence in their detection capabilities.

2   Background

Before we describe our approaches to determine the effects of cybersecurity awareness and decision-making under risk, we briefly provide the background information necessary to follow our methods.

2.1   Human User

Several papers describe the users of information processing systems as the weakest link [38], and human-centered cybersecurity is getting more and more attention nowadays. Cybersecurity decision-making is similar to other kinds of decisions, but cybersecurity decisions have distinctly other features. Security and Risk in themselves are intangible concepts, especially in the cyber domain invisible to users. As Schneier states: "*Security is both a feeling and a reality. And they are not the same*" [39]. For example, the presence of a TLS warning is often not enough to stop users from visiting a website anyway [2]. *Human Behavior and its Economics.* Behavioral Economics is the combination of psychology and economics. It considers human limitations and complications and determines what happens if these humans make decisions within a market [29]. It is important to note, that these models extend the predominant equilibrium and rational choice models [18]. Conventional decision-making, however, describes the trade-off between expected return and risk by combining risk and return calculations [27]. This result translates to the following: A decision maker in cybersecurity (user) will invest in cybersecurity if it yields a positive return under rational risks, chances, and returns. While this has various implications for the actual cost management of security [13] and investments into security [5, 7] it also has implications for cybersecurity decisions that users make. As Pfleeger and Caputo point out, security is *"rarely the primary task of those who use information infrastructure"* [33]. Kahnemann and Tversky defined prospect theory in 1991 [42]. The central argument of the prospect theory is, that humans do not have the underlying objective probabilities by which they measure gains and losses. They weigh the gains and losses (or their value) with a nonlinear transformation of these probabilities. The general assumption of loss aversion theory, as described by Tversky et al. [42], is that losses and/or disadvantages have a greater impact on preference than potential gains and advantages.

2.2   Cybersecurity Awareness

Awareness, in general, can be defined as *"knowledge that something exists, or understanding of a situation or subject at present based on information or experience"* according to the Cambridge Dictionary [6]. Cybersecurity awareness can thus be seen as the level of understanding, knowledge, and timely appreciation of cybersecurity aspects by an individual or a group. By researching websites of cybersecurity awareness providers and conducting a literature review, we manually identified the following four measures to raise awareness.

- **Live Hacking:** Live Hacking is an event format. One to two hackers perform various pre-planned attacks on stage to raise awareness of threats on the Internet. The duration ranges from 30 to 90 min.
- **Phishing Simulation Campaign:** A phishing simulation campaign is a cybersecurity activity characterized by the fact that participants are not necessarily aware of the activity. Selected employees are sent phishing emails at irregular intervals. Afterward, the type of phishing emails the employees could recognize, best or worst, is evaluated.
- **Seminar/course/workshop:** This measure involves the active development of learning content. For this, a group size of approx. 15–30 participants are advised, as all participants should actively partake. Often, this measure lasts one or more days, with the possibility of receiving a certificate or similar.
- **eLearning:** By eLearning, we mean a digital learning platform with short videos that explain various topics are available. Short tests and quizzes at the

end of the different lessons test the participants' knowledge: In and consolidate the other contents.

3   Methodology

In this section, we introduce the two user studies that we conducted in our work. We conducted an online survey (see Sect. 3.1) and an experiment with a follow-up survey (see Sect. 3.2) to obtain results about behavioral changes in cybersecurity awareness and decisions under risks. We wanted to elicit factors that contribute to cybersecurity perception and behavior accordingly. Our studies are based on the *Protection Motivation Theory* (PMT) [26]. PMT clarifies the cognitive processes that emphasize protective behavior in the event of threats. There are two approaches when facing a threat: (1) Focusing on the threat itself and (2) mitigation options (threat appraisal and coping appraisal). Based on the outcome of this assessment, humans adapt their behavior. Our survey focuses on the threat appraisal and the user study (see Sect. 3.2) on the coping appraisal.

3.1   Online User Survey

Our quantitative online user study aimed to understand how awareness measures and stress might affect risk tolerance and whether this decision is manipulated if framed otherwise. For most questions, we used 5-point Likert scales, and for the remainder, we used single and multiple-choice or open-ended questions. We provided 5-point Likert scales (ranging from "5 – Strongly Agree" to "1 – Strongly Disagree") [24] and an "I prefer not to answer" answer option. For further reference, we uploaded the survey online[1]. We conducted a pre-study ($n = 99$) in Germany. The goal of the pre-study was to understand which awareness measures users know. We used the study's results to compile a list of the four best-known awareness-raising measures (see Sect. 2.2). The participants in this sample can be seen as representatives of the general population. Based on the pre-study results, we dropped questions that were too detailed, e.g., regarding specific awareness measures that were not known to the participants in the pre-study. Further, we clarified the wording of the questions. Moreover, we dropped questions in which the participants did not provide meaningful insights, i.e., the respondents all selected "I do not know" or "I am not sure". Based on the feedback from the pre-study, we structured our final survey in three blocks (I–III). Block I focuses on "General information about awareness-raising measures"and the participants' knowledge of different awareness measures. We surveyed how well-known these measures are after introducing four different awareness-raising measures (see Sect. 2.2). Afterward, we asked the participants whether they already participated in such a measure. If not, they were forwarded to block IIa and otherwise to block IIb. Block IIa aims to determine why participants did not take part in an awareness training and how the measures would need to be changed so

---

[1] https://github.com/awareseven/scisec2024/tree/main.

that the participants would take part. Furthermore, we asked participants to provide their stress level when they use the Internet privately or professionally and to estimate the level of cybersecurity their employer has. Block IIb tries to determine how participants perceive the effectiveness of the awareness-raising measures they participated in. After they answered this, they were asked the same questions as block IIa. Block III elicits each participant's assessments of cybersecurity. Finally, demographic data of the participants is collected. To analyze the participants' willingness to take risks without any previous experience, we asked them two hypothetical questions concerning their decision-making under risk. Half of all participants (randomized by the survey tool) received the response options as a single-choice question worded to emphasize risk. The other half receives response options that emphasize the chance of not being compromised. The two questions on risk-taking differ in terms of time because the first question, which each participant receives, would have immediate consequences. The second question, however, would have the consequences occur after a certain period. The different time frames allow us to evaluate whether the timing of the possible consequences influences the participant's risk-taking. The following two scenarios were used:

Scenario 1: Imagine the following situation: You receive mail that looks like it is from your bank. There is a link in it. You click on the link and see a website that looks familiar. You are prompted to enter your username and password. Your primary goal is to check your account balance. What do you do now?

Answers for group 1: a) Do not enter username and password and do not check balance; b) Enter username and password and take a 35% risk of data being stolen immediately.

Answers group 2: a) Do not enter username and password and do not check balance; b) Enter the username and password and take a 65% chance that the data will not be stolen.

Scenario 2: Imagine the following situation: You work in a human resources department and receive a direct email with an application attached from a potential employee. If you do not open it, you risk a candidate who applied to the company dropping out and facing consequences because you failed as a recruiter. Your productivity is measured mainly by the number of candidates who reach the second step of the HR process.

Answers for group 1: a) Do not open the attachment and do not evaluate the applicant; b) Open the attachment and take a 25% chance that the attachment will endanger your job at a later time.

Answers for group 2: a) Do not open the attachment and do not evaluate the applicant; b) Open the attachment and take a 75% chance that the attachment will not endanger your job at a later time. We recruited participants using Amazon's *Mechanical Turk* (MTurk). We only accepted participants who are at least 18 years old with more than 100 tasks completed and high task completion rates ($\geq 75\%$) from around the globe. We asked for their consent to participate in our survey and disclosed our names, affiliations and all sponsors. We used Google Forms to conduct the survey and an instance of Google Workspace where the data location is set to EU. The workers received $1.40 for completing the study, and it took them, on average, 7 min (median: 6 min) to complete the survey. We saved all answers pseudonymously using MTurk's random unique string to pay the workers. Afterward, we deleted the string to increase the participants' level of anonymity.

3.2 Experiment with Follow-up Survey

To get a deeper understanding of how stress affects behavior and decision-making in situations that directly influence cybersecurity, an experiment with a follow-up survey was performed. We especially wanted to understand the impact of stress on security-relevant activities like categorizing whether an email is malicious or not. This scenario was the focus of Scenario 2 (cf. Block II, see Sect. 3.1) Therefore, this experiment explores how subjects identify potentially harmful emails when under stress. Our objective was to assess users' accuracy in detecting such emails without any prior indication of their malicious nature and how stress conditions influence this accuracy.

Study Design. The participants were all located in Germany. The study presented them with an email inbox simulation resembling real-world interfaces (Thunderbird). It was conducted online using a video tool (Zoom), with researchers guiding participants through the tasks. Four unique email datasets and a corresponding questionnaire were utilized for this purpose. The questionnaire, maintained by the researcher guiding the task, ensured a consistent interview process. *Introduction and Methodology.* Initially, to establish a baseline understanding for all participants, we briefed them on identifying malicious emails, referencing the German Federal Office for Security in Information Technology's "3 second check" [15], which includes checking the sender, subject, and attachments. These markers to detect phishing in an email are also recommended by the UK's National Cybersecurity Council [30], the French Agence nationale de la sécurité des systémes d'information [1], and the French Commission Nationale de l'Informatique et des Libertés [10]. To simulate a stressful environment, participants were given only five seconds to classify each email in the datasets as malicious or not. This time constraint was based on findings by Li et al. [23] and supported by the work of Van der Heijden and Jalali [20,43] on stress and email management. *Survey Process.* The process began with a Likert scale evaluation of participants' confidence in their ability to detect malicious emails, both with and without technological assistance. Following this, participants were asked to classify emails from each dataset as malicious or not, using a binary "Yes/No" system. *Datasets.* The four datasets, each containing ten emails, were designed to assess different scenarios.

- **Dataset 1:** A standard inbox with unmarked emails.
- **Dataset 2:** Emails marked according to a color scheme, differentiating between malicious and legitimate emails.

- **Dataset 3:** With false positives (legitimate emails misclassified as phishing).
- **Dataset 4:** With false negatives (phishing emails misclassified as legitimate).

Afterward, participants also evaluated the utility of the color scheme and their trust in protection against malicious emails. *Participant Recruitment and Data Handling.* We recruited participants via social networks associated with the authors and their institutions. Eligible participants were over 18 and consented to participate in the study. The study was unpaid, with an average completion time of 15 min. Data was stored pseudonymously to maintain confidentiality. *Data Analysis.* Data analysis was conducted using the Matthews Correlation Coefficient (MCC), as recommended by Powers [34] and Chicco [8]. The MCC was selected due to its effectiveness in handling imbalanced datasets and its capacity to simultaneously minimize false positives and negatives while maximizing true positives and negatives. Throughout the paper, we use the *Analysis of variance* (ANOVA) test with a 95% confidence interval ($\alpha = 0.5$) to find statistically significant differences between the measures of independent groups.

4   Results

In this section, we provide an overview of the results. First, we present the result of the online user study (see Sect. 4.1), and then we discuss the findings of the quantitative interview study (see Sect. 4.2).

4.1   Online Survey Results

In March 2021, 208 participants participated in our survey, which we recruited via Amazon's Mechanical Turk. The following describes the main results.

Table 1. Demographic overview of the online user study.

Region	Participants		Age	Participants	
North America	90	43%	Under 25	15	7%
Asia/Oceania	68	33%	25–35	139	67%
South America	27	13%	36–46	32	15%
Europe	14	7%	47–57	11	5%
Africa	4	2%	Over 57	11	5%
Others	3	1%			
Not specified	2	1%			

Demographics. Table 1 shows the demographics of the survey participants. The analysis of the localization of the 208 participants shows a strong predominance of North America and Asia/Oceania. Male participants are over-represented in our study, which is common if participants are recruited via MTurk. Many studies have examined gender distribution. One highly regarded study, which analyzed 24 European countries over 18 years, showed that there are approximately 105–107 male newborns for every 100 female newborns [9]. Also, common for MTurk studies, the age distribution shifts towards 25–35-year-olds (66.8%). Our recruitment approach has primarily reached participants who are strongly confronted with IT. More than three-fourths (82%) of the participants state that they are firmly or even very intensely involved with IT professionally. Therefore, the results of this survey are shifted to the professional sector.

Block I. In this block, we want to know how familiar our participants are with different cybersecurity measures. We ask what measures they know and how often they participated in any of those events. For all four queried event formats (i.e., live hacking show, course/seminar/workshop, phishing campaign, eLearning), an average of 63% indicated that they had a rough or even a firm idea of the format. The detailed numbers are as follows: Live Hacking (mean: 3.57; median: 4; SD: 0.84), Phishing Campaign (mean: 3.67; median: 4; SD: 0.84), Seminar/Course/Workshop (mean: 3.75; median: 4; SD: 0.92), and eLearning (mean: 3.92; median: 4; SD: 0.96). eLearning, in particular, is familiar to participants, with 73% (151) indicating that they have at least a rough idea of this awareness-raising measure. The most unknown measure in this survey is live hacking. 116 participants (56%) answered with the statement that they had at least a rough idea. Although various awareness events are very well known, only 14% (28 out of 208) have participated in more than two of such events. Only 38 participants (18.3%) have not yet taken part in any awareness-raising event, mainly because they did not receive an offer. This high participation rate shows that cybersecurity is gaining popularity worldwide. By far, most participants (142; 68%) stated that they had taken part in one or two such measures.

Block IIa. This section addresses the participants who indicated in the first section that they did not attend an awareness event. *Perception of security in the workspace.* Participants rate the cybersecurity of their own company as secure (71%), although some participants rate it as extremely poor (11%) or extremely good (11%) (mean: 3.29; median: 3; SD: 1.14). *Awareness of participants whether they clicked on malicious mail.* The high level of interest and self-assessment about knowledge in the area of cybersecurity is precise, as almost every participant who indicated in the first section that they did not attend an awareness event is sure whether he/she has already clicked on a malicious or fraudulent email (see Fig. 3; 1: "No, definitely not; 5: "Yes, fully"; mean: 3.29; median: 2; SD: 1.45). The standard deviation of almost 1.5 shows that most participants tend strongly in one direction, whether fraudulent emails have already been clicked.

Stress when Using the Internet. The stress rating shows that stress perception is equally distributed across the 38 participants' private and professional Internet use. It is noticeable that the Internet's professional use hardly influences the participants' stress levels. *Decision-making under risk.* The participants' willingness to take risks shows a clear difference in how the question or scenario is formulated. As explained in Sect. 3.1, all participants were asked the same question. The difference is that half of them are framed on the opportunity and the other half on the risk. In total, 38.2% of the participants would accept the presented risk and open the unknown email attachment. However, this is split between the two groups with the differently worded questions. The first group received the questions focusing on the risk, while the second group received the questions focusing on the chance that no bad result would occur. This formulation affects the users as 72% would not take the risk, and thus only 28% would accept the risk. The other half of the participants received a similar question, but it was formulated to focus on the chance that no damage would occur. As a result of this formal change, more than half of the participants, 53%, would now accept the risk. This change is significant ($t = 2.774$, $p < 0.01$). Thus, rephrasing the question to focus on the chance of no harm rather than the risk has a significant effect: Participants are more inhibited from taking risks. If we want to persuade other people not to take a particular risk, the risk should be addressed, not the chance that nothing will happen.

Block IIb. This section addresses only those participants who indicated in the first survey section that they had attended at least one awareness event. In total, this section has 170 participants. *Perception of instructional and entertaining aspects of awareness measures.* In particular, eLearning was rated as "very instructive" by 44% of the participants. The detailed numbers are as follows: Live Hacking (mean: 4.02; median: 4; SD: 0.96), Phishing Campaign (mean: 4.04; median: 4; SD: 0.95), Seminar/Course/Workshop (mean: 4.11; median: 4; SD: 0.91), and eLearning (mean: 4.27; median: 4; SD: 0.86). The assessments of the individual measures can also be seen in Fig. 1.

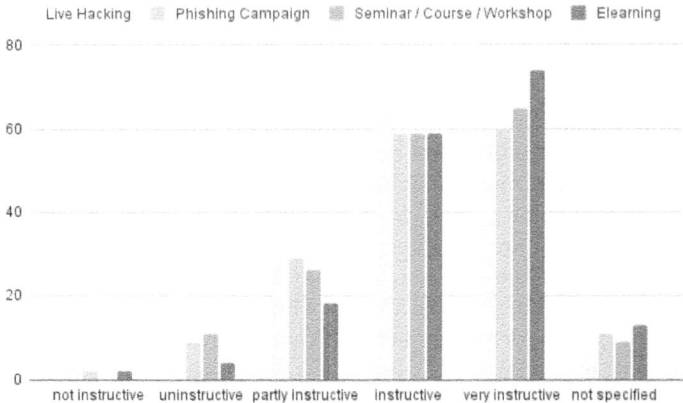

Fig. 1. Perception of instructiveness of the various cybersecurity awareness measures

Perception of Cybersecurity in the Workplace. The participants assess the cybersecurity of their own company as secure (71%), whereas no participant assesses it as extremely poor (1: "very poor", 5: "very good"; mean: 3.94; median: 4; SD: 0.81). Let us compare the views of participants who have already taken part in at least one cybersecurity measure with the participants' without cybersecurity measures. It is noticeable that their interest and the company's cybersecurity are assessed as better/higher. The participants who have already participated in a cybersecurity event rate their interest and the company's security better or higher. This observation shows the positive perception of such events among the participants. The estimated company security difference is statistically significant ($t = 4.13$, $p < 0.00001$). Likewise, the estimate about one's interest in cybersecurity is statistically significant ($t = -2.77$, $p < 0.01$).

Stress when Using the Internet. Fig. 2 shows the stress level in a private and professional context. It is noticeable that the participants are slightly stressed and that this assessment hardly changes between the professional and private contexts. This difference is *not* statistically significant ($t = 0.042$, $p = 0.966$).

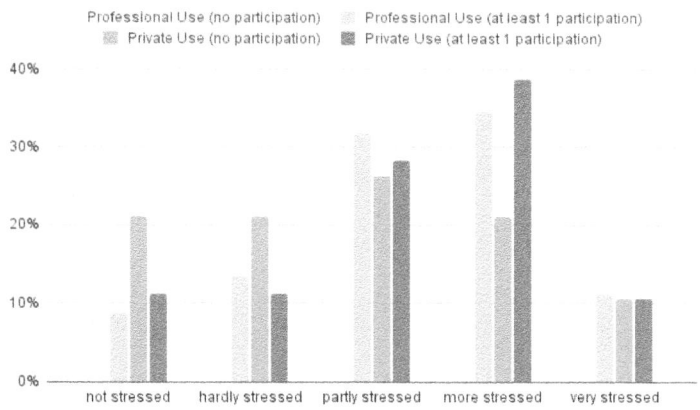

Fig. 2. Perception of stress in the private and professional context

The stress perception between the participants without participation in a cybersecurity measure and those of the participants with at least one participation shows that the most significant proportion is rather stressed. This perception is also seen in Fig. 2 and is consistent with a study from European Neuropsychopharmacology, from June 2020 [44]. We found a correlation between "stressed" participants and the assessment of clicking on fraudulent emails. According to their judgment, participants who are more stressed in their private or professional Internet use are significantly more likely to click on fraudulent emails ($t = 3.39$, $p < 0.001$); see Fig. 3.

Decision-Making Under Risk. The participants' willingness to take risks shows a clear difference in how the question or scenario is formulated. As explained in Sect. 3, the participants were asked the same question, but the answers were framed differently: either on opportunity or risk. In our survey, 59.1% of the participants would accept the presented risk and open the unknown email attachment. The difference between the participants who did not participate in any event is vast. Only 38.2% of these participants would take this risk and are thus much more cautious than the experienced participants. However, 59.1% is divided between the two groups with different phrasing. The first group received the questions focusing on risk. This rewording also affects these participants because now only around 55.6% would take the risk. The other half of the participants were asked a similar question with the difference that it focused less on the risk and was phrased so that it addressed the high chance that no damage would occur. As a result of this formal change, shown in Table 2, significantly more than half of the participants (62.4%) would accept the risk.

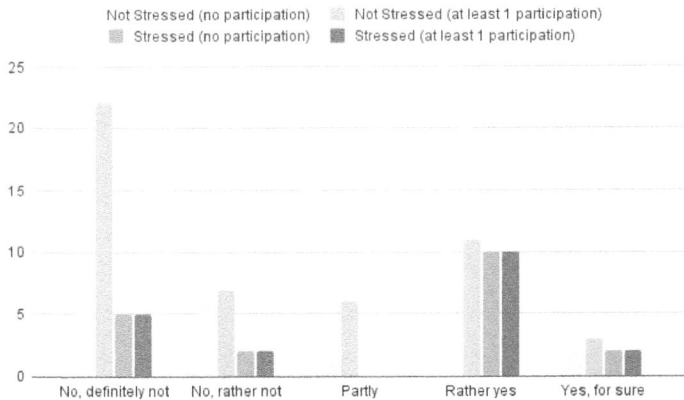

Fig. 3. Participant assessment whether they clicked on fraudulent emails. Differentiating stressed and non-stressed participants.

Combination of Blocks IIa and IIb. Here, we summarize the answers from all participants, regardless of which subgroup they were in. *Connection between stress and perception of security.* The stress level of the participants has an impact on the perception of the cybersecurity posture. The difference between private and professional use is not statistically significant ($t = 0.042$, $p = 0.966$). The participants who indicated they were "stressed" (i.e., rated their stress as 4 or 5) about using the Internet in either a personal or professional context rated their company's cybersecurity as follows: very bad: 0%; bad: 2.4%; average: 28.8%; good: 45.3%; very good: 23.5%.

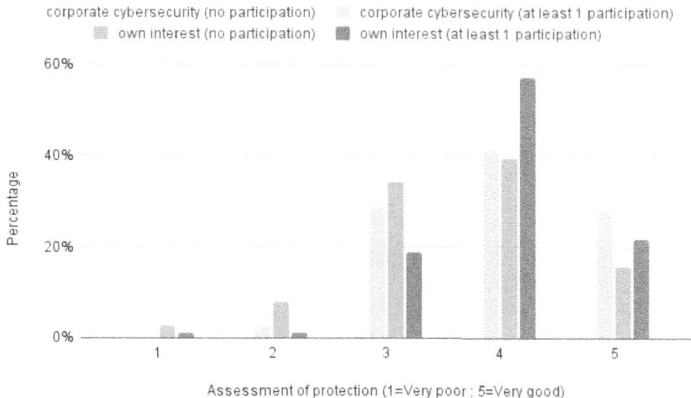

Fig. 4. Perception of company's cybersecurity

Figure 4 shows that participants who have already taken part in a cyber-security event rate the company's security as well as their interest better or higher. The difference in the estimated company security is statistically significant ($t = 4.13$, $p < 0.0001$). Likewise, the estimate about the participants' interest in IT security is statistically significant ($t = -2.77$, $p < 0.01$).

User Risk Perception. In summary, 55.3% of all participants would accept the risk. This refers to both groups of participants, with more critical questions about the risk and with less critical questions. If we consider only the group with phrasing focusing on the risk, half of the participants (49.25%) would take the risk presented. This difference from all participants' risk-taking is insignificant ($t = 1.42$, $p = 0.15$). Rephrasing the question focusing on the high chance that no harm will occur, led to 61.1% willing to take this risk (see Table 2). However, there is no added risk compared to the question that we framed otherwise. This difference is insignificant compared to all participants' risk-taking ($t = 1.31$, $p = 0.191$). As a result, the phrasing focusing on the opportunity or the risk affects the participants' willingness to take risks. If, for example, we as an employer would like our employees to show the lowest possible willingness to take risks, we should choose formulations focusing on risk since this formulation has shown that the participants would take a lower risk.

The different periods in which the possible consequences can occur have no statistically relevant effect on the participants' willingness to take risks. We want to assess if employees recognize this issue and are unwilling to take risks. We can distinguish between four user types depending on whether they participated in an awareness measure and whether they are willing to take a risk. For this comparison, the ANOVA test is suitable: With an f-value of 5.94 and a p-value of < 0.001, we can assume that the tested groups show different behavior on average. Accordingly, we can prove with this survey that the participants'

Table 2. Differences between answers that focus on risks or chances

	Take Risk	Avoid Risk
Focus on Risk	49,25%	50,75%
Focus on Chance	61,1%	38,9%
Total	55,3%	44,7%

behavior in terms of risk-taking depends on the phrasing of the scenario and the participants' previous experiences.

4.2 Qualitative User Study Results

Here we summarize the study results we introduced in Sect. 3.2. It was conducted in July 2020 with 31 participants using in-person interviews.

Demographics. 61% of the participants are male, and the majority (58%) are between 25 and 37 years old. Table 3 shows the participants' demographics and the industry in which the participants work. The majority of participants (38.71%) work in an IT-related domain.

Table 3. Participant demographics of the interview study.

		Participants		Field of work	Participants	
Gender	Male	19	61.3%	IT	12	38.7%
	Female	12	38.7%	Education	7	22.6%
	non-binary	0	0%	Consulting	4	12.9%
Age	18–37	18	58.1%	Healthcare	4	12.9%
	37–49	7	22.6%	Trade	2	6.5%
	49–65	6	19.4%	Politics	1	3.2%
				Construction	1	3.2%

Users' Ability to Detect Malicious Emails. Half of the participants (51.6%) rank themselves at least somewhat confident (4 or 5) in their ability to detect malicious emails. The mean is 3.3 for all participants with SD = 1.12.

Users' Trust in Technical Measures. The minority of participants (29.0%) rank technical measures to protect them from malicious emails as mostly sufficient. However, none of the participants believed that technical measures completely protect them from malicious emails. The majority of participants (71.0%) feel neutral or are skeptical about whether they should trust technical measures with their protection. The mean is below 3 with 2.87 and SD $= 0.92$.

Unmarked Dataset. Participants binary classified each email in the dataset based on whether they believed it was malicious. The average MCC is 0.650 (min: -0.327; max: 1; SD: 0.275), and five participants correctly identified every malicious email as malicious and every legitimate email as legitimate, scoring MCC $= 1$. An MCC of -1 would indicate that the participant did not flag any malicious email as malicious and flagged all legitimate emails as malicious. An MCC of 0 would indicate that the results resemble a completely random guess. Only one participant scored a negative MCC. However, 30 participants scored MCC > 0 on the first dataset, meaning they took at least *educated guesses*. Five participants (16.1%) misjudged malicious emails as benevolent. Lastly, 26 participants (83.9%) misjudged at least one legitimate email as malicious.

Correctly Marked Dataset. The average MCC is 0.761 (min: 0.218; max: 1; SD: 0.231), and 12 participants (38.7%) correctly identified every malicious email as malicious and every legitimate email as legitimate, scoring MCC $= 1$. The mean MCC is the highest among the four datasets. No participant scored MCC ≤ 0. Only 2 participants (6%) misjudged a maliciously marked email as benevolent. And 19 participants (61%) misjudged a legitimate email as malicious although it was flagged as legitimate. The correctly classified mean is 86.46% (SD $= 0.12$).

Dataset with One False Positive. The average MCC is 0.714 (min: 0,102; max: 1; SD: 0.244) and 9 participants (29.0%) correctly identified every malicious email as malicious and every legitimate email as legitimate scoring MCC $= 1$. No participant scored ≤ 0. Three participants (9.7%) misjudged a maliciously marked email as benevolent. 21 participants (67.7%) misjudged one of the legitimate emails as malicious, even though it was flagged as legitimate. We see correctness < 0.7 for the overall classification correctness in this specific dataset.

One email in our dataset was only identified correctly by 61.3% of the participants as legitimate, although it is tagged as legitimate. One approach to explaining why the email is classified as malicious by so many participants could be that the email is very generic. There is no introductory note, and it starts directly with linked images and bold typefaces. Email number 8 is the false positive item that was marked as malicious by the authors. The vast majority (96.8%) of participants detected the email as legitimate, although it was colored as a malicious one.

Dataset with one False Negative. This dataset contains the worst possible error with an email indicating "legit" even though it is *malicious*. The average MCC is 0.744 (min: 0.357; max: 1; SD: 0.205), and eleven participants (35.5%) correctly identified every malicious email as malicious and every legitimate email as legitimate, scoring MCC $= 1$. No participant scored MCC ≤ 0. Four participants (12.9%) misjudged a maliciously marked email as benevolent. Eighteen participants (58.1%) misjudged one of the legitimate emails as malicious, although it was flagged as legitimate. Email number 5 is the false negative item the authors placed wrongly marked in the dataset. The vast majority (93.5%) of the participants detected that email as malicious even though it was considered legitimate. This might be due to the raised awareness of participants during the whole experiment. Dataset 4 contains the most misclassified email of the whole experiment. It is legitimate and marked as legitimate. However, only a bit more than half of the respondents (58.1%) classified the email correctly as legit. Another email in our dataset was identified by only 58.1% of the participants correctly as legitimate, although it is tagged as legitimate. The email uses a generic address in the *From* field, but there is a *reply-to* field that might look suspicious. Further, the email is very generic, and there is no personalized introduction. These factors probably lead to the comparable low correctness rate of only 58.1%, which is significantly lower than the overall correctness rate of 85.4%. The average MCCs for all four datasets are summarized in Table 4. We list the rates of how many participants correctly identified the wrongly marked emails in the datasets with one false positive or negative, respectively.

Table 4. Average MCCs for the different datasets.

Dataset	MCC	False marking recognized
Unmarked	0.650	—
Correctly marked	0.761	—
One false positive	0.714	96.8%
One false negative	0.744	93.5%

Perception of Color Marking as Helpful. Most users (71.0%) rank the color marking as at least *somewhat helpful* with a 4 or 5 on the Likert scale when asked if they believe that marking emails with a color in the inbox helps detect malicious emails. The mean is 3.74 with SD = 0.93 for the 31 participants. Only one respondent evaluates the color marking as "not helpful". *Perception of color marking to increase confidence in own detection capabilities.* Most participants (65%) rank the color marking at least *somewhat helpful* (4 or 5) when asked if they believe that it increases their confidence in their detection capabilities (mean: 3.52; SD: 1.06). Only 4 respondents (13%) believe that the color marking would be counterproductive for their detection capabilities.

5   Related Work

Farahmand analyzes the decision weights of underwriters and corporate managers for cybersecurity [12]. He concludes that they overweigh low probability cybersecurity events and underweigh high probability cybersecurity events. He furthermore shows that the value function changes if an organization experiences a breach. Fineberg states that current cyber strategies are still operated as if the actors in cyberspace are acting rationally [14]. He extends the work of Herley, who showed that users reject security advice rationally [17]. Lahcen et al. also point out that research in behavioral economics intersecting with cybersecurity needs to be done, especially with the emerging importance of humans as an integral part of cybersecurity strategies [25]. Bada et al. examined cybersecurity awareness campaigns and found that they mostly fail because the simple transfer of knowledge is not enough. Positive cybersecurity behaviors need to be enforced so that thinking becomes a habit and part of organizational culture [4]. Current research primarily focuses on the poor design of security systems and policies, but not on the behavioral aspect and individual decision-making [19,31]. Jalali et al. find in a cybersecurity game experiment that decision-making is profoundly entrenched, and management decision alone is not helping in decision-making compared to inexperienced players [21]. Qu et al. investigate another approach by applying prospect theory to security decisions and show that those in a "disadvantage" situation are more likely to be persuaded to make better security decisions [35]. These findings are supported by the work of Amador et al. who investigated password selection processes and also found that intervention guided by prospect theory is causing 25% of users to improve their password strength [3]. To face this challenge from a research perspective, the field of *nudging* is more researched. Peer et al. show that personalized nudges increase the effect of nudging in choosing a solid password [32]. Zimmermann and Renaud show that the poor understanding of nudges in cybersecurity is currently hindering effective nudging. The *hybrid nudge*, consisting of a nudge and information provision, is a practical decision helper in some context [47]. Furthermore, warnings for malware and phishing exist in various contexts, such as browser toolbars, pop-up screens from firewalls, or browsers [11,16]. Warnings try to prevent users from entering sensitive information or executing attachments. Warnings should be understandable, authoritative, primarily accurate, and not just passive warnings that can be easily clicked away or ignored [28,41]. Qu et al. analyze that the interaction of framing and timing is important when nudging users towards an improvement in their cybersecurity decisions [36].

6   Discussion

Economic phenomena generally involve distributing scarce resources in combination with human behavior. In cybersecurity, this scarce resource is *attention*. Getting this attention is all about communication, and our survey reveals that even slightly different wording can lead to different outcomes. When communicating about cyber risks and building awareness, it is essential not to monger fear among users because this is likely to backfire [46]. This is another indication that it is not just about informing the user but also focusing on details of effectively influencing users to make decisions [37] that favor their cybersecurity. However, our research empirically shows that communicating risks directly hinders potential chances that arise from digitalization. Besides, the risk appetite of the participants is vast. Some participants took all the risks presented, but

212 J. Hörnemann et al.

other participants rejected all risks. Through the differently formulated scenarios in which the willingness to take risks was analyzed, it could be determined that the formulation of the situation exerts an effect on the willingness to take risks. The risk in the scenarios where the wording focused on the risk arising was taken significantly less often than in the scenarios where the wording focused on the chance of no harm arising. This finding is analog to the results of Tversky et al. [42], where they describe the fact that risks have a more significant effect on decisions than focusing on gains. This behavior shows that framing and economic principles are essential factors when dealing with cybersecurity risks. According to our results, using the correct wording without exuding fear can lead to an effective behavioral change. The majority of the participants stated being stressed in their private and professional Internet use. This feeling of stress affects their company's IT security because the stressed participants assess the IT security of their own company as secure. However, our survey did not reveal any statistically significant connection between the respondents' feelings of stress and their willingness to take risks, although this connection seemed natural. In our survey, stress affects clicking on fraudulent or harmful emails. These measured effects suggest that, in reality, there is an effect between stress perception and risk-taking, as risk-taking is often a trade-off between risk and benefit, so stressed people may be inclined to take the risk to avoid building up further stress. Our results from the experiment with the follow-up survey in Sect. 4.2 indicate that placing a nudge before opening a potentially malicious mail is helping users detect potentially malicious emails. To some extent, the participants were stressed because they had a time limit of five seconds for their decision, which is a realistic setting as discussed in Sect. 3.2. Section 4 shows that the average MCC and the total number of participants who score correctly when detecting malicious emails in a binary classification scheme are significantly improving for nudged emails. This study indicates that a color marking is a nudge. Just one participant answered perfectly and detected every malicious mail as malicious and every legitimate mail as legitimate and has MCC = 1. The average MCC for all participants is 0.697 (SD = 0.17), and no participant scores a negative MCC. The average MCC of the nudged dataset 2 is 17% better than the MCC in the non-nudged dataset 1. Even the average MCC over all datasets is 10% higher than in the first dataset 1, which also shows that the color marking helps users to make profound decisions if nudged. Another important factor is the increased confidence in the user's detection capabilities. Confident humans are more confident in detecting deceit and can discriminate between accurate and inaccurate lie detection [40, 45].

7 Limitations

We decided to provide the questionnaire only in English. Therefore, there might be a language bias because users did not receive the questionnaire in their native language. Confirming or rejecting this correlation is essential from the perspective of cybersecurity, as it can prevent risk-taking and, thus, possible damage in the long term. While we conducted the user study, we were given the feedback that the red and green marked mails are challenging to differentiate for people with red-green colorblindness. This needs to be addressed in the future. Furthermore, we put users under one specific stress (time), which is just one of many ways to exert stress on humans [22]. This study does not consider participants' characteristics (i.e., personality traits) or contextual factors (e.g., daytime/nighttime). A more diverse participant pool should be studied if other research groups try to improve our approach.

7.1 Ethical Consideration

For this study, we gathered responses from various participants. Our research institution does not require approval for this type of study, nor does it provide an Institutional Review Board (IRB). Nevertheless, we took strict ethical considerations into account. We never collected any PII like *name* or other information that would make the identification of a single respondent possible.

8 Conclusion

It is apparent from our survey that the majority of participants are optimistic about the awareness measures currently used in the industry. This positive attitude relates to both the perception of instructiveness and the entertaining nature of the various measures. We studied behavior change when nudging users towards an absolute security-relevant decision and hypothetical risk-taking. In our user study, we see that nudging provides an effective way to influence users' decisions and, thus, cope with the situation better. Even when users experience stress, we show empirically that nudging has a positive effect on the decision-making of our participants. On the other hand, our survey shows that the threat appraisal is different, depending on the situation's framing. To our knowledge, this is the first work to focus on empirically showing this behavioral economic effect in cybersecurity. Focusing on communicating the risks of data losses is an effective way to raise awareness, yet it is crucial not to foment fear, but to generate changes in behavior.

Acknowledgments. The authors gratefully acknowledge funding from the *Federal Ministry of Education and Research* (16KIS1648 "DigiFit", 16KIS1628K & 16KIS1629 "UbiTrans").

References

1. Agence nationale de la sécurité des systémes d'information: Signalement d'un contenu inadéquat (2023). https://web.archive.org/web/20240826204507/. https://cyber.gouv.fr/signalement-dun-contenu-inadequat. Accessed 28 Aug 2024
2. Akhawe, D., Felt, A.P.: Alice in warningland: a large-scale field study of browser security warning effectiveness. In: 22nd USENIX Security Symposium (USENIX Security 13) (2013)
3. Amador, J., Ma, Y., Hasama, S., Lumba, E., Lee, G., Birrell, E.: Prospects for improving password selection. In: Nineteenth Symposium on Usable Privacy and Security (SOUPS 2023), pp. 263–282 (2023)
4. Bada, M., Sasse, A.M., Nurse, J.R.C.: Cyber security awareness campaigns: why do they fail to change behaviour? (2019)
5. Butler, S.A.: Security attribute evaluation method: a cost-benefit approach. In: Proceedings of the 24th International Conference on Software Engineering (2002)
6. Cambridge University Press: Cambridge Dictionary. https://web.archive.org/web/20230516113857/. https://dictionary.cambridge.org/dictionary/english/awareness. Accessed 21 Mar 2021
7. Chai, S., Kim, M., Rao, H.R.: Firms' information security investment decisions: stock market evidence of investors' behavior. Decis. Support Syst. (2011)
8. Chicco, D., Jurman, G.: The advantages of the Matthews correlation coefficient (MCC) over F1 score and accuracy in binary classification evaluation. BMC Genomics (2020)
9. Coale, A.J.: Excess female mortality and the balance of the sexes in the population: an estimate of the number of "missing females". Popul. Dev. Rev. 517–523 (1991)
10. Commission Nationale de l'Informatique et des Libertés: Phishing : détecter un message malveillant (2017). https://web.archive.org/web/20240717151542/. https://www.cnil.fr/fr/phishing-detecter-un-message-malveillant. Accessed 28 Aug 2024
11. Egelman, S., Cranor, L.F., Hong, J.: You've been warned: an empirical study of the effectiveness of web browser phishing warnings. In: Proceedings of the SIGCHI Conference on Human Factors in Computing Systems (2008)
12. Farahmand, F.: Applying behavior economics to improve cyber security behaviors. Technical report, Georgia Institute of Technology Atlanta United States (2018)
13. Fielder, A., Panaousis, E., Malacaria, P., Hankin, C., Smeraldi, F.: Decision support approaches for cyber security investment. Decis. Support Syst. (2016)
14. Fineberg, V.: BEC: applying behavioral economics to harden cyberspace. J. Cybersecur. Inf. Syst. (2014)
15. German Federal Office for Information Security: Drei Sekunden für mehr E-Mail-Sicherheit (2020). https://web.archive.org/web/20240724031330/. https://www.bsi.bund.de/DE/Themen/Verbraucherinnen-und-Verbraucher/Informationen-und-Empfehlungen/Onlinekommunikation/E-Mail-Sicherheit/e-mail-sicherheit_node.html. Accessed 24 Aug 2024
16. Gupta, B.B., Tewari, A., Jain, A.K., Agrawal, D.P.: Fighting against phishing attacks: state of the art and future challenges. Neural Comput. Appl. (2017)
17. Herley, C.: So long, and no thanks for the externalities: the rational rejection of security advice by users. In: Proceedings of the 2009 Workshop on New Security Paradigms Workshop, pp. 133–144 (2009)
18. Ho, T.H., Lim, N., Camerer, C.F.: Modeling the psychology of consumer and firm behavior with behavioral economics. J. Mark. Res. (2006)

19. Iuga, C., Nurse, J.R.C., Erola, A.: Baiting the hook: factors impacting susceptibility to phishing attacks. HCIS **6**(1), 1–20 (2016). https://doi.org/10.1186/s13673-016-0065-2
20. Jalali, M.S., Bruckes, M., Westmattelmann, D., Schewe, G.: Why employees (still) click on phishing links: investigation in hospitals. J. Med. Internet Res. (2020)
21. Jalali, M.S., Siegel, M., Madnick, S.: Decision-making and biases in cybersecurity capability development: evidence from a simulation game experiment. J. Strateg. Inf. Syst. (2019)
22. Keay, K.A., Bandler, R.: Parallel circuits mediating distinct emotional coping reactions to different types of stress. Neurosci. Biobehav. Rev. (2001)
23. Li, X., Lee, C.J., Shokouhi, M., Dumais, S.: Characterizing reading time on enterprise emails. arXiv preprint arXiv:2001.00802 (2020)
24. Likert, R.: A technique for the measurement of attitudes. Arch. Psychol. (1932)
25. Maalem Lahcen, R.A., Caulkins, B., Mohapatra, R., Kumar, M.: Review and insight on the behavioral aspects of cybersecurity. Cybersecurity **3**(1), 1–18 (2020). https://doi.org/10.1186/s42400-020-00050-w
26. Maddux, J.E., Rogers, R.W.: Protection motivation and self-efficacy: a revised theory of fear appeals and attitude change. J. Exp. Soc. Psychol. (1983)
27. March, J.G., Shapira, Z.: Managerial perspectives on risk and risk taking. Manag. Sci. (1987)
28. Modic, D., Anderson, R.: Reading this may harm your computer: the psychology of malware warnings. Comput. Hum. Behav. (2014)
29. Mullainathan, S., Thaler, R.H.: Behavioral economics. Technical report, National Bureau of Economic Research (2000)
30. National Cyber Security Centre: Quick Guide: Phishing (2022). https://web.archive.org/web/20240622151113/. https://www.ncsc.gov.ie/pdfs/NCSC_Quick_Guide_Phishing.pdf. Accessed 28 Aug 2024
31. Nurse, J.R., Creese, S., Goldsmith, M., Lamberts, K.: Guidelines for usable cybersecurity: past and present. In: 2011 Third International Workshop on Cyberspace Safety and Security (CSS). IEEE (2011)
32. Peer, E., Egelman, S., Harbach, M., Malkin, N., Mathur, A., Frik, A.: Nudge me right: personalizing online security nudges to people's decision-making styles. Comput. Hum. Behav. (2020)
33. Pfleeger, S.L., Caputo, D.D.: Leveraging behavioral science to mitigate cyber security risk. Comput. Secur. (2012)
34. Powers, D.M.: Evaluation: from precision, recall and f-measure to roc, informedness, markedness and correlation. J. Mach. Learn. Technol. (2011)
35. Qu, L., Wang, C., Xiao, R., Hou, J., Shi, W., Liang, B.: Towards better security decisions: applying prospect theory to cybersecurity. In: Extended Abstracts of the 2019 CHI Conference on Human Factors in Computing Systems, pp. 1–6 (2019)
36. Qu, L., Xiao, R., Shi, W.: Interactions of framing and timing in nudging online game security. Comput. Secur. **124**, 102962 (2023)
37. Rogers, R.W.: Attitude change and information integration in fear appeals. Psychol. Rep. (1985)
38. Sasse, M.A., Brostoff, S., Weirich, D.: Transforming the 'weakest link'-a human/computer interaction approach to usable and effective security. BT Technol. J. (2001)
39. Schneier, B.: The psychology of security. In: International Conference on Cryptology in Africa. Springer (2008)
40. Smith, A.M., Leach, A.M.: Confidence can be used to discriminate between accurate and inaccurate lie decisions. Perspect. Psychol. Sci. (2019)

41. Stembert, N., Padmos, A., Bargh, M.S., Choenni, S., Jansen, F.: A study of preventing email (spear) phishing by enabling human intelligence. In: 2015 European Intelligence and Security Informatics Conference. IEEE (2015)
42. Tversky, A., Kahneman, D.: Loss aversion in riskless choice: a reference-dependent model. Quart. J. Econ. **106**, 1039–1061 (1991)
43. Van Der Heijden, A., Allodi, L.: Cognitive triaging of phishing attacks. In: 28th USENIX Security Symposium (USENIX Security 19) (2019)
44. Vinkers, C.H., et al.: Stress resilience during the coronavirus pandemic. Europ. Neuropsychopharmacology (2020)
45. Vrij, A., Baxter, M.: Accuracy and confidence in detecting truths andlies in elaborations and denials: truth bias, lie bias and individual differences. Expert Evid. (1999)
46. Witte, K.: Fear control and danger control: a test of the extended parallel process model (EPPM). Commun. Monogr. (1994)
47. Zimmermann, V., Renaud, K.: The nudge puzzle: matching nudge interventions to cybersecurity decisions. ACM Trans. Comput.-Hum. Interact. (TOCHI) (2021)

SVSM-KMS: Safeguarding Keys for Cloud Services with Encrypted Virtualization

Benshan Mei[1,2], Wenhao Wang[1,2(✉)], and Dongdai Lin[1,2]

[1] Key Laboratory of Cyberspace Security Defense, Institute of Information Engineering, Chinese Academy of Sciences, Beijing, China
`wangwenhao@iie.ac.cn`
[2] School of Cyber Security, University of Chinese Academy of Sciences, Beijing, China

Abstract. In recent years, numerous instances of data breaches have emerged due to the inadvertent or intentional disclosure of cryptographic keys. To address this issue, this paper proposes SVSM-KMS, which utilizes AMD's latest Encrypted Virtualization technology (AMD SEV-SNP) to deliver an efficient and seamless integrated secure key management service. We realized multilayered defense by integrating our mechanism within a privileged layer of a confidential virtual machine (CVM), thereby minimizing the trusted computing base (TCB) to prevent key leakage from compromised CVMs. Notably, we incorporated a zero-copy mechanism between the most privileged service module and the least privileged user applications, eliminating redundant data copies. To facilitate seamless integration, we propose a proxy server for existing cloud services. A prototype of SVSM-KMS has been developed based on the latest AMD SEV-SNP hardware platform. Evaluation results indicate that the performance of the Encrypted Virtualization-empowered SVSM-KMS is on par with Hadoop KMS, highlighting the practicality.

Keywords: Confidential Computing · Trusted Execution Environment · Encrypted Virtualization · Secure Virtual Machine Service Module · Key Management Systems

1 Introduction

Cloud computing transforms the way organizations store, process, and share data, offering numerous benefits such as scalability and cost-efficiency. However, it also introduces new security challenges, and data breaches have become a recurring threat. These breaches occur when unauthorized individuals gain access to sensitive data stored in the cloud, potentially leading to severe consequences such as financial loss, reputation damage, and privacy violations. Key management is one of the most critical services on the cloud [21,26]. If the keys are leaked, all relevant data associated with the keys are at risk of leaking [16,19]. A centralized key management system (KMS) is often deployed as

© The Author(s), under exclusive license to Springer Nature Singapore Pte Ltd. 2025
J. Zhao and W. Meng (Eds.): SciSec 2024, LNCS 15441, pp. 217–235, 2025.
https://doi.org/10.1007/978-981-96-2417-1_12

an integrated approach for generating, distributing, and managing cryptographic keys for devices and applications [6]. However, the TCB can be significant large in a centralized KMS, leading to a potential single point of failure [26]. Although a decentralized KMS may be desired, it still incurs substantial costs to maintain coherence among multiple nodes [12,14].

Motivations. Secure key management in the cloud presents several challenges that need to be addressed. One major concern revolves around the cloud service provider's extensive control over the platform, potentially leading to data inspection for their own business interests or due to insider threats. Furthermore, network latency in a centralized KMS can greatly impede service efficiency. Moreover, it can lead to key exposure if the hosting system is compromised.

To tackle these challenges, we focus on a defense-in-depth design, safeguarding sensitive keys in the cloud from untrusted cloud service providers (CSPs) and guest OS. Specifically, we utilize CVM to securely host our system. The CVM is isolated from each other and the hypervisor with hardware-based memory encryption. This ensures that the data remains confidential even if the CSP is untrustworthy. However, it is not sufficient for secure key management due to the large TCB introduced by the guest OS. The key still get leaked if the guest OS gets compromised. We observe that the recently proposed Virtual Machine Privilege Level (VMPL) hardware mechanism can be very applicable to further secure the keys used for encryption and decryption. We safeguard the keys from unauthorized access or tampering by integrating the system within the highest VMPL of a CVM. This ensures that even if the guest OS is compromised, the cryptography keys remain secure. In addition to these security measures, we devote to ensuring seamless integration with existing systems on the cloud, thereby minimizing the deployment effort.

Design. To achieve this, we place the key management service within the highest VMPL over the guest OS. Our design consists of the following aspects: access control, sealed storage, attestation, configuration and seamless integration. The majority of the components, including the access control, are contained within the service module, which runs in the highest VMPL. This means that even if the guest OS is compromised, the keys will remain secure. In order to facilitate seamless integration, we design a proxy server for Hadoop clients, which forwards requests and responses and does not interfere with access control. It consults the service module for token authentication, and securely persists the in-memory storage with sealing service provided by the service module. Unlike a decentralized KMS [12,14], we do not need to maintain distributed consistency, reducing data synchronization overhead. The performance is further improved by a zero-copy design between the proxy server and the service module, reducing extra data copies between those components.

To showcase the performance of our design, we have developed a prototype based on the latest AMD SEV-SNP platform. It consists of about 7000 lines of codes in total, and is composed of a service module, a kernel module and a proxy server. The evaluation consists of requesting service from KMS deployed on local and remote machine. Our experimental results demonstrate that our approach offers comparable performance to existing Hadoop KMS in both local and remote service scenarios. In the local service scenario, our KMS outperforms Hadoop KMS in term of service latency on most operations. While in serving Transparent Data Encryption (TDE) of Hadoop Distributed File System (HDFS), our system performs comparably in read and write test.

Contributions. The contributions of the paper are summarized as follows.

- We introduce multilayered defense for secure key management based on the latest feature of Encrypted Virtualization, preventing key leakage from vulnerable guest OS within CVM.
- We incorporate a zero-copy design in our system to improve the performance, allowing efficient service delivery.
- We introduce a proxy server to enable seamless integration, ensuring compatibility with traditional KMS.
- We implement a prototype of the design based on the latest linux-svsm project, and evaluate the performance of our system under realistic scenarios, showing the practical aspect of our design.

2 Background

2.1 AMD SEV

Over the past few years, establishing mutual trust between customers and Cloud Service Providers (CSPs) has been a persistent challenge. It's crucial to mitigate the reliance on CSP trust within the public cloud computing market. AMD's Secure Encrypted Virtualization (SEV) represents a significant leap forward, leveraging hardware-assisted virtualization technology to address this challenge through memory-encryption enhanced isolation [23,34]. To counter potential threats from malicious hypervisors, AMD introduced successive advancements like SEV-ES (Encrypted State), which encrypts memory pages and private register contents of Virtual Machines (VMs) using distinct keys [35,36]. However, a lingering vulnerability persists: the hypervisor retains control over nested paging, potentially allowing SEV VM pages to be mapped to other VMs or even the hypervisor itself [40]. Despite the encryption of VM's private status and pages under separate keys, SEV/SEV-ES lacks integrity protection, leaving room for exploits such as memory replay attacks by the hypervisor.

In 2020, AMD introduced SEV-SNP (Secure Nested Paging), a technology designed to fortify the protection of CVM against malicious hypervisor [41]. In SEV-SNP, a malicious hypervisor cannot map an encrypted physical page to multiple owners. This mechanism is implemented through the use of a Reverse Mapping Table (RMP). The RMP is a metadata table controlled by the AMD Platform Security Processor (AMD PSP). It keeps track of the ownership of every system physical page and specifies read, write, and execution permissions for each VMPL. Physical memory page access is restricted by configuring each page's VMPL in the RMP. On each nested-page table walk, the RMP is consulted to determine permission and ownership for each system physical memory page. A nested page fault (#NPF) is raised upon illegal access to a physical page, which can be captured and handled by the hypervisor. The hypervisor manages the VM Saved Areas (VMSAs) assigned to four VMPLs. With hypervisor assistance, the vCPU can operate in different VMPL by switching the corresponding VMSAs. A higher VMPL context can be seen as the secure world in the ARM TrustZone [39].

SVSM. The Secure Virtual-Machine Service Module (SVSM) is a newly proposed framework that aims at providing essential security services for the guest OS [9]. Both the SVSM and the guest OS share the guest physical address space. However, the SVSM's address space, with a higher VMPL, is inaccessible to a guest OS with a lower VMPL. The SVSM, operating in VMPL0, launches before the BIOS and guest OS in VMPL1. It occupies a fixed number of contiguous physical memory pages before transferring control flow to the BIOS and guest OS. These occupied memory pages remain unused by the guest OS. A dedicated secret page shared with the guest OS allows the guest OS to discover and request the service from SVSM.

The transitions between the guest OS and SVSM are handled by the hypervisor, while the communication protocols between the VM and hypervisor, e.g. passing parameters, and negotiating the shared memory pages, are specified in the Guest-Host Communication Block (GHCB) protocol [10]. The GHCB is a per-cpu guest physical page shared with the hypervisor. It is marked as unencrypted by clearing the C-bit of the page table entry of the guest. The communication between the VM and hypervisor is mediated through a new exception called VM Communication Exception (#VC). This exception can be triggered by Non-Automatic Exit (NAE) events.

A user application can request a service from SVSM by triggering a hypervisor trap through the VMGEXIT NAE event within the VM. Subsequently, the hypervisor schedules the SVSM. Alternatively, the guest OS can also request an SVSM service via the Model Specific Register (MSR) protocol. By creating a request in the GHCB MSR, the guest OS can ask the hypervisor to schedule the SVSM. In both scenarios, the Calling Area (CA) is used to pass the protocol version and call ID to the SVSM. The CA is a shared memory space, and its guest physical address is established through negotiation between SVSM and guest OS. Upon completion of the SVSM service routine, the hypervisor schedules the guest OS once again in accordance with the GHCB protocol.

2.2   Key Management System (KMS)

The KMS is an integrated approach for generating, distributing, and managing cryptographic keys for devices and applications [6]. It consists of various components and plays a vital role in securely managing cryptographic keys and secrets. It typically includes functionalities such as key's generation, storage, rollover and access control. The components of a KMS may include key servers, cryptographic hardware modules, APIs and management interfaces. KMS is essential for protecting sensitive information, ensuring secure communication, enabling data encryption, and meeting compliance requirements [32]. By providing centralized and controlled management of keys, KMS helps organizations maintain the confidentiality, integrity and availability of their data and systems, serving as a crucial foundation for secure operations and safeguarding against unauthorized access and data breaches.

2.3   Threat Model

Our goal is to develop a seamlessly integrated key protection scheme that operates within the highest VMPL inside a confidential VM empowered by AMD's SEV-SNP technology. Our threat model follows the standard assumptions of confidential computing, where any area outside the protected boundaries of the confidential VM is deemed to be under the control of potential attackers. We assume the framework's implementation follows the GHCB protocol reliably. However, we cannot place trust in the proxy server or kernel module. Even if these components compromised, the protected keys remains secure and inaccessible to unauthorized entities.

Ensuring client authentication is crucial for secure service delivery. In the case of SVSM-KMS, it plays a vital role in protecting the system from unauthorized access by malicious clients. To address this risk, SVSM-KMS only accepts requests from authenticated clients. However, the authentication mechanism is not included in the current design. By relying on a trusted third-party authentication mechanism, we can focus on building a strong key protection system while leveraging the security measures provided by the authentication service.

In addition, side channels and hardware attacks are out of the scope. It is assumed that the underlying hardware operates as described in the official documentation. The memory encryption and integrity protection are in place, adding an extra layer of security.

3   Design

3.1   Overview

Figure 1 illustrates the SVSM-KMS framework, comprising of the SVSM-KMS service module, kernel module, and proxy server. The service module resides in the highest VMPL area. Within the guest OS, the kernel module serves as an intermediary between the user application and the service module, managing shared memory and relaying requests from the proxy server. Consequently, user applications can access the SVSM-KMS service through the proxy server.

The service module of SVSM-KMS is integrated into the SVSM framework. Compared to normal KMS, it is running in an area with the highest VMPL, guarded against the tampering of the malicious hypervisor and guest OS with the hardware security mechanism provided from AMD SEV-SNP. Since the transition between the guest OS and SVSM is mediated by the hypervisor, the VM trap/resume events are inevitable, incurring performance impact. It is not easy

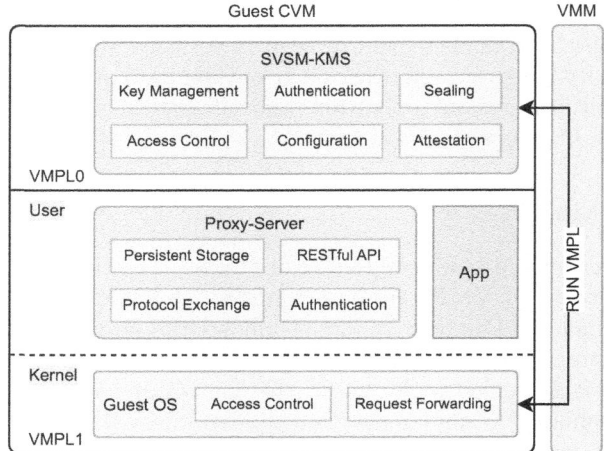

Fig. 1. An overview of the SVSM-KMS framework.

to deliver service from the most privileged SVSM-KMS to the least privileged user applications. Challenges present in integrating the key management service into the SVSM framework securely and efficiently, which outlined as follows.

Security. To ensure secure key protection throughout the entire life-cycle, the service module runs in the highest VMPL, i.e., VMPL0. The access control should not rely too much on the untrusted guest OS as much as possible. In other words, the authentication and authorization mechanism should be integrated into the highest VMPL. To prevent key leakage from untrusted guest OS, the keys should be sealed outside of the service module as well.

Performance. Considering the underlying SVSM framework, it is crucial to evaluate the performance impact, as each service request incurs at least two VM trap/resume events, potentially increasing service latency. These requests can originate from the kernel or traverse intermediary layers like the proxy server, kernel module, and hypervisor before reaching the service module. Given these indirect layers, we should mitigate the impact of these events and ensure efficient service delivery for user applications.

Compatibility. To enable key management services for various clients, it is necessary to bridge the gap between the SVSM-KMS service protocol and other network protocols. A proxy server is critical for seamless integration of our service with no code change, reducing the deployment effort.

The SVSM-KMS primarily serves the guest OS on the local VM. Therefore, it is not a pressing need to manage a large number of keys. According to the design of the Calling Area (CA) in the SVSM specification, the service request is handled on vCPUs one by one. At most, N requests can be handled in parallel, where N is the number of vCPUs. Due to the lack of high concurrency

requirements, a set of shared memory regions is used for each vCPU for simplicity. However, for better availability and scalability needs, we refer to existing
methods that can handle unexpected events such as power-offs and shutdowns
or managing a significant amount of keys [28,29].

To facilitate efficient and seamlessly integrated key protection, our design
consists of the following aspects: zero-copy mechanism, access control, sealed
storage, attestation and seamless integration, which are elaborated in the following subsections. In Sect. 3.2, we introduce how zero-copy design minimize
the performance impact caused by VMPL context switching. The access control
is introduced in Sect. 3.3. Section 3.4 introduces the sealed storage and attestation mechanisms, which establish a foundation of trust. Lastly, a proxy server is
introduced in Sect. 3.5 for seamless integration of existing web services.

3.2   Zero-Copy Design

Between the most privileged service module and the proxy server, user requests
go through two rounds of forwarding: first, they're passed through the proxy
server to the kernel module, and then forwarded from the kernel to the service
module via shared memory. So reducing cross-border data replication is crucial
for boosting service efficiency.

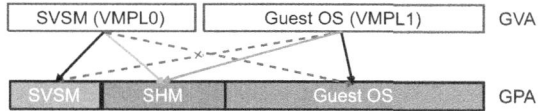

Fig. 2. An overview of the shared and private memory access.

Our system employs a zero-copy mechanism based on shared memory, allowing efficient communication between the most privileged service module and least
privileged user applications. To reduce the cost of mapping pages, the guest OS
shares memory pages with the SVSM-KMS in the full life-cycle of the service.
An overview of the shared memory mechanism is shown in Fig. 2. The SHM
is the shared memory. The dashed line with a cross indicates that the private
memory space of SVSM can not be accessed by the guest OS. The dashed line
without a cross indicates that the memory of the guest OS can be accessed by
mapping those memory pages in SVSM on demand. The solid arrow lines show
which physical memory space can be accessed by SVSM and guest OS.

The shared-memory regions allocated by the guest OS is mapped both in service module and the guest OS by issuing a shared-memory service request to the
service module. A user application can map those shared-memory regions continuously to his virtual memory space. The shared-memory keeps alive until
we remove the kernel module. The proxy server manages those regions for
each service request from each vCPU, enabling fully utilization of the hardware resources. In this way, a zero-copy mechanism between the user application

and the secure service module is realized and the overall communication overhead can be reduced to a minimum. In addition to the performance benefits, our innovative design also includes security measures for using shared memory. Since the shared memory can only be mapped once by a user application at a time, there is no intermediary who can intercept the communication between the proxy server and the service module. Furthermore, the shared memory is divided into request and response pages. The service module applies VMPL permission to the shared memory, ensuring the confidentiality of the request payload and the integrity of the response. This additional security layer enhances the overall system security.

3.3 Authentication and Access Control

To address security concerns, the majority of security measures, including access control, are integrated within the service module itself rather than the web proxy. These mechanisms operate within the highest VMPL layer over the guest OS, rendering them resistant to external circumvention. The access control mechanism is referenced from the Hadoop KMS and encompasses simple token authentication (STA) and access control list (ACL) mechanisms.

Authentication. A user application can request the token authentication service with user group information. The SVSM-KMS grants a legal user an access token signed with a secret. Then a user application can request our service with this token. The secret used to sign the token can be the same as Hadoop KMS for compatibility. A secret obtained inside the SVSM-KMS can also be used for signing for security. As such, the signed token cannot be forged outside of the service module. The algorithm used to sign the token is HMAC(SHA256) for compatibility with Hadoop KMS. Since the token authentication is moved to the SVSM-KMS service module, the proxy server consults the service module for a signed token on each authentication request. The granted token is put into the Cookie on each successful authentication for later user requests.

Access Control List. The ACL mechanism includes system-level and key-level ACLs. Authenticated users have restricted access to specific operations and keys based on the ACL configuration. The service module verifies user permissions for operations or services, and also check permissions for each individual key. The ACL are also maintained within the service module.

3.4 Sealed Storage and Attestation

The AMD PSP exposes the firmware service to the VM through the GHCB protocol. To request an AMD PSP firmware service, the VM issues an SNP Guest Request using the GHCB protocol [11]. The hypervisor mediates the communication between the firmware and the VM, ensuring protection against malicious attackers through AES-GCM authenticated encryption. There are two services can be used in our system, i.e., *key derivation* and *attestation report* services. According to the SEV-SNP firmware ABI specification [11], the guest can request a key from the firmware, derived from a root key. This key can be utilized by the guest for various purposes, including sealing keys or establishing communication with external entities. Meanwhile, the guest can request the firmware to generate an attestation report, which external entities can utilize to verify the guest's identity and security configuration. The guest generates attestation reports for VMPLs that meet or exceed the current VMPL, with the desired VMPL specified in the request message. Since both the *key derivation* and *attestation report* request parameters include the VMPL level, which must be greater than or equal to the current VMPL. Therefore, we can generate a dedicated export key and authentication report for SVSM-KMS running in the highest VMPL.

Sealed Storage. Concerning secure storage, we integrate a sealing service to facilitate the transfer of sealed data between the service module and local storage. The data stored in memory can be secure protected outside of the trusted service module for long-term storage. Compared to traditional KMS, our system persistently stores key data to the external world using a sealing key. This sealing key is obtained through a *key derivation* service request to the AMD PSP firmware within the service module and remains inaccessible to external entities, thus ensuring the confidentiality of the sealed data.

The proxy server utilizes SVSM-KMS's sealing service to store its in-memory data persistently. Before starting the web service, the proxy server transmits an unsealing request to upload the sealed data to SVSM-KMS. When the web service is about to stop, a sealing request is dispatched to locally save the sealed data. This guarantees a secure storage of the in-memory data. By proactively invoking the sealing service whenever there are updates to the in-memory keys, the proxy server not only ensures performance but also improves fault tolerance. For future improvements, it may be worth considering the implementation of an incremental backup or synchronization mechanism.

Attestation. The security status of the initial service module image can be measured and attested by remote attestation mechanism. The SVSM Attestation Protocol [9] is specified as a standard way to obtain an attestation report of the SVSM. By sending an *attestation report* service request to the firmware, the report for the service module can be obtained, enabling the attestation of its security state. As a result, a separate design for this service is unnecessary.

3.5   Seamless Integration

To enable seamless integration of our system into actual applications, we introduce a proxy server tailored for Hadoop applications, such as Hadoop KMS and DFS clients. In fact, other applications can also be supported by integrating the Key Management Interoperability Protocol (KMIP) [7]. Its main functionality is to redirect user requests to the secure SVSM-KMS and transmit the service response to HTTP clients. The proxy server consists of the following components.

Protocol Exchange. Since the service module follows standard SVSM protocol, protocol exchange is require for the proxy server to request service from the service module. The request and response of the service module are packed into an appropriate form before forwarding, i.e., JSON. Compared to standard JSON-RPC protocol [5], the request and response consist of message *header* and *body*. The *header* is transferred through general registers, while the *body* is stored in the shared memory. The user request is routed to the corresponding service handler according to the message header. The message body are in the form of JSON for simplicity, except for the sealing service where encrypted data is transferred between local and in-memory storage. The status code from the service module is mapped to the corresponding HTTP status code. In this way, we avoid parsing the message header and simplify the protocol design.

RESTful API. The RESTful API is based on the Hadoop KMS RESTful API[1] for compatibility, allowing our SVSM-KMS to support various Hadoop applications, especially the HDFS. The Hadoop KMS is used to assign encryption zone keys (EZK) to encryption zones automatically for transparently encrypting/decrypting data stored in HDFS [28]. Here, we demonstrate the seamless substitution of Hadoop-KMS with SVSM-KMS in the TDE of HDFS[2], showcasing the flexibility of our solution within existing Hadoop infrastructures.

Configuration. The configuration of the service module can be modified via the configuration service, allowing for the import of various configurations. We support the Hadoop KMS configuration format for compatibility. Presently, our system solely permits updating ACL and log configurations, along with the ability to adjust the log level for debugging purposes. Additionally, this service also enables the persistence of in-memory configurations, sealed locally with secure authenticated encryption.

4   Implementation

To showcase our design, we developed a prototype based on the infrastructure available from AMD[3]. Our project comprises three parts: the service module (over 4000 lines of Rust code), the kernel module (around 1000 lines of C code), and the proxy server (about 2000 lines of C++ code). The implementation details are elaborated in the following sections.

4.1   SVSM-KMS Service Module

Presently, we have developed the service module based on the linux-svsm project[4]. The service module provides functionalities encompassing key management, authentication, configuration, sealing, and more. Additionally, the

---

[1] https://hadoop.apache.org/docs/current/hadoop-kms/index.html.

[2] https://hadoop.apache.org/docs/current/hadoop-project-dist/hadoop-hdfs/ TransparentEncryption.html.

[3] https://github.com/AMDESE.

[4] https://github.com/AMDESE/linux-svsm/, commit ID: 2ea8eeab.

recently open-sourced coconut-svsm project[5] offers an alternative infrastructure for deploying our system within CVM. Both frameworks are written in Rust, ensuring memory safety. The Rust language ecosystem also greatly facilitates the development of SVSM-KMS with various crates for encryption and persistent storage.

Key Management. We provide key management services for Hadoop KMS Client and HDFS. These services primarily involve managing key metadata and performing tasks like creating, updating, encrypting, and decrypting key versions. Each service has a status code that complies with the official SVSM specification. The requests for these services originate from the proxy server and are indirectly derived from the HTTP clients. These core services are implemented within the service module as service handlers, offering support for managing AES and SM4 encryption keys. To ensure secure generation of encryption keys and initialization vectors (IV), the *rdseed* instruction is utilized to retrieve hardware-generated random values, which are then used to initialize the *ChaCha* Random Number Generator (RnG). By leveraging the hardware-provided RnG, the system ensures robust and secure generation of cryptographic material.

Configuration and Sealing. The configuration parameters for the service module are primarily obtained from the configuration files of Hadoop KMS. The ACL configuration from Hadoop KMS is also utilized in our service module. Furthermore, the sealing service within our implementation employs AES-GCM encryption for sealing operations.

Compared to Hadoop KMS, our implementation does not require frequent local storage access. All requests are managed using in-memory storage and operations. As a prototype, the in-memory data is organized in a HashMap data structure. Moreover, our service module runs directly in the VM, while Hadoop KMS runs in a JVM (Java VM). This significantly reduces the impact on performance. Meanwhile, we have not integrated certain important security mechanisms into our current system, such as Kerberos Authentication and Audit Logs supported in the Hadoop KMS. Since they are not the main focus of our current work and have little impact on performance evaluation, they can be integrated in the future. Furthermore, as in-memory storage space is limited, a swap-memory mechanism can be introduced as a future enhancement.

4.2 SVSM-KMS Kernel Module

Currently, there is no official example of calling an SVSM service in user-space. Here, we implement a custom kernel module for requesting the service. This kernel module is a *misc* device driver, exposing a /dev/svsm device file in the file system. It also provides *mmap* and *ioctl* interfaces to user applications, allowing service request through the *ioctl* syscall. The kernel module then forwards these user requests directly to the service module using the GHCB protocol. To prevent invalid and illegal service requests from being forwarded to the service module, a simple service filtering mechanism is introduced within the forwarding logic.

---

[5] https://github.com/coconut-svsm.

The request and response headers of the SVSM-KMS include the value of five general registers. These registers are used by the SVSM Core Protocol as well. The user applications wrap the value of these registers into a structure. The kernel module then extracts those values from the structure to the corresponding registers before issuing an GHCB Protocol request for scheduling the service within VMPL0. After the completion of the service, the kernel module copies those values back to the user-space so that the user application can receive a complete response from the service module.

Following above design, the kernel module does nothing but prepare shared memory and forward user request and service response across the whole lifetime. Such flexible design allows providing more security services for the user applications without major changes to the kernel module, ensuring extensibility.

4.3 SVSM-KMS Proxy Server

The proxy server is implemented with the open-sourced libhv framework[6]. It maps the shared memory to the private virtual memory space with the *mmap* syscall and forwards user requests to the service module with the *ioctl* syscall. It is capable of uploading configurations to SVSM-KMS, supporting the Hadoop KMS configuration format for compatibility. Prior to uploading, the configuration is transformed into JSON format. Afterwards, SVSM-KMS updates the corresponding configurations. Both configurations and sealed data are uploaded to the service module before accepting user requests. By doing so, we ensure the compatibility and seamless integration of a wider range of applications with our SVSM-KMS solution. Currently, two clients have undergone testing, namely the Hadoop KMS Client and HDFS (TDE). These clients can request services successfully via the RESTful API offered by SVSM-KMS.

While the SVSM-KMS is indeed leverages the VMPL security features specific to AMD SEV-SNP, it is not limited to deployment only on AMD SEV-SNP servers. Given that other mainstream CPU architectures, such as Intel Trust Domain Extensions (TDX) and ARM Confidential Computing Architecture (CCA), have introduced similar security features like td-partitioning [27] and ARM Planes [15], the SVSM-KMS can also be adapted to these environments. In fact, the coconut-svsm framework, which is similar to linux-svsm, already supports Intel's td-partitioning. This means that SVSM-KMS can be deployed on Intel TDX platforms, and potentially on ARM CCA as well, provided the necessary adaptations are made. Therefore, the SVSM-KMS is versatile and can be deployed across various CVM architectures beyond just AMD SEV-SNP.

5   Evaluation

In this section, we evaluate the performance of SVSM-KMS and Hadoop KMS. The tests are performed on a dual-socket 3rd Gen AMD EPYC processor (codenamed Milan) with 128 logical cores and 64 GB RAM, supporting the SEV-SNP security feature. The host system operates QEMU 6.1.50 on Ubuntu 22.04

---
[6] https://github.com/ithewei/libhv.

(kernel version 6.1.0-rc4-snp-host), while the VM is allocated 8 vCPUs and 8GB RAM, running Ubuntu 22.04 (kernel version 6.2.0-snp-guest). Throughout the experiments, Hadoop-3.3.4 is deployed on both the VM and a remote machine for evaluation. Simple Token Authentication is configured for both Hadoop KMS and SVSM-KMS. Additionally, the ACL is identical for both KMS.

5.1 Micro-benchmarks

For the micro-benchmarks, we evaluate the service latency of SVSM-KMS and Hadoop-KMS. Both KMS are running on the same VM, with SVSM-KMS on VMPL0 and Hadoop KMS on VMPL1. The experiments are performed on both local and remote scenarios. In total, we evaluate thirteen RESTful APIs, excluding *Invalidate Cache*, which is not supported in our system currently. We make 150 requests to each service and record the average elapsed time for each request. To request the service in both local and remote scenarios, we utilize the popular *requests* Python package.

The latency of the service is evaluated in both local and remote situations. When calling a service of SVSM-KMS, the support of the hypervisor is necessary, thereby inevitably increasing the service latency. However, the service module operates on a real machine, not a JVM like Hadoop KMS. This latency represents the only impact that cannot be reduced through software means. In remote cases, there is network latency. The impact of the SVSM framework can be somewhat neglected. The evaluation is presented in Fig. 3. The terms Hadoop and SVSM correspond to the implementation of KMS. The L and R denote local and remote scenarios, respectively.

While differences exist between SVSM-KMS and Hadoop KMS, it's essential to acknowledge that SVSM-KMS serves as a prototype and does not encompass all the optimizations featured in Hadoop KMS. Nonetheless, our system demonstrates comparable performance to Hadoop KMS in both local and remote service scenarios. Due to implementation issues, it shows significant performance degradation on specific operations, such as GenerateEncryptedKeys, GetKeyNames, and GetKeysMetadata in the remote scenario. It is noteworthy that these operations are infrequently utilized in a deployed cloud platform. Therefore, their impact on cloud service performance is negligible. Conversely, SVSM-KMS holds a notable advantage in the DecryptEncryptedKey operation, serving as a hot function in HDFS (TDE).

The SVSM-KMS outperform Hadoop KMS in local service scenarios due to the zero-copy design, which helps mitigate the performance impact of the SVSM framework. While not incorporating all optimizations, SVSM-KMS show the potential of providing efficient and secure key protection services within CVM. By integrating key-cache and warm-up mechanisms into SVSM-KMS, further improvements in performance can be achieved, highlighting the scalability and extensibility of our design.

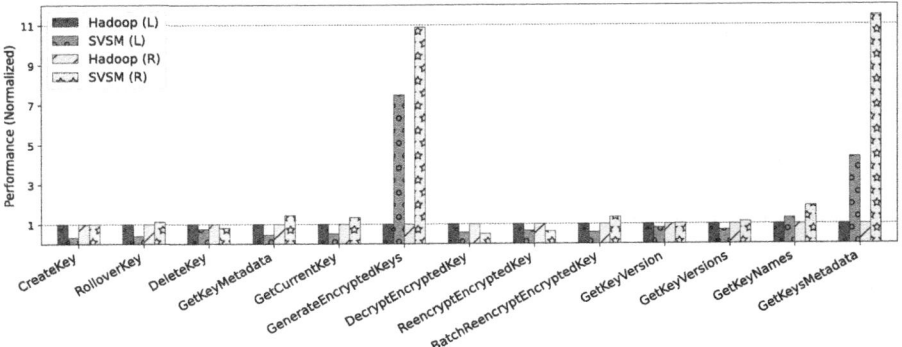

Fig. 3. The service latency of the proposed SVSM-KMS and Hadoop KMS in serving local and remote machines.

5.2   Macro-Benchmarks

We perform read and write operations on an encrypted zone of HDFS ten times, with each operation resulting in a service request to the KMS. The average elapsed time of each operation is recorded. To evaluate the performance impact on HDFS, file sizes range from 32 KB to 4 MB. The experiments consists of both local and remote scenarios. In the local scenario, the HDFS and KMS are deployed in the same VM. Conversely, in the remote scenario, the HDFS is deployed on another machine, sending service requests to a local or remote deployed KMS. The evaluation on HDFS (TDE) is illustrated in Fig. 4.

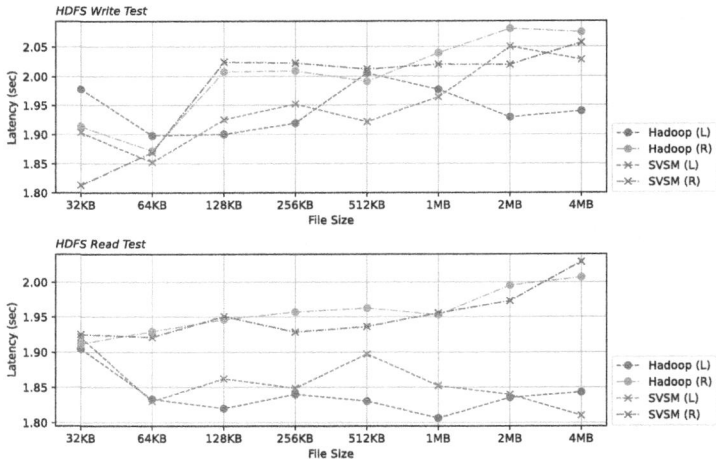

Fig. 4. The service latency (seconds) evaluation between SVSM-KMS and Hadoop-KMS in serving HDFS (TDE) under local and remote scenarios.

Figure 4 shows the average time consumption of read and write operations in different scenarios. Since performing read and write operations in the encryption zone is time-consuming, the performance impact caused by the KMS can be largely ignored. Although each operation requires sending a request to the KMS, the average time consumption shows little difference on both SVSM-KMS and Hadoop KMS when serving HDFS (TDE). Whether in local or remote scenarios, as the file size increases, the performance impact on the HDFS (TDE) can be largely ignored. Overall, these experimental results demonstrate the effectiveness and efficiency of our prototype system, indicating its potential to provide comparable or better performance than existing solutions, especially in local service scenarios and TDE (HDFS).

6   Related Works

Microsoft Azure KeyVault [8], AWS KMS [1], and Google Cloud KMS [2] are cloud-based solutions for centralized management of keys, certificates, and secrets. Cloudera Navigator Key Trustee Server [3] offers key management for securing data in Hadoop, while Ranger KMS provides centralized key management for secure encryption in Apache Hadoop. HashiCorp Vault [4] offers secure storage and dynamic management of secrets. While these systems are secure, they are often deployed in VM where vulnerable guest OS can compromise their security. Their TCB is still too large to hosting such services. Therefore, further privilege separation within CVM is critical to the security of KMS on the cloud.

The Trusted Execution Environment (TEE) can effectively mitigate the risks of key compromise and unauthorized access from the untrusted OS. One advancement in improving KMS security is the STYX framework [44], which incorporates Intel Software Guard eXtension (SGX) [25]. By leveraging the capabilities of Intel SGX, cloud system KMS security is greatly enhanced [20,30], while also considering aspects such as performance and high availability. In [43], a novel technique for key management was presented to enhance security, privacy, and confidentiality. The proposed approach, combining Intel SGX and FIWARE components, is scalable and suitable for safeguarding IoT data. Another notable work proposes a key management solution based on Intel SGX for data-centric networking [38]. MultiSGX-KMS is a decentralized system designed to safeguard user keys during exchange and ensure fault tolerance through secret sharing and Intel SGX [12], which can be leveraged to realize a decentralized design of the SVSM-KMS framework. Researchers also introduced RansomClave, a ransomware family that securely manages cryptographic keys using an enclave [18]. Similarly, TZ-KMS is introduced for joint cloud computing based on ARM Trust-Zone [33]. Samsung's Keymaster is a key management system for mobile platforms based on ARM TrustZone [42]. In general, TEEs provide a secure and isolated environment where sensitive operations, such as key generation, storage, and cryptographic operations, can be performed with high assurance. However, these Intel SGX and ARM TrustZone based proposals are not multi-layered defense as SVSM-KMS in the realm of confidential computing.

Previous works have surveyed traditional key management across various domains, including the smart grid [22], wireless sensor networks [24] in Internet of Things (IoT) scenarios [17], as well as cloud security contexts [13, 19, 37]. A comparative analysis of cryptographic key management systems is detailed in [31]. Recently, researchers surveyed three popular key management systems, namely AWS CloudHSM, Keyless SSL, and STYX [26]. However, their threat model clearly contrasts to SVSM-KMS, where the guest OS is considered vulnerable.

7   Conclusion

SVSM-KMS is a secure key protection mechanism that leverages the latest features from AMD SEV-SNP for cloud services. It integrates a series of mechanisms to enhance performance and establishes a layered defense. A zero-copy mechanism is introduced to optimize overall performance, utilizing shared memory throughout the entire life-cycle to eliminate redundant data transfers between the service module and user applications. To realize multilayered defense, we integrate our core service into the SVSM framework, ensuring that critical key management service operates at the highest VMPL, shielded from vulnerable guest OS. The system's security is further fortified through sealed storage and the capability to measure and attest the security state using the standard SVSM Attestation Service. Meanwhile, hardware instructions are leveraged for the secure generation of keys, effectively addressing security and privacy concerns associated with centralized and decentralized KMS. We focus on seamless integration by introducing a proxy-server for Hadoop clients, streamlining the deployment of our system. Evaluation results demonstrate the performance of the SVSM-KMS across local and remote scenarios, including the compatibility as a drop-in replacement for Hadoop KMS in HDFS (TDE), demonstrating comparable performance across various usage scenarios.

In conclusion, the SVSM framework can be leveraged to provide security critical services (e.g., key management) for a variety of applications. Future works would be the key-cache and warm-up mechanisms for performance. A decentralized system design of the SVSM-KMS can be considered for security.

Acknowledgment. This work was supported by National Natural Science Foundation of China (Grant No. 62272452).

References

1. Amazon Key Management Service | Manage encryption keys | Amazon Web Services (2024). https://www.amazonaws.cn/en/kms/
2. Cloud Key Management | Google Cloud (2024). https://cloud.google.com/security-key-management?hl=zh-cn
3. Cloudera Navigator Key Trustee Server Overview | CDP Private Cloud (2024). https://docs.cloudera.com/cdp-private-cloud-base/7.1.3/security-key-trustee-server/topics/cm-security-kts-overview.html

4. HashiCorp Vault - Manage Secrets & Protect Sensitive Data (2024). https://www.hashicorp.com/products/vault
5. JSON-RPC 2.0 specification (2024). https://www.jsonrpc.org/specification
6. Key management (2024). https://en.wikipedia.org/wiki/Key_management
7. Key management interoperability protocol specification version 2.0 (2024). https://docs.oasis-open.org/kmip/kmip-spec/v2.0/os/kmip-spec-v2.0-os.html
8. Key Vault | Microsoft Azure (2024). https://azure.microsoft.com/en-us/products/key-vault
9. Secure VM service module for SEV-SNP guests (2024). https://www.amd.com/content/dam/amd/en/documents/epyc-technical-docs/specifications/58019.pdf
10. SEV-ES guest-hypervisor communication block standardization (2024). https://www.amd.com/content/dam/amd/en/documents/epyc-technical-docs/specifications/56421.pdf
11. SEV secure nested paging firmware ABI specification (2024). https://www.amd.com/system/files/TechDocs/56860.pdf
12. Abdulsalam, Y.S., Bouamama, J., Benkaouz, Y., Hedabou, M.: Decentralized SGX-based cloud key management. In: International Conference on Network and System Security, pp. 327–341. Springer (2023)
13. AlBelooshi, B., Damiani, E., Salah, K., Martin, T.: Securing cryptographic keys in the cloud: a survey. IEEE Cloud Comput. 3(4), 42–56 (2016)
14. An, H., Choi, R., Kim, K.: Blockchain-based decentralized key management system with quantum resistance. In: Information Security Applications: 19th International Conference, WISA 2018, Jeju Island, Korea, 23–25 August 2018, Revised Selected Papers 19, pp. 229–240. Springer (2019)
15. Arm: Evolution of the arm confidential compute architecture (2024). https://www.youtube.com/watch?v=1AsvIt7bSLY
16. Arul Oli, S., Arockiam, L.: A framework for key management for data confidentiality in cloud environment. In: 2015 International Conference on Computer Communication and Informatics (ICCCI), pp. 1–4 (2015)
17. Bartsch, W., et al.: Design rationale for symbiotically secure key management systems in IoT and beyond. In: ICISSP, pp. 583–591 (2023)
18. Bhudia, A., O'Keeffe, D., Sgandurra, D., Hurley-Smith, D.: RansomClave: ransomware key management using SGX. In: Proceedings of the 16th International Conference on Availability, Reliability and Security, pp. 1–10 (2021)
19. Buchade, A.R., Ingle, R.: Key management for cloud data storage: methods and comparisons. In: 2014 Fourth International Conference on Advanced Computing & Communication Technologies, pp. 263–270. IEEE (2014)
20. Chakrabarti, S., Baker, B., Vij, M.: Intel SGX enabled key manager service with openstack barbican. arXiv preprint arXiv:1712.07694 (2017)
21. Chandramouli, R., Iorga, M., Chokhani, S.: Cryptographic key management issues and challenges in cloud services. Secure Cloud Comput. 1–30 (2013)
22. Ghosal, A., Conti, M.: Key management systems for smart grid advanced metering infrastructure: a survey. IEEE Commun. Surv. Tutor. 21(3), 2831–2848 (2019)
23. Hashimoto, M.: Overview of memory security technologies. In: 2021 International Symposium on VLSI Technology, Systems and Applications (VLSI-TSA), pp. 1–2 (2021)
24. He, X., Niedermeier, M., De Meer, H.: Dynamic key management in wireless sensor networks: a survey. J. Netw. Comput. Appl. 36(2), 611–622 (2013)
25. Hu, X., et al.: STYX: a hierarchical key management system for elastic content delivery networks on public clouds. IEEE Trans. Dependable Secure Comput. 18(2), 843–857 (2021)

26. Huang, X., Chen, R.: A survey of key management service in cloud. In: 2018 IEEE 9th International Conference on Software Engineering and Service Science (ICSESS), pp. 916–919. IEEE (2018)
27. Intel: Intel Trust Domain Extensions (2020). https://cdrdv2-public.intel.com/773039/intel-tdx-module-1.5-td-partitioning-spec-354807001.pdf
28. Jin, W., Geng, K., Yu, M., Guo, Y., Li, F.: Efficiently managing large-scale keys in HDFS. In: 2021 IEEE 23rd International Conference on High Performance Computing & Communications; 7th International Conference on Data Science & Systems; 19th International Conference on Smart City; 7th International Conference on Dependability in Sensor, Cloud & Big Data Systems & Application (HPCC/DSS/SmartCity/DependSys), pp. 353–360. IEEE (2021)
29. Karande, V., Bauman, E., Lin, Z., Khan, L.: SGX-Log: securing system logs with SGX. In: Proceedings of the 2017 ACM on Asia Conference on Computer and Communications Security, pp. 19–30 (2017)
30. Kuan, S.: Improving the security of KMS on a cloud platform using trusted hardware. Master's thesis, Aalto University, School of Science (2018)
31. Kuzminykh, I., Ghita, B., Shiaeles, S.: Comparative analysis of cryptographic key management systems. In: International Conference on Next Generation Wired/Wireless Networking, pp. 80–94. Springer (2020)
32. Lei, S., Zishan, D., Jindi, G.: Research on key management infrastructure in cloud computing environment. In: 2010 Ninth International Conference on Grid and Cloud Computing, pp. 404–407. IEEE (2010)
33. Luo, S., Hua, Z., Xia, Y.: TZ-KMS: a secure key management service for joint cloud computing with arm trustzone. In: 2018 IEEE Symposium on Service-Oriented System Engineering (SOSE), pp. 180–185. IEEE (2018)
34. Mattioli, M.: Rome to Milan, AMD continues its tour of Italy. IEEE Micro **41**(4), 78–83 (2021)
35. Mofrad, S., Zhang, F., Lu, S., Shi, W.: A comparison study of Intel SGX and AMD memory encryption technology. In: Proceedings of the 7th International Workshop on Hardware and Architectural Support for Security and Privacy. HASP '18. Association for Computing Machinery, New York (2018)
36. Ning, Z., Zhang, F., Shi, W., Shi, W.: Position paper: challenges towards securing hardware-assisted execution environments. In: Proceedings of the Hardware and Architectural Support for Security and Privacy. HASP '17. Association for Computing Machinery, New York (2017)
37. Oruganti, R., Churi, P.: Systematic survey on cryptographic methods used for key management in cloud computing. In: International Conference on Innovative Computing and Communications: Proceedings of ICICC 2021, vol. 2, pp. 445–460. Springer (2022)
38. Park, M., et al.: An SGX-based key management framework for data centric networking. In: International Workshop on Information Security Applications, pp. 370–382 (2019)
39. Pinto, S., Santos, N.: Demystifying arm trustzone: a comprehensive survey. ACM Comput. Surv. (CSUR) **51**(6), 1–36 (2019)
40. Qin, H., et al.: Protecting encrypted virtual machines from nested page fault controlled channel. In: Proceedings of the Thirteenth ACM Conference on Data and Application Security and Privacy, pp. 165–175 (2023)
41. AMD SEV-SNP: strengthening VM isolation with integrity protection and more. White Paper, p. 8 (2020)

42. Shakevsky, A., Ronen, E., Wool, A.: Trust dies in darkness: shedding light on Samsung's TrustZone keymaster design. In: 31st USENIX Security Symposium (USENIX Security 22), pp. 251–268 (2022)
43. Valadares, D.C.G., da Silva, M.S.L., Brito, A.E.M., Salvador, E.M.: Achieving data dissemination with security using FIWARE and Intel software guard extensions (SGX). In: 2018 IEEE Symposium on Computers and Communications (ISCC), pp. 1–7. IEEE (2018)
44. Wei, C., Li, J., Li, W., Yu, P., Guan, H.: STYX: a trusted and accelerated hierarchical SSL key management and distribution system for cloud based CDN application. In: Proceedings of the 2017 Symposium on Cloud Computing, pp. 201–213 (2017)

GNNexPIDS: An Interpretation Method for Provenance-Based Intrusion Detection Based on GNNExplainer

Ziyang Yu[1,2], Wentao Li[1,2], Xiu Ma[1,2], Baorui Zheng[1,2], Xinbo Han[1,2],
Ning Li[1(✉)], Qiujian Lv[1], and Weiqing Huang[1]

[1] Institute of Information Engineering, Chinese Academy of Sciences, Beijing, China
{yuziyang,lining01}@iie.ac.cn
[2] School of Cyber Security, University of Chinese Academy of Sciences,
Beijing, China

Abstract. With the growing menace of Advanced Persistent Threats
(APTs) in recent years, provenance graphs have become the focus of
studies on various techniques to analyze APTs using Graph Neural Networks (GNNs) for automated host intrusion detection. However, a bar-
rier to the practical adoption of GNN-based intrusion detection systems
(IDS) is the lack of interpretation. Although some studies on the interpretation of GNNs have been proposed, they have not been applied to
IDS and do not focus on how to interpret anomalies.

To overcome these limitations, we introduce an interpreter called
GNNexPIDS to interpret provenance-based intrusion detection systems
(PIDSes) by adapting and extending GNNExplainer, a generic GNN
model interpretation tool. Since GNNExplainer only implements the
interpretation on nodes, we extend GNNExplainer to interpret both
nodes and edges. By this extension, our GNNexPIDS can interpret PIDS
that use link prediction and node classification tasks. Additionally, we
introduce a graph reduction method based on anomaly status to optimize interpretation results, aiding in finding interactive behaviors leading to abnormal alerts. Experimental results on two datasets CADETS
E3 and THEIA E3 demonstrate fidelity and stability of our proposed
method, while comparing to the baselines, e.g., LIME and GraphLime.
Our method outperforms baselines on two datasets for 5% to 20% fidelity
and 10% to 30% stability.

Keywords: Security Applications · GNN Interpretability ·
Provenance Graphs · Provenance-based Intrusion Detection ·
Interpretation of Intrusion Detection

1 Introduction

In recent years, the rise of Advanced Persistent Threats (APTs) has significantly
challenged cybersecurity efforts [1]. Effective and prompt detection methods to

J. Zhao and W. Meng (Eds.): SciSec 2024, LNCS 15441, pp. 236–253, 2025.
https://doi.org/10.1007/978-981-96-2417-1_13

counteract APTs are crucial. Intrusion detection is crucial for identifying deviations that may signal threats [2]. Utilizing provenance graphs has emerged as an innovative approach to enhancing intrusion detection capabilities.

Provenance graphs encapsulate a system's operational history through structured audit logs, which are critical for intrusion detection [3–8]. Originating from system-level logs, they detail interactions among kernel objects, providing a comprehensive narrative of the system. Despite their detail, they present challenges in modeling complex relationships and features [1,9,10]. To overcome these challenges, Graph Neural Networks (GNNs) are employed, leveraging their capacity to integrate neighboring information and effectively capture both structural and feature data [5–7,11–13]. Research basically splits into two approaches: one focuses on identifying anomalous system objects via node classification [5,6], and the other on detecting abnormal interactions through link prediction [7,11].

Despite their success in lab settings, GNN models' lack of transparency and interpretability hinders their real-world application in security [14,15]. The "black-box" nature of these models challenges the adoption of GNN-based Provenance Intrusion Detection Systems (PIDSes) due to difficulty in trusting system decisions without clear justifications [18]. These systems often yield binary outcomes (abnormal or normal) without explaining the influence of specific edges or nodes, increasing the security operators' workload. Thus, there's a critical need for interpreters that can clarify the impact of particular nodes or edges, significantly easing the experts' analysis efforts. However, existing related research are facing the following challenges:

– Existing methods for interpreting GNN models help clarify the importance of nodes, edges, and features in making predictions. Techniques like SA [16] and CAM [17] use gradients or feature activations to assess input significance. GraphLime [19] focuses on the importance of node features for classification, while LRP [21] links model parameters with input-output features. However, works above are limited by GNNs' structure, tasks or the type and scale of the input graphs. The complexity and scale of provenance graphs add to the challenge, making it harder for security operators to verify the accuracy of anomalies detected by PIDS. Thus, there's is a need to develop a generic GNN interpreter for both link prediction and node classification. And this interpreter should work on provenance graphs.
– Most prior methods have been designed for non-security domains, primarily to understand the mechanisms of GNNs, whereas security operators are more concerned with quickly understanding PIDS alerts and responding to them. The main function of IDS is to provide alarms for anomalies, and the related behaviors that cause alarms often have a high tendency be anomalous. In order to help security operators quickly handle alarms, interpretation result should not only consider the relevance to the explained node or edges but also the features' tendency to be anomalous.
– Research indicates that prior methods have failed to be adopted in security domains due to poor performance [22,23]. The interpretation results of these methods often contain a large number of features irrelevant to security. This issue leads to poor performance in the fidelity and stability of the interpreters.

To address these issues, this paper proposes an interpreter called GNNex-PIDS. GNNexPIDS aims at enhancing decision transparency of GNN-based PIDSes by extending the generic GNN interpreter named GNNExplainer [25] to provenance graphs. Moreover, GNNexPIDS broadens the scope of GNNExplainer from focusing on node interpretation to interpreting edges. Additionally, it reduces irrelevant parts from the interpretation results based on anomaly status, thereby enhancing the decision credibility of GNN-based PIDSes. Given the primary tasks of PIDSes are node classification and link prediction, we focus on threaTrace and Kairos, due to their significant contributions to these areas [6,7]. This study's main contributions are threefold:

- We apply GNNExplainer to provenance graphs by expanding the neighborhood search depth to interpret the provenance graph. Additionally, we extend GNNExplainer from its original node interpretation to further encompass edge interpretation. With this extension, our interpreter can operate not only on link prediction tasks but also on node classification tasks.
- To obtain the interactive behaviors leading to abnormal alerts in PIDS, we introduce a graph reduction method based on anomaly status to further optimize the interpretation results from GNNExplainer. Through this method, our interpreter can focus more on anomalies and achieve higher performance. This result aids security operators in analyzing alerts generated by PIDS from both anomaly and causality perspectives.
- We evaluate GNNexPIDS on two datasets, CADETS E3 and THEIA E3. In comparison with the baselines such as LIME and GraphLime, our method surpasses these baselines by achieving 5% to 20% higher fidelity and 10% to 30% greater stability on these datasets. These metrics show the faithfulness of GNNexPIDS. Furthermore, by contrasting the fidelity with and without graph reduction, we demonstrate the effectiveness of our graph reduction method.

2   Related Work

2.1   Interpreters on Intrusion Detection Systems

While there has been extensive research in the AI domain on various interpretability techniques for a wide range of deep learning models, interpretation of various anomaly detection systems in the security domain still requires further investigation.

Wei et al. propose xNIDS [33], a framework for interpreting deep learning-based network intrusion detection systems (DL-NIDS) and generating actionable defense rules. XNIDS identifies historical inputs that lead to predictions near the original prediction and utilizes sparse group lasso to capture feature dependencies in structured data. However, due to the multitude of system call types and system entity types, interactions are more complex compared to the network side. Consequently, the computational complexity of PIDS and the consideration of causal relationships between entities are higher compared to DL-NIDS, and neural network models are more complex. Therefore, the methods employed by XNIDS cannot be directly applied to address the interpretability issues of GNN-based PIDS.

Han et al. present DeepAID [34], a framework for interpreting and enhancing DL-based anomaly detection in security applications. DeepAID includes two key techniques: Interpreter and Distiller. Interpreter formulates anomaly interpretation as an optimization problem and proposes techniques to ensure interpretations meet security domain concerns. Distiller extends Interpreter by distilling high-level heuristics and expert feedback into a simplified model, facilitating easier understanding and modification by security practitioners. However, DeepAID briefly discusses interpretation methods for simple homogeneous graphs and provides only one case study focusing on network lateral movement attacks. DeepAID leaves metrics such as fidelity and stability on graphs open for further discussion. Further research is required to determine whether DeepAID can be applied to GNN-based PIDS.

2.2   Explanations for Graph Neural Networks

Interpretability methods in graph models provide insights into the importance of various graph components, crucial for improving predictive accuracy. These methods aim to explain graph input features and are categorized into four main types based on their approach:

Gradients/Features-Based Methods: These methods use gradients or hidden feature activations to estimate the importance of inputs. For example, SA uses squared gradient values as importance metrics [16], while CAM identifies key nodes by mapping features from the last layer back to the input space, requiring specific GNN layers [17]. However, their applicability is limited by GNN structure and task-specific constraints.

Surrogate Methods: This approach uses a simpler model to approximate the complex model's predictions, helping explain the significance of node features in tasks like classification. GraphLime extends the LIME algorithm to graphs [19], and PGM-Explainer creates a probabilistic model for GNN explanations [20].

Decomposition Methods: These methods break down predictions to reflect the importance of input features, often delving into model parameters. LRP provides interpretations based on model parameters but focuses only on node importance [21], not addressing broader graph structures.

Perturbation-based Methods: By generating masks for input graphs, these methods highlight critical features across different tasks (e.g., nodes, edges). PGExplainer learns edge masks for explanations [24], while GNNExplainer optimizes masks for edges and node features, aiding in interpreting GNN predictions in feature-rich graphs [25].

Most of these interpretability approaches, while offering valuable insights, have limitations, such as being model or task-specific, not fully applicable to link prediction, or only addressing node importance. Adapting and deploying new models remains a challenge due to these constraints.

2.3   Provenance-Based Intrusion Detection

Methods that identify anomalies in host behaviors through provenance data typically analyze kernel audit logs to establish normal patterns, treating deviations as potential threats [3,8]. Within this framework, GNNs are utilized for two main tasks: node classification and link prediction [26].

StreamSpot employs hash computations on subgraphs to generate a fixed-size feature vector [32]. SIGL employs a graph autoencoder with GNNs to compress provenance graph features into a vector, analyzing software installations [8]. Unicorn divides the provenance graph into temporal snapshots, using graph kernel methods for feature extraction [3].

Significant advancements include threaTrace and Kairos [6,7]. threaTrace uses GNNs for node classification within provenance graphs to detect anomalies [6]. Kairos focuses on link prediction, using GNNs to predict connections and identify security risks [7]. These efforts demonstrate the versatility of GNNs in improving detection in provenance-based systems.

3   Method

3.1   Approach Overview

For a given anomaly detection system based on a provenance graph, denote as $G_P = (V_P, E_P)$, we aim to interpret its anomaly detection results rather than simply obtaining binary outcomes, such as anomaly or normal. When the system generates an anomaly alert, which corresponds to a specific edge e or node v flagged as anomalous in the provenance graph (depending on whether the PIDS task is node classification or link prediction), our objective is to use the interpreter to explain the anomalous node or edge and obtain the subgraph $G_S = (V_S, E_S)$ that led to it being identified as anomalous by the PIDS.

By analyzing the subgraph causing the anomaly, security operators can determine whether the anomaly is a true positive (TP) or false positive (FP), enabling targeted responses to anomalies. If the anomaly is a TP, security operators can identify attacker techniques, tactics, and attack types based on the interpretation subgraph, facilitating swift responses. Conversely, if the anomaly is a FP, security operators can identify and summarize the reasons for the false alarm, and further refine the anomaly detection system to improve performance and reduce the probability of generating alerts.

Figure 1 provides an overview of our approach. Nodes in the graph are system entities where rectangles are processes, ovals are files, and diamonds are sockets. Edges among nodes represent system calls. GNNexPIDS contains three parts to interpret a given PIDS and provenance graph: Anomaly detection, subgraph generation and graph reduction based on anomaly status. As shown in Fig. 1 (a), a provenance graph is sent into a PIDS to detect which edges or nodes are anomalous and the anomalous ones will be labeled as 'positive'. Every edge or node will also have an anomaly score. The provenance graph G_P will get a threshold $TH(G_P)$. The PIDS model, an anomaly edge or node and the

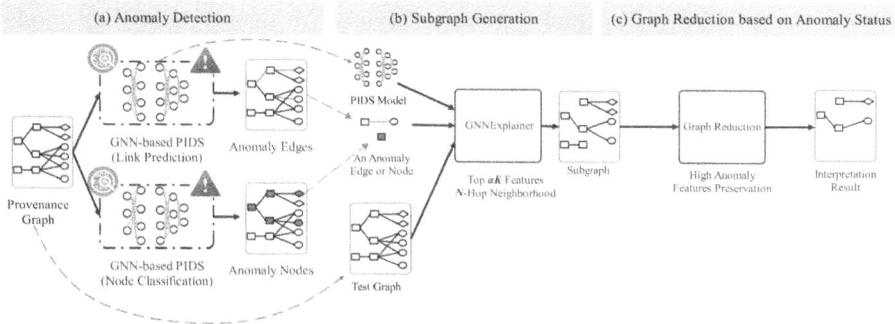

Fig. 1. An overview of our approach. We first obtain three inputs for the interpreter by anomaly detection. Then we generate subgraphs through GNNExplainer. Finally, we perform a reduction on the subgraph based on their anomalous status to produce the final interpretation results

provenance graph will be the input for the next part. Then in Fig. 1 (b), we use GNNExplainer to obtain a subgraph. This subgraph stands for several interactions within the system that have a close relationship with the anomaly edge or node. The interpreter will search for potential features within the N-hop neighborhood of the interpreted node or edge and select top αK features to generate subgraph, where α is a hyperparameter and $\alpha > 1$. K means the final desired number of features. Finally, in Fig. 1 (c), we employ a graph reduction method based on abnormal status to obtain the top K anomalous features among the subgraph, thus enhancing the performance and credibility of interpretation.

3.2 Anomaly Detection

Anomaly Detection is used to detect which nodes or edges are anomalous and provide essential inputs and threshold of the provenance graph for the next two parts. This subsection will show how we get these.

Provenance Graph. The constructed provenance graph serves as input not only for the PIDS but also for the interpreter. To construct a provenance graph $G_P = (V_P, E_P)$, where V_P represents a set for all nodes in G_P and E_P represents a set for all edges, we need retrieve audit logs from running hosts and converts them into provenance graphs. The methods used for processing audit logs and constructing the graph depend on the PIDS. For example, in Kairos [7], a hierarchical feature hashing technique is utilized to encode features for source and destination nodes, and a one-hot vector encoding is employed for edge types. Then, these data are input into the PIDS for anomaly detection.

PIDS Model. PIDS are usually trained through a graph neural network using numbers of provenance graphs composed entirely of normal behaviors, in order to learn the patterns of normal system behavior. We denote the PIDS model for link prediction as $PIDS_{lp}$ and the PIDS model for node classification as $PIDS_{nc}$. Through the PIDS model, we can detect anomaly entities or behaviors in provenance graph. PIDS model will also be an input for subgraph generation part of GNNexPIDS.

Detection Results. The PIDS model learns the relationships between entities and provides rich contextual information. This allows PIDS to identify patterns of benign behavior. Based on these patterns, PIDS detects anomalies in the system. It calculates an anomaly score for each node or edge. A threshold is determined for the entire provenance graph.

Given a provenance graph $G_P = (V_P, E_P)$ and a $PIDS_{lp}$ that detect this graph, we can get anomaly detection results $label_{lp}(e_i)$ and anomaly score $AS_{lp}(e_i)$ for edge $e_i \in E_P$. Similarly, we can get anomaly detection results $label_{nc}(v_i)$ and anomaly score $AS_{nc}(v_i)$ for node $v_i \in V_P$ through $PIDS_{nc}$. We also have the threshold $TH_{lp}(G_P)$ or $TH_{nc}(G_P)$ for provenance graph G_P.

After thorough training and anomaly detection in provenance graphs, we obtain crucial inputs for the interpreter, which are shown in Fig. 1: the trained model $PIDS$ and a positive node v or an edge $e = (u, v)$ and provenance graph G_P.

3.3   Subgraph Generation

After determining the three inputs for the interpreter, we use GNNExplainer to find a set number of features and generate them to a subgraph $G_S = (V_S, E_S)$. These features have a close relationship with the node v or edge $e = (u, v)$ being explained. This relationship helps security operators understand the reasons behind the anomaly detection for this instance. Additionally, we limit the neighborhood search depth N when generating masks and subgraphs. This limitation is due to the complexity and large scale of the provenance graphs. The large scale makes searching the entire graph difficult, and searching only within a 1-hop distance results in information loss. Thus, it is crucial to define an appropriate search range. This subsection will first introduce the overall structure of GNNExplainer, followed by a detailed explanation of how GNNExplainer generates feature masks when explaining nodes. Subsequently, we will describe how we utilize GNNExplainer to explain edges.

The Architecture of GNNExplainer. As shown in Fig. 2, for the input graph G_P, X represents the set of feature vectors for each node in the graph, while A represents the adjacency matrix of the provenance graph. Masks M are obtained to indicate important input features. Different types of masks are generated based on different interpretation tasks, such as node masks, edge masks, and node feature masks. Then, the generated masks are combined with the input graph to obtain a new graph containing essential input information. Finally, the new graph is input into the pre-trained PIDS model to obtain new predictions, which are compared with the original predictions to evaluate the masks and update the mask generation algorithm. Intuitively, the important input features captured by the masks should convey critical semantic meanings, resulting in new predictions that are essentially similar to the original predictions.

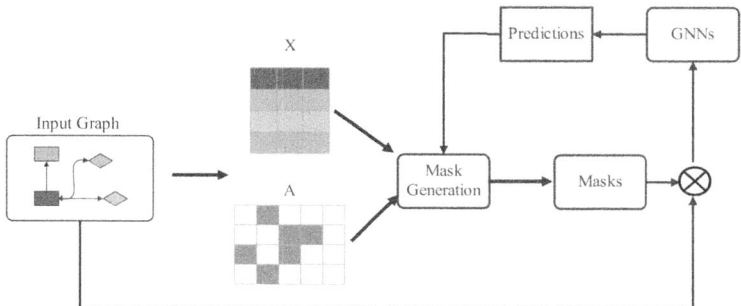

Fig. 2. The architecture of GNNExplainer. Given the input graph, we generate masks essential for highlighting critical input features. Then masks are integrated with the graph for new predictions in the PIDS model, ensuring the updated outputs closely reflect the original interpretations

Subgraph for a Node. This subsubsection introduces the formulation of the "Mask Generation" part in Fig. 2. For a node v detected as positive by PIDS, our goal is to find the important subgraphs $G_S \subseteq G_P$ and the important feature subset $X_s = x_j \mid v_j \in G_S$ that are crucial for the positive result $\hat{y}$. G_S and X_S are both limit to the N hop neighborhood of v. By utilizing mutual information (MI), it can be represented as below.

$$\max_{G_s} MI(Y, (G_S, X_S)) = H(Y) - H(Y|G = G_S, X = X_S) \tag{1}$$

For instance, when $v_j \in G_P(v_i), v_j \neq v_i$, if removing node v from $G_P(v_i)$ significantly decreases the abnormal score for anomaly detectiont, then we can consider v_j as an important node for v_i. Similarly, if $(v_j, v_k) \in G_C(v_i), v_j, v_k \neq v_i$, and removing the relationship (v_j, v_k) from $G_P(v_i)$ significantly reduces the abnormal score, then we consider this relationship important for v_i.

Returning to the previous Eq. 1, once a GNN-based PIDS model is trained, the entropy $H(Y)$ becomes deterministic. Thus, the equation above becomes minimizing the conditional entropy $H(Y|G = G_S, X = X_S)$ as below.

$$H(Y|G = G_S, X = X_S) = -\mathbb{E}_{Y|G_S, X_S}[log P\Phi(Y|G = G_S, X = X_S)] \tag{2}$$

Then, we can optimize Eq. 2 as follows:

$$\min_{\mathcal{G}} \mathbb{E}_{G_S \sim G} H(Y|G = G_S, X = X_S) \tag{3}$$

Then, Jensen's inequality allows us to derive Eq. 4 from Eq. 3.

$$\min \mathcal{G} H(Y|G = \mathbb{E}_{\mathcal{G}}[G_S], X = X_S) \tag{4}$$

Clearly, GNN-based PIDS models do not satisfy the assumption of convex functions. However, it has been found in the literature that by combining the objective function described above with regularization terms, the learned local optimal solutions already exhibit good explanatory power [25].

For the expectation $\mathbb{E}\mathcal{G}[GS]$, it is implemented through a mask, denoted as $A_c \odot \sigma(M)$, $where M \in \mathbb{R}^{n \times n}$, meaning that what the actual interpreter needs to learn is the mask M.

Furthermore, in general, we are interested in understanding "why a particular sample is predicted as a certain class" rather than understanding the global model. Therefore, we further modify the objective function to:

$$\min_{M} - \sum_{p=1}^{P} 1[y = p] \log P_\Phi(Y = y|G = A_p \odot \sigma(M), X = X_p) \tag{5}$$

Then, we compute the element-wise multiplication of $\sigma(M)$ and A_p and remove low values in M through thresholding to arrive at the explanation G_S for the GNN-based PIDS model's prediction $\hat{y}$ at positive node v.

Finally, we generate the subgraph based on the obtained final mask M, where the size of the subgraph is determined by the hyperparameters α and the final output feature number K.

Subgraph for an Edge. Although the native GNNExplainer only provides methods for generating masks for nodes, we can extend these techniques to develop our own approach for explaining a specific edge based on GNNExplainer.

For an edge $e = (u, v)$ detected as positive by PIDS, where u and v represent the source and destination nodes of edge e, our goal is to find the important subgraphs $G_S \subseteq G_P$ in N hop neighborhood of e. This subgraph G_S are crucial for the positive result $\hat{y}$. Considering edge e is anomalous, nodes u and v would also be considered anomalous in security analysis. Therefore, we compute the masks for u and v separately using the formulas from the previous subsection. We denote these masks as M_u and M_v respectively. Afterward, we compute the mask for the edge based on these two masks. Since the masks for u and v represent features correlated with the nodes' causality. This correlation means the features are closely related to the edge. Therefore, we need to consider both node u and v. Additionally, to ensure that the dimensions to be masked are non-zero, we take the final edge mask as Eq. 6.

$$M_e = max(M_u, M_v) \tag{6}$$

Finally, we generate the subgraph based on the obtained final mask M_e, where the size of the subgraph is determined by the hyperparameters α and the final output feature number K.

3.4   Graph Reduction Based on Anomaly Status

Although we obtained a subgraph for an anomalous node or edge using GNNExplainer in the Subsect. 3.3, having just one such subgraph is not sufficient. Several reasons contribute to this inadequacy, including the following:

– Although the subgraphs generated in Subsect. 3.3 with the assistance of GNNExplainer focus solely on the causal relationship between the explained node or edge and the explanation subgraph, they do not consider the abnormal states of individual nodes and edges within the subgraph. Security operators are particularly interested in identifying entities and behaviors that show a higher propensity for anomalies in the subgraph, thereby enabling quicker decision-making.
– The subgraphs contain numerous system behaviors and entities with strong causal relationships to the explained entity or behavior. However, they may also include underlying system behaviors that are not directly related to the anomaly but are correlated with the explained entity or behavior. For example, in a Windows system, a simple process of writing a file, whether anomalous or not, may also trigger numerous processes accessing .dll files simultaneously. These strongly causal behaviors or entities with a lower tendency for anomalies can significantly dominate the features found in Subsect. 3.3, leading to poor performance and reliability of the interpreter. Moreover, it poses significant challenges for security operators to interpret the alerts based on the interpretation results.

Hence, we propose a graph reduction method based on the anomaly status of each node or edge in the explanation subgraph.

Most PIDS for nodes and edges provide anomaly scores for each node or edge in the provenance graph, denote as $AS(v)$ or $AS(e)$. We can obtain the anomaly status of nodes or edges in the subgraph. This is done by retrieving the anomaly scores provided by the PIDS. These values are then compared with the threshold $TH(G_P)$. Then, we can apply the following rules to prune the graph based on the anomaly status:

1. Preserve all anomalous features. Given that most APT attacks involve multiple steps and employ various techniques and tactics, it is crucial for security analysts to analyze other related anomalous behaviors if one anomalous behavior is identified. Thus, for an edge $e_i \in G_S$ or a node $v_i \in G_S$, if $label_{lp}(e_i) =$'positive', or $label_{nc}(v_i) =$'positive', then e_i or v_i should be retained. However, observation shows that due to the sparsity of the provenance graph and the limited depth of neighborhood search, the probability of other anomalous nodes or edges appearing in the explanation subgraph is usually low. Even if they do appear, they typically represent a very small proportion of the subgraph, usually ranging from 0 to 2 anomalous nodes or edges.
2. Preserve a portion of normal features with anomaly scores close to the threshold. Since security analysts are more interested in behaviors that lean towards

anomalies in the explanation subgraph, we can retain a certain proportion of nodes that lean towards anomalies to reduce the size of the subgraph. Additionally, for behaviors or entities with strong causal relationships but with a lower tendency for anomalies, their anomaly scores usually fall within the normal range and are far from the threshold. Thus, these behaviors or entities and their corresponding edges or nodes can be effectively filtered out. In this step, the proportion of features to be retained depends on the hyperparameter α. Since the number of αK features were generated in Subsect. 3.3, we need to retain the final features around K. Therefore, approximately $1/\alpha$ of the normal features are retained, and these retained features are closer to the threshold compared to the absorbed features.

The specific algorithm for graph reduction is shown in Algorithm 1.

Algorithm 1. Graph Reduction Algorithm

Require: Subgraph G_{old}, anomaly scores of each node $AS(v_i)$, threshold $TH(G_P)$, detection results $label(v_i)$ (positive/negative)
1: hyperparameter α, output limit K
2: Initialize an empty subgraph G_{new}
3: **for** node v_i, $i < \alpha K$ in G_{old} **do**
4:     **if** $label(v_i) = $ 'positive' **then**
5:         Add the node v_i to G_{new}
6:     **end if**
7: **end for**
8: **for** the remaining nodes v_j in G_{old} **do**
9:     $\text{Score}(v_j) = |AS(v_j) - TH(G_P)|$
10: **end for**
11: Sort $\text{Score}(v)$
12: Select the top $1/\alpha$ nodes with the smallest Score
13: Add them to G_{new}
14: Establish connections between nodes in G_{new} based on the connections in G_{old}
15: **Output:** G_{new}

---

4 Evaluations

4.1 Environment Setup and Dataset

Here, we briefly introduce the implementation and experimental setup. All experiments are performed on a server running Ubuntu 20.04 with 3.20 GHz 32-core Intel Xeon Silver 4215R CPU, 128 GB of memory and Nvidia GeForce RTX 3090.

PIDSes & Datasets. We use two GNN-based provenance intrusion detector systems. ThreaTrace pinpoints only anomalous nodes that might be involved in the attack [6]. Kairos [7], on the other hand, focuses on the anomaly edges in the attack chains. These two works respectively employ node classification and link prediction as tasks for the decoder.

Table 1. Summary of the experimental datasets

Dataset	# of Nodes	# of Edges(in millions)	# of Attack Edges	% of Attack Edges
DARPA-E3-CADETS	178,965	10.1	1248	0.012%
DARPA-E3-THEIA	690,105	32.4	3119	0.010%

We obtain our experimental datasets from DARPA [27]. They are the few open-source datasets widely used in evaluating provenance-based systems [3,4, 6,7]. Table 1 summarizes the statistics of the graphs in those datasets.

Implementation of Our Interpreter and Baselines. We implemented our interpreter in Python. Firstly, we use scikit-learn [28] and PyG [29], a PyTorch-based graph neural networks library, to reproduce the anomaly detection work of ThreatTrace and Kairos. After we gained the trained model and data, we use PyG to adapt the GNNExplainer and the baseline interpreters. For the interpreters, we fix the interpretation results into K-dimension and set neighborhood scale $N = 5$ by default and evaluate its effect below. The hyperparameter α, which determines the retention proportion of the subgraph, is set to $\alpha = 3$. For baselines, we choose LIME [31], a classic linear explanation method in deep learning, and GraphLime [19], a gradients-based method on graph neural networks which is derived from LIME.

4.2   Fidelity-Conciseness Evaluation

Assessing the quality of interpretation results is essential, relying on two methods: comparison with Ground Truth and fidelity evaluation. Ground Truth comparison measures the match between interpretation results and a known correct interpretation, while fidelity evaluates the consistency of the model's predictions between the interpreted results and the original graph. Fidelity focuses on whether essential predictive elements are maintained in the subgraph. Due to the lack of interpretation Ground Truths for Darpa Datasets, fidelity is used to evaluate the quality of interpretation results.

To assess the fidelity of our interpretations, we adopt an indicator inspired by [30], termed the Label Flipping Rate (LFR). LFR measures the proportion of data labeled as 'abnormal' that is reclassified as 'normal' following the application of interpretation results. This metric assumes that interpreters of higher fidelity will exhibit higher LFR values, as they more accurately identify the crucial features responsible for the data's abnormal classification. It is important to note, however, that an increase in LFR, driven by including more features in the interpretation, may inadvertently compromise the conciseness of the interpretation.

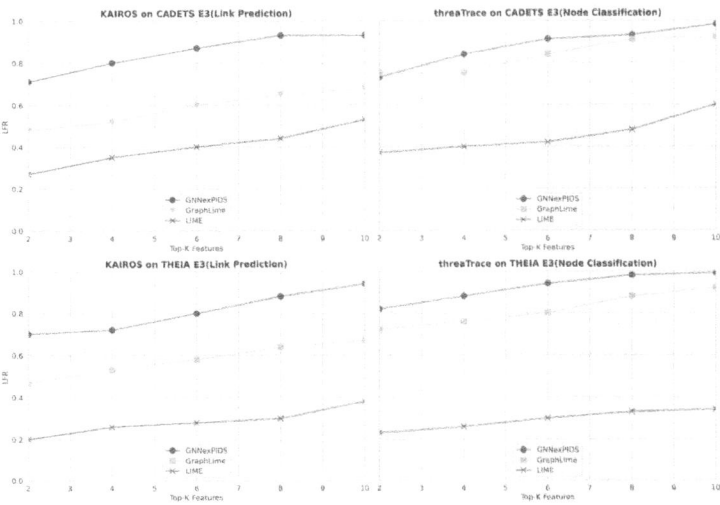

Fig. 3. Fidelity evaluation of interpreters(*higher is better*)

Our evaluation, depicted in Fig. 3, explores the fidelity-conciseness trade-off. The x-axis ranks the top-K important features, illustrating that, with constant interpreter, interpreted model, and dataset, the Label Flipping Rate (LFR) escalates in tandem with K—the number of top-K significant features determined by the interpreter. Despite this trend, GNNExplainer consistently outperforms both GraphLime and LIME in fidelity assessments for any given interpreted model and dataset configuration. This superiority underscores GNNexPIDS's efficacy in not only identifying pivotal features with greater precision but also in maintaining a balance between fidelity and conciseness of interpretations.

4.3   Stability Evaluation

Stability is essential for interpreters, ensuring consistent explanations for similar inputs. It verifies that minor changes to the input don't significantly alter the interpretation, crucial for the reliability of the insights provided. Thus, stability is a key metric for the effectiveness and trustworthiness of interpretation methods.

To evaluate the stability of our interpretation methods, we focus on the consistency of interpretations for identical samples across different executions. Drawing inspiration from [22, 23], we concentrate on the indexes of important feature dimensions, disregarding their specific values. For any two sets of interpretation vectors, v_1 and v_2, representing important dimension indexes obtained from consecutive runs under identical settings, we employ the *Jaccard Similarity* (**JS**) to quantify the similarity between these sets. JS is calculated as the size of the intersection divided by the size of the union of the two sets: $|set(v_1) \cap set(v_2)|/|set(v_1) \cup set(v_2)|$. By measuring and averaging the similarity of interpretation results for each anomaly across multiple runs, we gain insights into the interpreters' stability.

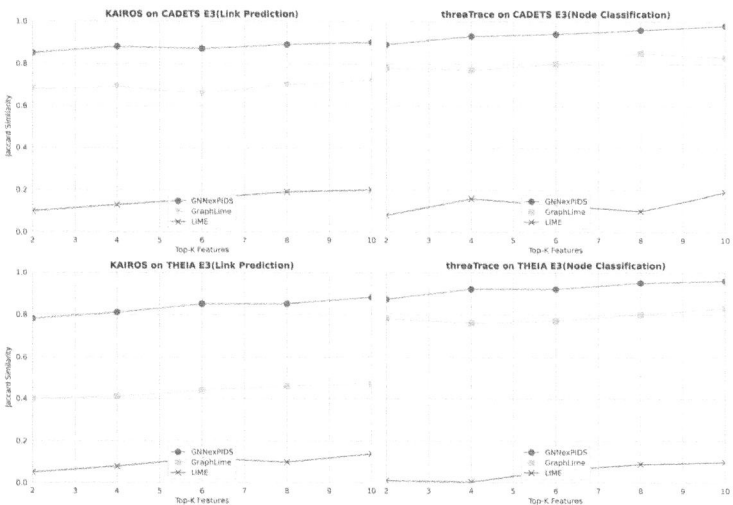

Fig. 4. Stability evaluation of interpreters(*higher is better*)

The findings, illustrated in Fig. 4, reveal that GNNExplainer consistently exhibits superior stability across different models and their respective datasets when compared to GraphLime, and markedly surpasses LIME. LIME's lower performance in this context underscores its limitations for graph data, highlighting the importance of employing interpretation methods like GNNexPIDS that are specifically designed for such data structures.

4.4 Discussions on the Graph Reductions

In this subsection, we primarily investigate the impact of our anomaly-based graph reduction method on the interpretation results. As a metric for assessing interpretation quality, we will explore the fidelity of the interpreter when using and not using this method. Since graph reduction is based on the subgraph determined by GNNExplainer, stability is not expected to vary significantly, hence it will not be discussed. Other baseline methods are not included in this discussion. We solely focus on exploring the fidelity performance of GNNexPIDS with or without graph reduction.

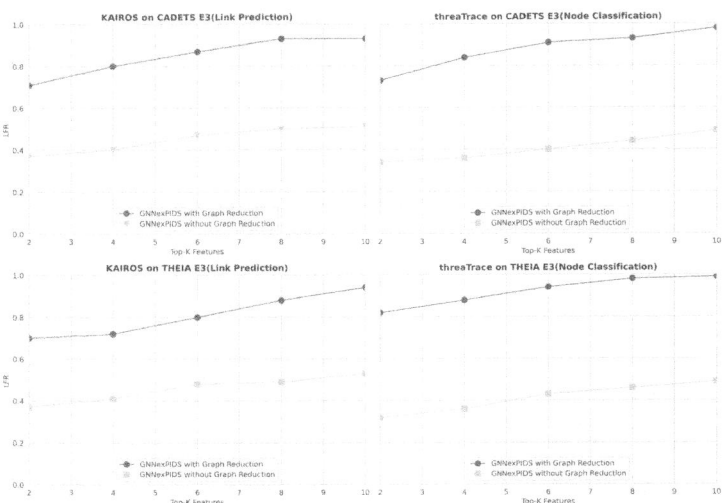

Fig. 5. Fidelity with or without Graph Reductions (*higher is better*)

As shown in Fig. 5, for the node classification and link prediction anomaly detection tasks on CADETS E3 and THEIA E3, the fidelity of the interpreter with graph reduction is significantly improved compared to the native GNNExplainer. This indicates that our method effectively filters out redundant parts from the features generated by GNNExplainer, thereby enhancing the performance and credibility of the interpreter. This demonstrates the effectiveness of our graph reduction approach.

5   Conclusions

This paper introduces a novel interpretation methodology named GNNexPIDS for Provenance Intrusion Detection Systems (PIDSes) based on the GNNExplainer framework. We adapt GNNExplainer to work on provenance graphs. Following the basic methodology of node interpretation in GNNExplainer, we extend it to enable edge interpretation. This enhancement makes our interpreter applicable to any GNN-based Provenance Intrusion Detection System (PIDS) tasked with node classification or link prediction. Subsequently, we propose an effective graph reduction method based on anomaly status to adapt to PIDS application scenarios, meet the interpretation needs of security operators, and improve the performance of the interpreter in explaining anomalies. Our work enables the precise identification of the features most significantly influencing the model's decisions. Our evaluation focuses on the interpretation of two seminal PIDS models within the DARPA dataset, employing both fidelity and stability as metrics for assessment. Through comparative analysis with GraphLime and LIME, our findings affirm the effectiveness and superiority of our method.

GNNexPIDS, while effective, has limitations and areas for improvement. Stability and fidelity in link prediction remain lower compared to node classification, indicating potential for future enhancements. This suggests that there is still room for improvement in our approach of extending node interpretation to edge interpretation. There is likely a more accurate and efficient method to replace ours. We will further explore this in future research endeavors. Additionally, the impact of increasing neighborhood search depth on interpretability has not been explored. From a security perspective, relying solely on interpreter outputs is insufficient; manual analysis of these results still imposes significant workload on security operators. Future work should investigate integrating expert knowledge to automate further analysis of interpretability outcomes. We aim to address these shortcomings in our subsequent research.

Acknowledgments. This research has been partially funded by the Institute of Information Engineering, Chinese Academy of Science, Project E4V01511G3 and is supported by the University of Chinese Academy of Science.

References

1. Wang, R., Nie, K., Wang, T., Yang, Y., Long, B.: Deep learning for anomaly detection. In: Proceedings of the 13th International Conference on Web Search and Data Mining, pp. 894–896 (2020)
2. Chandola, V., Banerjee, A., Kumar, V.: Anomaly detection: a survey. ACM Comput. Surv. (CSUR) **41**(3), 1–58 (2009)
3. Han, X., Pasquier, T., Bates, A., Mickens, J., Seltzer, M.: UNICORN: runtime provenance-based detector for advanced persistent threats. In: Network and Distributed Systems Security (NDSS) Symposium 2020, pp. 1–18. Internet Society (2020)
4. Milajerdi, S.M., Gjomemo, R., Eshete, B., Sekar, R., Venkatakrishnan, V.N.: HOLMES: real-time APT detection through correlation of suspicious information flows. In: 2019 IEEE Symposium on Security and Privacy (SP), pp. 1137–1152. IEEE (2019)
5. Zengy, J., et al.: SHADEWATCHER: recommendation-guided cyber threat analysis using system audit records. In: 2022 IEEE Symposium on Security and Privacy (SP), pp. 489–506. IEEE (2022)
6. Wang, S., et al.: THREATRACE: detecting and tracing host-based threats in node level through provenance graph learning. IEEE Trans. Inf. Forensics Secur. **17**, 3972–3987 (2022)
7. Cheng, Z., et al.: KAIROS: practical intrusion detection and investigation using whole-system provenance. In: 2024 IEEE Symposium on Security and Privacy (SP), p. 5. IEEE Computer Society (2023)
8. Han, X., et al.: SIGL: securing software installations through deep graph learning. In: 30th USENIX Security Symposium (USENIX Security 21), pp. 2345–2362 (2021)
9. Zhang, Z., Cui, P., Zhu, W.: Deep learning on graphs: a survey. IEEE Trans. Knowl. Data Eng. **34**(1), 249–270 (2020)
10. Zhou, J., et al.: Graph neural networks: a review of methods and applications. AI Open **1**, 57–81 (2020)

11. Zhang, M., Chen, Y.: Link prediction based on graph neural networks. In: Advances in Neural Information Processing Systems 31 (2018)
12. Ying, Z., You, J., Morris, C., Ren, X., Hamilton, W., Leskovec, J.: Hierarchical graph representation learning with differentiable pooling. In: Advances in Neural Information Processing Systems 31 (2018)
13. Hamilton, W., Ying, Z., Leskovec, J.: Inductive representation learning on large graphs. In: Advances in Neural Information Processing Systems 30 (2017)
14. Adadi, A., Berrada, M.: Peeking inside the black-box: a survey on explainable artificial intelligence (XAI). IEEE access **6**, 52138–52160 (2018)
15. Guidotti, R., Monreale, A., Ruggieri, S., Turini, F., Giannotti, F., Pedreschi, D.: A survey of methods for explaining black box models. ACM Comput. Surv. (CSUR) **51**(5), 1–42 (2018)
16. Baldassarre, F., Azizpour, H.: Explainability techniques for graph convolutional networks. In: International Conference on Machine Learning (ICML) Workshops, 2019 Workshop on Learning and Reasoning with Graph-Structured Representations (2019)
17. Pope, P.E., Kolouri, S., Rostami, M., Martin, C.E., Hoffmann, H.: Explainability methods for graph convolutional neural networks. In: Proceedings of the IEEE/CVF Conference on Computer Vision and Pattern Recognition, pp. 10772–10781 (2019)
18. Yuan, H., Yu, H., Gui, S., Ji, S.: Explainability in graph neural networks: a taxonomic survey. IEEE Trans. Pattern Anal. Mach. Intell. **45**(5), 5782–5799 (2022)
19. Huang, Q., Yamada, M., Tian, Y., Singh, D., Chang, Y.: GraphLIME: local interpretable model explanations for graph neural networks. IEEE Trans. Knowl. Data Eng. (2022)
20. Vu, M., Thai, M.T.: PGM-Explainer: probabilistic graphical model explanations for graph neural networks. In: Advances in Neural Information Processing Systems 33, pp. 12225–12235 (2020)
21. Baldassarre, F., Azizpour, H.: Explainability techniques for graph convolutional networks. arXiv preprint arXiv:1905.13686 (2019)
22. Fan, M., Wei, W., Xie, X., Liu, Y., Guan, X., Liu, T.: Can we trust your explanations? Sanity checks for interpreters in Android malware analysis. IEEE Trans. Inf. Forensics Secur. **16**, 838–853 (2020)
23. Warnecke, A., Arp, D., Wressnegger, C., Rieck, K.: Evaluating explanation methods for deep learning in security. In: 2020 IEEE European Symposium on Security and Privacy (EuroS&P), pp. 158–174. IEEE (2020)
24. Luo, D., et al.: Parameterized explainer for graph neural network. In: Advances in Neural Information Processing Systems 33, pp. 19620–19631 (2020)
25. Ying, Z., Bourgeois, D., You, J., Zitnik, M., Leskovec, J.: GNNExplainer: generating explanations for graph neural networks. In: Advances in Neural Information Processing Systems 32 (2019)
26. Zipperle, M., Gottwalt, F., Chang, E., Dillon, T.: Provenance-based intrusion detection systems: a survey. ACM Comput. Surv. **55**(7), 1–36 (2022)
27. Keromytis, A.D.: Transparent computing engagement 3 data release (2018). https://github.com/darpa-i2o/Transparent-Computing/blob/master/READMEE3:md
28. scikit-learn: machine learning in Python (2021). https://scikit-learn.org/
29. Fey, M., Lenssen, J.E.: Fast graph representation learning with PyTorch Geometric. In: ICLR Workshop on Representation Learning on Graphs and Manifolds (2019)

30. Guo, W., Mu, D., Xu, J., Su, P., Wang, G., Xing, X.: LEMNA: explaining deep learning based security applications. In: Proceedings of the 2018 ACM SIGSAC Conference on Computer and Communications Security, pp. 364–379 (2018)
31. Ribeiro, M.T., Singh, S., Guestrin, C.: "Why should i trust you?" Explaining the predictions of any classifier. In: Proceedings of the 22nd ACM SIGKDD International Conference on Knowledge Discovery and Data Mining, pp. 1135–1144 (2016)
32. Manzoor, E., Momeni, S., Venkatakrishnan, V., et al.: Fast memory-efficient anomaly detection in streaming heterogeneous graphs. In: International Conference on Knowledge Discovery and Data Mining (KDD'16) (2016)
33. Wei, F., Li, H., Zhao, Z., Hu, H.: xNIDS: explaining deep learning-based network intrusion detection systems for active intrusion responses. In: 32nd USENIX Security Symposium (USENIX Security 23), pp. 4337–4354 (2023)
34. Han, D., et al.: DeepAID: interpreting and improving deep learning-based anomaly detection in security applications. In: Proceedings of the 2021 ACM SIGSAC Conference on Computer and Communications Security, pp. 3197–3217(2021)

Performance Evaluation of Lightweight Cryptographic Ciphers on ARM Processor for IoT Deployments

Mohsin Khan[(✉)], Dag Johansen, and Håvard Dagenborg

UiT The Arctic University of Norway, Tromsø, Norway
{mohsin.khan,dag.johansen,havard.dagenborg}@uit.no

Abstract. The security mechanisms used to protect Internet-of-things and embedded systems often depend on cryptographic tools designed specifically to operate in resource-constrained environments. Several lightweight cryptographic ciphers have been proposed for such purposes, providing various properties and tradeoffs that balance security and performance. This paper provides a comprehensive evaluation of key lightweight ciphers, evaluating their throughput, processing efficiency, memory footprint, ROM utilization, and energy consumption. We focus on the software implementation and performance of these ciphers on a representative single-core ARM processor and propose an *E-Rank* metric that combines essential performance characteristics with resource consumption into a single comparative metric. The metric aids in identifying lightweight ciphers that achieve an optimal balance of efficiency by normalizing the trade-offs between performance, memory usage, and energy consumption.

Keywords: Lightweight cryptography · block cipher · stream cipher · performance evaluation · internet-of-things

1 Introduction

The widespread use of Internet of Thing (IoT) devices across various technological landscapes has emphasized the need for effective and sufficient security measures. One problem domain is securing the IoT devices themselves, where we previously have investigated secure enclave technologies supporting confidential computing [8,20]. Next, we have focused on securing the external IoT communication channels, where we conjecture that conventional cryptographic algorithms, such as, for instance, AES and RSA, are secure but less suitable for IoT applications due to their high computational and resource demands.

IoT devices generally operate with limited processing power, energy consumption, and memory, making it a challenge to balance security and efficiency. To address security challenges in IoT systems, Lightweight Cryptography (LWC) has become essential for providing integrity and confidentiality.

J. Zhao and W. Meng (Eds.): SciSec 2024, LNCS 15441, pp. 254–272, 2025.
https://doi.org/10.1007/978-981-96-2417-1_14

Lightweight Cryptographic Ciphers (LWCCs) are carefully designed to meet the unique demands and constraints of IoT environments, providing effective and adequate security solutions that protect devices without overburdening their limited computational resources or disrupting their operational efficiency. This strategic adaptation ensures that security implementations are both practical and effective for IoT applications.

This study offers an in-depth assessment of LWCCs across two primary categories: Lightweight Block Cipher (LWBC) and Lightweight Stream Cipher (LWSC) [13]. LWBCs, such as PRESENT [10], XTEA [19], CLEFIA [24], SIMON, and SPECK [5], are known for their ability to encrypt data blocks in constrained environments. On the other hand, LWSCs, including Mickey [2], Salsa [7], Sosemanuk [6], Grain-v1 [16], and an optimized version of Grain-128a [18], are designed for efficient encryption of continuous data streams.

The experiment is conducted on a Raspberry Pi Zero W, characterized by its single-core ARM11 processor. This preference is motivated by the device's widespread use in IoT applications due to its compact size, low power consumption, and ability to integrate with various sensors and actuators, thereby presenting a typical resource-constrained IoT environment. To measure the performance of each selected LWCC precisely, the study employs a custom benchmarking framework that evaluates throughput, Cycles per Byte (CpB), memory footprint, ROM utilization, and energy consumption. An Arduino UNO, coupled with a power measurement sensor, is used to measure the energy consumption during cipher operations. This provides insights into the energy consumption and efficiency of each lightweight cryptographic solution. Moreover, the study employs the *Rank* metric [5] and develops *E-Rank* that aggregates performance, resource usage, and energy consumption metrics into a single comparative value. This metric allows for a thorough evaluation of ciphers, aiding informed decision-making based on overall software performance.

The subsequent sections of this paper are structured as follows. The section on background will provide a concise overview of the selected LWCCs. The next section on research methodology explains our approach, including tool development, cipher selection criteria, hardware considerations, and evaluation metrics. The results and discussion section presents our findings and explores their implications. The related work section explains existing benchmarking tools briefly, highlighting their limitations and inapplicability to our research. Finally, the conclusions section summarizes key insights and suggests avenues for future research.

2   Background

2.1   Lightweight Block Ciphers

Significant progress has been made on LWBCs to accommodate the constraints of resource-limited environments. The Tiny Encryption Algorithm (TEA) [26] is characterized by its simplicity and the use of minimal cryptographic operations such as XOR, ADD, and SHIFT. However, vulnerabilities identified in TEA related to key-specific attacks [23] led to the development of an improved version, XTEA [19], which improves security through dynamic key arrangement and delayed diffusion mechanisms. The National Security Agency (NSA) has contributed to this field as well by developing the SIMON and SPECK families of ciphers [4]. SIMON utilizes simple bitwise operations like AND, XOR, and circular shifts to achieve secure encryption in hardware environments. On the other hand, SPECK is designed to balance security and performance in software applications by using operations like modulo addition, XOR, and left and right circular shifts.

In 2007, the PRESENT [10] and CLEFIA [24] LWBCs were introduced and later standardized under ISO/IEC 29192 [17]. The PRESENT cipher has a compact design, achieved through the use of scan flip-flops and a single 4×4-bit S-Box with a linear mixing layer for permutation. Sony developed CLEFIA using a Diffusion Switching Mechanism (DSM), resulting in an increased number of active S-Boxes. The DSM operates by rotating the diffusion layers during encryption, increasing the number of active S-Boxes, and altering the interconnection patterns.

2.2   Lightweight Stream Ciphers

A significant initiative for LWSCs was the eSTREAM project, launched in 2004 under the European Network of Excellence for Cryptology (ECRYPT) [22]. Initially, 34 ciphers were presented, but only 7 were selected as the finalists. The seven ciphers were evaluated thoroughly in three phases and divided into two profiles: Profile 1 focused on software-oriented applications requiring high throughput, while Profile 2 focused on hardware-oriented applications with limited resources.

Profile 1 includes four cryptographic algorithms namely Rabbit [9], Salsa [7], HC-128 [27], and Sosemanuk [6]. Salsa operates by utilizing a pseudorandom function that is based on a hash function core. It uses a set of operations, including addition modulo, bitwise XOR, and constant distance rotation of 32 bits. Salsa20 is structured around a 20-round process that uses an invertible quarter-round function as its primary building block. Sosemanuk combines the elements of both synchronous stream ciphers and block ciphers, utilizing a Linear Feedback Shift Register (LFSR) and a Finite State Machine (FSM) to generate a keystream. The LFSR in Sosemanuk is responsible for providing a fast and efficient mechanism for generating bits, while the FSM handles the non-linear transformation of the state.

Profile 2 consists of Trivium [12], Grain [16], and Mickey [2]. Trivium uses three shift registers and a small amount of combinatorial logic. The cipher operates in two phases. During the initialization phase, the cipher loads the key and initialization vector (IV) into a 288-bit internal state, which then evolves over 1152 cycles. In the keystream generation phase, output bits are produced using XOR and AND operations on specific bits from the internal registers. Grain-v1 also features a two-phase operation. First, the Non-linear Feedback Shift Register (NLFSR) is loaded with a key, and the LFSR is initialized with an IV. In the second phase, feedback from both NLFSR and LFSR is used to generate output bits through XOR operation with the input. Grain-128a is an improvement over Grain-v1, as it increases the key size to 128 bits and the internal state to 256 bits, making it more secure against vulnerabilities. Finally, the Mickey cipher improves randomness by clocking shift registers irregularly. Two registers are used, each comprising 80 stages, to create a keystream generator, with each stage controlling one bit.

3   Research Methodology

This study presents a structured approach for evaluating the software performance of LWCCs on Raspberry Pi Pico's ARM11 processor. Our methodology relies on a custom benchmarking tool to select, execute, and precisely measure cryptographic encryption and decryption operations for selected LWCCs. This tool evaluates the performance of ciphers using specific key metrics, including throughput, CpB, memory footprint, flash memory, and energy consumption. These metrics provide essential insights into the processor speed, memory efficiency, and energy utilization of the LWCCs, which are vital for their effective deployment in resource-constrained environments. During the selection process of the eleven ciphers, we carefully considered their relevance to IoT security challenges and their prominence within the lightweight cryptographic community.

3.1   Development of Benchmarking Tool

The benchmarking tool was developed using a systematic approach to ensure that it is lightweight, adaptable, and user-friendly. The tool was programmed using Python, which was extended with the C implementation of the ciphers through shared object files. This combination offers simple, flexible, and fast performance for both cryptographic operations and system monitoring. The following sections elaborate on the development process, providing a detailed account of each stage and methodology employed.

Requirement Analysis: The initial phase involved a thorough analysis of the requirements and objectives of the benchmarking tool. This included defining the desired features, input parameters, and output metrics. Based on the requirement analysis, a comprehensive design was developed for the benchmarking tool. The overview of the design of the tool and its associated workflow is illustrated in Fig. 1. The design of the tool involves defining its structure, including modules for input processing, cryptographic operations, and metrics related to performance measurement.

Implementation: The implementation phase is focused on translating the design and architecture into functional code. The core structure of the tool integrates the C implementations of the selected LWCCs with Python using the *ctypes* library [21], which enables the calling of C-implemented shared object

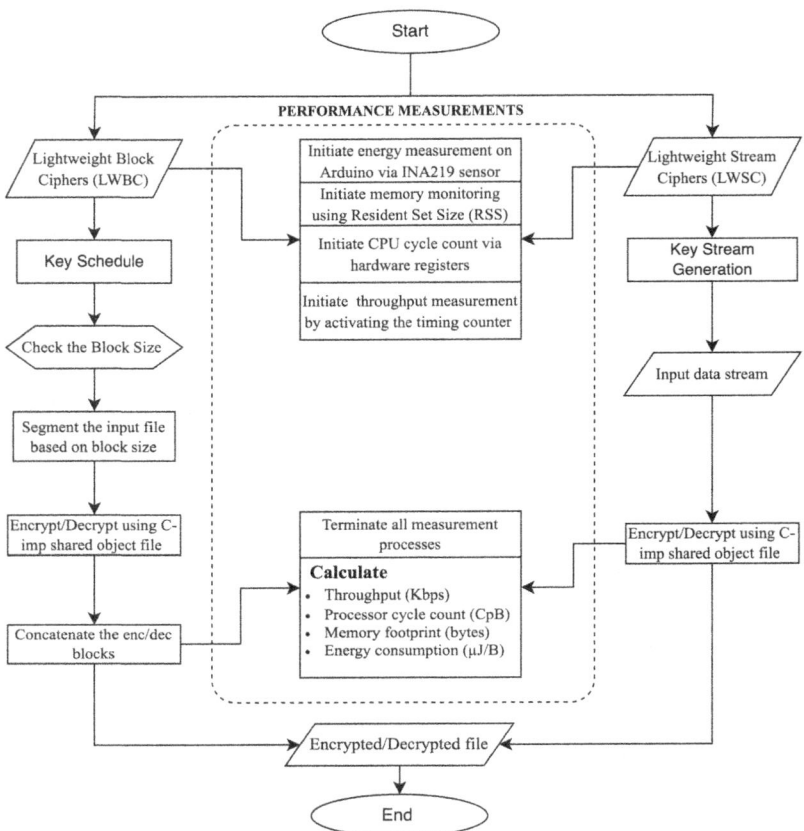

Fig. 1. Flowchart demonstrating basic encryption/decryption steps and associated initialization of performance tracking.

functions directly from Python. This hybrid approach leverages the speed and efficiency of C language for cryptographic operations while utilizing Python's robust capabilities for orchestrating the system procedures and data handling necessary to capture key performance metrics. Python was chosen for taking performance measurements due to its simplicity, extensive library ecosystem, and proven track record in rapid prototyping, significantly accelerating the development process. During this phase, several key functionalities were implemented:

– **Input Processing:** The program is designed to receive user inputs such as the block size, key size, and input file for encryption and decryption.
– **Cryptographic Operations:** Encryption and decryption are performed using specified parameters through the integration of a C implementation of selected LWCCs.
– **Performance Measurement:** The benchmarking tool incorporates libraries, functions, and system calls to measure key performance metrics like

throughput, CpB, memory usage, and energy consumption. This tool enables a comprehensive comparison of selected ciphers by evaluating the quantitative performance of the ciphers under standard conditions.

– **Result Analysis:** The data gathered from the benchmarking tool is used to generate bar graphs that demonstrate the performance of each LWCC across multiple metrics. To ensure the accuracy of these visual representations, the confidence interval is calculated at a 95% confidence level by measuring the performance of each cipher repeatedly across several iterations. This statistical approach provides a better understanding of the performance consistency and potential variability, allowing for a more fine interpretation of the data. By including these confidence intervals in the bar graphs, the analysis not only highlights average performance metrics but also provides insights into the stability and predictability of each cipher.

Testing and Debugging: During this phase, a set of tests were conducted. The tests include regression testing to ensure that any new changes made did not interfere with the existing functionality. Statistical testing was also conducted to verify the accuracy of data handling and analysis. Lastly, load testing was carried out to evaluate how the system would perform under various stress conditions. The tool was thoroughly validated against established benchmarks and reference implementations to ensure its accuracy and effectiveness. Feedback collected during this phase was vital in identifying and resolving any problems, bugs, or discrepancies, resulting in improved overall functionality and performance of the tool.

Optimization and Performance Tuning: Optimization techniques were applied to enhance both the efficiency and performance of the benchmarking tool. This process involved using only efficient, valid C implementations of LWCCs to execute core cryptographic operations with maximum speed and low-level access. Furthermore, to enhance the performance measurement and minimize resource usage, we have taken a careful approach to selecting and utilizing essential libraries, thereby reducing unnecessary overhead. These optimizations improved the tool's accuracy and responsiveness while keeping its footprint lightweight. This is particularly important when evaluating performance in environments with limited resources, ensuring reliable data delivery.

3.2   Selection Criteria for Lightweight Cryptographic Ciphers

In this study, eleven LWCCs were selected for evaluation, primarily due to their relevance to embedded system security, their prominence within the research community, and their availability in open-source cryptographic libraries. The ciphers selected include PRESENT, XTEA, CLEFIA, SIMON, and SPECK among LWBC, and Grain-v1, Grain-128a, Trivium, Mickey, Salsa, and Sosemanuk among LWSC.

PRESENT and CLEFIA were chosen based on their recognition in the lightweight cryptography community and their compliance with the ISO/IEC 29192 standard [17]. XTEA was included due to its simple and compact structure and its development at the Cambridge Computer Laboratory, marking it as one of the pioneering ciphers with a small footprint suitable for constrained environments. SIMON and SPECK were selected because of their development by the researchers at the National Security Agency (NSA).

Grain-v1, Trivium, Mickey, Salsa, and Sosemanuk were all finalists in the ECRYPT Stream Cipher Project, highlighting their cryptographic robustness and efficiency. We also chose a speed-optimized version of Grain-128a, which includes enhanced features such as added authentication, to evaluate potential performance improvements.

This diverse selection of LWCCs allows for a comprehensive analysis of cryptographic techniques applicable to IoT devices in terms of their computational efficiency and performance evaluation.

3.3   Implementation Considerations for ARM-Processor

The benchmarking tool has been specifically optimized for deployment on the Raspberry Pi Zero W. This device is equipped with a single-core ARM1176JZF-S processor [1]. The processor was chosen due to its specific computational capabilities and inherent memory constraints. These constraints are important in simulating the performance of LWCCs in constrained environments.

The Raspberry Pi Zero W features 512 MB of RAM, which provides efficient memory management to prevent bottlenecks during the execution of selected ciphers. The 40-pin header on the Raspberry Pi Zero W is utilized strategically, with designated pins specifically used for supplying power to the device and for data transfer. This setup facilitates the use of the transfer (Tx) line to enable communication with the Arduino, which is required for signaling the start and stop of power measurement during the cryptographic benchmarking process. This configuration ensures that the Arduino precisely calculates and tracks the energy consumption of the Raspberry Pi as it executes the selected LWCCs. More detailed information about this connection and its role in the overall benchmarking framework will be elaborated in the upcoming section on metrics.

3.4   Evaluation Metrics

The following metrics are used in our evaluation: throughput, CpB, code size, memory footprint, energy consumption, and *E-Rank*.

Throughput refers to the rate at which an LWCC can process data and is measured in kilobits per second (Kbps). Higher throughput values indicate rapid encryption or decryption speeds, offering a significant advantage for applications requiring swift data processing.

Cycles per Byte (CpB) indicates the average number of CPU cycles required to process one byte of data. Here, we used a specialized method to improve the precision of CpB measurements for LWCCs on the ARM11 processor. Our strategy involves direct access to the ARM timer registers, and the system timer registers. These registers are essential for managing various timing and scheduling operations within the Raspberry Pi's operating system. By interfacing directly with these hardware timers, our program can meticulously measure processor cycles by capturing timer values straight from the hardware. This method ensures a highly accurate assessment of the computational effort each LWCC demands, providing valuable insights into their efficiency on constrained devices.

To ensure accurate measurement of CpB for the selected LWCCs without being influenced by background processes, the system first calculates the CpB during a baseline period in which only background processes are running. This baseline CpB is then averaged over a specific time period. During the execution of LWCCs, the CpB is determined by subtracting the baseline CpB from the observed CpB at each instant. These differences are then summed to obtain the total processor cycles utilized during the cipher's execution. This approach improves measurement precision and mitigates the processor cycles of the background processes. Lower values of CpB indicate faster LWCC performance, requiring fewer processing cycles for encryption or decryption.

Code size refers to the amount of memory used in the non-volatile memory, which is the flash memory of the target device. In the case of the Raspberry Pi Zero W, the code size quantifies the storage space occupied by the LWCC, measured in bytes.

Memory footprint is the amount of memory used by a process during its execution. In our approach, we compute the memory footprint of LWCCs by assessing the volatile memory and subsequently dividing it by the size of the input. This computation yields the amount of volatile memory required (in bytes) for processing each byte of the data input. The benchmarking tool carefully monitors and records the amount of memory footprint of each selected LWCC implementation. The memory footprint assessment relies on the Resident Set Size (RSS), which represents the portion of memory reserved for a process in the main memory (RAM). In this context, RSS represents the portion of memory held by an LWCC within the RAM, excluding any data swapped out to disk. It contains memory allocated for important segments such as the code segment, data segment, heap, and stack, offering a comprehensive view of the actual memory usage of the LWCC. This provides valuable insights into the memory footprint of the ciphers and their impact on system resources.

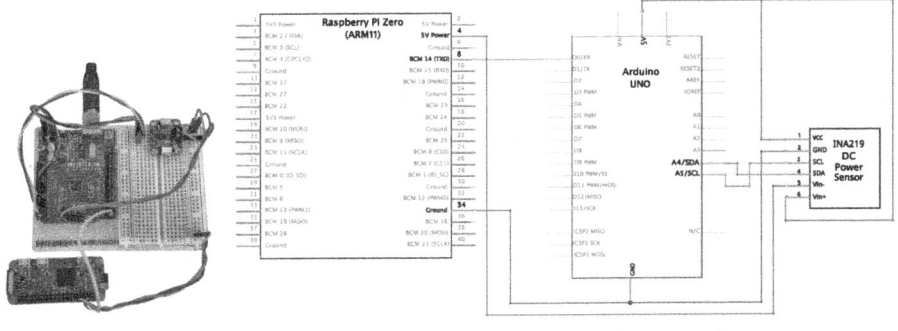

(a) Physical Configuration   (b) Schematic Diagram of the Connection Setup

Fig. 2. Connection setup between a Raspberry Pi and an Arduino UNO for power and energy measurement using an INA219 module.

Energy consumption refers to the quantity of power consumed by a process during its execution over a specified period of time. In this context, it refers to the energy consumption of LWCCs on the ARM11 processor, calculated in microjoules per byte ($\mu J/B$). Due to limitations in directly measuring energy consumption via software, a hardware-based approach was devised to obtain precise power measurements. As illustrated in Fig. 2, the setup involves a circuit schematic specifically designed for power measurement and energy calculation. The Raspberry Pi board is powered by an Arduino UNO, which supplies power through GPIO pin-4 (5V power supply) and pin-6 (ground). A digital current and power monitor sensor developed by Texas Instruments, known as the INA219 sensor [25], is integrated into the power line from the Arduino to the Raspberry Pi. This sensor is capable of measuring the power consumption of Raspberry Pi with precision, enabling it to directly monitor and record electrical parameters.

During the initialization, the sensor monitors the power consumption of the Raspberry Pi while it executes background processes for a set duration. This data is then averaged to determine the typical power consumption during idle operation. Further, GPIO pin-14 on the Raspberry Pi is connected to pin-0 on the Arduino, establishing a one-way serial communication channel from the Raspberry Pi to the Arduino. This channel is required to signal the start and end of power and energy measurements. When encryption begins, the Raspberry Pi sends a signal to the Arduino, indicating the cipher's name along with its block and key size. The Arduino then activates the INA219 to start logging the power usage. Upon successful completion of the encryption/decryption process, another signal is transmitted to the Arduino to halt the power and energy measurement for that specific LWCC. To ensure that energy consumption is accurately calculated for the specific LWCC and not affected by background processes, the average power consumption calculated during the initialization is subtracted from the current power readings. The subtraction isolates the power used exclusively by the cipher during each discreet time interval. This approach mitigates the impact of background processes, thereby enhancing the precision and accuracy of energy consumption measurements for the LWCCs.

Rank measures the overall software efficiency of LWCCs and is defined as the ratio of the amount of data processed per second (throughput) to the aggregate amount of volatile and non-volatile memory used during the execution of cipher, as seen in Eq. 1. The overall software performance metric (*Rank*) was introduced by Beaulieu et al. [4], but in our work, we use a modified *Rank* metric (*E-Rank*) based on the Figure of Merit (FoM), which is used to evaluate the overall hardware performance. FoM is shown in Eq. 2 and was formulated by Badel et al. [3] to evaluate their LWBC called ARMADILLO. The authors highlighted the absence of power dissipation considerations in their FoM metric. This omission resulted from the knowledge that dynamic power, which is linked to the total switched capacitance, maintains a direct proportionality with area and, consequently, gate count. As a result, power is conventionally associated with the number of gates within the hardware context. As can be seen in Eq. 1, the *Rank* metric relies on aggregated memory utilization for calculation. Unlike FoM, memory utilization does not have direct proportionality with power dissipation due to no direct relation with the switched capacitance. Thus, integrating energy consumption into the *Rank* formulation would provide a comprehensive and more precise view of the overall software performance, referred to as *E-Rank*. Equation 3 provides the *E-Rank* formula, where accuracy is upheld through unit normalization, ensuring the uniform expression of all metrics in bytes.

$$Rank = \frac{\text{Throughput}}{\text{ROM} + 2 \times \text{RAM}} \tag{1}$$

$$FoM = \frac{\text{Throughput}}{\text{Clock freq} \times \text{Gate count}^2} \tag{2}$$

$$E\text{-}Rank = \frac{\text{Throughput}}{(\text{ROM} + 2 \times \text{RAM}) \times \text{Energy consumption}} \tag{3}$$

4 Experimental Results

4.1 Throughput Analysis

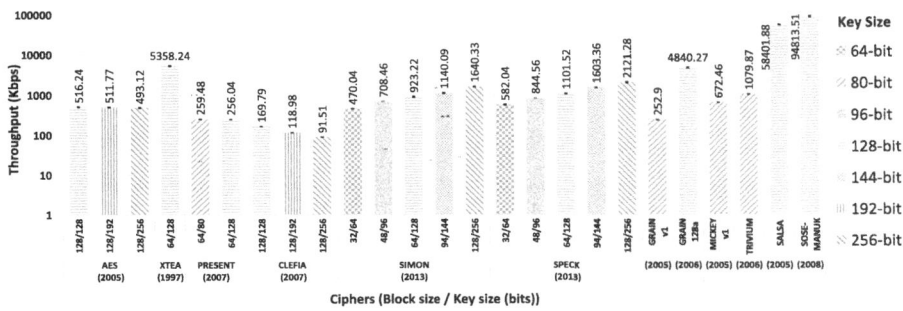

Fig. 3. Throughput of selected LWCCs.

Among the LWBCs, as depicted in Fig. 3, XTEA demonstrates the highest throughput, followed by SPECK and SIMON. SPECK is recognized as a software-oriented cipher and thus exhibits better throughput than SIMON. CLEFIA has a lower throughput compared to PRESENT, while AES has a higher throughput than both.

Regarding LWSCs, as demonstrated in Fig. 3, these ciphers collectively exhibit significantly better throughput than that of block ciphers, with the exception of Grain-v1. Sosemanuk leads the group with the highest throughput, followed by Salsa and then Grain-128a (optimized version). Trivium demonstrates better throughput compared to Mickey, followed by Grain-v1.

4.2   Cycles per Byte (CpB) Analysis

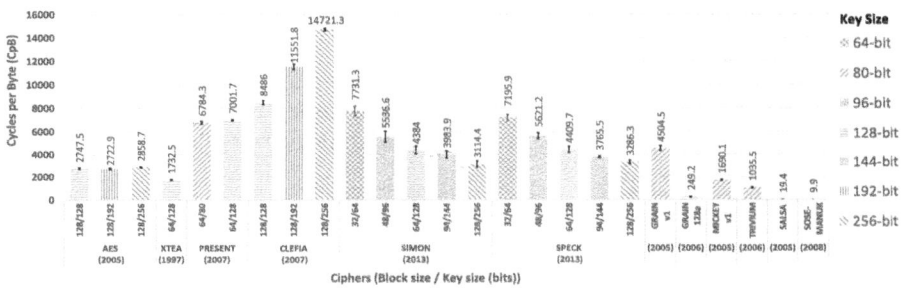

Fig. 4. Processor cycles of selected LWCCs.

Among LWBCs, as represented in Fig. 4, CLEFIA demonstrates the highest processor cycle count (CpB), followed by PRESENT. Contrarily, XTEA exhibits the lowest consumption, trailed by AES, SIMON, and SPECK. Note that the exception of 64-bit and 144-bit key sizes of SPECK show lower CpB than SIMON. AES displays lower CpB consumption compared to both SIMON and SPECK.

Regarding LWSCs, illustrated in Fig. 4, the overall processor cycle count is remarkably lower than that of LWBCs. Grain-v1 consumes the most among LWSCs, whereas Sosemanuk, Salsa, and Grain-128a (optimized version) exhibit successively lower CpB consumption. Trivium shows lower CpB utilization compared to Mickey.

4.3 Memory Consumption Analysis

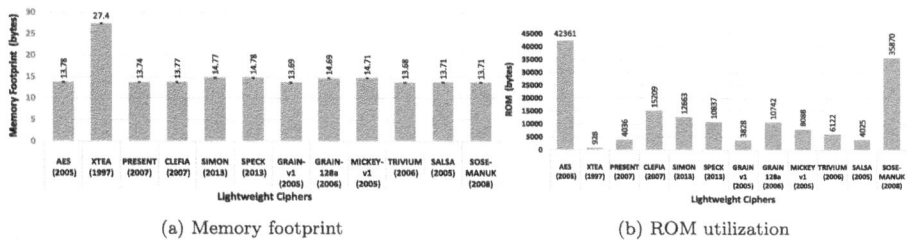

(a) Memory footprint   (b) ROM utilization

Fig. 5. Memory footprint and ROM utilization of selected LWCCs.

The memory footprint, as illustrated in Fig. 5a, demonstrates that XTEA consumes the highest memory footprint, followed by SPECK and SIMON among LWBCs. Contrarily, the lowest footprint is observed with PRESENT, CLEFIA, and AES. In the LWSCs, Mickey-v1 and Grain-128a exhibit the highest memory footprints, while Trivium, Grain-v1, Salsa, and Sosemanuk show successively lower footprints.

The analysis of code size, illustrated in Fig. 5, shows that XTEA boasts the smallest code size, while AES exhibits the largest among LWBCs. XTEA leads with the smallest code size, followed by PRESENT, SPECK, SIMON, and then CLEFIA in ascending order. In the LWSCs, Sosemanuk has the largest code size, while Grain-v1 has the smallest. The Grain-v1 is followed by Salsa, Trivium, Mickey, and Grain-128a in successive order of increasing code size.

4.4 Energy Consumption Analysis

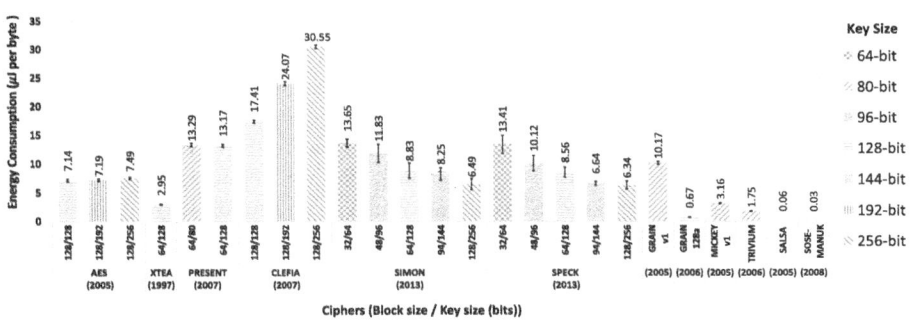

Fig. 6. Energy consumption of selected LWCCs.

Among LWBCs, as shown in Fig. 6, CLEFIA demonstrates the highest energy consumption, followed by PRESENT. Contrarily, XTEA exhibits the lowest energy consumption, followed by AES, SPECK, and SIMON. AES indicates lower energy consumption compared to both SIMON and SPECK, except for the 256-bit key size.

Among LWSCs, illustrated in Fig. 6, the overall energy consumption of LWSCs is remarkably lower than that of LWBCs. Grain-v1 consumes the most energy among LWSCs, while Sosemanuk, Salsa, and Grain-128a (optimized version) demonstrate successively lower energy consumption. Trivium shows lower energy consumption compared to Mickey.

4.5   E-Rank Analysis

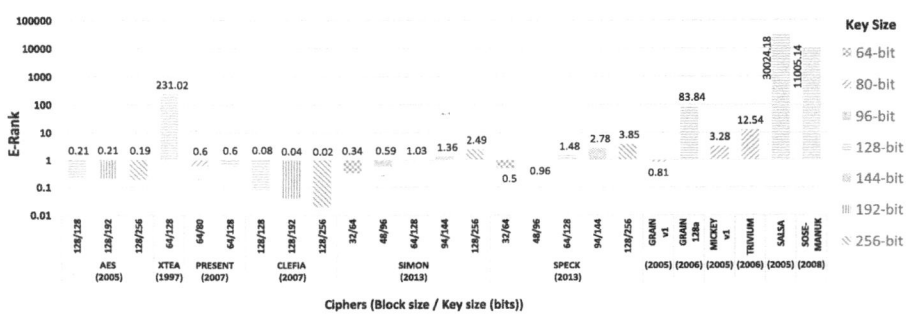

Fig. 7. E-Rank of selected LWCCs.

In our analysis of LWBCs, XTEA emerged as the best LWCC in overall software performance (*E-Rank*), as represented in Fig. 7, with CLEFIA demonstrating the least favorable results. XTEA's performance is followed by SPECK and then SIMON. Although AES displayed better throughput and energy consumption than PRESENT from our previous results, the *E-Rank* metric favored PRESENT, which is in line with its classification as a LWCC. Moreover, SPECK demonstrated better *E-Rank* utilization in comparison to SIMON, which indicates its orientation toward software implementation.

In the context of LWSCs, Fig. 7 shows that while Sosemanuk demonstrated substantial throughput and energy consumption, Salsa demonstrated better performance in terms of *E-Rank*, followed by Sosemanuk. On the other hand, Grain-v1 presented the least favorable *E-Rank*, while the optimized version of Grain-128a demonstrated better *E-Rank* than Mickey and Trivium. Trivium shows better overall software performance than Mickey.

5   Discussion

This study has systematically evaluated the performance and resource usage of LWCCs through a comprehensive benchmarking framework. Our analysis covered all the essential metrics, such as processor cycles, throughput, energy consumption, and total memory utilization, which includes both memory footprint and flash memory.

The performance measurement of LWBCs in Table 1 illustrates that XTEA demonstrates leading performance, exhibiting the best throughput, lowest CpB, and minimal energy consumption. This efficiency can be attributed to XTEA's simple and compact structure, which utilizes basic operations like XOR, AND, and SHIFT within a Feistel Network. This network splits the plaintext into two equal blocks, performing cryptographic operations on these divided segments. Following XTEA, SPECK and SIMON were notable for their throughput, with SPECK's performance surpassing SIMON's due to its Addition Rotation XOR (ARX) based structure, tailored for software optimization, compared to SIMON's hardware-oriented Feistel structure. PRESENT and CLEFIA, which are based on the Substitution Permutation Network (SPN) and Generalized Feistel Network (GFN), respectively, demonstrated inferior software performance compared to AES due to their hardware-oriented design. In the SPN design framework, block ciphers are created by applying multiple rounds of substitution and permutation to perform confusion and diffusion, respectively. Unlike Feistel Networks, the GFN splits the plaintext into sub-blocks of greater than or equal to $2n$ bits.

Among LWSCs, Sosemanuk and Salsa have shown remarkable throughput, lowest CpB, and energy efficiency. This is due to their software-oriented structures. Sosemanuk combines elements of synchronous stream and block ciphers, enhancing its initialization speed and overall performance, while Salsa employs a complex structure of addition modulo, bitwise XOR, and constant distance rotation, optimizing it for software implementation. On the contrary, hardware-oriented LWSCs like Grain, Mickey, and Trivium exhibited lower performance efficiency, except Grain-128a, which showed the best performance among them due to its speed-optimized implementation.

The analysis of resource usage among LWBCs reveals that XTEA, despite its best performance efficiency, demands the highest memory footprint, whereas AES requires the largest code size. Following closely, SIMON and SPECK demonstrate considerable memory usage while maintaining an optimal code size, achieving a balance between performance and resource consumption. PRESENT, on the other hand, is the most resource-efficient. Among LWSCs, Sosemanuk leads in resource usage in terms of code size, followed by Grain-128a and Mickey-v1. This analysis reveals a consistent trend: LWCCs that deliver higher performance efficiency also tend to require more resources, indicating a trade-off that necessitates careful cipher selection based on a balanced consideration of both performance and resource constraints to ensure optimal system functionality.

Integrating performance efficiency with resource usage is effectively accomplished through the *E-Rank* metric, which provides a comprehensive measure-

Table 1. Software implementation results of LWCCs on ARM processor, along with E-RANK

Ciphers	Block Length (bits)	Key Size (bits)	Code Size (bytes)	Memory Footprint (bytes)	Processor Cycles (CpB)	Throughput (Kbps)	Energy (μ J/B)	E-RANK
AES	128	128	42361	13.78	2747.5	516.24	7.14	0.21
		192			2722.9	511.77	7.19	0.21
		256			2858.7	493.12	7.49	0.19
XTEA	64	128	928	27.4	1732.5	5358.24	2.95	231.02
PRESENT	64	80	4036	13.74	6784.3	259.48	13.29	0.6
		128			7001.7	256.04	13.17	0.6
CLEFIA	128	128	15209	13.77	8486	169.79	17.41	0.08
		192			11551.8	118.98	24.07	0.04
		256			14721.3	91.51	30.55	0.02
SIMON	32	64	12663	14.77	7731.3	470.04	13.65	0.34
	48	96			5536.6	708.46	11.83	0.59
	64	128			4384	923.22	8.83	1.03
	96	144			3983.9	1140.09	8.25	1.36
	128	256			3114.4	1640.33	6.49	2.49
SPECK	32	64	10837	14.78	7195.9	582.04	13.41	0.5
	48	96			5621.2	844.56	10.12	0.96
	64	128			4409.7	1101.52	8.56	1.48
	96	144			3765.5	1603.36	6.64	2.78
	128	256			3286.3	2121.28	6.34	3.85
Grain-v1	–	80	3828	13.69	4504.5	252.9	10.17	0.81
Grain-128a	–	128	10742	14.69	249.2	4840.27	0.67	83.84
Mickey-v1	–	80	8088	14.71	1690.1	672.46	3.16	3.28
Trivium	–	80	6122	13.68	1035.5	1079.87	1.75	12.54
Salsa	–	128	4025	13.71	19.4	58401.88	0.06	30024.18
Sosemanuk	–	128	35870	13.71	9.9	94813.51	0.03	11005.14

ment of the overall software performance of LWCCs. As illustrated in Eq. 3, the *E-Rank* yielded several notable insights. In the category of LWBCs, XTEA achieved the highest *E-Rank* due to its balanced code size, memory footprint, and throughput, followed by SPECK, which surpassed SIMON. On the contrary, hardware-oriented ciphers such as SIMON, CLEFIA, and PRESENT demonstrated lower *E-Rank* values, with PRESENT outperforming AES due to its better resource management. In the selected set of LWSCs, software-oriented ciphers like Salsa achieved the highest *E-Rank* due to their efficient resource usage, despite Sosemanuk's better performance efficiency. Hardware-oriented LWSCs such as Grain-v1, Mickey, and Trivium are ranked lower, with Trivium showing comparatively better *E-Rank* due to more balanced performance and resource consumption. Additionally, Grain-128a, being an optimized version, naturally exhibited a higher *E-Rank* than Grain-v1, Mickey, and Trivium, highlighting the benefits of optimization in enhancing the cipher's overall software performance. So, *E-Rank* assists in identifying cryptographic solutions that offer an optimal balance between performance and resource efficiency, customized to the specific needs of resource-constrained environments.

6   Related Work

We started our investigation by examining various benchmarking tools designed to evaluate the performance of LWCC on resource-constrained devices. Although we considered several tools that offered certain performance metrics, there were specific limitations or functionality gaps in those tools that were important to our research objectives. We were motivated by the need for a comprehensive performance analysis covering all the major performance metrics like memory consumption, processor utilization, execution speed, and energy consumption. As a result, we decided to develop a new benchmarking framework from scratch rather than utilizing the existing platforms. The decision to create a new benchmarking framework was necessary due to various challenges posed by the existing frameworks, such as platform interdependency issues, code complexity, and inherent limitations to add new LWCCs.

In the BLOC project [11], a total of 16 LWBCs were evaluated using an MSP430 microcontroller. However, the accompanying C library lacked a standardized interface for all ciphers, causing integration difficulties with new platforms. It is important to note that the project only measured the performance of LWBCs in terms of memory and processor cycle count without providing any insights on energy consumption.

Dinu et al. [14] designed and developed a framework with the aim of achieving fair and consistent benchmarking of LWBCs across 8, 16, and 32-bit processors. This tool enables the analysis and comparison of the execution time, RAM requirements, and code size of lightweight primitives across various embedded platforms. The framework, while versatile, focused exclusively on block cipher primitives, providing performance information only in terms of memory. The associated research paper provides detailed documentation of the research methodology and findings.

The ECRYPT Benchmarking of Cryptographic Systems (eBACS) [15] is a part of the ECRYPT II project and measures cryptographic primitives' speed. It covers public-key systems (eBATS), stream ciphers (eSTREAM), and hash functions (eBASH). The project is built on the SUPERCOP (System for Unified Performance Evaluation Related to Cryptographic Operations and Primitives) framework, which offers an extensive range of cryptographic primitive implementations. The framework's source code is open and written in C with inline assembly, Bash, and Python, making it easy to add new LWCC implementations across the research community. While it is valuable for speed assessment, it lacks any insights into energy and memory consumption.

7   Conclusions

In this paper, we introduced a benchmarking framework optimized for ARM processors, designed to evaluate the performance of selected LWCCs across multiple dimensions: processor speed, memory efficiency, and energy utilization. To accurately measure energy consumption, we enhanced the framework with a hardware extension comprising an Arduino and a power measurement sensor.

In the category of LWBCs, XTEA stood out by offering the lowest CpB, highest throughput, and lowest energy consumption, indicating faster execution. However, it also exhibited the highest total memory consumption among the LWBCs. Similarly, in the LWSCs, Sosemanuk demonstrated the lowest CpB, highest throughput, and lowest energy consumption, yet it also recorded the highest total memory consumption among LWSCs. These findings highlight the trade-offs between performance efficiency and resource usage that must be considered when selecting cryptographic solutions for resource-constrained environments. The *E-Rank* metric has demonstrated significant practical insights, effectively providing an optimum balance between performance, energy, and resource utilization. For instance, PRESENT performed better than AES in terms of *E-Rank* despite AES demonstrating efficient performance. This is because PRESENT has better resource management, which is an essential feature that is expected from LWCC. Similarly, among the selected LWSCs, Salsa ranked highest in overall software performance due to its more effective resource usage, even though Sosemanuk exhibited better performance efficiency. These findings validate the effectiveness of the *E-Rank* metric in providing an optimum and balanced evaluation of the overall software performance of LWCCs.

The future work involves measuring the LWCCs in real-world, resource-intensive applications, such as live video streaming via a camera module, and evaluating the performance of encryption and decryption processes within these applications.

Acknowledgments. This research received no external funding.

Disclosure of Interests. The authors have no conflicts of interest to declare.

References

1. ARM: ARM1176JZF-S Technical Reference Manual r0p7. ARM Limited (2009). https://developer.arm.com/documentation/ddi0301/h
2. Babbage, S., Dodd, M.: The MICKEY stream ciphers. In: Robshaw, M., Billet, O. (eds.) New Stream Cipher Designs. LNCS, vol. 4986, pp. 191–209. Springer, Heidelberg (2008). https://doi.org/10.1007/978-3-540-68351-3_15
3. Badel, S., et al.: Armadillo: a multi-purpose cryptographic primitive dedicated to hardware. In: Mangard, S., Standaert, F.X. (eds.) Cryptographic Hardware and Embedded Systems, CHES 2010. CHES 2010. LNCS, vol. 6225, pp. 398–412. Springer, Berlin, Heidelberg (2010). https://doi.org/10.1007/978-3-642-15031-9_2
4. Beaulieu, R., Shors, D., Smith, J., Treatman-Clark, S., Weeks, B., Wingers, L.: The simon and speck block ciphers on avr 8-bit microcontrollers. In: Lightweight Cryptography for Security and Privacy: Third International Workshop, LightSec 2014, Istanbul, Turkey, 1–2 September 2014, Revised Selected Papers 3, pp. 3–20. Springer, Berlin, Heidelberg (2015). https://doi.org/10.1007/978-3-319-16363-5_1
5. Beaulieu, R., Shors, D., Smith, J., Treatman-Clark, S., Weeks, B., Wingers, L.: The simon and speck lightweight block ciphers. In: Proceedings of the 52nd Annual Design Automation Conference, pp. 1–6 (2015). https://doi.org/10.1145/2744769.2747946

6. Berbain, C., et al.: Sosemanuk, a fast software-oriented stream cipher. New Stream Cipher Designs: The eSTREAM Finalists, pp. 98–118 (2008). https://doi.org/10.1007/978-3-540-68351-3_9

7. Bernstein, D.J.: The salsa20 family of stream ciphers. In: Robshaw, M., Billet, O. (eds.) New Stream Cipher Designs: The eSTREAM Finalists, LNCS, vol. 4986, pp. 84–97. Springer, Berlin, Heidelberg (2008). https://doi.org/10.1007/978-3-540-68351-3_8

8. Birrell, E., Gjerdrum, A., van Renesse, R., Johansen, H., Johansen, D., Schneider, F.B.: Sgx enforcement of use-based privacy. In: Proceedings of the 2018 Workshop on Privacy in the Electronic Society, pp. 155–167 (2018). https://doi.org/10.1145/3267323.3268954

9. Boesgaard, M., Vesterager, M., Pedersen, T., Christiansen, J., Scavenius, O.: Rabbit: a new high-performance stream cipher. In: Johansson, T. (ed.) FSE 2003. LNCS, vol. 2887, pp. 307–329. Springer, Heidelberg (2003). https://doi.org/10.1007/978-3-540-39887-5_23

10. Bogdanov, A., et al.: PRESENT: an ultra-lightweight block cipher. In: Paillier, P., Verbauwhede, I. (eds.) CHES 2007. LNCS, vol. 4727, pp. 450–466. Springer, Heidelberg (2007). https://doi.org/10.1007/978-3-540-74735-2_31

11. Cazorla, M., Marquet, K., Minier, M.: Survey and benchmark of lightweight block ciphers for wireless sensor networks. In: 2013 International Conference on Security and Cryptography (SECRYPT), pp. 1–6. IEEE (2013)

12. De Canniere, C., Preneel, B.: Trivium. In: Robshaw, M., Billet, O. (eds.) New Stream Cipher Designs: The eSTREAM Finalists, LNCS, vol. 4986, pp. 244–266. Springer, Berlin, Heidelberg (2008). https://doi.org/10.1007/978-3-540-68351-3_18

13. Dhanda, S.S., Singh, B., Jindal, P.: Lightweight cryptography: a solution to secure iot. Wirel. Pers. Commun. **112**(3), 1947–1980 (2020). https://doi.org/10.1007/s11277-020-07134-3

14. Dinu, D., Biryukov, A., Großschädl, J., Khovratovich, D., Le Corre, Y., Perrin, L.: Felics-fair evaluation of lightweight cryptographic systems. In: NIST Workshop on Lightweight Cryptography, vol. 128. NIST (2015)

15. Eisenbarth, T., et al.: Compact implementation and performance evaluation of block ciphers in ATtiny devices. In: Mitrokotsa, A., Vaudenay, S. (eds.) AFRICACRYPT 2012. LNCS, vol. 7374, pp. 172–187. Springer, Heidelberg (2012). https://doi.org/10.1007/978-3-642-31410-0_11

16. Hell, M., Johansson, T., Meier, W.: Grain: a stream cipher for constrained environments. Int. J. Wirel. Mob. Comput. **2**(1), 86–93 (2007). https://doi.org/10.1504/IJWMC.2007.013798

17. ISO Central Secretary: Information security – lightweight cryptography part 2: Block ciphers. Standard ISO/IEC 29192-2:2019, International Organization for Standardization, Geneva, CH (2019). https://www.iso.org/standard/78477.html

18. Martin, Hell, M., Johansson, T., Meier, W.: Grain-128a: a new version of grain-128 with optional authentication. Int. J. Wirel. Mob. Comput. **5**(1), 48–59 (2011). https://doi.org/10.1504/IJWMC.2011.044106

19. Needham, R.M., Wheeler, D.J.: Tea extensions. Report (Cambridge University, Cambridge, UK, 1997) (1997)

20. Pettersen, R., Johansen, H.D., Johansen, D.: Secure edge computing with arm trustzone. In: IoTBDS, pp. 102–109 (2017). https://doi.org/10.5220/0006308601020109

21. Python Software Foundation: Python ctype Library Documentation (2024). https://docs.python.org/3/library/ctypes.html. Accessed 14 May 2024

22. Robshaw, M.: The eStream Project. In: New Stream Cipher Designs: The eSTREAM Finalists, pp. 1–6. Springer, London (2008). https://doi.org/10.1007/978-1-4471-5079-4_10
23. Saarinen, M.J.: Cryptanalysis of block tea, October 1998. Unpublished manuscript
24. Shirai, T., Shibutani, K., Akishita, T., Moriai, S., Iwata, T.: The 128-bit blockcipher clefia. In: Biryukov, A. (ed.) Fast Software Encryption: 14th International Workshop, FSE 2007, Luxembourg, Luxembourg, 26-28 March 2007, Revised Selected Papers 14, FSE 2007. LNCS, vol. 4593, pp. 181–195. Springer, Berlin, Heidelberg (2007). https://doi.org/10.1007/978-3-540-74619-5_12
25. Texas Instruments: Ina219 zerø-drift, bidirectional current/power monitor with i2c interface (2021). https://www.ti.com/product/INA219
26. Wheeler, D.J., Needham, R.M.: Tea, a tiny encryption algorithm. In: Preneel, B. (eds.) Fast Software Encryption: Second International Workshop Leuven, Belgium, 14–16 December 1994, Proceedings 2, FSE 1994. LNCS, vol. 1008, pp. 363–366. Springer, Berlin, Heidelberg (1995). https://doi.org/10.1007/3-540-60590-8_29
27. Wu, H.: The stream cipher HC-128. In: Robshaw, M., Billet, O. (eds.) New Stream Cipher Designs. LNCS, vol. 4986, pp. 39–47. Springer, Berlin, Heidelberg (2008). https://doi.org/10.1007/978-3-540-68351-3_4

AutoCRAT: Automatic Cumulative Reconstruction of Alert Trees

Eric Ficke[1], Raymond M. Bateman[2], and Shouhuai Xu[3]($\boxtimes$)

[1] The University of Texas at San Antonio, San Antonio, TX, USA
[2] U.S. Army Research Laboratory South - Cyber, San Antonio, TX, USA
[3] University of Colorado Colorado Springs, Colorado Springs, CO, USA
sxu@uccs.edu

Abstract. When a network is attacked, cyber defenders need to precisely identify which systems (i.e., computers or devices) were compromised and what damage may have been inflicted. This process is sometimes referred to as *cyber triage* and is an important part of the incident response procedure. Cyber triage is challenging because the impacts of a network breach can be far-reaching with unpredictable consequences. This highlights the importance of *automating* this process. In this paper we propose AutoCRAT, a system for quantifying the breadth and severity of threats posed by a network exposure, and for prioritizing cyber triage activities during incident response. Specifically, AutoCRAT automatically reconstructs what we call *alert trees*, which track network security events emanating from, or leading to, a particular computer on the network. We validate the usefulness of AutoCRAT using a real-world dataset. Experimental results show that our prototype system can reconstruct alert trees efficiently and can facilitate data visualization in both incident response and threat intelligence analysis.

Keywords: Cyber Triage · Alert Tree · Alert Path · Threat Score · Alert Prioritization · Incident Response · Intrusion Detection · Cyber Attack

1 Introduction

In cyber incident response, the defender needs to precisely identify what happened to the network in question, including: how did the attacker propagate through the network, what was the attacker's intent, and where and how much damage did the attacker inflict? Since attackers may target a large portion of a network, the defender must quickly and effectively determine the scope of their impact. Specifically, the defender must isolate the routes that the attacker may have used to enter and propagate through the network. These are referred to as *alert paths*, and may be aggregated into so-called *alert trees*.

Isolating alert paths turns out to be a difficult task for two reasons. First, for any amount of incoming alerts, the number of paths that need to be examined

J. Zhao and W. Meng (Eds.): SciSec 2024, LNCS 15441, pp. 273–294, 2025.
https://doi.org/10.1007/978-981-96-2417-1_15

grows quadratically. This is the problem of *efficiency*. Second, without examining all possible alert paths, it is possible that the defender will overlook the actual attack path. This is the problem of *coverage*, which is closely related to false negatives in intrusion detection systems. These problems have serious implications on incident response, as the average time to contain threats can be as high as 85 days [50]. The same report suggests that security automation can reduce this by 26% (to 63 days). In order to help defenders effectively and efficiently respond to cyber incidents, the research community needs to investigate principled solutions to tackling this problem. This motivates the present study, which aims to facilitate *cyber triage* by automatically identifying the potential scope of an attack.

Our Contributions. In this paper, we make three contributions. First, we formalize a suite of concepts to automate cyber triage, including: *alert graph*, which presents a Graph-Theoretic representation of alerts triggered on a network; *alert path*, which indicates a sequence of potentially-related attack steps across multiple computers; *alert tree*, which represents a set of attack paths (as identified by alerts) with respect to a computer of interest. These concepts allow us to represent alerts in a structural fashion to facilitate various analysis and reasoning cyber triage tasks of interest.

Second, we present a novel system called AutoCRAT to reconstruct alert trees from the output of a Network Intrusion Detection System (NIDS). Auto-CRAT's key features can be characterized as follows. (i) It can continuously process streams of alerts reported by security devices. This is important for real-world usage where security devices constantly produce alerts. (ii) It can quickly reconstruct alert trees on demand. Our asymptotic analysis shows that in the worst-case scenario, graph maintenance scales cubically with the number of alerts, while tree reconstruction scales quadratically or log-linearly, depending on the type of tree. (iii) It can model *multi-step attacks*, which include lateral movements secondary to a compromise. This is important because attacker objectives often cannot be accomplished after a single compromise. (iv) It can model *muti-target attacks*, in which a single attacker produces multiple attack paths with different targets. This is important because it does not require the assumption that an attacker has a single target. (v) It can model *multi-source attacks*, in which a single victim is targeted by different attackers or by an attacker with multiple access points. This is important because it does not require the assumption that only a single attacker is active on the network.

Third, we demonstrate the usefulness of alert trees and the alert tree reconstruction method by conducting a case study using a dataset which is collected from a realistic cyber attack testbed, namely CSE-CIC-IDS2018, as published by the University of New Brunswick's Canadian Institute for Cybersecurity. Our experimental results show that our methods were able to analyze the data within the timeframe during which the data was collected, while also incorporating a larger portion of the data than previous works.

Related Work. We divide related prior studies into four categories: *attack modeling*, *attack reconstruction*, *alert prioritization*, and *intrusion detection*. In terms of *attack modeling*, the problem has been approached using alert correlation [22, 70] and clustering [13]. The present study moves a step further by making sense of alerts through the notion of alert paths, which are more comprehensive than alert correlation because they explicitly model temporal-spatial relationships. This means they may be useful in mapping the attack to a particular model (e.g., [28, 37, 52]), which may accommodate explicit happens-before dependencies. Similarly, attack paths (not to be confused with alert paths) have been studied for their usefulness in predicting attack outcomes [6, 8, 25, 26, 40, 41, 46]. These are distinct from alert paths, which are instead retrospective in nature and based on observed attacker activity rather than potential vulnerabilities. Attack graphs have also been used to model the potential propagation of attacks through a computer network [12, 21, 27, 30, 31, 42]. Like attack paths, attack graphs are focused on static network evaluation or attack prediction, rather than reconstruction of observed attack patterns.

In terms of *attack reconstruction* for multi-step attacks, one work that reconstructs multi-step attacks is MAAC [56]. This model assumes a four-stage attack model, where steps of an attack operation can only be assembled if stage 3 alerts are found. This means that false negatives in alert production are more likely to produce false negatives in the model. Although MAAC assembles an alert graph, it only identifies paths of length one, which is likely an attack step within a host rather than across hosts. Another reconstruction method is MIF [38], which uses supervised machine learning to reduce graph size and then produces a risk-state graph to track network attacks. Edges are ordered by start times only, which may induce false positives when two paths overlap. The model uses a recursive depth-first-traversal to build paths. On LLDoS 2000 (containing 60 hosts), MIF processes one million (upsampled) flows in 3 m 24 s. It processed CICIDS2017 for accuracy but did not give runtimes. Yet another reconstruction method is APIN [19], which builds an alert graph using raw alerts and extracts alert paths with respect to a particular node using a chronological traversal. It includes complexity analysis and runtimes for the DARPA 1999 intrusion detection evaluation dataset and the CSE-CIC-IDS2018 dataset.

In terms of *alert prioritization*, the concept of alert trees benefits from relevant studies of alert prioritization [3, 4, 17, 48]. This process ranks alerts according to their severity or associated risk. These rankings are useful to constructing alert trees because they enable more intuitive tree interpretation. For example, visualizing the colors of nodes in a tree based on the ranking (i.e., prioritization) can help defenders identify hotspots in the network. An innovative method is presented in [17] to achieve better visualization of alert tress.

In terms of *intrusion detection*, alert trees depend on input from intrusion detection systems (IDSs), including network-based (NIDS) and host-based (HIDS). IDSs have been criticised heavily on account of the base-rate fallacy, poorly-representative environments, limited attack scope, and weak ground truth [5, 11, 54, 55]. IDSs are ineffective in practice because of alert volumes, false positives and alert interpretability [16, 18, 20, 29].

Paper Outline. Section 2 formalizes the problem. Section 3 presents Auto-CRAT. Section 4 presents the results of applying AutoCRAT to a dataset. Section 5 discusses limitations of AutoCRAT. Section 6 concludes the paper. Table 1 summarizes common terms used in the paper and shows the symbols used to represent them.

Table 1. Common terms and symbols used throughout the paper

Term	Symbol	Meaning/Usage
Alert/Arc	$\alpha \in A$ (or $\mathcal{A}$)	Event indicating the presence of an attack. By convention, $\mathcal{A}$ denotes the set of all alerts in a dataset, while $A \subseteq \mathcal{A}$
Alert Graph	$G(\mathcal{A})$	The set of vertices, arcs, maps and labels used to model a set of alerts representing network attacks
Endpoint pair	$e \in E$	The ordered pair $(source, destination)$, which is used to group certain sets of arcs
Alert Path	$p \in P(\mathcal{A})$	A sequence of vertices and accompanying arc sets derived from $G(\mathcal{A})$ with partial happens-before ordering
Origin	v_1^p	First node in a path p
Target	v_n^p	Last node in a path p of length n
Alert Tree	$T_{fwd}(\mathcal{A}, \hat{v})$, $T_{bwd}(\mathcal{A}, \hat{v})$	An aggregation of alert paths with common origin or target, respectively
Threat Score	$TS(A)$, $ETS(e)$, $PTS(p)$	Metric used to rank sets of alerts by relative threat to the network. Also used for endpoints and paths, respectively
AutoCRAT		The proposed model for tracking network events

2 Problem Statement

Informal Problem Statement. Consider an enterprise network, which consists of computer systems and security devices (e.g., NIDSs), and is managed by the *defender*. Note that the concept of an enterprise network is generally applicable to many types of computer networks, including IoT networks and mobile networks. Computers on the enterprise network may be the target of cyber attacks from some *attacker*, which may come from inside or outside the network. Network traffic is monitored by security devices, which produce alerts when an attack is observed by a security device. A successful attacker may conduct secondary attacks (known as *lateral movements*) against other computers within the network. The term *multi-step attack* refers to such a sequence of attacks.

We aim to develop an understanding of multi-step attacks against a network. This leads to three informal questions: (i) What routes could an attacker have taken to get from one computer to another? (ii) What is the scope of a given attack operation, as represented by an alert or group of alerts? (iii) How may the attacker have used lateral movements to traverse the network and finally set up a particular attack? To answer them, we first need to formalize them.

Formalization: Alert and Alert Graph/Path/Tree. We begin with the input of a stream of alerts, denoted by $\mathcal{A}$, generated by network security/defense devices (e.g., NIDSs).

Definition 1 (Alert). *An alert α is generated by a network security device corresponding to a communication between a source computer and a destination computer and can be described as a tuple $\alpha = (source, destination, time, ID)$, where source and destination are the endpoints of the alert, time represents the timestamp at which the communication begins, and ID is the alert identifier given by the security device (e.g., signature identifier or remote logon type).*

Having defined alerts, we construct more complex objects, namely *alert graphs*, *alert paths* and *alert trees*. Given a set of alerts, we can construct an *alert graph*, a labeled multidigraph where multiple arcs may exist between a pair of nodes (e.g., an attack may make multiple connections before it succeeds, resulting in multiple arcs). Moreover, vertices and arcs can have common labels because alert (i.e., arc) IDs belong to a pre-defined set with specific cybersecurity meanings.

Definition 2 (Alert Graph). *Given a set of alerts $\mathcal{A}$, an alert graph is a labeled multidigraph $G(\mathcal{A}) = (\Sigma_V, \Sigma_A, V, A, s, t, \ell_V, \ell_{ID}, \ell_{time})$, where vertex $v \in V$ represents a computer on the network, arc $a \in A$ represents an alert produced by a security device, Σ_V denotes the set of computer labels (such as IP addresses), Σ_A denotes the set of alert labels (i.e., α's ID), $s : A \to V$ maps arcs to their source vertex (i.e., $\alpha.source$), $t : A \to V$ maps arcs to their target vertex (i.e., $\alpha.destination$), $\ell_V : V \to \Sigma_V$ maps vertices to their labels, $\ell_{ID} : A \to \Sigma_A$ maps arcs to their alert labels, and $\ell_{time} : A \to \mathbb{N}$ maps alerts to the set of natural numbers, representing the time at which they occurred.*

Given an alert graph $G(\mathcal{A})$, the concept of *alert path* describes a sequence of vertices through which an attack is observed (as indicated by alerts). Since alerts representing a multi-step attack may consist of multiple repeated arcs between a pair of vertices, a path is associated with a set of arcs where multiple arcs may link the same pair of vertices. This arc set is further divided into a sequence of smaller arc sets, grouped by the endpoints corresponding to pairs of vertices in the path. Within these arc sets, there must be a valid sequence of arcs, one each from consecutive arc sets, such that they appear in chronological order.

Definition 3 (Alert Path). *Given an alert graph $G(\mathcal{A})$, an alert path is defined as $p = (V^p, A^p)$, where $V^p = (v_1^p, \ldots, v_n^p)$ is a sequence of unique vertices in V; A^p is a corresponding set of arcs in A, for which there must exist a sequence of arc sets $(A_1^p, \ldots, A_{n-1}^p)$ such that $A_i^p = \{\alpha \in A : s(\alpha) = v_i^p \wedge t(\alpha) = v_{i+1}^p\}$, and $A^p = \bigcup_{i=1}^{n-1} A_i^p$, and there must exist some sequence $(\alpha_1, \ldots, \alpha_{n-1})$ such that for each $i \in [1, \ldots, n-1]$, $\alpha_i \in A_i^p \wedge i < n-1 \to \ell_{time}(\alpha_i) < \ell_{time}(\alpha_{i+1})$.*

Given an alert path p with $V^p = (v_1^p, \ldots, v_n^p)$, we say v_1^p is the *origin* and v_n^p is the *target*. Given $\mathcal{A}$, let $P(\mathcal{A})$ denote the set of all alert paths in $G(\mathcal{A})$. A path $p \in P(\mathcal{A})$ may be uniquely identified by V^p.

An alert tree represents a set of alert paths $P(A)$ (where $A \subseteq \mathcal{A}$) with a common *reference vertex*, which is either the origin or the target, but not both. Trees with a common origin are called *forward trees* and trees with a common target are called *backward trees*, because of the happens-before relationship of alerts in the tree. Because the graph $G(\mathcal{A})$ is a labeled multidigraph, it is possible that cyclical subgraphs will result in sets of paths $P(A)$ such that they cannot naturally be combined to form trees. For example, the graph constructed from the set of paths $P = \{(a, b, c), (a, c, b)\}$ contains a cycle (b, c, b) and is therefore not a tree. However, if we manipulate the set of paths by changing duplicate nodes into unique nodes with the same identifier (in Σ_V), we can artificially prevent this loop formation. Consider the previous example. We can change the second path to (a, c', b') such that $\ell_V(c') = \ell_V(c)$ and $\ell_V(b') = \ell_V(b)$. This results in a tree with arcs $(a, b), (b, c), (a, c'), (c', b')$ such that the modified graph forms a tree, which can be used to extract the relevant computer identifiers. The advantage of the tree structure over an arbitrary graph is that one can visualize the *temporal dependence* of the arcs based on their height within the tree. Since temporal dependence is an important feature of alert paths, it is natural to visualize sets of alert paths using alert trees rather than alert graphs.

Definition 4 (Alert Tree). *Given an alert graph $G(\mathcal{A})$, a reference vertex $\hat{v} \in V$, and a set of paths $P' = \{p \in P(\mathcal{A}) : v_1^p = \hat{v}\}$, a forward alert tree is a labeled digraph denoted $T_{fwd}(\mathcal{A}, \hat{v}) = \{\Sigma_{V,T}, V_T, A_T, \ell_{V,T}\}$, where $\Sigma_{V,T} = V$, $V_T = V \times P'$, $A_T = (v, v') \in V_T$ such that $\exists p \in P', i \in [1, \ldots, |V^p|]$ for which $v = v_i^p \land v' = v_{i+1}^p$, and $\ell_{V,T} : V_T \rightarrow \Sigma_{V,T}$ maps vertices to their corresponding nodes in V. A forward alert tree is rooted at reference vertex $\hat{v}$ and represents paths $p \in P'$, such that every vertex in p has a corresponding vertex $v \in V_T$ for which all of the ascendants of v (denoted $asc(v)$) are labeled with the vertices in the path. Specifically, $\forall i \in [1, \ldots, |asc(v)|], \ell_{V,T}(asc(v)_i) = v_i^p$. A backward alert tree T_{bwd} is a reversed forward tree in the following sense: reference vertex $\hat{v}$ is the target rather than the origin of all of the corresponding alert paths, while the vertex identifiers must match vertex descendants rather than ascendants.*

Equipped with the preceding definitions, we can now formalize the afore-mentioned informal questions as the following research questions (RQs).

- **RQ1**: This question asks for the set of all alert paths with a specified origin and target. Formally, given a set of alerts $\mathcal{A}$, a known attack origin v_{origin}, and a known attack target v_{target}, produce the set of attack paths $P' = \{p \in P(\mathcal{A}) : v_1^p = v_{origin} \land v_{|V^p|}^p = v_{target}\}$.
- **RQ2**: This question asks for the depth and breadth of access that an attacker achieved corresponding to an alert (or attack). Formally, given a set of alerts $\mathcal{A}$ and a known attack origin v_{origin}, produce the forward alert tree $T_{fwd}(\mathcal{A}, v_{origin})$.

– **RQ3**: This question asks for the set of multi-stage attacks that may have led to the compromise of a computer. Formally, given a set of alerts $\mathcal{A}$ and a known attack target v_{target}, produce the backward alert tree $T_{bwd}(\mathcal{A}, v_{target})$.

3   The AutoCRAT System

Here we propose the AutoCRAT system to address the RQs mentioned above. Figure 1 highlights its architecture, which has two main components: *database* and *core* functions. The system receives inputs in the form of *management* and *data retrieval*, which make calls to the core functions to change or retrieve data from the database, respectively. This section discusses the *database* and *core* functions of AutoCRAT.

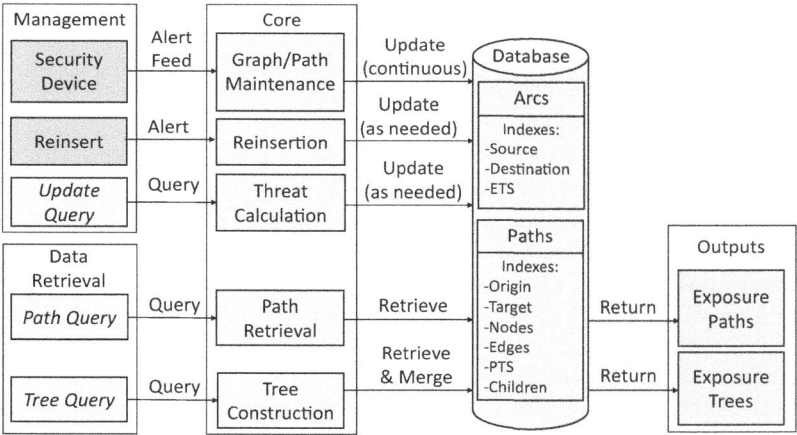

Fig. 1. Architecture of the AutoCRAT system

3.1   Database

The database stores the arcs (i.e., alerts) and paths derived from G as follows. First, the arc collection of the database stores pairs of arc endpoints. Arcs are stored as annotations to the arc endpoints. This enables more efficient retrieval of alerts, which are always considered in the context of the other alerts with the same endpoints. Second, alert paths are stored in their own collection. Each path $p \in P(\mathcal{A})$ contains a list of nodes, the corresponding node pairs (to facilitate retrieval via pairwise indexing), the path's PTS, and a list of child vertices (denoted $Children^p$) that have been used to produce lengthened copies of the path, $\{v \in V : (\exists p' \in P(\mathcal{A}), \forall v_i^p \in V^p)v_i^p = v_i^{p'} \wedge |V^{p'}| = |V^p| + 1 \wedge v = v_{|V^{p'}|}^{p'}\}$.

This list is used to facilitate path validation, as discussed later in this section. Every time a path is lengthened, the new path is added as its own object in the database, in order to facilitate indexing as described below.

Indexing. Each collection is indexed so that its objects may be efficiently found and retrieved from the database. Our approach uses multiple types of indices, including individual indexes (on a single field), compound indexes (on two or more fields) and multikey indexes (on a set or sequence). Some of these index types are only available in certain kinds of databases, such as MongoDB, which is used in our implementation. This limits the interoperability of our approach, or risks compromising the efficiency of accessing some objects in the database.

The arc collection is indexed by arc endpoints (each individually as well as together in a compound index), and by the arc's threat score (discussed later in this section). The path collection has individual indices on the origin, the target, and the path's threat score (also discussed later). It also has multikey indexes on the vertex sequence and the arc endpoint sequence. Additionally, the path collection has one compound index, covering the target and the list of child vertices (as a multikey index).

Duplicate paths are prevented by validating new arcs against this list of children in the path's so-called parent (i.e., $p' \in P$ such that $V^{p'} = V^p - (v^p_{|V^p|}) \wedge v^p_{|V^p|} \in Children^{p'}$. Subpaths are stored as their own objects because this allows them to be more efficiently indexed. Specifically, we index the path's origin and target, because multikey indexes do not preserve order. This means that, if a multikey index was used on the path nodes, then an attempt to retrieve paths with a certain origin and target would need to parse all paths containing both the origin and target in any position, then trim the path appropriately. In this case, the runtime optimization of multikey indexing of path vertices trades off with storage complexity (i.e., storing more paths).

On the other hand, it is efficient to use a multikey index to index to children of a particular path, because queries that search for children need only find a single child (i.e., the *destination* of the alert being inserted). This means that the ordering of children in a multikey index is unimportant.

3.2   Core

The core has five modules: *alert graph/path maintenance, reinsertion, threat calculation, path retrieval,* and *tree construction.* These modules are invoked by the respective management services.

Algorithm 1. Alert Graph / Path Maintenance

Input: $\alpha, G(\mathcal{A}) = (\Sigma_V, \Sigma_A, V, A, s, t, \ell_V, \ell_{ID}, \ell_{time}), P, Children^{p \in P}$
Output: $G(\mathcal{A} \cup \{\alpha\}), P, Children^{p \in P}$
1: **if** $\not\exists \alpha' \in A : s(\alpha) = s(\alpha') \wedge t(\alpha) = t(\alpha')$ **then**
2:     $V^{\hat{p}} \leftarrow (s(\alpha), t(\alpha))$ ▷ Create $\hat{p}$ via $V^{\hat{p}}$
3:     $Children^{\hat{p}} \leftarrow \emptyset$
4:     $P \leftarrow P \cup \{\hat{p}\}$
5: **end if**
6: $A \leftarrow A \cup \{\alpha\}$
7: $V \leftarrow V \cup \{s(\alpha), t(\alpha)\}$
8: $Candidate\_Paths \leftarrow \{p \in P : V^p_{|V^p|} = s(\alpha)\}$ ▷ Find paths to lengthen using "end" index
9: **for** $c \in Candidate\_Paths$ **do**
10:     **if** $t(\alpha) \notin V^c$ **then**   ▷ Prohibit cycles
11:         **if** $t(\alpha) \notin Children^c$ **then**   ▷ Prohibit identical twins
12:             $p' \leftarrow \{V^c, A^c\}$
13:             $V^{p'} \leftarrow V^{p'} \cup (t(\alpha))$
14:             $A^{p'} \leftarrow A^{p'} \cup \{\alpha\}$
15:             $Children^{p'} \leftarrow \emptyset$
16:             $Children^c \leftarrow Children^c \cup \{t(\alpha)\}$
17:             $P \leftarrow P \cup \{p'\}$
18:         **end if**
19:     **end if**
20: **end for**
21: **return** $G = (V, A), P, Children^{p \in P}$

Core Module 1: Alert Graph/Path Maintenance. As new security events arrive, AutoCRAT incorporates them into the relevant databases according to Algorithm 1, which proceeds as follows: (i) It first checks if the new arc includes existing endpoints. If so, it adds the arc to the endpoint object as an annotation; otherwise, it inserts a new endpoint object annotated with the arc (Lines 1–13). (ii) It queries the database to find paths which end at the source of the event (Line 14). (iii) The algorithm copies the paths found, appending the alert's *destination* onto the copies. If the new paths already exist in the database or are cyclical, they are discarded. The original paths are annotated with a list of children to facilitate this check on future inserts, and the new paths are inserted (Lines 15–27).

In the worst-case scenario, every insertion of a new alert α will add a new endpoint object and path, and lengthen the set of existing paths that terminate at the source of the inserted arc. This will give us a storage complexity in $\mathcal{O}(|E| + |A| + |P|)$. Now suppose one inserts the first arc, with endpoints (v_1, v_2), into a database. This will produce a single path: (v_1, v_2). A second insertion of (v_2, v_3) adds a new path and lengthens the existing path, resulting in three paths: (v_1, v_2), (v_2, v_3), and (v_1, v_2, v_3). Clearly, each arc adds at most one new endpoint object (exactly one in this case), leading to $|E|$ or $\mathcal{O}(|\mathcal{A}|)$. This means that each new arc adds up to $|E|$ new paths of length $|E|$ or less, leading to $|P| \in \mathcal{O}(|\mathcal{A}|^2)$. This is demonstrated in Table 2.

Table 2. The worst-case storage complexity after i arc insertions. As each new arc α_i can add as many as i paths, the worst-case number of paths is $|P| = \frac{1}{2}|\mathcal{A}|^2 + \frac{1}{2}|\mathcal{A}|$ or $\mathcal{O}(|\mathcal{A}|^2)$.

Arcs	1	2	3	4	$|\mathcal{A}|$
Endpoint pairs	1	2	3	4	$|\mathcal{A}|$
Paths	1	3	6	10	$\sum_{i=1}^{|\mathcal{A}|}(i)$

Under the same worst-case scenario, the insert algorithm must access every path in the database, $|P|$, copy each, adding one endpoint, and insert the copies, plus one new path with two nodes. Then the worst-case runtime (in terms of the number of accesses to the database) is $2 \cdot |P| + 1$ or $\mathcal{O}(|P|)$.

Core Module 2: Reinsertion. In some cases, the defender may need to find paths that were unavailable at the time of data collection (e.g., if the NIDS produces a false negative which is later corrected). Since alerts are otherwise inserted chronologically, we need another algorithm to perform such retroactive insertions ("reinsert"). For this purpose we propose Algorithm 2.

Algorithm 2. Alert Reinsertion

Input: $\alpha, G(\mathcal{A}), P$
Output: $G(\mathcal{A} \cup \{\alpha\}), P$
1: $P^{pre} \leftarrow \{p \in P : V_{|V^p|}^p = s(\alpha)\}$
2: $P^{pre} \leftarrow P^{pre} \setminus \{p \in P^{pre} : t(\alpha) \in V^p\}$
3: $P^{post} \leftarrow \{p \in P : V_1^p = t(\alpha)\}$
4: $P^{post} \leftarrow P^{post} \setminus \{p \in P^{post} : s(\alpha) \in V^p\}$
5: $V \leftarrow V \cup \{s(\alpha), t(\alpha)\}$
6: $A \leftarrow A \cup \{\alpha\}$
7: $V^{\hat{p}} \leftarrow (s(\alpha), t(\alpha))$ ▷ Create $\hat{p}$ via $V^{\hat{p}}$
8: $P \leftarrow P \cup \{\hat{p}\}$
9: **for** $p \in P^{pre}$ **do**
10:     **for** $p' \in P^{post}$ **do**
11:         **if** $V^p \cap V^{p'} = \emptyset$ **then**   ▷ Prohibit Cycles
12:             $V^{New\_Path} \leftarrow V^p \cup V^{p'}$ ▷ Create New_Path via V^{New_Path}
13:             **if** New_Path is valid given A **then**
14:                 $P \leftarrow P \cup \{New\_Path\}$
15:             **end if**
16:         **end if**
17:     **end for**
18: **end for**
19: **return** G, P

The storage complexity is the same as for the regular maintenance function, $\mathcal{O}(|\mathcal{A}|^2)$. The asymptotic runtime is dominated by $|p^{pre}| \cdot |p^{post}|$. Because cycles about α are removed (lines 2,4), $P^{pre} \cap P^{post} = \emptyset$. This is bounded by $(\delta \cdot |P|) \cdot ((1 - \delta) \cdot |P|)$, where $\delta < 1$ is the proportion of the larger of the two path sets relative to $|P|$. This simplifies to $(\delta - \delta^2) \cdot |P|^2$, for which the maximum value of the first term (when $\delta = \frac{1}{2}$) is $\frac{1}{4}$. The worst-case runtime is $\frac{1}{4}|P|^2$ or $\mathcal{O}(|P|^2)$.

Core Module 3: Threat Calculation. We propose measuring a given set of alerts $A \subset \mathcal{A}$ according to its threat score (TS), which can also be used during alert visualization. For example, nodes in a graph may be colored from black to red based on their TS in ascending order, to guide the viewer to network hotspots. We use this approach in our examples to follow.

There can be many definitions for TS, and identifying the "best" definition is orthogonal to the focus of the present paper. Since any "good" definitions can be incorporated into AutoCRAT in a plug-and-play fashion, we will use one example definition to demonstrate the idea. In this example definition of TS, we define it as the geometric mean of the number and diversity of alerts in a set, in order to estimate the risk posed by the associated attacks. This definition can be naturally extended to specify the TS of a pair of endpoints (Endpoint Threat Score — ETS) or a path (Path Threat Score — PTS). Specifically, we have:

Definition 5 (Threat Score (TS)). *The TS of a set of alerts is defined as:*

$$TS(A) = \sqrt{|\{\sigma \in \Sigma_A : (\exists \alpha \in A), \sigma = \ell_{ID}(\alpha)\}| \cdot |\{\alpha \in A\}|}. \quad (1)$$

The ETS of an endpoint pair (source, destination) can be calculated as $TS(\{\alpha \in \mathcal{A} : s(\alpha) = source \wedge t(\alpha) = destination\})$. The PTS of a path p can be calculated as $TS(A_p)$.

In order to identify the endpoints and paths in the database with the highest threat, AutoCRAT must periodically update ETS and PTS for endpoints and paths, respectively. In order to do this, the defender submits an update query on demand. The update function is simple: first, it calculates ETS for every endpoint object, then it calculates PTS for every path. To reduce the number of accesses to the database, a copy of each endpoint object's arc annotations is cached during the ETS calculations, in order to facilitate the PTS calculations.

The runtime of this approach is based on the following: (i) calculating ETS for all endpoints requires the parsing of each arc (which each belong to exactly one endpoint annotation), meaning the runtime of updating is in $\Omega(|\mathcal{A}|)$; (ii) calculating PTS for all paths requires the parsing of each endpoint object a number of times equal to the number of paths in which it appears. This is analyzed in Table 3, which shows that the worst case number of appearances of an endpoint object in the set of paths is in $\mathcal{O}(|E^2|)$. Since there are $|E|$ endpoint objects, this results in a worst-case combined runtime in $\mathcal{O}(|\mathcal{A}| + |E|^3)$.

Table 3. The worst-case number of paths that contain a particular endpoint. The ith endpoint can only belong to $i \cdot (|E| - i + 1) = i \cdot |E| - i^2 + i$ paths. Then the worst-case is in $\mathcal{O}(|E|^2)$, and the worst-case number of endpoint objects across the set of all paths is $\sum_{i=1}^{|E|} i \cdot (|E| - i + 1) = \frac{1}{6}|E|(|E| + 1)(|E| + 2)$ or $\mathcal{O}(|E|^3)$.

Endpoint i	Insertion	1	2	3	4	5	$|E|$	
1	1		2	3	4	5	$|E|$	
2	0		2	4	6	8	2 · (|E| − 1)	
3	0		0	3	6	9	3 · (|E| − 2)	
4	0		0	0	4	8	4 · (|E| − 3)	
5	0		0	0	0	5	5 · (|E| − 4)	
$|E|$	0		0	0	0	0	$|E|$	
i	0		0	0	0	0	$i \cdot (|E| - i + 1)$	

ETS is also used in the visualization of attack paths. Specifically, the color of an alert tree node n represents the normalized ETS of $(parent(n), n)$ for forward trees or $(n, parent(n))$ for backward trees. Vertex colors range from red to black, where the vertex in the tree with the highest ETS is red and the root is black (along with any other vertices with $ETS = 1$). In RGB (i.e., hexadecimal) notation, colors range from 0x000000 (black) to 0xFF0000 (red), such that colors may be compared ordinally to mimic the comparison of ETS. This is demonstrated in Fig. 2, showing the value of the ETS measurement in presenting salient information to the defender.

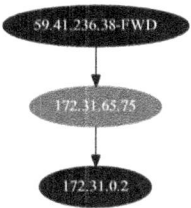

Fig. 2. Example alert tree coloring, where the root is black, the child is red (ETS 179.10), and the grandchild is very nearly black (color code 0x0D0000 and ETS 10.49). (Color figure online)

Core Module 4: Path Retrieval. Path retrieval requires that the graph and path databases are properly maintained. The path retrieval process then involves a simple query to the database, which stores each path individually. This query leverages the $(source, destination)$ compound index, which efficiently retrieves the appropriate paths from the database. The retrieval algorithm executes with a runtime efficiency of $\mathcal{O}(|P|)$, where $P \subseteq P(\mathcal{A})$ is the set of paths to be retrieved.

Core Module 5: Tree Reconstruction. Tree reconstruction is to reassemble paths into a tree which represents either the attack surface exposed to an attacker or the attack vectors exposing a target, depending on the type of query. In either case, the user must specify a node to act as the root of the tree and the tree's direction, and the module handles the rest. The tree reconstruction algorithm takes as input a reference node to act as the tree's root, a direction (i.e., forward or backward), G, and P. It begins by selecting all paths which have the reference node as their origin (for forward trees) or target (for backward trees), using the approach described above. It then parses each path, adding each edge as a node of the tree if it was not already added from a previous path.

The worst-case scenario for the tree reconstruction algorithm is the same as the insert algorithm as discussed above, in which each subsequent alert produces a new endpoint which extends that of the previous alert. Specifically, we have at most $|E|$ paths rooted at a given node, and the maximum path length is $|E|$. This results in a final worst-case runtime of $\mathcal{O}(|E|^2)$.

4   Case Study

Our experiments were run using Ubuntu 20.04 with 192 GB of RAM, 2 cores of an Intel Xeon Gold 6242 CPU @2.80 GHz, and a 200 GB HDD. These resources were shared among AutoCRAT functions and the corresponding MongoDB database, which was installed on the same computer to eliminate variability imposed by network conditions.

4.1   Dataset

The dataset was published by the University of New Brunswick, and is referred to as CSE-CIC-IDS2018 [51]. It contains data collected over the course of 9 days, during which multiple distinct attack scenarios were executed against the network. The environment was connected to the internet during the experiments, thus real-world attacks can also be observed in the data. We preprocessed the packet capture (PCAP) files in the dataset using Suricata 4.0 [1] with the corresponding Emerging Threats signature set [2], to produce a set of 3,323,426 alerts, of which 19,921 were strictly internal to the target network as defined by the dataset authors. We converted the alerts into JSON objects to conform to AutoCRAT's expected format, and sorted them chronologically before feeding them into the database.

4.2   Experimental Results

Database construction for CSE-CIC-IDS2018 took 35h14m07s, resulting in 1,053,710 edges and 3,591,217 paths. Efficiency and accuracy (in terms of graph coverage, which may impose false negatives) are analyzed relative to an existing model, APIN [19]. Path selection for APIN was done using its relevant heuristics. The comparison is shown in Table 4.

Table 4. Comparison of the proposed AutoCRAT model with an existing model [19]. Query runtimes are the average of 10 runs. Improvements shown in bold. * [19] only ranks nodes, while AutoCRAT scores endpoints and paths. †The models do not directly rank these objects, so they are selected based on applicable node and edge rankings. These inherited rankings may not be accurate with respect to other metrics but offer a reasonable baseline.

	[19]	AutoCRAT
Build DB	29m43s	13h42m41s
Rank Objects*	49s	1h00m29s
Top 100 paths	52s†	**32ms**
Top 20 trees	52s†	2m42s†
Coverage (nodes)	99.6%	**100%**
Coverage (events)	3.4%	**100%**
DB size	637 MB	1.1 GB

Note that [19] sacrifices coverage in order to improve runtime. This is necessary because its tree retrieval time suffers extraordinary slowdown in the presence of highly connected nodes. Even though [19] only excludes .4% of the vertices in the graph, 96.6% of the arcs in the graph are adjacent to these vertices, and are effectively blacklisted from tree reconstruction, meaning that the set of paths (and trees) that can be reconstructed is incomplete. In AutoCRAT, the connectedness slowdown problem is solved by shifting the bulk of the work into the pre-processing stage. This results in a much faster path retrieval time and full graph coverage at the cost of maintenance time. However, this pre-processing time remains feasible in practice since 9 days of data (constituting approximately 164 hours of activity) are processed in under 14 hours. Graph coverage is important because low coverage induces false negatives in the path and tree reconstruction. This means that the model in [19] is vulnerable to DoS attacks, which may allow an attacker to conceal their attack paths by creating so much traffic around a key node that the algorithm removes it from the analysis entirely.

In order to compare the two models under comparable conditions, we ran both models again using a reduction of the original dataset, which filtered nodes outside the network as well as edges which crossed the border of the network. This resulted in only 1,994 edges and 3,019 paths, as shown in Table 5. This resulted in a coverage of 0.6% for both nodes and paths, much closer to that of [19] (i.e., roughly 1/6). Given the intuition that internal nodes are far more relevant to defenders, we do not consider the loss of node coverage important in this case (although some of the nodes filtered by [19] were in fact internal nodes). This reduction greatly improved the performance of AutoCRAT, resulting in runtimes and storage efficiency that far outperformed [19].

The above discussion demonstrates trade-offs between pre- and post- processing times and between processing time and accuracy.

Table 5. Comparison of the proposed AutoCRAT model with an existing model [19], using only the internal events. Query runtimes are the average of 10 runs. Improvements shown in bold. * [19] only ranks nodes, while AutoCRAT ranks endpoints and paths. †The models do not directly rank these objects, so they are selected based on applicable node or endpoint rankings. These inherited rankings may not be consistent with respect to other metrics but offer a reasonable baseline.

	[19]-internal	AutoCRAT-internal
Build DB	9 s	35 s
Rank Objects*	0.28 s	5 s
Top 100 paths	3 s†	**23ms**
Top 20 trees	3 s†	**1.97s†**
Coverage (nodes)	0.6%	0.6%
Coverage (events)	0.6%	0.6%
DB size	2.9 MB	**2.4 MB**

Insight 1. *To attain the same accuracy, AutoCRAT (relative to [19]) front-loads its processing time into building and maintaining paths so that when it comes time to retrieve them, it can do so more quickly. For both models, reducing the volume of data processed improves both processing time and database size, but sacrifices accuracy (as measured by coverage).*

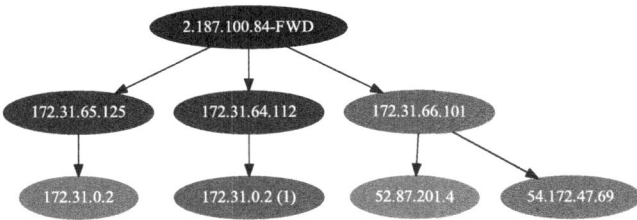

Fig. 3. A forward tree containing eight vertices, colored red to black based on normalized ETS, descending. Of the four leaves (i.e., path targets), the reddest vertex, which represents the endpoints (172.31.66.101, 52.87.201.4) scored an ETS of 5.92 with 35 alerts sharing a single *ID*. (Color figure online)

To demonstrate how the alert tree structure and threat score heuristic may be useful in practice, we include example forward and backward trees in Figs. 3 and 4, respectively. These trees were selected for their size; some trees produced had well over 1000 vertices and would not be legible in the present format. This problem is a challenge that we leave to future work, as the present focus is efficiency of reconstruction.

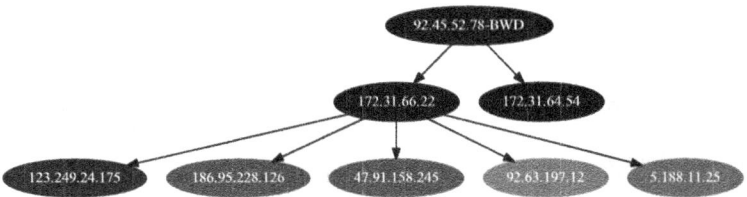

Fig. 4. A backward tree containing eight verticess, colored red to black based on normalized ETS, descending. Of the six leaves (i.e., path origins), the reddest vertex, which represents the endpoints (92.63.197.12, 172.31.66.22) scored an ETS of 7.35 with 54 alerts sharing a single ID. (Color figure online)

Answering RQ1. In order to answer RQ1, we must retrieve a set of paths corresponding to a known attack origin and a known attack target. Assuming the database maintenance has kept up with the alert stream, this can be accomplished with a query to the database utilizing the $(source, destination)$ index. Because subpaths are also stored in the database, we can be confident that we need only retrieve paths that start and end with the origin and target, respectively. This efficiently returns a list containing exactly the required corresponding paths, without the need to parse the paths to truncate them at the proper destination.

Answering RQ2. To answer RQ2, we must reconstruct the forward alert tree corresponding to a particular origin. This process begins with retrieving all of the paths beginning at the specified node, leveraging the *source* index. We then pass these paths to the tree reconstruction function, which arranges them based on their relationships to each other.

Answering RQ3. In order to answer RQ3, we must reconstruct the backward alert tree corresponding to a particular target. Similar to RQ2, this process retrieves the appropriate paths from the database and passes them to the tree reconstruction function.

5   Limitations

We identify six limitations that should be addressed in future studies. First, AutoCRAT depends on IDS correctness, meaning that IDS false positives and false negatives can result in errors in AutoCRAT's path and tree reconstruction. However, in the case where errors occur along paths with true positives, the penalty will only reduce the accuracy of the threat score calculations. It is important to build metrics to quantify the impact of IDS (in)correctness on the trustworthiness of the results, in a fashion similar to [7]. In particular, it would be exciting to establish a systematic quantitative methodology that can be seamlessly incorporated into the Cybersecurity Dynamics framework [62,66] to enable not only reactive defenses but more importantly proactive and adaptive defenses [23, 24, 32, 33, 61, 63–65, 72, 73], by possibly leveraging data-driven cyber threats forecasting techniques [15, 44, 45, 53, 59, 60, 68, 69].

Second, AutoCRAT assumes that attacks are always initiated by a malicious node. This assumption may be violated in client-side attacks, such as drive-by downloads [47, 57, 58]. In such cases, an advanced security device may be able to reverse the order of nodes in the alert during preprocessing, preserving attack semantics. It is important to investigate how to extend AutoCRAT to accommodate cyber social engineering attacks and defenses as they are often used as a means to penetrate into a network, especially in relation to the psychological aspect [34–36, 49]. It is also interesting to extend AutoCRAT to smart homes, which is an emerging field especially from the cyber insurance perspective [71].

Third, AutoCRAT depends on the accuracy of threat scores. This limitation may result in rankings that do not reliably show the importance of a path or tree relative to the mission at hand. However, it does not affect the accuracy of path or tree reconstruction. Regardless, AutoCRAT can be easily adapted to incorporate better ranking methods, which may include asset values and risk scores. The problem of ranking alerts also remains an open research problem independent of alert path and tree modeling. Thus, we need to develop a systematic family of metrics [9, 10, 14, 39, 43, 67].

Fourth, AutoCRAT's path maintenance algorithm assumes that events are inserted sequentially. This means that it may be difficult to parallelize its execution. Since parallelization is a powerful tool of efficiency, this problem may impact viability of the methods in practice. However, it may be possible to parallelize some alert insertions if the adjacent nodes are disparate relative to sequential alerts. We leave this investigation to future work.

Fifth, AutoCRAT assumes that each computer has a single, unique, and static address. It may be extensible to accommodate computers with multiple IP addresses (e.g., by aliasing node names before inserting them into the database or when preprocessing alerts). In the case of a segmented network with private subnets, some computers on disparate subnets may have matching addresses (e.g., 192.168.1.1). This case may be harder to accommodate.

Sixth, the efficiency of the path maintenance algorithm depends on the assumption that the database is capable of efficiently indexing elements of an array. This restricts the interoperability of the framework to certain kinds of databases.

6   Conclusion

We have introduced the AutoCRAT system for modeling and tracking multi-step network attacks as indicated by alerts generated by security devices. The key concepts behind AutoCRAT are those of alert graphs, alert paths, and alert trees. The technical contributions include data structures and algorithms for efficiently representing and constructing alert paths and alert trees, as well as asymptotic storage and runtime complexity analysis. This study is useful to cyber defenders because it quantifies threats against a network and its components and presents them in an intuitive form that is easy to understand. Our case study based on an implementation of AutoCRAT and a research dataset shows that AutoCRAT can efficiently reproduce alert paths and trees, keeping pace with alerts produced on a testbed network. The paper is a significant step towards automating cyber triage with and risk quantification, which remains an important and elusive problem. More research needs to be conducted with real-world datasets, including cyber-physical systems such as smart homes and hospitals.

Acknowledgment. We thank the reviewers for their comments. This research was supported in part by NSF Grant #2115134 and Colorado State Bill 18-086. This research used the Chameleon testbed.

References

1. Suricata | open source ids / ips / nsm engine (2018). https://suricata-ids.org/download/
2. Welcome to the emerging threats rule server (2019). https://rules.emergingthreats.net/
3. Alsubhi, K., Aib, I., Boutaba, R.: Fuzmet: a fuzzy-logic based alert prioritization engine for intrusion detection systems. Int. J. Netw. Manag. **22**(4), 263–284 (2012)
4. Apruzzese, G., Pierazzi, F., Colajanni, M., Marchetti, M.: Detection and threat prioritization of pivoting attacks in large networks. IEEE Trans. Emerg. Top. Comput. **8**(2), 404–415 (2017)
5. Axelsson, S.: The base-rate fallacy and the difficulty of intrusion detection. ACM Trans. Inf. Syst. Secur. (TISSEC) **3**(3), 186–205 (2000)
6. Chen, C.M., Guan, D., Huang, Y.Z., Ou, Y.H.: Attack sequence detection in cloud using hidden Markov model. In: 2012 Seventh Asia Joint Conference on Information Security, pp. 100–103. IEEE (2012)
7. Chen, H., Cho, J., Xu, S.: Quantifying the security effectiveness of firewalls and dmzs. In: Proceedings of the HoTSoS'2018, pp. 9:1–9:11 (2018)
8. Chen, Y., Boehm, B., Sheppard, L.: Value driven security threat modeling based on attack path analysis. In: 2007 40th Annual Hawaii International Conference on System Sciences (HICSS'07), p. 280a. IEEE (2007)
9. Cho, J., Hurley, P., Xu, S.: Metrics and measurement of trustworthy systems. In: Proceedings of the IEEE MILCOM (2016)
10. Cho, J.H., Xu, S., Hurley, P.M., Mackay, M., Benjamin, T., Beaumont, M.: Stram: measuring the trustworthiness of computer-based systems. ACM Comput. Surv. **51**(6), 128:1–128:47 (2019)
11. Chou, D., Jiang, M.: Data-driven network intrusion detection: a taxonomy of challenges and methods. arXiv preprint arXiv:2009.07352 (2020)
12. Cinque, M., Della Corte, R., Pecchia, A.: Contextual filtering and prioritization of computer application logs for security situational awareness. Futur. Gener. Comput. Syst. **111**, 668–680 (2020)
13. De Alvarenga, S.C., Barbon, S., Jr., Miani, R.S., Cukier, M., Zarpelão, B.B.: Process mining and hierarchical clustering to help intrusion alert visualization. Comput. Secur. **73**, 474–491 (2018)

14. Du, P., Sun, Z., Chen, H., Cho, J.H., Xu, S.: Statistical estimation of malware detection metrics in the absence of ground truth. IEEE T-IFS **13**(12), 2965–2980 (2018)
15. Fang, Z., Xu, M., Xu, S., Hu, T.: A framework for predicting data breach risk: leveraging dependence to cope with sparsity. IEEE T-IFS **16**, 2186–2201 (2021)
16. Fernandez, G.C., Xu, S.: A case study on using deep learning for network intrusion detection. In: 2019 IEEE Military Communications Conference (MILCOM'2019), pp. 1–6 (2018)
17. Ficke, E., Bateman, R.M., Xu, S.: Reducing intrusion alert trees to aid visualization. In: Yuan, X., Bai, G., Alcaraz, C., Majumdar, S. (eds.) Network and System Security - 16th International Conference, NSS 2022, Denarau Island, Fiji, 9–12 December 2022, Proceedings. LNCS, vol. 13787, pp. 140–154. Springer (2022)
18. Ficke, E., Schweitzer, K.M., Bateman, R.M., Xu, S.: Analyzing root causes of intrusion detection false-negatives: methodology and case study. In: Proceedings of the IEEE MILCOM'2019 (2019)
19. Ficke, E., Xu, S.: Apin: automatic attack path identification in computer networks. In: IEEE ISI 2020 (2020)
20. FireEye: The numbers game: How many alerts is too many to handle? (2015). https://www.trellix.com/advanced-research-center/
21. Frigault, M., Wang, L.: Measuring network security using bayesian network-based attack graphs. In: 2008 32nd Annual IEEE International Computer Software and Applications Conference, pp. 698–703, July 2008. https://doi.org/10.1109/COMPSAC.2008.88
22. Haas, S., Fischer, M.: GAC: graph-based alert correlation for the detection of distributed multi-step attacks. In: Proceedings of the 33rd Annual ACM Symposium on Applied Computing, pp. 979–988 (2018)
23. Han, Y., Lu, W., Xu, S.: Characterizing the power of moving target defense via cyber epidemic dynamics. In: HotSoS, pp. 1–12 (2014)
24. Han, Y., Lu, W., Xu, S.: Preventive and reactive cyber defense dynamics with ergodic time-dependent parameters is globally attractive. IEEE TNSE **8**(3), 2517–2532 (2021)
25. Hossain, M.N., et al.: {SLEUTH}: real-time attack scenario reconstruction from {COTS} audit data. In: 26th {USENIX} Security Symposium ({USENIX} Security 17), pp. 487–504 (2017)
26. Howard, M., Pincus, J., Wing, J.M.: Measuring relative attack surfaces. In: Computer security in the 21st century, pp. 109–137. Springer (2005)
27. Hu, H., Zhang, H., Yang, Y.: Security risk situation quantification method based on threat prediction for multimedia communication network. Multimed. Tools Appl. **77**(16), 21693–21723 (2018). https://doi.org/10.1007/s11042-017-5602-0
28. Hutchins, E.M., Cloppert, M.J., Amin, R.M.: Intelligence-driven computer network defense informed by analysis of adversary campaigns and intrusion kill chains. In: 2011 International Conference on Information Warfare and Security (2011)
29. Khraisat, A., Gondal, I., Vamplew, P., Kamruzzaman, J.: Survey of intrusion detection systems: techniques, datasets and challenges. Cybersecurity **2**(1), 1–22 (2019). https://doi.org/10.1186/s42400-019-0038-7
30. Lee, S., Kim, S., Choi, K., Shon, T.: Game theory-based security vulnerability quantification for social internet of things. Futur. Gener. Comput. Syst. **82**, 752–760 (2018)
31. Leitold, F., Arrott, A., Hadarics, K.: Quantifying cyber-threat vulnerability by combining threat intelligence, it infrastructure weakness, and user susceptibility. In: 24th Annual EICAR Conference, Nuremberg, Germany (2016)

32. Li, X., Parker, P., Xu, S.: A stochastic model for quantitative security analyses of networked systems. IEEE TDSC **8**(1), 28–43 (2011)
33. Lin, Z., Lu, W., Xu, S.: Unified preventive and reactive cyber defense dynamics is still globally convergent. IEEE/ACM ToN **27**(3), 1098–1111 (2019)
34. Longtchi, T., Rodriguez, R.M., Al-Shawaf, L., Atyabi, A., Xu, S.: Internet-based social engineering psychology, attacks, and defenses: a survey. Proc. IEEE **112**(3), 210–246 (2024)
35. Longtchi, T., Xu, S.: Characterizing the evolution of psychological factors exploited by malicious emails. In: Proceedings of International Conference on Science of Cyber Security (SciSec'2024) (2024)
36. Longtchi, T., Xu, S.: Characterizing the evolution of psychological tactics and techniques exploited by malicious emails. In: Proceedings of International Conference on Science of Cyber Security (SciSec'2024) (2024)
37. Mandiant: Apt1 report (2013). https://www.fireeye.com/content/dam/fireeyewww/services/pdfs/mandiant-apt1-report.pdf
38. Mao, B., Liu, J., Lai, Y., Sun, M.: MIF: a multi-step attack scenario reconstruction and attack chains extraction method based on multi-information fusion. Comput. Netw. **198**, 108340 (2021)
39. Mireles, J., Ficke, E., Cho, J., Hurley, P., Xu, S.: Metrics towards measuring cyber agility. IEEE Trans. Inf. Forensics Secur. **14**(12), 3217–3232 (2019)
40. Ning, P., Cui, Y., Reeves, D.S.: Constructing attack scenarios through correlation of intrusion alerts. In: Proceedings of the 9th ACM Conference on Computer and Communications Security, pp. 245–254 (2002)
41. Ning, P., Xu, D.: Learning attack strategies from intrusion alerts. In: Proceedings of the 10th ACM Conference on Computer and Communications Security, pp. 200–209 (2003)
42. Ou, X., Boyer, W.F., McQueen, M.A.: A scalable approach to attack graph generation. In: Proceedings of the 13th ACM Conference on Computer and Communications Security, pp. 336–345 (2006)
43. Pendleton, M., Garcia-Lebron, R., Cho, J.H., Xu, S.: A survey on systems security metrics. ACM Comput. Surv. **49**(4), 62:1–62:35 (2016)
44. Peng, C., Xu, M., Xu, S., Hu, T.: Modeling and predicting extreme cyber attack rates via marked point processes. J. Appl. Stat. **44**(14), 2534–2563 (2017)
45. Peng, C., Xu, M., Xu, S., Hu, T.: Modeling multivariate cybersecurity risks. J. Appl. Stat. 1–23 (2018)
46. Phillips, C., Swiler, L.P.: A graph-based system for network-vulnerability analysis. In: Proceedings of the 1998 Workshop on New Security Paradigms, pp. 71–79 (1998)
47. Pritom, M., Xu, S.: Supporting law-enforcement to cope with blacklisted websites: framework and case study. In: IEEE CNS'2022 (2022)
48. Ramaki, A.A., Rasoolzadegan, A., Bafghi, A.G.: A systematic mapping study on intrusion alert analysis in intrusion detection systems. ACM Comput. Surv. (CSUR) **51**(3), 1–41 (2018)
49. Rodriguez, R.M., Xu, S.: Cyber social engineering kill chain. In: Proceedings of International Conference on Science of Cyber Security (SciSec'2022), pp. 487–504 (2022)
50. Security, I.: Cost of a data breach report 2021 (July 2021). https://www.ibm.com/downloads/cas/OJDVQGRY
51. Sharafaldin, I., Lashkari, A.H., Ghorbani, A.A.: Toward generating a new intrusion detection dataset and intrusion traffic characterization. In: ICISSP, pp. 108–116 (2018)

52. Strom, B.: Att&ck 101: cyber threat intelligence (2018). https://www.mitre.org/capabilities/cybersecurity/overview/cybersecurity-blog/attck-101
53. Sun, Z., Xu, M., Schweitzer, K., Bateman, R., Kott, A., Xu, S.: Cyber attacks against enterprise networks: characterization, modeling and forecasting. In: Proceedings of the SciSec'2023 (2023)
54. Thakkar, A., Lohiya, R.: A review of the advancement in intrusion detection datasets. Procedia Comput. Sci. **167**, 636–645 (2020)
55. Vasilomanolakis, E., Cordero, C.G., Milanov, N., Mühlhäuser, M.: Towards the creation of synthetic, yet realistic, intrusion detection datasets. In: NOMS 2016-2016 IEEE/IFIP Network Operations and Management Symposium, pp. 1209–1214. IEEE (2016)
56. Wang, X., Gong, X., Yu, L., Liu, J.: Maac: novel alert correlation method to detect multi-step attack. In: 2021 IEEE 20th International Conference on Trust, Security and Privacy in Computing and Communications (TrustCom), pp. 726–733. IEEE (2021)
57. Xu, L., Zhan, Z., Xu, S., Ye, K.: An evasion and counter-evasion study in malicious websites detection. In: IEEE CNS, pp. 265–273 (2014)
58. Xu, L., Zhan, Z., Xu, S., Ye, K.: Cross-layer detection of malicious websites. In: ACM CODASPY'13, pp. 141–152 (2013)
59. Xu, M., Hua, L., Xu, S.: A vine copula model for predicting the effectiveness of cyber defense early-warning. Technometrics **59**(4), 508–520 (2017)
60. Xu, M., Schweitzer, K.M., Bateman, R.M., Xu, S.: Modeling and predicting cyber hacking breaches. IEEE T-IFS **13**(11), 2856–2871 (2018)
61. Xu, M., Xu, S.: An extended stochastic model for quantitative security analysis of networked systems. Internet Math. **8**(3), 288–320 (2012)
62. Xu, S.: The cybersecurity dynamics way of thinking and landscape (invited paper). In: ACM Workshop on Moving Target Defense (2020)
63. Xu, S., Lu, W., Xu, L.: Push- and pull-based epidemic spreading in networks: thresholds and deeper insights. ACM TAAS **7**(3) (2012)
64. Xu, S., Lu, W., Xu, L., Zhan, Z.: Adaptive epidemic dynamics in networks: thresholds and control. ACM TAAS **8**(4) (2014)
65. Xu, S., Lu, W., Zhan, Z.: A stochastic model of multivirus dynamics. IEEE Trans. Dependable Secure Comput. **9**(1), 30–45 (2012)
66. Xu, S.: Cybersecurity dynamics: a foundation for the science of cybersecurity. In: Lu, Z., Wang, C. (eds.) Proactive and Dynamic Network Defense, vol. 74, pp. 1–31. Springer Nature Switzerland AG (2019)
67. Xu, S.: SARR: a cybersecurity metrics and quantification framework. In: Third International Conference on Science of Cyber Security (SciSec'2021), pp. 3–17 (2021)
68. Zhan, Z., Xu, M., Xu, S.: Characterizing honeypot-captured cyber attacks: statistical framework and case study. IEEE Trans. Inf. Forensics Secur. **8**(11), 1775–1789 (2013)
69. Zhan, Z., Xu, M., Xu, S.: Predicting cyber attack rates with extreme values. IEEE Trans. Inf. Forensics Secur. **10**(8), 1666–1677 (2015)
70. Zhang, K., Zhao, F., Luo, S., Xin, Y., Zhu, H.: An intrusion action-based ids alert correlation analysis and prediction framework. IEEE Access **7**, 150540–150551 (2019)
71. Zhang, X., Xu, M., Xu, S.: Smart home cyber insurance pricing. In: Proceedings of International Conference on Science of Cyber Security (SciSec'2024) (2024)
72. Zheng, R., Lu, W., Xu, S.: Active cyber defense dynamics exhibiting rich phenomena. In: Proceedings of the HotSoS (2015)

73. Zheng, R., Lu, W., Xu, S.: Preventive and reactive cyber defense dynamics is globally stable. IEEE TNSE **5**(2), 156–170 (2018)

FAF-BM: An Approach for False Alerts Filtering Using BERT Model with Semi-supervised Active Learning

Dan Du[1,2], Yunpeng Li[1,2], Yiyang Cao[1,2], Yuling Liu[1,2], Guozhu Meng[1,2], Ning Li[1,2], Dongxu Han[1,2(✉)], and Huamin Feng[3]

[1] Institute of Information Engineering, Chinese Academy of Sciences, Beijing, China
{dudan,liyunpeng,caoyiyang,liuyuling,mengguozhu,
lining6,handongxu}@iie.ac.cn
[2] School of Cyber Security, University of Chinese Academy of Sciences, Beijing, China
[3] Beijing Electronic Science and Technology Institute, Beijing, China
fenghm@besti.edu.cn

Abstract. In the field of cybersecurity, the deluge of alerts presents a significant challenge to human review capabilities. Despite existing solutions, there is still an urgent need for more advanced methods to improve the effectiveness and accuracy of false alerts filtering. In this paper, we propose FAF-BM, a cutting-edge approach that integrates the BERT model, semi-supervised learning and active learning to enhance alert filtering capabilities. FAF-BM leverages the fine-tuned BERT model to fully exploit the deep semantics of alerts without being constrained by the format of the alerts. Subsequently, the semi-supervised learning is dedicated to mining the hidden potential within unlabeled data, thus expanding the learning scope beyond the confines of labeled datasets. In addition, the active learning strategically utilizes the expertise of security professionals to guide the learning process, ensuring that the approach adapts to the evolving threat landscape. Through a series of experiments, it has been demonstrated that FAF-BM not only improves the effectiveness of the filter but also enhances the generalization ability of dealing with the heterogeneity of alerts.

Keywords: False alerts filtering · BERT model · Semi-supervised learning · Active learning

1 Introduction

With the rapid development of cybersecurity, the number of alerts generated by security monitoring devices such as IDS has surged [8], with a significant portion being false positives. False positives occur when normal behavior is incorrectly identified as intrusion [20]. This issue of false positives is prevalent across all types of IDS and constitutes a high percentage. A study [3] from the perspective of SOC

J. Zhao and W. Meng (Eds.): SciSec 2024, LNCS 15441, pp. 295–312, 2025.
https://doi.org/10.1007/978-981-96-2417-1_16

security analysts suggests that the rate of false positives in security alerts can reach as high as 99%. The study stated that some larger organizations received over 100,000 alerts per day. Verifying the authenticity of alerts is a tedious process, and the continuous validation of false positives can lead to analyst fatigue, causing security analysts to overlook real security issues. Against this backdrop, effectively filtering out false positives has become an urgent issue to address.

To deal with the issue of false positives, several techniques have been proposed. Abouabdalla et al. [1] conducted a review of false alerts reduction techniques, proposing a variety of methods including signature-based enhancement, state signatures, vulnerability signatures, alert mining, and alert correlation. These techniques aim to reduce false alerts by analyzing the context of alerts, employing data mining techniques, and constructing attack graphs. Hubballi and Suryanarayanan [13] conducted a comprehensive review of false alerts filtering techniques, which included various methods to reduce false positives, such as alert correlation, alert validation, and flow analysis. Their work laid the foundation for understanding the strengths and limitations of different false alerts filtering techniques. In response to the volume of alerts, Wang et al. [33] proposed an alert prioritization technique that scores and ranks alerts based on factors such as alerts features and analyst feedback.

Despite the progress made by existing research, the filtering of false alerts still faces some challenges as follows: (i) Given the diversity in the sources and structures of alerts, attempts to standardize them may inadvertently lead to the loss of critical information inherent in heterogeneous alerts. (ii) Considering the limited time and energy of security analysts, only a small fraction of the extensive volume of alerts are manually labeled. This presents a significant obstacle to the construction of a high-performing classifier using supervised learning methodologies. (iii) Verifying whether an alert is a false positive depends on the actual business scenario and the expertise of the analysts themselves. Consequently, the specialized knowledge and experience of security analysts play an essential role in the filtering of false alerts.

To address these problems, designing and constructing an appropriate false alerts filter is a promising solution. The literature [5] proposes the use of AI and machine learning techniques to alleviate alert fatigue. In this paper, we propose FAF-BM, it is a false alerts filtering approach that combines BERT (Bidirectional Encoder Representations from Transformers) model, semi-supervised learning and active learning. While BERT model can be employed to filter false alerts by leveraging its ability to extract deep semantic insights from alerts without being constrained by the format of the alerts, semi-supervised learning attempts to exploit the potential of unlabeled data, active learning effectively utilizes the knowledge of security experts, which enhances the precision and efficiency of the false alerts filtering.

The contributions of this paper are the followings:

– We present FAF-BM, a novel approach for filtering false alerts. FAF-BM can rapidly improve the performance of false alerts filtering in the presence of sparsely labeled alerts, significantly saving security analysts' time and effort.

- We innovatively apply the BERT model to the field of false alerts filtering by leveraging the deep semantics of the alerts. We treat alerts from different sources and formats as textual data for fine-tuning the BERT model, thereby significantly contributing to filter false alerts without being constrained by alert heterogeneity.
- We conduct a series of experiments to evaluate FAF-BM's superior performance. Results demonstrate its advantages in the field of false alerts filtering, with FAF-BM outperforming popular algorithms, achieving an AUC of 97.89% under 15% labeled alert conditions on benchmark datasets.

We organize the remainder of this paper as follows. We survey some related works about false alerts filtering in Sect. 2. We then present the design of our proposed approach in Sect. 3, including Fine-tuning BERT Model, semi-supervised learning and active learning algorithms. In Sect. 4, we evaluate our approach and analyze the experimental results. Finally, Sect. 5 concludes this paper and discusses future work of FAF-BM.

2   Related Work

False alerts filtering was addressed by many studies using different techniques and methods like Mahmoud [14] proposed the solution to reduce the false alert rate by using fuzzy cognitive maps. Law and Kwok [17] designed a false positive filter based on the KNN classifier. [2] reduced false positives and improved the detection rates based on both clustering and Naive Bayes classifier. Meng and Kwok [25] proposed a scheme of hash-based contextual signatures that combined the original intrusion detection signatures with contextual information and hash functions to filter out non-critical alerts. [22] proposed an algorithm of reducing the false positives in IDS based on correlation Analysis. [32] proposed an effective model for false positives identification using gradient boosting tree models based on the analysis of security features of the IDS alerts. The aforementioned research methodologies typically demand a substantial amount of labeled data, which is often scarce in real-world application scenarios. Semi-supervised learning and active learning algorithms have shown great potential in handling limited labeled data and a vast amount of unlabeled data [30,35], while BERT is good at handling heterogeneous text formats and grasping subtle relationships and differences in context.

Semi-supervised Learning in False Alerts Filtering. Semi-supervised learning is to make use of the vast amounts of unlabeled data that are often readily available, while still benefiting from the guidance provided by the smaller set of labeled examples. Chiu et al. [7] proposed a semi-supervised learning mechanism that significantly reduced false positives by using TCP connection information and related algorithms. Meng and Kwok [23] further applied a disagreement-based semi-supervised learning algorithm to reduce false positives by constructing an alert filter. Ede et al. [11] introduced the DEEPCASE system, which utilized semi-supervised learning methods to automatically correlate security events. Li et al. [20] proposed a multi-view learning method to reduce false positives in NIDS by extracting features from different perspectives and jointly optimizing them. [29] introduced a semi-supervised learning scheme called SAFI, aimed at addressing alert fatigue.

Active Learning in False Alerts Filtering. Active learning is a machine learning paradigm that reduces annotation costs by selecting the most informative samples for expert annotation. For instance, Settles et al. [28] proposed an active learning framework for malware detection. [4] explored the application of active learning in intrusion detection. Pietraszek [27] proposed a model called Adaptive Learner for Alert Classification that considered feedback from analysts to identify irrelevant alerts. Meng and Kwok [24] designed a simple but efficient pool-based active learning algorithm in a false alert filter. Li and Guo [18] designed an active learning-based TCM-KNN algorithm that can effectively detect anomalies while maintaining a low false positives rate. The AlertPro framework by [33] used the active learning model, combined with expert feedback, it reorganized alerts to prioritize the display of high-risk alerts.

BERT in False Alerts Filtering. BERT model is a pre-trained language representation model based on the Transformer architecture, proposed by Google in 2018 [9]. It provides sophisticated tools for grasping contextual nuances and has delivered remarkable outcomes across a spectrum of language understanding tasks such as natural language understanding, machine translation, question-answering systems, sentiment analysis, and named entity recognition [6,12,21, 36]. However, it has not yet been applied to the field of false alerts filtering.

Above all, despite extensive research that have proposed or adopted semi-supervised learning or active learning for the detection of false alerts where labeled data is scarce, but they address the same problem from opposite directions, while semi-supervised learning exploits what the learner thinks it already knows about the unlabeled data, active learning attempts to explore the unknown aspects [28], no scholar has yet suggested combining these for filtering false alerts. Moreover, BERT model is capable of fully exploiting the contextual information of alerts, which makes it as an ideal base model for semi-supervised active learning. Inspired by this, an approach that combines BERT model with semi-supervised active learning could be employed for filtering false alerts.

3   Proposed Approach

The nature of false alerts filtering is fundamentally a classification issue, in this section, we discuss the approach for filtering false alerts called FAF-BM. An overall approach of FAF-BM is provided in Fig. 1.

Initially, Fine-tuning BERT model module involves preprocessing alerts and establishing a classifier based on labeled alerts called BERT-AlertClassifier during steps 1 and 2, serving as the foundation of semi-supervised learning. Subsequently, semi-supervised learning module is engaged to leverage the wealth of unlabeled data, augmenting the training set and retaining the classifier. This is achieved by employing the classifier to predict the probability of pseudo-labels upon unlabeled alerts in step 3, and then filtrate the most-uncertain alerts and high-confidence alerts in step 4. Alerts with high-confidence pseudo-labels are subsequently incorporated into the expanded labeled alerts as training set during steps 6, enabling iterative training of the classifier in step 8. In active learning module, active learning is utilized to query batches of alerts that are most uncertain in step 5, those alerts are annotated by experts and are added into the expanded labeled alerts set in step 7. Thereby repeat steps 3 to 8 until either the stability of the classifier's performance or unlabeled alert set is empty for enhancing the classifier's optimization. The false alerts to be filtered out include those in the expanded labeled alerts set, and additionally, if there are still unlabeled alerts, these are filtered out through the well-iterated classifier. We will now introduce each module of FAF-BM in details.

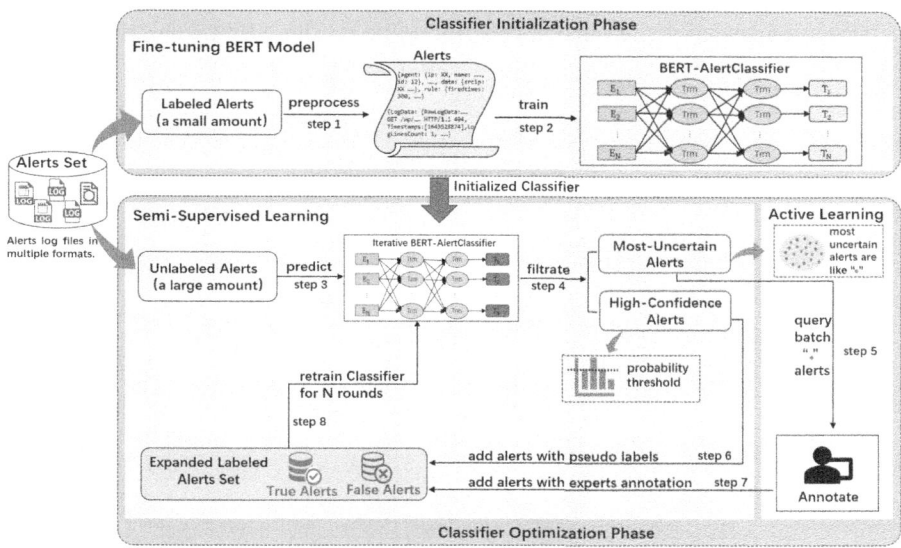

Fig. 1. An overall of FAF-BM approach. It includes two main phases: the classifier initialization phase and the classifier optimization phase. During the classifier initialization phase, the alert classifier is constructed primarily by Fine-tuning BERT Model module. In the classifier optimization phase, the performance of the classifier is enhanced through the utilization of a vast amount of unlabeled alerts via semi-supervised learning module and active learning module.

3.1 Fine-Tuning BERT Model

In order to initialize the classifier, we should first preprocess the alerts that are used for fine-tuning BERT model. In this paper, we remove duplicate alerts within limited time and then remove several fields deemed less significant from dataset as preprocessing operations, yet the alerts still keep a heterogeneous text of formats.

In this study, we choose bert-base-uncased [9] as the base model to construct a classifier tailored for false alerts filtering. The architecture of fine-tuned BERT as BERT-AlertClassifier is shown in Fig. 2. In brief, the model takes alerts which have been preprocessed as sequences of tokens as input, the tokens will be further divided into sub-tokens and add [CLS],[SEP] tokens, then the sequence of sub-tokens is fed into the BERT model. The BERT model processes the input sub-tokens through multiple layers of the Transformer structure to produce context-aware representations for each sub-token. These representations capture complex deep semantics within the alert text, such as sequential relationships and long-range dependencies. A dense layer, which is newly introduced during the fine-tuning process, is used to convert the output of BERT into predictions for the classification task. The dense layer receives the output from the final layer of the BERT model (usually the output corresponding to the [CLS] token) and learns to map these representations to two category labels (true alert and false alert).

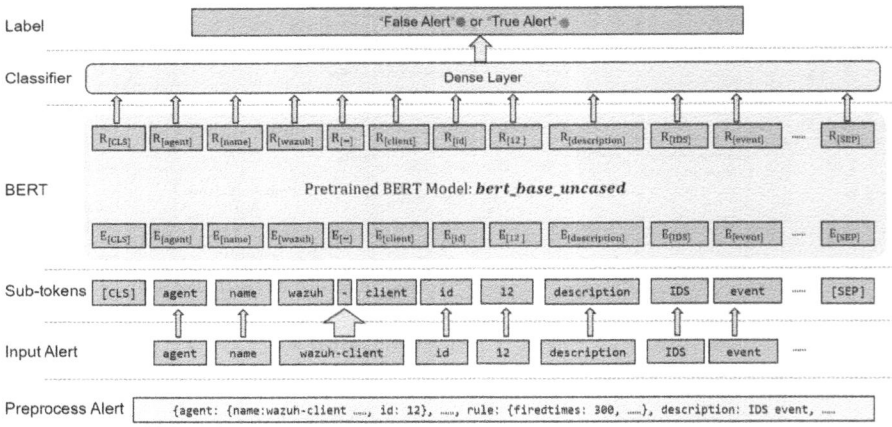

Fig. 2. The architecture of fine-tuned BERT as BERT-AlertClassifier.

During fine-tuning, the newly added dense layer is trained using a dataset of alerts with labels. When training, the model adjusts the weights of the dense layer by minimizing the difference between the outputs and their corresponding ground truth labels, to measure the difference between the predicted probability distribution and the true distribution, we employ the BCE loss as the objective function, which is calculated as Eq. 1, here w is the weight of the alert, y_i is the true label of the i alert, typically 0 or 1, $logits_i$ is the model's predicted probability of the true alert class for the i alert.

$$loss = -\sum_i w \times y_i \times log\left(logits_i\right) + \left(1 - y_i\right) \times log\left(1 - logits_i\right) \qquad (1)$$

The fine-tuning module is used to craft a well-calibrated initial classifier that lays the groundwork for the subsequent phases of semi-supervised active learning.

3.2 Semi-supervised Learning

In the filtering of false alerts, the challenge often lies in the scarcity of labeled alerts compared to the abundance of unlabeled alerts. This scenario is particularly well-suited for semi-supervised learning methods.

In our study, we adopt the self-training methodology for semi-supervised learning to refine our classifier. This approach involves a single supervised classifier that undergoes iterative training on both the labeled alerts and those that have been assigned pseudo-labels in preceding iterations of the algorithm. Initially, during the Fine-tuning BERT Model module Sect. 3.1, a supervised classifier BERT-AlertClassifier is trained exclusively on the labeled alerts. This classifier is then utilized to generate predictions for the unlabeled alerts. Subsequently, the high-confidence alerts above threshold are incorporated into the expanded labeled dataset. The supervised classifier BERT-AlertClassifier is then retrained on this combined set of labeled and newly pseudo-labeled alerts.

To enhance the performance of semi-supervised learning, the threshold adjustment mechanism from the literature [34] is referenced to adaptively adjust the high-confidence threshold in a manner that is tailored to each class, taking into account the diversity within each class as well as the distinctions between classes.

The class-specific learning status is estimated by computing the expected predictions $\tilde{p}_t(c)$ of the classifier on each class c. At the initialization, the training step $t = 0$, then $\tilde{p}_t(c)$ is initialized to $1/C$. In subsequent training steps, the expected predictions $\tilde{p}_t(c)$ are updated using the Exponential Moving Average (EMA) as follows:

$$\tilde{p}_t(c) = \begin{cases} \frac{1}{C}, & \text{if } t = 0, \\ \lambda \tilde{p}_{t-1}(c) + (1 - \lambda)\frac{1}{\mu B} \sum_{i=1}^{\mu B} q_i(c), & \text{otherwise.} \end{cases} \quad (2)$$

where C is the total number of classes, λ is the decay parameter of EMA, μ is the ratio of unlabeled data batch size to labeled data batch size, B is the batch size and $q_i(c)$ is the classifier's predicted probability for class c in the batch of unlabeled data.

The global threshold τ_t is combined with the class-specific expectation $\tilde{p}_t(c)$, and the final self-adaptive threshold $\tau_t(c)$ is calculated using MaxNorm. The MaxNorm function ensures that $\tau_t(c)$ maintains relative fairness across classes.

$$\tau_t(c) = \text{MaxNorm}(\tilde{p}_t(c)) \cdot \tau_t = \frac{\tilde{p}_t(c)}{\max \tilde{p}_t(c) : c \in [C]} \cdot \tau_t \quad (3)$$

The False Alerts Filter Iterative Process of the FAF-BM, which integrates both semi-supervised learning and active learning modules, is shown in the Algorithm 1. This iterative process continues until unlabeled dataset U is empty or the classifier's performance is stable. The `SelectBatchUnlabeledAlerts` function is employed to choose batches of unlabeled alerts, while `SelectHighConfidence` is used to select alerts with high-confidence pseudo-labels. `GetExpertAnnotations` is responsible for filtering a batch of most uncertainty alerts and requesting annotations from experts. The `TrainClassifier` function is used to update the classifier based on the current classifier and the labeled set L, `UpdateThresholdStrategy` is adaptively adjust the high-confidence threshold according to the Eq. (3).

Algorithm 1. False Alerts Filter Iterative Process of FAF-BM.

Input: Labeled dataset L , Unlabeled dataset U, The original $classifier$ from Fine-tuning BERT Model, Threshold τ
Output: $classifier$, L
1: **while** ¬stopping_criteria_met($classifier, U$) **do**
2:     $selected\_unlabeled\_alerts \leftarrow$ SELECTBATCHUNLABELEDALERTS(U)
3:     $pseudo\_labeled\_alerts \leftarrow classifier.predict(selected\_unlabeled\_alerts)$
4:     $high\_confidence\_alerts \leftarrow$ SELECTHIGHCONFIDENCE($pseudo\_labeled\_alerts, \tau$)
5:     $annotated\_alerts \quad\leftarrow\quad$ GETEXPERTANNOTATIONS($pseudo\_labeled\_alerts$, $high\_confidence\_alerts, expert\_annotations$)
6:     $L \leftarrow L \cup annotated\_alerts \cup high\_confidence\_alerts$
7:     $U \leftarrow U \setminus (annotated\_alerts \cup high\_confidence\_alerts)$
8:     $classifier \leftarrow$ TRAINCLASSIFIER($classifier, L$)
9:     $\tau \leftarrow$ UPDATETHRESHOLDSTRATEGY(τ)
10: **end while**
11: **return** $classifier$

3.3   Active Learning

Active learning constitutes an interactive machine learning approach, as studies such as [10,19] have demonstrated its efficacy in proactively selecting a limited yet 'valuable' subset of samples. By engaging with experts to obtain ground truth annotations for these samples, active learning incrementally enhances model performance. Central to active learning is the query strategy, which identifies the most informative data points as 'valuable' candidates for manual annotation. This strategy encompasses various methods, including uncertainty sampling, query committees, random sampling, and others. Uncertainty sampling operates on the premise that labeling instances where the model exhibits the greatest uncertainty will provide more substantial information compared to instances where the model is highly confident. Traditional active learning, which selects one valuable sample per iteration for expert supervision, can incur higher time costs. Therefore, in the context of filtering false alerts, we propose a query strategy that selects a batch of uncertain alerts from the unlabeled pool for manual annotation in each querying round.

The objective of the query strategy is to select a batch of alerts S from U to augment the semi-supervised training as active learning advances, thereby achieving a higher level of classifier generalization with the least number of queries. This is accomplished by utilizing the expanded training set $S \cup L$, with the constraint that $S \subset U$ and $|S| = b$, where b is the batch size. For binary classification datasets, we employ the probability divergence margin to assess the certainty criterion concerning whether x_i belongs to a given label, defined by

$$c(x_i) = |p(y_0|x_i) - p(y_1|x_i)| \tag{4}$$

where $| \cdot |$ denotes the absolute value function, while $p(y_0|x_i)$ is the probability that the given alert x_i belongs to category y_0, 0 represents false alerts and 1 represents true alerts. Equation (4) clearly indicates the informativeness of sample x_i based on labeled alerts. A smaller $c(x_i)$ value signifies greater uncertainty of classifier regarding the class of x_i, thus allowing for the most significant refinement of the classifier. For a candidate batch $S \subset U$, we define $C(S)$ as the uncertainty accumulation of all samples within S, leading to the following optimization problem:

$$\min_{S \subset U, |S|=b} C(S) = \sum_{i=1}^{b} c(x_i), c(x_i) < q \tag{5}$$

As shown in Eq. (5), in each iteration, We first sort all unlabeled alerts in ascending order based on the value of $c(x_i)$, and then select the top b alerts. Subsequently, from the top b alerts, we select those with a $c(x_i)$ value less than q to constitute the current optimal batch. Security analysts then annotate these alerts.

Algorithm 2. FAF-BM for Expert Annotations.

Input: pseudo_labeled_alerts, high_confidence_alerts, batchSize b, value q, expert_
    annotations
Output: annotated_alerts
 1: $annotated\_alerts \leftarrow \emptyset$
 2: $pseudo\_labeled\_alerts \leftarrow pseudo\_labeled\_alerts \setminus high\_confidence\_alerts$
 3: $uncertainty\_alerts \leftarrow$ SELECTUNCERTAINTY$(pseudo\_labeled\_alerts, b, q)$
 4: **for** $alert \in uncertainty\_alerts$ **do**
 5:     $expert\_annotation \leftarrow$ ASKEXPERT$(alert)$
 6:     $annotated\_alerts \leftarrow annotated\_alerts \cup \{alert, expert\_annotation\}$
 7: **end for**
 8: **return** $annotated\_alerts$

FAF-BM for expert annotations procedure is shown in the Algorithm 2, filtering the most uncertain alerts through Function SelectUncertainty, and then label the information through Function AskExpert. This approach effectively improves the learning model's performance within a constrained annotation budget.

4   Evaluation

In this section, we evaluate FAF-BM through a series of experiments and analyze the experimental results.

4.1   Dataset Description

We use two publicly available datasets to explore the performance of FAF-BM: AIT Alert dataset [16] in 2023 and DARPA LLDOS1.0 in 2000 [26].The advantages of these two datasets are as follows:

– AIT Alert dataset is that it contains many false positives caused by normal user behavior (e.g., user login attempts or software updates), heterogeneous alert formats, collection of alerts from diverse log sources (application logs and network traffic) and all phases in the network (mail server, web server, DNS, firewall, file share, etc.), and labels for attack phases.
– DARPA LLDOS1.0 dataset is a well-established and widely recognized dataset within the field of intrusion detection. Despite being over a decade old and lacking the characteristics of newer attack patterns, it is renowned for its comprehensive study, meticulous documentation, and public availability.

Table 1. Statistical Analysis of Alerts from Two Public Datasets

Statistical Dimensions	Santos of AIT ADS [16]	DARPA LLDOS1.0 [26]
Total Alert Count	130,779	4,608
True Alert Count	13,004	323
False Alert Count	117,775	4,285
Alert Formats	Heteregeneous	Homogeneous
Alert Sources	Network Traffic, Applicaiton Logs	Network Traffic
Alert Duration	4 days	9 h

AIT Alert Dataset. AIT Alert dataset is forensically generated from the AIT Log Data Set V2 (AIT-LDSv2) [15] and origin from three open-source intrusion detection systems, namely Suricata (a network-based and signature-based IDS), Wazuh (a host-based and signature-based IDS), and AMiner (a host-based and anomaly-based IDS). The data sets comprise eight scenarios independently executed in eight networks, we choose Santos scenario as experimental datasets which reflects the challenges commonly found in actual environments. Santos scenario information is shown in Table 1.

DARPA LLDOS1.0 Dataset. The DARPA intrusion detection evaluation dataset, sponsored by the Defense Advanced Research Projects Agency and the Air Force Research Laboratory and managed by the MIT Lincoln Laboratory

[26]. We utilized the DARPA LLDOS1.0 dataset in our experimental analysis, processed the DARPA network traffic by employing Snort version 2.9.15 to generate alerts. The details are shown in Table 1.

4.2 Experiment Setup

We implement our experiments using one P100 GPU,choosing Python as the primary programming language. In the experiments, we preprocess alert data and configure the experimental parameters which are crucial in establishing a robust framework for our study.

Preprocess Alerts. In order to improve BERT-AlertClassifier performance, we removed duplicate alerts within one minute and then removed several fields deemed less significant from the dataset as preprocessing operations, yet the alerts still keep a heterogeneous range of formats. E.g., we removed "predecoder, description, previous_output, input, groups, location" from Wazuh alerts.

Configure Parameters. In our study, We employ the *bert-base-uncased* version of the model, with the input's maximum length set to 512. The learning rate is set as $3e - 5$, and a dropout probability of 0.1 is applied to mitigate overfitting. The batch size for training is set to 24. Throughout the training process, the Adam optimizer is utilized for adjusting the network weights.

4.3 Evaluation Metrics

In order to evaluate FAF-BM, we employ common metrics include Accuracy, Recall-Positive, Precision-Negative and the Area Under the Receiver Operating Characteristic curve (AUC). Accuracy quantifies the proportion of total correct classifications, Recall-Positive emphasizes the model's ability to identify true alerlts, Precision-Negative measures the fraction of identified positive samples that are actually true alerlts. AUC represents the model's overall discrimination capability across varying thresholds, thus it is a dependable measure for assessing the quality of classifier.

4.4 Experiments and Results

We conducted three experiments to substantiate the efficacy and advancement of FAF-BM in filtering false alerts. The first experiment compared FAF-BM with popular supervised and semi-supervised algorithms on two public datasets, confirming that FAF-BM possesses superior false alerts filtering capabilities. The second experiment demonstrated that FAF-BM is capable of handling heterogeneous alerts and exhibits commendable classification performance. The third experiment employed an ablation study approach, demonstrating that each module of FAF-BM contributes to the enhancement of the classifier's performance.

Experiment 1. To evaluate the effectiveness of FAF-BM, we selected the proven and excellent supervised algorithm GBDT [31] and semi-supervised algorithm MVPSys [20] as comparative algorithms. The GBDT achieved the highest final score in *IEEE BigData Cup: Suspicious Network Event Recognition* for identifying false alerts. The BERT-AlertClassifier, serving as a crucial module within FAF-BM approach, is inherently a supervised learning algorithm. When subjected to a comparative analysis with GBDT under equivalent conditions, it can accurately reveal the outstanding performance of the foundational module that supports FAF-BM. Subsequently, FAF-BM demonstrates its advanced integration of semi-supervised active learning when compare with BERT-AlertClassifier, GBDT and MVPSys, thereby proving its innovative superiority.

Table 2. Selected different proportions of labeled dataset from two public datasets and compared the performance of GBDT, BERT-AlertClassifier(BERT-AC), MVPSys, FAF-BM in terms of Accuracy(ACC), Recall-Positive (RP), Precision-Negative (PN), and AUC.

Dataset	Labeled Alerts	Type	Algorithm	ACC	RP	PN	AUC
AIT ADS	1%	Supervised	GBDT [31]	0.9741	0.7913	0.9821	0.8905
			BERT-AC	0.9720	0.8329	0.9863	0.9116
		Semi-Supervised	MVPSys [20]	0.9504	0.8412	0.9869	0.9033
			FAF-BM	**0.9813**	**0.9077**	**0.9926**	**0.9626**
	5%	Supervised	GBDT [31]	**0.9856**	0.8098	0.9847	0.9043
			BERT-AC	0.9797	0.8981	0.9918	0.9377
		Semi-Supervised	MVPSys [20]	0.9614	0.9023	0.9920	0.9421
			FAF-BM	0.9852	**0.9527**	**0.9962**	**0.9718**
	15%	Supervised	GBDT [31]	0.9867	0.8276	0.9867	0.9134
			BERT-AC	0.9843	0.9431	0.9954	0.9641
		Semi-Supervised	MVPSys [20]	0.9684	0.9141	0.9930	0.9508
			FAF-BM	**0.9868**	**0.9581**	**0.9966**	**0.9789**
	30%	Supervised	GBDT [31]	0.9878	0.8309	0.9871	0.9154
			BERT-AC	**0.9895**	**0.9592**	**0.9967**	**0.9776**
LLDOS1.0	30%	Supervised	GBDT [31]	**0.9872**	0.8273	0.9864	0.9403
			BERT-AC	0.9646	0.9858	0.9905	0.9913
		Semi-Supervised	MVPSys [20]	0.9697	0.9345	0.9946	0.9702
			FAF-BM	0.9823	1	1	**0.9985**

This comparison is conducted using different proportions of labeled alerts from two public datasets. For the AIT ADS dataset, we select four different proportions alerts, i.e. {proportion of labeled alerts, proportion of unlabeled alerts} from Wauzh of Santos, specifically {1%, 29%}, {5%, 25%}, {15%, 15%}, and {30%, –} (where the '–' indicates that the unlabeled alerts is null, therefore, this set of data is only applicable to supervised learning algorithms), to train four different algorithms, an additional 10% alerts is allocated for testing. In the LLDOS1.0 dataset, for the comparative experiments, we allocate 30% as the labeled alerts for training the classifier, 40% as the unlabeled alerts for iterative optimize the classifier, with the remaining 30% reserved for testing.

Table 2 shows alert classification performances of different algorithms. FAF-BM outperforms all the other algorithms on the two public benchmark datasets. Additionally, we have the following observations.

(i) **Regarding the comparison of different algorithms**, FAF-BM consistently performs better than supervised learning algorithms GBDT,BERT-AlertClassifier and semi-supervised learning algorithm MVPSys in terms of Recall-Positive, Recall-Positive and AUC performance. In most cases, the Accuracy of FAF-BM is higher than that of GBDT, BERT, and MVPSys. Only in a few cases, the Accuracy of GBDT is higher than that of FAF-BM, but the Recall-Positive value of GBDT is more than 15% lower than that of FAF-BM. This is because the proportion of false alerts in datasets is high, which leads GBDT to erroneously classify alerts as false positives. This tendency underscores the inferior performance of GBDT in comparison to FAF-BM. Notably, FAF-BM achieves an AUC of 0.9626 with only 1% of the labeled alerts, significantly surpassing GBDT, BERT-AlertClassifier and MVPSys, even when these algorithms are provided with a higher proportion of labeled alerts, such as 5% or 15%. This shows the FAF-BM's capability to effectively leverage unlabeled alerts and expert knowledge when dealing with a small number of labeled alerts, thereby evidencing its advanced nature.

(ii) **Regarding the comparison of different datasets**, FAF-BM exhibits a consistent performance. On the LLDOS 1.0 dataset, mirroring its efficacy on AIT ADS, FAF-BM surpasses GBDT, BERT-AlertClassifier and MVPSys in terms of AUC. These results indicate that FAF-BM not only excels on a single dataset but also possesses a robust generalization capability across various datasets and business scenarios.

(iii) **Regarding the comparison of different data proportions**, FAF-BM showcases its broad applicability. Across different proportions of labeled alerts at 1%, 5%, and 15%, FAF-BM maintains high levels of Accuracy, Recall-Positive, Precision-Negative and AUC. Notably, FAF-BM achieves an AUC of 97.89% under 15% labeled alert conditions on AIT ADS. At labeled alert proportions of 5% and 15%, the performance of FAF-BM is comparable to that of the supervised learning model BERT-AlertClassifier with a 30% labeled alert proportion. This indicates that FAF-BM can achieve equivalent false alerts filtering performance with significantly fewer labeled alerts than are required for supervised learning.

Upon a comprehensive synthesis of the aforementioned analysis, it is evident that FAF-BM algorithm exhibits a high degree of efficacy in critical evaluative metrics. In light of these observations, it is reasonable to infer that the FAF-BM is an efficient and dependable classifier for alert filtration, with applicability across diverse datasets and business scenarios, thereby portending a promising landscape for its utilization.

Experiment 2. To validate the performance of FAF-BM on heterogeneous alerts, we selected alerts in various formats from the Santos scenario (heterogeneous alert formats are shown in Fig. 3), thereby establishing a heterogeneous alert dataset, with detailed information as presented in Table 3. Considering the balance of the dataset, we constructed a training set comprising 3% of Wauzh alerts and 80% of AMiner alerts, totaling 7552 alerts. Additionally, we formed a test set consisting of 1% of Wauzh alerts and 20% of AMiner alerts, which amounts to 1667 alerts.

{"agent": {"ip": "172.21.131.50", "name": "wazuh-client", "id": "22"}, "manager": {"name": "wazuh.manager"}, "rule": {"mail": false, "level": 3, "description": "Dovecot Authentication Success.", "groups": ["dovecot", "authentication_success"], "nist_800_53": ["AU.14", "AC.7"], "firedtimes": 119, "mitre": {"technique": ["Valid Accounts"], "id": ["T1078"], "tactic": ["Defense Evasion", "Persistence", "Privilege Escalation", "Initial Access"]}, "id": "9701", "gpg13": ["7.1", "7.2"]}, "decoder": {"parent": "dovecot", "name": "dovecot"}, "full_log": "Jan 14 12:16:03 mail dovecot: imap-login: Login: user=<vickie.sharp>, method=PLAIN, rip=172.21.131.50, lip=172.21.131.50, mpid=13761, TLS, session=<405JxInViOisFYMy>", "input": {"type": "log"}, "@timestamp": "2022-01-14T12:16:03.000000Z", "location": "/var/log/mail.info", "id": "1687219549.260202"}

{"AnalysisComponent": {"AnalysisComponentIdentifier": 8, "AnalysisComponentType": "EntropyDetector", "AnalysisComponentName": "AMiner: High entropy in Apache Access referer.", "Message": "Value entropy anomaly detected", "PersistenceFileName": "entropy_referer", "TrainingMode": true, "AffectedLogAtomPaths": ["/model/combined/combined/referer"], "AffectedLogAtomValues": ["-"], "CriticalValue": 0.0, "ProbabilityThreshold": 0.05}, "LogData": {"RawLogData": ["192.168.104.95 - - [14/Jan/2022:05:24:13 +0000] \"GET / HTTP/1.1\" 302 453 \"-\" \"Mozilla/5.0 (X11; Ubuntu; Linux x86_64; rv:86.0) Gecko/20100101 Firefox/86.0\""], "Timestamps": [1642137853], "DetectionTimestamp": [1642137853], "LogResources": ["/var/log/apache2/mail-access.log"]}, "AMiner": {"ID": "172.21.129.224"}}

(a) An example alert from Wazuh   (b) An example alert from AMiner

Fig. 3. Heterogeneous alert examples from different IDS (Wazuh and AMiner) within Santos of AIT ADS.

During the experiment, 30% alerts as labeled and 70% alerts as unlabeled which from training set of Heterogeneous Alert Dataset were utilized to train FAF-BM, and adopting the entire test set of Heterogeneous Alert Dataset for test. Furthermore, we selected the same number of alerts from the homogeneous alerts that were produced by Wauzh IDS from Santos to train and test FAF-BM. Moreover, the performance of FAF-BM on heterogeneous and homogeneous alerts across Accuracy, Recall-Positive (RP), Precision-Negative (PN) and AUC were evaluated.

Table 3. Heterogeneous Alert Dataset

DataSet	From Wauzh	From AMiner	Alert Count
Training set	3,784(3%)	3,768(80%)	7,552
Test set	1,253(1%)	414(20%)	1,667

The experimental results presented in Table 4 demonstrate that FAF-BM achieves high performance on heterogeneous alerts, with Accuracy exceeding 98%, Recall-Positive above 95%, Precision-Negative above 99%, and an AUC value surpassing 97%. These metrics are marginally lower than those observed on homogeneous alerts, underscoring FAF-BM's robust capability to effectively handle alerts across varying data types.

Table 4. Evaluation of FAF-BM Performance on Heterogeneous and Homogeneous Alerts Across Accuracy(ACC), Recall-Positive (RP), Precision-Negative (PN) and AUC.

Alert Source	Alert Format	ACC	RP	PN	AUC
Wauzh, AMiner	Heterogeneous	0.9865	0.9538	0.9962	0.9716
Wauzh	Homogeneous	0.9895	0.9538	0.9963	0.9741

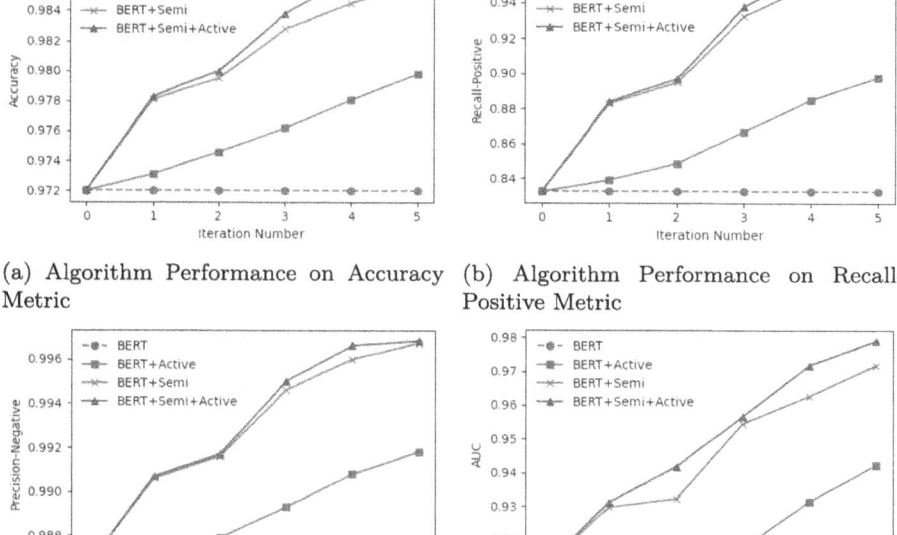

(a) Algorithm Performance on Accuracy Metric

(b) Algorithm Performance on Recall-Positive Metric

(c) Algorithm Performance on Precision-Negative Metric

(d) Algorithm Performance on AUC Metric

Fig. 4. The figure clearly illustrates the impact of different strategies-BERT, BERT+Active, BERT+Semi and BERT+Semi+Active-on the evaluation metrics of Accuracy, Recall-Positive, Precision-Negative and AUC, as well as their respective trends during model iterations. Serving as the ground truth benchmark, the BERT classifier itself was not subjected to iterative optimization, its performance remained relatively stable.

Experiment 3. In order to assess the contribution of each module within FAF-BM to its overall performance, we conduct ablation experiments.

Throughout the experiment, we compared the performance of using only the BERT-AlertClassifier of FAF-BM (referred to as BERT), along with variations incorporating semi-supervised learning (BERT+Semi), active learning (BERT+Active), and a combination of both (BERT+Semi+Active) within Santos scenario of the AIT ADS dataset in pursuit of optimal filtering performance. We selected 1% alerts as labeled for training, other 29% alerts as unlabeled for each iteration, and 10% alerts for test.

From Fig. 4, it is observable that with the increment of iterations, the models BERT+Active, BERT+Semi, and BERT+Semi+Active all exhibit improvements in the metrics of Accuracy, Recall-Positive, Precision-Negative, and AUC. This outcome confirms that each module within the FAF-BM-the Fine-tuning BERT Model, Semi-supervised Learning, and Active Learning modules-positively contributes to the enhancement of overall performance.

5   Conclusion

In the field of cybersecurity, the large number of alerts generated is a great challenge and often exceeds the capabilities of the human review process. This paper introduces FAF-BM, an iterative self-training false alerts filtering approach. Its ability to integrate BERT and semi-supervised active learning techniques not only optimizes the utilization of available alerts regardless of the format, but also minimizes the reliance on extensive manual labeling.

The future of false alerts filtering lies in the pursuit of approaches that minimize the reliance on manual annotation. Future work will focus on refining FAF-BM, i.e. constructing a superior classifier, exploring additional strategies for pseudo-labeling and active query, and expanding the framework to enhance these apprroachs' adaptability and generalizability within dynamic and complex network environments.

Acknowledgments. This study is funded by science and technology project of the headquarters of State Grid Corporation of China (Research and Application of Network Security Situation Awareness Technology Based on Knowledge Graph, Project Code: 5700-202352606A-3-2-ZN). It also receives support from the Key Laboratory of Network Assessment Technology at the Chinese Academy of Sciences, and the Beijing Key Laboratory of Network Security and Protection Technology.

References

1. Abouabdalla, O., El-Taj, H., Manasrah, A., Ramadass, S.: False positive reduction in intrusion detection system: a survey. In: 2009 2nd IEEE International Conference on Broadband Network & Multimedia Technology, pp. 463–466. IEEE, Beijing, China (2009)

2. Abu Afza, A.J.M., Uddin, M.S.: Intrusion detection learning algorithm through network mining. In: 16th International Conference on Computer and Information Technology, pp. 490–495. IEEE, Khulna (2014)

3. Alahmadi, B.A., Axon, L., Martinovic, I.: 99% false positives: a qualitative study of soc analysts' perspectives on security alarms, pp. 2783–2800 (2022)

4. Almgren, M., Jonsson, E.: Using active learning in intrusion detection. In: Proceedings. 17th IEEE Computer Security Foundations Workshop, 2004, pp. 88–98. IEEE, Pacific Grove, CA, USA (2004)

5. Ban, T., Takahashi, T., Ndichu, E.A.: Breaking alert fatigue: AI-assisted SIEM framework for effective incident response. Appl. Sci. **13**(11), 6610 (2023)

6. Behera, S.K., Dash, R.: Fine-tuning of a BERT-based uncased model for unbalanced text classification. In: Mohanty, M.N., Das, S. (eds.) Advances in Intelligent Computing and Communication, pp. 377–384. Springer Nature, Singapore (2022)

7. Chiu, C.-Y., Lee, Y.-J., Chang, C.-C., Luo, W.-Y., Huang, H.-C.: Semi-supervised learning for false alarm reduction. In: Perner, P. (ed.) ICDM 2010. LNCS (LNAI), vol. 6171, pp. 595–605. Springer, Heidelberg (2010). https://doi.org/10.1007/978-3-642-14400-4_46

8. De Alvarenga, S.C., Barbon, S., Miani, R.S., et al., C.: Process mining and hierarchical clustering to help intrusion alert visualization. Comput. Secur. **73**, 474–491 (2018)

9. Devlin, J., Chang, et al.: BERT: pre-training of deep bidirectional transformers for language understanding. In: Proceedings of the 2019 Conference of the North American Chapter of the Association for Computational Linguistics: Human Language Technologies, vol. 1 (Long and Short Papers), pp. 4171–4186. Association for Computational Linguistics, Minneapolis, Minnesota (2019)

10. Doak, J.E., Ingram, J., Shelburg, J., Johnson, J., Rohrer, B.R.: Active learning for alert triage. In: 2013 12th International Conference on Machine Learning and Applications, pp. 34–39. IEEE, Miami, FL, USA (2013)

11. Ede, T.V., et al.: DEEPCASE: semi-supervised contextual analysis of security events. In: 2022 IEEE Symposium on Security and Privacy (SP), pp. 522–539. IEEE, San Francisco, CA, USA (2022)

12. Fang, Y., et al.: EVA: Exploring the Limits of Masked Visual Representation Learning at Scale, pp. 19358–19369 (2023)

13. Hubballi, N., Suryanarayanan, V.: False alarm minimization techniques in signature-based intrusion detection systems: a survey. Comput. Commun. **49**, 1–17 (2014)

14. Jazzar, M., Jantan, A.B.: Using fuzzy cognitive maps to reduce false alerts in SOM-based intrusion detection sensors. In: 2008 Second Asia International Conference on Modelling & Simulation (AMS), pp. 1054–1060. IEEE (2008)

15. Landauer, M., Skopik, F., Frank, M., Hotwagner, W., Wurzenberger, M., Rauber, A.: Maintainable log datasets for evaluation of intrusion detection systems. IEEE Trans. Dependable Secure Comput. **20**(4), 3466–3482 (2023)

16. Landauer, M., Skopik, F., Wurzenberger, M.: Introducing a New Alert Data Set for Multi-Step Attack Analysis, August 2023. http://arxiv.org/abs/2308.12627

17. Law, K.H., Kwok, L.F.: IDS false alarm filtering using KNN classifier. In: Lim, C.H., Yung, M. (eds.) WISA 2004. LNCS, vol. 3325, pp. 114–121. Springer, Heidelberg (2005). https://doi.org/10.1007/978-3-540-31815-6_10

18. Li, G., Yan, Z., Fu, Y., Chen, H.: Data fusion for network intrusion detection: a review. Secur. Commun. Netw. **2018**, 1–16 (2018)

19. Li, H., et al.: Learning adaptive criteria weights for active semi-supervised learning. Inf. Sci. **561**, 286–303 (2021)

20. Li, W., Meng, W., Luo, X., Kwok, L.F.: MVPSys : toward practical multi-view based false alarm reduction system in network intrusion detection. Comput. Secur. **60**, 177–192 (2016)
21. Lin, Z., Akin, H., Rao, R., Hie, B., Zhu, E.A.: Evolutionary-scale prediction of atomic-level protein structure with a language model. Science **379**(6637), 1123–1130 (2023)
22. Liu, J., Li, S., Zhang, R.: Algorithm of reducing the false positives in IDS based on correlation analysis. In: IOP Conference Series: Materials Science and Engineering, vol. 322, p. 062016 (2018)
23. Meng, Y., Kwok, L.: Intrusion detection using disagreement-based semi-supervised learning: detection enhancement and false alarm reduction. In: Xiang, Y., Lopez, J., Kuo, C.-C.J., Zhou, W. (eds.) CSS 2012. LNCS, vol. 7672, pp. 483–497. Springer, Heidelberg (2012). https://doi.org/10.1007/978-3-642-35362-8_36
24. Meng, Y., Kwok, L.-F.: Enhancing false alarm reduction using pool-based active learning in network intrusion detection. In: Deng, R.H., Feng, T. (eds.) ISPEC 2013. LNCS, vol. 7863, pp. 1–15. Springer, Heidelberg (2013). https://doi.org/10.1007/978-3-642-38033-4_1
25. Meng, Y., Kwok, L.F.: Adaptive non-critical alarm reduction using hash-based contextual signatures in intrusion detection. Comput. Commun. **38**, 50–59 (2014)
26. MIT Lincoln Laboratory: Darpa lldos 1.0 (2000). https://www.ll.mit.edu/r-d/datasets/2000-darpa-intrusion-detection-scenario-specific-datasets. Accessed 07 Apr 2024
27. Pietraszek, T.: Using adaptive alert classification to reduce false positives in intrusion detection. In: Jonsson, E., Valdes, A., Almgren, M. (eds.) RAID 2004. LNCS, vol. 3224, pp. 102–124. Springer, Heidelberg (2004). https://doi.org/10.1007/978-3-540-30143-1_6
28. Settles, B.: Active learning literature survey (2009)
29. Shon, H.G., Lee, Y., Yoon, M.: Semi-supervised alert filtering for network security. Electronics **12**(23), 4755 (2023)
30. Tharwat, A., Schenck, W.: A survey on active learning: state-of-the-art. Pract. Chall. Res. Dir. Math. **11**(4), 820 (2023)
31. Vu, Q.H., Ruta, D., Cen, L.: Gradient boosting decision trees for cyber security threats detection based on network events logs. In: 2019 IEEE International Conference on Big Data, pp. 5921–5928. IEEE, Los Angeles, CA, USA (2019)
32. Wang, T., Zhang, C., Lu, Z., Du, D., Han, Y.: Identifying truly suspicious events and false alarms based on alert graph. In: 2019 IEEE International Conference on Big Data (Big Data), pp. 5929–5936. IEEE, Los Angeles, CA, USA (2019)
33. Wang, X., Yang, X., Liang, X., Zhang, X., Zhang, W., Gong, X.: Combating alert fatigue with AlertPro: context-aware alert prioritization using reinforcement learning for multi-step attack detection. Comput. Secur. **137**, 103583 (2023)
34. Wang, Y., Chen, H., Heng, Q., et al.: FreeMatch: self-adaptive thresholding for semi-supervised learning (2023). http://arxiv.org/abs/2205.07246
35. Yuan, Z., et al.: DualTeacher: bridging coexistence of unlabelled classes for semi-supervised incremental object detection (2023). http://arxiv.org/abs/2401.05362
36. Zhao, X., Greenberg, J., An, Y., Hu, X.T.: Fine-tuning BERT model for materials named entity recognition. In: 2021 IEEE International Conference on Big Data (Big Data), pp. 3717–3720. IEEE, Orlando, FL, USA (2021)

Smart Home Cyber Insurance Pricing

Xiaoyu Zhang[1], Maochao Xu[2], and Shouhuai Xu[3]($\boxtimes$)

[1] Jiangsu Normal University, Xuzhou, Jiangsu, China
[2] Illinois State University, Normal, IL, USA
[3] University of Colorado Colorado Springs, Colorado Springs, CO, USA
sxu@uccs.edu

Abstract. Our homes are increasingly employing various kinds of Internet of Things (IoT) devices, leading to the notion of smart homes. While this trend brings convenience to our daily life, it also introduces cyber risks. To mitigate such risks, the demand for smart home *cyber insurance* has been growing rapidly. However, there are no studies on analyzing the *competency* of smart home cyber insurance policies offered by cyber insurance vendors (i.e., insurers), where 'competency' means the insurer is profitable and smart home owners are not overly charged with premiums and/or deductibles. In this paper, we propose a novel framework for pricing smart home cyber insurance, which can be adopted by insurers in practice. Our case studies show, among other things, that insurers are over charging smart home owners in terms of premiums and deductibles.

Keywords: Smart home cyber insurance · cyber risk · pricing strategy

1 Introduction

The omnipresence of Internet of Things (IoT) technology allows average households to transform into smart homes to provide more convenient and comfortable living environments. According to a report by the Zion Market Research [5], the global smart home market is likely to reach US$137.9 billion by 2026. This explains why smart homes are attracting increasing attention from cyber insurance vendors (i.e., insurers).

Figure 1 illustrates the smart devices in a smart home. As illustrated in Fig. 1, there are a variety of IoT devices in a smart home [34], including thermostats, TVs, laptops, cameras, locks, sensors, smart home appliances, and alarms. These devices may collect and exchange data with each other via a local network. Moreover, there is a possibility of a home gateway that serves as a hub, and a cloud server may be involved in for storing bulk data for the long term.

While ushering in great convenience, these IoT devices, like many other new technologies, contain vulnerabilities that can be exploited by attackers to cause damages to smart home owners [6,24]. For instance, an attacker can wage eavesdropping attacks to steal personal information, or exploit vulnerabilities in smart cameras to extract private videos for cyber extortion purposes [10]. As another example, the 2024 IOT Security Landscape Report [4], which is based on the threat intelligence sampled from 3.8 million smart homes around the

J. Zhao and W. Meng (Eds.): SciSec 2024, LNCS 15441, pp. 313–333, 2025.
https://doi.org/10.1007/978-981-96-2417-1_17

Fig. 1. Illustration of a smart home with 9 devices (excluding cloud server).

world that are protected by the NETGEAR Armor powered by Bitdefender, shows that about 50 million IoT devices generate more than 9.1 billion security events related to vulnerabilities. These highlight that smart home networks have become a target of cyber attackers.

To mitigate smart home cyber risks, insurance companies have expanded their services to include smart home cyber insurance. The current smart home insurance market typically provides coverage for smart home cyber insurance as an add-on to a standard home insurance policy. For instance, a home safety insurance [9] offers smart home cyber insurance coverage, including data breaches, computer and home systems attacks, cyber extortion, and online fraud, while coverage limit tops out at $50,000 and a $500 deductible; State Farm [9] offers a personal cyber insurance add-on to its standard home insurance policy by covering smart home cyber risks, including cyber attacks and extortion, with a coverage limit of $15,000; AIG [9] offers an insurance coverage up to $250,000 at a $1,652 annual premium.

However, we are not aware of any study in the public literature the analyzes the *competency* of smart home cyber insurance, where competency indicates that smart home insurers are profitable and smart home owners do not overpay in premiums and/or deductibles. This is crucial to foster a viable and sustainable smart home cyber insurance market because home owners have no idea whether they are charged and covered appropriately, while noting that insurers would never make their proprietary pricing methods public. This highlights the importance of studying principled solutions to the problem of smart home cyber insurance pricing. To our knowledge, there are essentially no studies on smart home cyber insurance despite the many studies on smart home cyber security (e.g., [6,24,28,29,34]). In this paper, we make a first step to fill the void.

Our Contributions. In this paper, we make two contributions. First, we propose a novel framework for tackling the smart home cyber insurance problem. The framework aims to identify *competent* smart home cyber insurance policies where insurers are profitable and smart home owners do not overpay in premiums and/or deductibles. The framework considers smart home cyber risks in terms of *business lines*, which are often used by the insurance sector. This framework can be adopted by insurers to price smart home cyber insurance premiums and deductibles. Second, we demonstrate the usefulness of the framework by conducting case studies. Our findings include: (i) the current smart home cyber insurance requiring deductibles overly charges smart home owners; (ii) the current smart home cyber insurance requiring no deductibles is not profitable; and (iii) our framework leads to more competent smart home cyber insurance market, meaning that insurers remain profitable while home owners do not overpay in premiums or deductibles.

Related Work. To our knowledge, the present study is the first on smart home cyber insurance, which has unique aspects, such as prevalence of IoT devices and risks of cyber extortion (via private video), online fraud, and property theft. This is true despite the many studies on smart home devices (e.g., [6,14,15,24,28, 29,42]) and cybersecurity risk management and dynamics (e.g., [12,36,45–48]). Nevertheless, cyber insurance has been studied in other contexts (than smart homes) in two categories: *model-driven* vs. *data-driven*. Model-driven studies include: modeling the cyber insurance market and pricing [8] via the *expectation principle* (one of the 4 principles we will consider); insurance pricing [44] via the *standard deviation* principle (another of the 4 principles we will consider); we refer to [7,13,22] for excellent surveys. The present study falls into this category, but initiating the investigation of smart home cyber insurance. Moreover, the present study appears to be the first that investigates cyber risks based on *business lines*, making academic research one step closer to insurance practice. On the other hand, data-driven cyber insurance pricing has been investigated in [16,33,39,40].

Table 1. Summary of the major notations used in this paper.

Notation	Description
n	Number of vulnerabilities
M	Number of risk business lines
V	Set of nodes(vertices) with each representing a vulnerability
$\mathbf{S}$	Set of random variables with each representing a vulnerability state
$\mathbb{S}$	Set of all possible vulnerability states
E	Set of directed edges where $(i,j) \in E$ indicating vulnerability i can exploit j
$G(V,E)$	Directed vulnerability graph consist of V and E
$\mathbf{pa}_j$	Parent node set of vulnerability $j \in V$
$\mathrm{L_m}$	Loss in business lines $m \in \{1,\dots,M\}$ without insurance
$\mathcal{L}_m$	Subset of V whose exploitation can incur loss $\mathrm{L_m}, m \in \{1,\dots,M\}$
X_m	Loss in business line $m \in \{1,\dots,M\}$ with insurance
TL	Total loss in M business lines
(d,C)	Deductible and coverage limit in an insurance product
θ	Parameter reflecting the risk attitude of the insurer
p_i	Exploitation probability(EPSS score) of entry point $i \in V$
e_{ij}	Conditional probability that vulnerability $i \in V$ exploited $j \in V$

Paper Outline. Section 2 presents the framework. Section 3 conducts case studies on pricing. Section 4 concludes the paper with future research directions. For the easy of reference, Table 1 summarizes the main notations used in this paper.

2   Framework for Smart Home Cyber Insurance Pricing

Our framework consists of four steps: (i) defining smart home cyber insurance business lines; (ii) identifying and representing smart home cyber risks; (iii) modeling smart home cyber risks in terms of business lines; and (iv) determining smart home cyber insurance premiums and deductibles.

2.1   Defining Smart Home Cyber Insurance Business Lines

For cyber insurance purposes, an insurer needs to estimate the potential loss of a smart home incurred by cyber attacks in terms of *business lines*, which is an insurance term describing products offered by an insurer. In practice, the following six business lines are widely used and thus adopted in this study.

– *Data breach* (L_1). This business line of risk refers to the exposure of private data in a smart home, such as the home owner's daily activities, emotions, health conditions, audios, and videos. The data breach insurance covers attorney cost, IT professionals, and mitigation of damage.
– *Loss of use* (L_2). This business line of risk refers to the damage incurred by the unavailability of service. The insurance covers the cost associated with the recovery of service, including "cleaning up" compromised devices, data recovery, home applicant repair, and system restoration.
– *Ransomware* (L_3). This business line of risk refers to the loss incurred by ransomware, which encrypts the data on victims' devices. The insurance covers the ransom upon the approval of the insurer when no other methods can recover the data. Note that this risk is different from the *loss of use* risk because the only solution in this case is to pay the ransom.
– *Cyber extortion* (L_4). This business line of risk refers to when the attacker threatens to release sensitive personal data, activities, conversations, or videos of a victim, the insurance covers what is being demanded by the attacker.
– *Online Fraud* (L_5). This business line of risk refers to the financial loss incurred by cyber attacks that stole funds via unauthorized use of bank or credit cards, phishing schemes, and other types of fraud. The insurance covers the direct financial loss incurred by the attack.
– *Property theft* (L_6). This business line of risk refers to the loss incurred by cyber attacks against the cyber defense systems employed at a smart home. The insurance covers the failure of the employed cyber defense system. Note that a smart home that employed cyber defense tools but suffered from a data breach can make claims with respect to L_1 and L_6.

2.2   Identifying and Representing Smart Home Cyber Risks

In practice, vulnerabilities in smart home devices can be detected by vulnerability scanners, such as Nessus [2] and OpenVAS [3]. It would be ideal to patch all vulnerabilities in smart homes to mitigate risks. However, this is not always possible in practice (e.g., home owners do not have the technical expertise). As a consequence, vulnerabilities are present in smart home devices and can be exploited by attackers. Thus, we must deal with the presence of vulnerabilities in smart homes.

Each vulnerability is identified by a Common Vulnerabilities and Exposures (CVE) number and described by the Common Vulnerability Scoring Systems (CVSS) [1]. For each vulnerability, the base CVSS scores, also known as impact metrics, describe the impact when a vulnerability is exploited in terms of confidentiality, integrity, and availability; the Exploit Prediction Scoring System (EPSS) score [23] measures the probability that the vulnerability will be exploited within period of time after its public disclosure, independent of others. Thus, CVSS and EPSS together can describe the risk associated with a vulnerability, assuming their exploitations are independent of each other. That is, the EPSS score can be seen as the probability of exploiting a vulnerability as an entry point for penetrating into a smart home.

However, vulnerabilities are not necessarily exploited independent of each other; rather, the exploitation of one vulnerability often leads to the exploitation of another. Thus, we need to estimate the exploitation probability that goes beyond what is provided by EPSS. This requires us to describe how vulnerabilities may be exploited by attackers after gaining entry point into a smart home. For this purpose, we propose using a Graph-Theoretic representation as illustrated by the following example.

Suppose a smart home has three vulnerabilities, say CVE-2022-22667 (in iPhone), CVE-2018-3919 (in smart home hub), and CVE-2021-32934 (in smart camera). These vulnerabilities enable the attacker to conduct the following attack: the attacker first exploits CVE-2022-22667 in the iPhone, then pivots from the compromised iPhone to exploit CVE-2018-3919 to attack the smart home hub, then pivots from the compromised hub to exploit CVE-2021-32934 to attack the smart camera, and finally breaches videos taken by the camera.

The preceding intuitive discussion prompts us to propose the following steps for identifying and representing smart home cyber risks.

1. Scan vulnerabilities in smart home devices. The detected vulnerabilities are represented by their CVE numbers, CVSS scores, and EPSS scores.
2. Create a directed vulnerability graph $G = (V, E)$, where V is the set of nodes (i.e., vertices) with each representing a vulnerability and E is the set of arcs (i.e., directed edges) with $(i, j) \in E$ representing that the exploitation of vulnerability $i \in V$ could lead to the exploitation of vulnerability $j \in V$. For instance, the aforementioned attack scenario can be represented as a vulnerability graph G, where $V = \{1, 2, 3\}$ (respectively representing CVE-2022-22667, CVE-2020-27403, and CVE-2018-3919) and $E = \{(1, 2), (2, 3)\}$.

Smart home owners typically lack the expertise to create vulnerability graphs. Therefore, insurers should assess cyber risks by understanding attack paths and developing these vulnerability graphs. This can be achieved, for instance, by collaborating with a third-party smart home cybersecurity assessment service provider and/or utilizing an insurer's own cybersecurity team.

2.3   Modeling Cyber Risks in Terms of Business Lines

For cyber insurance purposes, we need to estimate the Total Loss (TL), which is a random variable and incurred by cyber attacks against devices in a smart home. Suppose there are M business lines of risk, represented by $\mathsf{L}_1, \ldots, \mathsf{L}_M$. In order to estimate TL or its distribution $\Pr(\mathsf{TL} \leq \ell)$, we need to estimate the random variable of loss per business line, namely $\Pr(\mathsf{L}_m \leq \ell_m)$ for $m \in \{1, \ldots, M\}$ where $\Pr(\mathsf{TL} \leq \ell) = \Pr\left(\sum_{m=1}^{M} \mathsf{L}_m \leq \ell\right)$. To estimate $\mathsf{L}_1, \ldots, \mathsf{L}_M$ and thus TL, we need to model the security states of smart home devices because the compromise of different devices may incur losses in different business lines. That is, we need to model the exploitation probability for *each and every* vulnerability $i \in V$ in the vulnerability graph $G = (V, E)$ produced in the previous step.

For a vulnerability that is exploited as an entry point into a smart home (e.g., vulnerability in the iPhone in the preceding example), the EPSS score, or a justifiable amendment of it, can be used as its exploitation probability. However, determining the exploitation probability of a non-entry vulnerability in the smart home would require a careful treatment. For instance, the exploitation probability of the smart home hub in the preceding example would conditionally depend on the exploitation of the iPhone, and the exploitation of the smart home camera would conditionally depend on the exploitation of the smart home hub. Given vulnerability graph $G = (V, E)$, there are methods to determine the exploitation probabilities of the non-entry nodes, such as the method that will be used in our case study.

To make our framework able to accommodate any relevant method, we only require that a suitable method, which is equivalent to estimating the security state of each node in a smart home, should take the following as input: (i) A vulnerability graph $G = (V, E)$, where V is the set of nodes with each representing a vulnerability and E is the set of arcs with $(i, j) \in E$ meaning that the exploitation of $i \in V$ could lead to the exploitation of $j \in V$; (ii) each $i \in V$ is annotated with a CVE number, the CVSS scores of i, and the EPSS score of i; and (iii) each arc $(i, j) \in E$ is annotated with the condition probability that $j \in V$ is exploited when $i \in V$ is exploited.

To demonstrate feasibility, Algorithm 1 simulates the distributions of L_m and TL on inputs: (i) the set V of smart home devices; (ii) the exploitation probability p_i of device $i \in V$; (iii) the subset $\mathcal{L}_m \subseteq V$ of nodes whose exploitation can incur loss L_m in business line m, where $m \in \{1, \ldots, M\}$; (iv) the conditional probability e_{ij} for $i, j \in \{1, \ldots, n\}$; and (v) the number R of simulation runs.

2.4   Determining Premiums and Deductibles

Given the estimated loss in business line L_m for $m \in \{1, \ldots, M\}$, this step determines the cyber insurance premium and deductible. Let C be the coverage

---

Algorithm 1: Simulating distributions of L_m and TL

Input: $V = \{1, \ldots, n\}$; the exploitation probability p_i (e.g., EPSS score) of
each possible entry point $i \in V$; $\mathcal{L}_m$; distribution of the L_m when
vulnerabilities in $\mathcal{L}_m$ are exploited, where $m \in \{1, \ldots, M\}$; conditional
probability e_{ij} for $i, j \in \{1, \ldots, n\}$; the number R of simulation runs

Output: Simulated distributions of L_m for $m \in \{1, \ldots, M\}$ and total loss TL

1 Draw $G = (V, E)$ where $V = \{1, \ldots, n\}$ and $(i, j) \in E$ means the exploitation of
vulnerability $i \in V$ can lead to the exploitation of $j \in V$

2 **for** $r = 1$ **to** R **do**

3   Generate Bernoulli vector $\boldsymbol{S}^{(r)} = (S_1, \ldots, S_n)$ based on the p_i's of the
possible entry points and e_{ij} for $i, j \in \{1, \ldots, n\}$

4   Determine the L_m's that are impacted by the exploitation of $i \in V$

5   **for** $m = 1$ **to** M **do**

6   Randomly generate loss $L_m^{(r)}$ according to the distribution of L_m when
the vulnerabilities in $\mathcal{L}_m$ are exploited

7   Record $L_m^{(r)}$ and $TL^{(r)} = \sum_{m=1}^{M} L_m^{(r)}$ for $m \in \{1, \ldots, M\}$ and $r \in \{1, \ldots, R\}$

8 **return** $L_m^{(r)}$ for $m \in \{1, \ldots, M\}$ and total loss $TL^{(r)}$ where $r \in \{1, \ldots, R\}$

---

limit (i.e., the maximum pay by cyber insurer to a smart home owner in the case
of successful cyber attacks against the smart home), which is often an input
parameter determined by the insurer. Let d denote the deductible of a smart
home owner, which is also an input parameter and determined by the insurer.
Since an insurer does not necessarily know in advance about what a suitable
deductible d should be, the insurer would need to try a range of candidate
deductible d's based on the following two pre-determined parameters, which are
widely used in the insurance industry [26]: the Profit to the insurer, defined as
Profit = Premium $-$ Claim, where Premium is the total amount of premiums
collected by the insurer and Claim is the total amount of claims paid to the smart
home owners by the insurer; and the Loss Ratio (LR) to the insurer, defined as
LR = Claim/Premium. For example, the permissible LR may be 40% to assure
that an insurer can make a profit.

To an insurer, the loss in business line m is defined as

$$X_m = \min\{(L_m - d)_+, C\}, \tag{1}$$

where $(L_m - d)_+ = L_m - d$ if $L_m > d$ and $(L_m - d)_+ = 0$ otherwise. In Actuarial
Science, there are many premium pricing principles [25]. In the present study,
we consider the following 4 popular premium pricing principles.

Expectation Principle [25]: Under this principle, the premium is defined as
$\rho_1(X_m) = (1 + \theta) E(X_m)$, where E is the expectation function, and θ reflects
the risk attitude of the insurer (i.e., $\theta = 0$ means risk-neutral, $\theta < 0$ means risk-
seeking, and $\theta > 0$ means risk-averse). The basic idea behind the principle is to
adjust the expected value of the loss by a factor that accounts for the insurer's
risk attitude. The principle is easy to understand and implement. However, it has
the disadvantage that it does not account for the variability in the distribution

of loss X_m, which is critical when dealing with highly uncertain events (e.g., which vulnerabilities will be exploited).

Standard Deviation Principle [25]: Under this principle, the premium is defined as $\rho_2(X_m) = E(X_m) + \theta \sqrt{\text{Var}(X_m)}$, where E is the same as above (i.e., expectation), Var is the variance function, and θ reflects the risk attitude of the insurer (same as above). The basic idea behind the principle is to adjust the expected value of the loss by a factor proportional to the standard deviation of the loss, thus accounting for the variability in the distribution of loss X_m. This leads to the advantage of accommodating uncertainty of the loss distribution, offering a more accurate pricing. Its disadvantage is that it assumes a linear relationship between the loss X_m and its standard deviation, which may not be true in some circumstance (e.g., when losses follow a highly skewed distribution).

Gini Mean Difference (GMD) Principle [19,20]: Under this principle, the premium is defined as $\rho_3(X_m) = E(X_m) + \theta E(|X_{m1} - X_{m2}|)$, where θ also reflect an insurer's risk attitude (as above), $E(|X_{m1} - X_{m2}|)$ is the Gini Mean Difference (GMD) that measures the statistical variability between a pair of independent realizations of X_m, denoted by X_{m1} and X_{m2}. The basic idea behind the principle is to adjust the expected value of the loss by a factor proportional to the GMD, which captures the average absolute difference between pairs of independent realization of the loss, thereby accounting for variability. The advantage is that it is sensitive to the shape of the loss distribution and captures more information about the variability of the loss than the standard deviation. The disadvantage is that it is somewhat involved to cope with.

Conditional Tail Expectation Principle [21,41]: Under this principle, the premium is defined as $\rho_4(X_m) = E(X_m | X_m \geq \text{VaR}_\beta)$, where VaR_β is the value-at-risk at level $\beta \in (0,1)$, namely $\text{VaR}_\beta = \min_\gamma \{\gamma : \Pr(X_m \leq \gamma) \geq \beta\}$. The basic idea behind the principle is to determine the premium based on the conditional expectation of the loss X_m, given that the loss exceeds the threshold that is defined by the value-at-risk at a desired confidence level β. Its advantage is that it focuses on the tail of the loss distribution, which is important for assessing extreme risks and ensuring that sufficient funds are available to cover significant losses. Its disadvantage is that it only considers losses beyond a certain threshold and ignores the distribution of losses below this threshold, which may offer an incomplete assessment of the overall risk.

Given that each of the preceding principles has its advantages and disadvantages, the overall guideline in selecting principles to guide pricing is the following: choose a principle that aligns with the specific risk management objective, the nature of the risks being insured, and the risk attitude of an insurer. For our case study, we compare all the principles to understand their implications and suitability for different scenarios in smart home insurance for the first time.

3   Case Studies

We inherit the 6 business lines defined in the framework. Thus, in what follows, we only focus on the other three steps.

3.1   Identifying and Representing Smart Home Cyber Risks

Our case study is based on the smart home illustrated in Fig. 1, which has 9 smart home devices. Suppose a vulnerability scanning process shows that 7 (out of the 9) devices contain the vulnerabilities listed in Table 2. Thus, the vulnerability graph $G = (V, E)$ has $V = \{1, 2, 3, 4, 5, 6, 7\}$, where node 1 supposedly represents vulnerability CVE-2022-22667 in the iPhone, node 2 for CVE-2020-27403 in the smart TV, node 3 for CVE-2018-3919 in the smart home hub, node 4 for CVE-2021-29438 in the smart sensor, node 5 for CVE-2021-32934 in the smart camera, node 6 for CVE-2019-7256 in the smart lock, and node 7 for CVE-2017-8759 in the laptop.

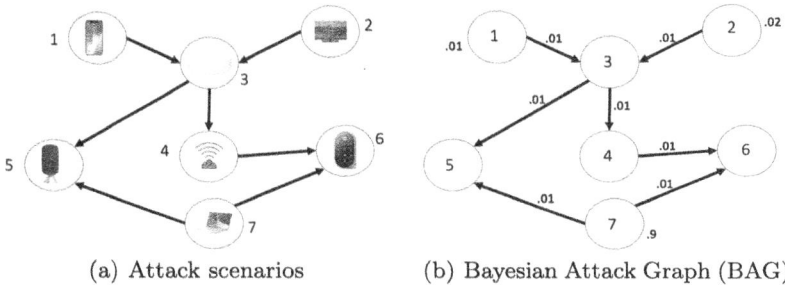

(a) Attack scenarios   (b) Bayesian Attack Graph (BAG)

Fig. 2. Graph-theoretic representation of vulnerabilities and attacks in a smart home. (a) Attack steps represented as arcs. (b) BAG with exploitation probabilities, such as $\Pr(S_7 = 1) = .9$ and $\Pr(S_5 = 1 | S_7 = 1) = .01$.

Figure 2(a) illustrates the Graph-Theoretic representation of vulnerabilities and attacks in the smart home based on the following assumptions. (i) Nodes 1, 2, and 7 are the entry points, meaning that they (i.e., the devices containing these vulnerabilities) can be exploited by the attacker from outside of the smart home. (ii) Arc $(i, j) \in E$ means that compromise of node $i \in V$ can cause the compromise of node $j \in V$. Note that there are 6 attack paths, namely $1 \rightarrow 3 \rightarrow 5$; $1 \rightarrow 3 \rightarrow 4 \rightarrow 6$; $2 \rightarrow 3 \rightarrow 5$; $2 \rightarrow 3 \rightarrow 4 \rightarrow 6$; $7 \rightarrow 5$; and $7 \rightarrow 6$. For example, attack path $1 \rightarrow 3 \rightarrow 5$ means the following: the attacker first compromises the iPhone, then pivots to compromise the smart home hub, and finally pivots to compromise the smart home camera. This attack may cause, for example, *data breach* (L_1), *online fraud* (L_5), and *cyber extortion* (L_4) because the attacker targets data stored in the smart home hub or recorded at the smart home camera. Similarly, attack path $1 \rightarrow 3 \rightarrow 4 \rightarrow 6$ can cause risks including *data breach* (L_1) and *property theft* (L_6) because the attacker can unlock the door; attack path $7 \rightarrow 5$ can cause risks including *loss of use* (L_2) and *cyber extortion* (L_4); and attack path $7 \rightarrow 6$ can cause risks including *data breach* (L_1), *ransomware* (L_3) and *property theft* (L_6);

In our case study we use the Bayesian Attack Graph (BAG) approach [27, 38] to model cyber risks because of its simplicity. At a high level, BAGs are graphical models representing information about network vulnerabilities and how they may be exploited in terms of attack paths. This leads to the BAG illustrated in Fig. 2(b), where entry nodes 1 (iPhone), 2 (smart TV), and 7 (laptop) are assumed to be exploited by the attacker from outside of the smart home with the probability that is specified by their respective EPSS score, .01, .02, and .9. The probability e_{ij} associated with arc $(i, j) \in E$ is defined as the conditional probability that the exploitation of $i \in V$ leads to the exploitation of $j \in V$. For simplicity, we assume $e_{ij} = .01$ for all $(i, j) \in E$.

To make the discussion below useful in general cases, we consider $V = \{1, \ldots, n\}$ and $|V| = n$, with $n = 7$ in the preceding example. Let $s_i = 1$ denote the fact (i.e., with probability 1 or certainty) that vulnerability $i \in V$ is exploited (i.e., the corresponding device is compromised) and $s_i = 0$ otherwise. Let S_i be the random variable denoting $i \in V$ is exploited, where $S_i = 1$ means i is exploited and $S_i = 0$ otherwise. With these notations, we can write $e_{ij} = \Pr(S_j = 1 | S_i = 1)$. Let $\mathbf{S} = (S_1, \ldots, S_n)$, $\mathbf{s} = (s_1, \ldots, s_n)$, and $\mathbb{S} = \{\mathbf{s} | \mathbf{s} \in \{0, 1\}^n\}$ or the set of all possible states. For $j \in V$, we denote its parent node set by $\mathbf{pa}_j$, which is the set of nodes that point to j, namely $\mathbf{pa}_j = \{i : (i, j) \in E\}$. For example, the parent node set of node 5 in Fig. 2 is $\mathbf{pa}_5 = \{3, 7\}$. Let $S_{\mathbf{pa}_j}$ denote the states of the parent nodes of node j.

Remark. We can use the preceding representation to enable "what if" analysis by accommodating zero-day vulnerabilities. For a hypothetical zero-day vulnerability, one needs to determine the probability it can be an entry point, and the conditional probability e_{ij}, where both i and j can be zero-day vulnerabilities.

3.2   Modeling Cyber Risks in Terms of Business Lines

Modeling Security States in a Smart Home. Table 2 summarizes the aforementioned impacts of the exploitation of the 7 vulnerabilities with respect to the 6 business lines, while recalling that the exploitation of one vulnerability can impact multiple business lines.

Table 2. Vulnerabilities in a smart home and losses caused by their exploitation, where ✓ indicates applicability, vulnerability $i \in \{1, \ldots, 7\}$, and business line $j \in \{1, \ldots, 6\}$.

Smart Home Device	Node Identity	Vulnerability	L_1	L_2	L_3	L_4	L_5	L_6
iPhone	1	CVE-2022-22667	✓			✓		
Smart TV	2	CVE-2020-27403	✓					
Smart Home Hub	3	CVE-2018-3919	✓	✓				
Smart Sensor	4	CVE-2021-29438	✓					
Smart Camera	5	CVE-2021-32934			✓	✓		
Smart Lock	6	CVE-2019-7256						✓
Laptop	7	CVE-2017-8759	✓		✓			

Let $\mathcal{L}_m$ denote the subset of nodes in V whose exploitation can incur loss in business line m, namely L_m, where $m \in \{1,\ldots,M\}$. According to Table 2, we have $\mathcal{L}_1 = \{1,2,3,4,7\}$, $\mathcal{L}_2 = \{3,5\}$, $\mathcal{L}_3 = \{7\}$, $\mathcal{L}_4 = \{5\}$, $\mathcal{L}_5 = \{1\}$, and $\mathcal{L}_6 = \{6\}$. Then, we have [27]:

$$\Pr(\mathbf{S} = \mathbf{s}) = \Pr(S_1 = s_1,\ldots,S_n = s_n) = \prod_{i=1}^{n} \Pr(S_i = s_i | S_{\mathbf{pa}_i}), \quad s_i \in \{0,1\}. \quad (2)$$

For the loss incurred in business line $m \in \{1,\ldots,M\}$, we have

$$\Pr\left(\mathsf{L}_m \le \ell_m\right) = \sum_{\mathbf{s} \in \mathbb{S}} \Pr\left(\mathsf{L}_m \le \ell_m | \mathbf{S} = \mathbf{s}\right) \Pr(\mathbf{S} = \mathbf{s})$$

$$= \sum_{\mathbf{s} \in \mathbb{S}} \Pr\left(\mathsf{L}_m \le \ell_m | \mathbf{S} = \mathbf{s}\right) \prod_{i=1}^{n} \Pr(S_i = s_i | S_{\mathbf{pa}_i}). \quad (3)$$

Then, we have

$$\Pr\left(\mathsf{TL} \le \ell\right) = \Pr\left(\sum_{m=1}^{M} \mathsf{L}_m \le \ell\right)$$

$$= \sum_{\mathbf{s} \in \mathbb{S}} \Pr\left(\sum_{m=1}^{M} \mathsf{L}_m \le \ell | \mathbf{S} = \mathbf{s}\right) \prod_{i=1}^{n} \Pr(S_i = s_i | S_{\mathbf{pa}_i}). \quad (4)$$

Note that we cannot compute $\Pr(\mathsf{L}_m \le \ell_m)$ and $\Pr(\mathsf{TL} \le \ell)$ according to Eqs. (3) and (4) without knowing both $S_{\mathbf{pa}_i}$ and the dependence among the L_m's. The issue of $S_{\mathbf{pa}_i}$ could be resolved if we can assume that the exploitation of j is independently incurred by its compromised parent nodes $i \in \mathbf{pa}_j$, which means that the exploitation probability of $j \in V$ can be computed as

$$\Pr(S_j = 1 | S_{\mathbf{pa}_j}) = \begin{cases} 0 & \forall i \in \mathbf{pa}_j, S_i = 0; \\ 1 - \prod_{i \in \mathbf{pa}_j, S_i = 1}(1 - e_{ij}), & \text{otherwise} \end{cases}. \quad (5)$$

However, the computation would be very complex due to the BAG structure. Moreover, Eq. (4) is still infeasible to compute because of the dependence among the L_m's. To tackle these two issues, we propose using Algorithm 1 to conduct the simulation to empirically derive $S_{\mathbf{pa}_i}$ and the total loss without computing the dependence among the L_m's.

Running Example of Modeling Security States in a Smart Home. Consider the example BAG in Fig. 2(b). Equation (2) says

$$\Pr(\mathbf{S} = \mathbf{s}) = \Pr(S_1 = s_1) \cdot \Pr(S_2 = s_2) \cdot \Pr(S_7 = s_7)$$
$$\cdot \Pr(S_3 = s_3 | S_2 = s_2, S_1 = s_1) \cdot \Pr(S_5 = s_5 | S_3 = s_3, S_7 = s_7)$$
$$\cdot \Pr(S_4 = s_4 | S_3 = s_3) \cdot \Pr(S_6 = s_6 | S_4 = s_4, S_7 = s_7). \quad (6)$$

Table 3. Probability (*Prob*) of the smart home in 8 example states

S_1	S_2	S_3	S_4	S_5	S_6	S_7	*Prob*
0	0	0	0	0	0	0	.097
0	0	0	0	0	0	1	.856
0	0	0	0	0	1	0	.000
0	0	0	0	0	1	1	.009
0	0	0	0	1	0	0	.000
0	0	0	0	1	0	1	.009
0	0	0	0	1	1	0	.000
0	0	0	0	1	1	1	.000

Suppose $S_1 = 1$ and $S_2 = 1$. Recall that $e_{23} = \Pr(S_3 = 1|S_2 = 1)$ and $e_{13} = \Pr(S_3 = 1|S_1 = 1)$. Then, we have

$$\Pr(S_3 = 1|S_2 = 1, S_1 = 1) = 1 - (1 - e_{13})(1 - e_{23}).$$

Since there are 7 vulnerabilities, there are 2^7 possible states in $\mathbb{S}$. Table 3 presents the probabilities of 8 (out of the 2^7) states, computed according to Eq. (6).

Now we show how to apply Algorithm 1 to empirically compute the L_m's and thus the TL based on reasonable assumptions. Suppose L_1 and L_2 follow the exponential distribution, $L_1 \sim \exp(\lambda_1)$ and $L_2 \sim \exp(\lambda_2)$, where $\lambda_1 = \sum_{i \in \mathcal{L}_1} a_1 S_i$, $\lambda_2 = \sum_{j \in \mathcal{L}_2} a_2 S_j$, $a_1 = (1/160, 1/32, 1/80, 1/80, 1/160)$ is the coefficients representing the rates of loss incurred by the exploitation of vulnerabilities in $\mathcal{L}_1 = \{1, 2, 3, 4, 7\}$ with the rate of loss incurred by the exploitation of vulnerability 1 being $1/160$ per year (i.e., the mean loss is \$160 per year), which is an input parameter that can be estimated based on historic data or domain expertise, and $a_2 = (1/640, 1/320)$ is the coefficients representing the rates of loss incurred by the exploitation of vulnerabilities in $\mathcal{L}_2 = \{3, 5\}$. The exponential distribution assumptions about L_1 and L_2 can be justified as follows: For a smart home, L_1 (data breach) and L_2 (loss of use) may not be extremely large losses, and thus can be modeled as an exponential distribution [11].

Suppose L_3 and L_4 follow lognormal distributions when the vulnerabilities in $\mathcal{L}_3 = \{7\}$ and $\mathcal{L}_4 = \{5\}$ are exploited, namely $L_3 \sim \text{Lognorm}(\mu_1, \sigma^2)$ and $L_4 \sim \text{Lognorm}(\mu_2, \sigma^2)$, where $(\mu_1, \mu_2, \sigma) = (4, 7, 1)$ based on historic data or domain expert's estimation. The lognormal distribution assumption can be justified as follows: For a smart home, L_3 (ransomware) and L_4 (cyber extortion) could be large losses and thus can be modeled via the heavy tail of the lognormal distribution [11].

Suppose L_5 and L_6 follow Gamma distributions when the vulnerabilities in $\mathcal{L}_5 = \{1\}$ and $\mathcal{L}_6 = \{6\}$ are exploited, namely $L_5 \sim \Gamma(\alpha_1, \beta)$ and $L_6 \sim \Gamma(\alpha_2, \beta)$, where $(\alpha_1, \alpha_2, \beta) = (1000, 2000, 1)$ based on historic data or domain expert's estimation. The Gamma distribution assumption can be justified as follows: For a smart home, the loss incurred by L_5 (online fraud) and L_6 (property theft) could vary significantly and thus can be modeled by the flexibility of the Gamma distribution as it has two parameters (i.e., the shape parameter and the scale parameter) [11].

Table 4 summarizes the simulation result of L_m for $m \in \{1, \ldots, 6\}$ and the resulting total loss TL, where $n = 7$ and $R = 10,000$. We observe that L_1 exhibits the highest mean value because vulnerability 7 is most likely exploited, that L_3 has the second highest mean value, and L_4 exhibits extreme values because cyber extortion could cause significant loss when the smart camera is exploited.

Table 4. Simulation result based on Algorithm 1 and parameters discussed in the text

	Min	Q_{25}	Median	Q_{75}	Q_{90}	Q_{95}	Q_{99}	$Q_{99.5}$	$Q_{99.9}$	Max	Mean	SD
L_1	.00	28.31	92.35	204.10	354.61	476.86	749.93	867.01	1143.49	1875.83	144.81	165.50
L_2	.00	.00	.00	.00	.00	.00	.00	199.73	704.06	1388.21	3.02	43.15
L_3	.00	21.26	48.13	99.59	191.60	273.29	57.05	727.44	1047.45	4597.62	83.16	123.43
L_4	.00	.00	.00	.00	.00	.00	.00	868.25	4867.59	15563.73	18.80	319.33
L_5	.00	.00	.00	.00	.00	.00	.00	998.13	1029.31	1069.61	9.46	96.69
L_6	.00	.00	.00	.00	.00	.00	.00	2004.76	2072.27	2150.52	19.70	198.09
TL	.00	87.22	181.37	332.07	539.64	771.11	2142.00	2458.68	5326.47	16521.96	278.95	465.62

3.3 Determining Cyber Insurance Premiums and Deductibles

Is the Current Real-world Smart Home Insurance Policy Competent? We use the term *competency* to indicate that a smart home cyber insurer can make a profit and a smart home owner does not overly pay premium or deductible. We consider two real-world insurers whose names are anonymized in this paper: Insurer A requiring deductibles vs. Insurer B not requiring deductibles.

Case 1: Insurer A requiring deductibles. In this case, the one-year policy has a $1,000 deductible and a $50,000 coverage limit. The coverage includes cyber extortion (L_4), data restoration, crisis management, and cyber bullying, where the last three are not considered in this paper. Table 5 summarizes the yearly premium per business line.

Table 5. Cyber insurance premium ($) offered by Insurer A.

Type	Cyber extortion	Data restoration	Crisis management	Cyber bullying	Total Premium
Premium	28	151	231	28	438

Since L_4 is common to Insurer A's policy and the present study, we use it as a baseline to determine the parameters in our pricing formulas. Specifically, we set the premium for L_4 to $28 per year, and the deductible and coverage limit to the same as that of Insurer A's. Then, we determine the parameters in ρ_1 to

ρ_4 based on the formulas given in the framework (Sect. 2.4) and the simulated losses. The resulting parameters are $\theta = .5, .03, .25$ for ρ_1, ρ_2, ρ_3, and $\beta = .34$ for ρ_4, respectively.

Table 6. Our premiums (\$) under the 4 pricing principles, where 'total premium' is the sum of all premiums.

Premium (\$)	ρ_1	ρ_2	ρ_3	ρ_4
L_1	217	150	185	211
L_2	5	4	5	5
L_3	125	87	107	120
L_4	**28**	**28**	**28**	**28**
L_5	14	12	14	14
L_6	30	26	29	30
Total premium	418	307	368	408

Table 6 shows the premium for each business line under the 4 premium principles described in the framework, where the total premium is the sum of individual premiums in the 6 business lines or $\sum_{m=1}^{6} \rho_j(L_m)$ where $j \in \{1, \ldots, 4\}$. We observe ρ_1 is the largest total premium (\$418) and ρ_2 is the smallest (\$307).

To assess the performance of premium pricing principles, we use the aforementioned two metrics: Profit, namely Profit = Premium−Claim; and Loss Ratio (LR), namely LR = Claim/Premium. Suppose the permissible LR is 40%, and 500 smart home owners purchase smart home insurance (i.e., a portfolio of 500 policyholders) where premiums are charged according to Table 6. We simulate the loss scenarios of the portfolio with 10,000 independent runs and consider the distribution of the loss scenarios of these 10,000 simulation runs.

Table 7. Summary statistics of Profits and LRs under the 4 pricing principles with \$1,000 deductible and \$50,000 coverage limit, where 'Profit' and 'LR' correspond to the loss of individual business lines (i.e., total premium in Table 6).

	Min	Q_1	Q_5	Q_{10}	Q_{15}	Q_{50}	Q_{75}	Max	Mean	SD
ρ_1 Profit	127,549	174,542	183,192	186,869	189,152	196,215	199,612	207,628	195,089	6,429
ρ_2 Profit	72,049	119,042	127,692	131,369	133,652	140,715	144,112	152,128	139,589	6,429
ρ_3 Profit	102,549	149,542	158,192	161,869	164,152	171,215	174,612	182,628	170,089	6,429
ρ_4 Profit	122,549	169,542	178,192	181,869	184,152	191,215	194,612	202,628	190,089	6,429

	Min	Q_{25}	Q_{50}	Q_{75}	Q_{90}	Q_{95}	$Q_{99.5}$	Max	Mean	SD
ρ_1 LR	.01	.04	.06	.08	.11	.12	.19	.39	.07	.03
ρ_2 LR	.01	.06	.08	.11	.14	.17	.26	.53	.09	.04
ρ_3 LR	.01	.05	.07	.09	.12	.14	.22	.44	.08	.03
ρ_4 LR	.01	.05	.06	.08	.11	.13	.19	.40	.07	.03

Table 7 shows the summary statistics of portfolio Profit and LRs based on the loss of individual business lines and the aggregated loss under the 4 pricing principles. We observe that all the profits are positive, with ρ_1 being the largest and ρ_2 being the smallest. Moreover, the mean LRs are small ($< .1$) for every pricing principle, the high quantiles of LRs (e.g., $Q_{99.5}$) are still smaller than 40%, but the worst-case scenarios for ρ_2 and ρ_3 are beyond the permissible LR of 40%. Note that all the standard deviations are the same because the coverage limits and deductibles are fixed.

Insight 1. *Insurer A, which requires deductibles, is too conservative, meaning that home owners are over charged for their smart home cyber insurance.*

Case 2: Insurer B not Requiring Deductibles. In this case, Insurer B offers a smart home policy covering cyber extortion and ransomware, cyber financial loss, and cyber personal protection with coverage limits and premiums shown in Table 8.

Table 8. Smart home insurance policy offered by Insurer B with 0 deductible and $50,000 coverage limit for the covered attacks.

Coverage limit				Premium
Cyber extortion	Cyber financial loss	Cyber personal protection	All covered events	
10,000	50,000	50,000	50,000	200

We apply the premium strategy of Insurer B to our simulated portfolio losses, and present the summary statistics of the Profit and LR in Table 9. We observe that under this premium strategy, Insurer B cannot make a profit, and the mean LR is 1.35 which is much larger than the permissible LR (40%).

Table 9. Summary statistics of profit and LR under premium practice of Insurer B.

	Min	Q_1	Q_5	Q_{10}	Q_{50}	Q_{95}	Q_{995}	Q_{999}	Max	Mean	SD
Profit	−106,801	−61,674	−51,710	−46,830	−34,010	−20,912	−13,866	−10,635	−4,540	−34,764	9,416
LR	1.05	1.16	1.21	1.24	1.34	1.52	1.65	1.74	2.07	1.35	.09

Insight 2. *Insurer B, which does not require deductibles, cannot make a profit and cannot survive.*

Our Proposals for Competent Smart Home Insurance Policies. Given that the current real-world smart home cyber insurance policies are not competent, we propose our insurance policy that would be more competent (i.e., benefit both insurers and smart home owners). Since requiring no deductible is not a standard practice and is not profitable in our analysis mentioned above, we only consider the case of requiring deductibles. We ask and address two questions: Given the same premium and coverage limit as Insurer A in practice, what is the competent (or affordable) deductible that makes the insurer profitable? Given the same deductible and coverage limit as Insurer A, what is the competent (or affordable) premium that makes the insurer profitable?

Seeking Small Yet Competent Deductibles. To search for competent insurance polices, or small deductibles, we start the $1,000 required by Insurer A while making the premiums and coverage limit the same as that of Insurer A. We consider two strategies: (i) the permissible mean LR is 40%; (ii) the permissible high quantile of LR, $Q_{99.5}$, is 40%. We consider these two strategies because they reflect an insurer's risk attitude from the policy-making perspective. Strategy (i) ensures that, on average, the insurer maintains a profitable margin while providing coverage. Strategy (ii) represents a risk-averse approach, ensuring that the insurer remains profitable even in extreme scenarios.

We start with a $500 deductible, which would be more attractive to smart home owners than Insurer A which requires a $1,000 deductible.

Table 10. Summary statistics of Profits and LRs with $500 deductible and $50,000 coverage limit.

	Min	Q_1	Q_5	Q_{10}	Q_{15}	Q_{50}	Q_{75}	Max	Mean	SD
ρ_1 Profit	112,217	157,355	166,918	170,996	173,536	181,815	186,350	199,217	180,880	7,743
ρ_2 Profit	56,717	101,855	111,418	115,496	118,036	126,315	130,850	143,717	125,380	7,743
ρ_3 Profit	87,217	132,355	141,918	145,996	148,536	156,815	161,350	174,217	155,880	7,743
ρ_4 Profit	107,217	152,355	161,918	165,996	168,536	176,815	181,350	194,217	175,880	7,743
	Min	Q_{25}	Q_{50}	Q_{75}	Q_{90}	Q_{95}	$Q_{99.5}$	Max	Mean	SD
ρ_1 LR	.05	.11	.13	.15	.18	.20	.27	.46	.13	.04
ρ_2 LR	.06	.15	.18	.21	.25	.27	.37	.63	.18	.05
ρ_3 LR	.05	.12	.15	.18	.21	.23	.31	.53	.15	.04
ρ_4 LR	.05	.11	.13	.16	.19	.21	.28	.47	.14	.04

Table 10 presents the resulting Profits and LRs with a deductible of $500. We observe that Profit of the insurer decreases when compared with that of Insurer A (Table 7), but this decrease is still acceptable because the mean LR is far below 40% for ρ_1 to ρ_4, which is required by strategy (i). The same conclusion applies to strategy (ii), which requires that $Q_{99.5}$ is 40%.

The fact that a $500 deductible is profitable in both strategies (i) and (ii) prompts us to search for the strategy that is profitable to insurers while requiring a deductible that is as small as possible. Specifically, we gradually reduce the quantity of deductibles under each strategy, as Eq. (1) indicates that the loss faced by the insurer decreases with the amount of deductible. Details follow.

First, consider a $250 deductible. Owing to space limit, we omit the details but highlight the result as follows. We find that the Profit decreases. However, under strategy (i), the mean LR is still less than 40%, indicating that we can continue to decrease the deductible. Under strategy (ii), the $Q_{99.5}$ of LR for ρ_1 is equal to 40%, meaning a \$250 deductible is sufficient; however, the $Q_{99.5}$ of LR for ρ_2, ρ_3, and ρ_4 exceeds 40%, meaning a \$250 deductible is not sufficient.

Second, consider a \$200 deductible. While omitting the details (owing to space limit), we highlight we observe a further decrease in Profit, which is expected. Under strategy (i), we observe the mean LR for ρ_2 surpasses 40%, meaning a \$200 deductible is not sufficient. Under strategy (ii), the $Q_{99.5}$ of LR under all the 4 pricing principle exceeds 40%, meaning a \$200 deductible is not enough.

Third, consider a \$150 deductible. While omitting details (owing to space limit), we highlight that the Profit further decrease. Under strategy (i), the mean LR for ρ_3 surpasses 40%, meaning a \$150 deductible is not sufficient; the mean LR for ρ_1 and ρ_4 is less than 40%, meaning a smaller deductible is possible. Under strategy (ii), a \$150 deductible is also not profitable.

Table 11. Summary statistics of Profits and LRs with a \$100 deductible and \$50,000 coverage limit.

	Min	Q_1	Q_5	Q_{10}	Q_{15}	Q_{50}	Q_{75}	Max	Mean	SD
ρ_1 Profit	44,965	89,249	99,248	104,001	107,026	116,589	122,186	144,683	115,821	9,223
ρ_2 Profit	-10,535	33,749	43,748	48,501	51,526	61,089	66,686	89,183	60,321	9,223
ρ_3 Profit	19,965	64,249	74,248	79,001	82,026	91,589	97,186	119,683	90,821	9,223
ρ_4 Profit	39,965	84,249	94,248	99,001	102,026	111,589	117,186	139,683	110,821	9,223
	Min	Q_{25}	Q_{50}	Q_{75}	Q_{90}	Q_{95}	$Q_{99.5}$	Max	Mean	SD
ρ_1 LR	.31	.42	.44	.47	.50	.53	.59	.78	.45	.04
ρ_2 LR	.42	.57	.60	.64	.68	.71	.81	1.07	.61	.06
ρ_3 LR	.35	.47	.50	.54	.57	.60	.67	.89	.51	.05
ρ_4 LR	.32	.43	.45	.48	.51	.54	.61	.80	.46	.05

Fourth, consider a \$100 deductible. Table 11 presents the resulting Profits and LRs. We observe that the Profit further decreases. Under strategy (i), the mean LR under all the 4 pricing principles, including ρ_1 and ρ_4, which are profitable with a \$150 deductible, surpass 40% under, meaning that a \$100 deductible is not sufficient. Under strategy (ii), a \$100 deductible is also not profitable.

Based on the preceding simulation results, we propose the deductibles under the 4 pricing principles in Table 12, while showing their mean Profits. When compared with the real-world Insurer A's insurance policy (Table 7), we observe that the mean Profits based on our insurance policies are reduced, but insurers are still profitable, while smart home owners only need to pay a significantly smaller amount of deductibles, which would attract more smart owners. More policyholders can increase the insurer's profit as they might face the same kinds of cyber risks.

Table 12. Our proposed deductibles, where Deductible 1 and Mean Profit 1 are derived based on strategy (i) or the permissible mean LR being smaller than 40%, Deductible 2 and Mean Profit 2 are derived based on strategy (ii) or the permissible 99.5th LR ($Q_{99.5}$) being smaller than 40%.

	Total premium	Coverage limit	Deductible 1	Mean Profit 1	Deductible 2	Mean Profit 2
ρ_1	Premium 418	50,000	150	131,809	250	154,670
ρ_2	Premium 307	50,000	250	99,170	500	125,380
ρ_3	Premium 368	50,000	200	119,583	500	155,880
ρ_4	Premium 408	50,000	150	126,809	500	175,880

Insight 3. *When compared with the current smart home cyber insurance practice (via Insurer A's policy), our framework leads to smaller mean profits to cyber insurers but significantly smaller deductibles to home owners.*

Seeking Small Yet Competent Premiums. We fix the deductible at $1,000 and the coverage limit at $50,000 as in the policy of Insurer A. We use the mean LR of 40% in strategy (i) and the $Q_{99.5}$ LR of 40% in strategy (ii), to determine the respective premiums.

Table 13. Summary statistics of Profits and LRs with $1,000 deductible and $50,000 coverage limit.

Premium		Min	Q_1	Q_5	Q_{10}	Q_{50}	Q_{95}	$Q_{99.5}$	$Q_{99.9}$	Max	Mean	SD
198	Profit	17,549	64,542	73,192	76,869	86,215	93,089	95,761	96,699	97,628	85,089	6,429
	LR	.01	.04	.06	.07	.13	.26	**.40**	.50	.82	.14	.06
70	Profit	-46,451	542	9,192	12,869	22,215	29,089	31,761	32,699	33,628	21,089	6,429
	LR	.04	.11	.17	.20	.37	.74	1.13	1.41	2.33	**.40**	.18

Table 13 presents the premiums corresponding to the summary statistics of Profits and LRs. We observe that if the permissible LR of $Q_{99.5}$ is 40%, a $198 premium leads to a $85,089 mean profit (i.e., the mean of the total profit of the insurer in serving 500 smart homes); if the permissible mean LR is 40%, a $70 premium leads to a $21,089 mean profit. Therefore, it is possible to charge a low premium ($70) for a decent coverage ($50,000), while remaining profitable.

Insight 4. *When compared with the current smart home cyber insurance practice (via Insurer A's policy), our framework leads to a lower premium while insurers remain profitable.*

4   Conclusion

We have presented a framework for smart home cyber insurance that can lead to competent policies to make insurers profitable while smart home owners pay a more affordable premium or deductible than their counterpart in the current practice. We conducted case studies to demonstrate the usefulness of the framework, while showing that the current smart insurance pricing can be further adjusted to offer more attractive policies (e.g., lower deductible or premium).

The present study has several limitations that need to be addressed in the future. First, the loss distributions and parameters used in our case study are assumed to be given. While they can be derived from experience and/or historic data in principle, this needs to be calibrated when real-world smart home cyber claim data are available. Second, our case study is based on BAG for analyzing the probability that a vulnerability in a smart home will be exploited. This method, or any method for the same purpose, needs to be validated with real-world smart home experiments. Third, we empirically searched for competent smart home cyber insurance deductibles and premiums. It is interesting to define optimal deductibles and premiums and solve such optimization problems analytically. Fourth, we do not consider the systemic risk that occurs when common vulnerabilities exist in multiple smart home networks. Fifth, we need to deepen our understanding of cyber risks in the business lines. For instance, characterizing and forecasting data breaches have been investigated at the enterprise or industry level [17,18,37,43] but not at the smart home level (L_2); characterizing the psychological aspects of cyber social engineering attacks has been conducted in a general context [30–32,35] but not the smart home context (L_5). Sixth, it is interesting to extend or adapt the present study to accommodate other settings, such as the financial service and healthcare sectors.

Acknowledgment. We thank the reviewers for their comments. This research was supported in part by NSF Grant #2115134 and Colorado State Bill 18-086. This research work is also a contribution to the International Alliance for Strengthening Cybersecurity and Privacy in Healthcare (CybAlliance, Project no. 337316).

References

1. Common vulnerability scoring system. http://www.rst.org/cvss/cvss-guide.html, Accessed 30 Dec 2021
2. Nessus home page. https://www.tenable.com/products/nessus
3. Openvas home page. https://www.openvas.org/
4. The 2024 IoT security landscape report. https://blogapp.bitdefender.com/hotforsecurity/content/files/2024/06/2024-IoT-Security-Landscape-Report_consumer.pdf
5. Zion market research report. https://www.zionmarketresearch.com/report/smart-home-market (2023)
6. Alrawi, O., Lever, C., Antonakakis, M., Monrose, F.: Sok: security evaluation of home-based IoT deployments. In: 2019 IEEE Symposium on Security and Privacy (SP), pp. 1362–1380 (2019). https://doi.org/10.1109/SP.2019.00013
7. Awiszus, K., Knispel, T., Penner, I., Svindland, G., Voß, A., Weber, S.: Modeling and pricing cyber insurance: idiosyncratic, systematic, and systemic risks. Eur. Actuar. J. **13**(1), 1–53 (2023)

8. Böhme, R., Schwartz, G.: Modeling cyber-insurance: towards a unifying framework. In Proceedings of the Workshop on the Economics of Information Security (2010)
9. Breiner, B.: What is personal cyber insurance? and how can homeowners buy a policy? https://www.valuepenguin.com/personal-cyber-home-insurance (2022)
10. Bugeja, J., Jacobsson, A., Davidsson, P.: Prash: a framework for privacy risk analysis of smart homes. Sensors **21**(19), 6399 (2021)
11. Casella, G., Berger, R.: Statistical Inference. CRC Press (2024)
12. Cho, J.H., Xu, S., Hurley, P.M., Mackay, M., Benjamin, T., Beaumont, M.: Stram: measuring the trustworthiness of computer-based systems. ACM Comput. Surv. **51**(6), 128:1–128:47 (2019)
13. Cremer, F., et al.: Cyber risk and cybersecurity: a systematic review of data availability. Geneva Papers Risk Insurance-Issues Pract. **47**(3), 698–736 (2022)
14. Das, S.R., Chita, S., Peterson, N., Shirazi, B.A., Bhadkamkar, M.: Home automation and security for mobile devices. In: 2011 IEEE International Conference on Pervasive Computing and Communications Workshops (PERCOM Workshops), pp. 141–146. IEEE (2011)
15. Denning, T., Kohno, T., Levy, H.M.: Computer security and the modern home. Commun. ACM **56**(1), 94–103 (2013)
16. Eling, M., Jung, K., Shim, J.: Unraveling heterogeneity in cyber risks using quantile regressions. Insur. Math. Econ. **104**, 222–242 (2022)
17. Fang, X., Xu, M., Xu, S., Zhao, P.: A deep learning framework for predicting cyber attacks rates. EURASIP J. Inf. Secur. **2019**, 5 (2019)
18. Fang, Z., Xu, M., Xu, S., Hu, T.: A framework for predicting data breach risk: leveraging dependence to cope with sparsity. IEEE T-IFS **16**, 2186–2201 (2021)
19. Furman, E., Kye, Y., Su, J.: Computing the Gini index: a note. Econ. Lett. **185**, 108753 (2019)
20. Furman, E., Wang, R., Zitikis, R.: Gini-type measures of risk and variability: Gini shortfall, capital allocations, and heavy-tailed risks. J. Banking Finance **83**, 70–84 (2017)
21. Hardy, M.R.: An introduction to risk measures for actuarial applications. SOA Syllabus Study Note **19** (2006)
22. He, R., Jin, Z., Li, J.S.H.: Modeling and management of cyber risk: a cross-disciplinary review. In: Annals of Actuarial Science, pp. 1–40 (2024)
23. Jacobs, J., Romanosky, S., Edwards, B., Adjerid, I., Roytman, M.: Exploit prediction scoring system (EPSS). Digit. Threat. Res. Pract. **2**(3), 1–17 (2021)
24. Jacobsson, A., Boldt, M., Carlsson, B.: A risk analysis of a smart home automation system. Futur. Gener. Comput. Syst. **56**, 719–733 (2016)
25. Kaas, R., Goovaerts, M., Dhaene, J., Denuit, M.: Modern actuarial risk theory: using R, vol. 128. Springer (2008)
26. Klugman, S.A., Panjer, H.H., Willmot, G.E.: Loss models: from data to decisions. Wiley (2012)
27. Koller, D., Friedman, N.: Probabilistic graphical models: principles and techniques. MIT Press (2009)
28. Kozlov, D., Veijalainen, J., Ali, Y.: Security and privacy threats in IoT architectures. In: Proceedings of the 7th International Conference on Body Area Networks, pp. 256–262 (2012)
29. Lee, C., Zappaterra, L., Choi, K., Choi, H.A.: Securing smart home: Technologies, security challenges, and security requirements. In: 2014 IEEE Conference on Communications and Network Security, pp. 67–72. IEEE (2014)

30. Longtchi, T., Rodriguez, R.M., Al-Shawaf, L., Atyabi, A., Xu, S.: Internet-based social engineering psychology, attacks, and defenses: a survey. Proc. IEEE **112**(3), 210–246 (2024)
31. Longtchi, T., Xu, S.: Characterizing the evolution of psychological factors exploited by malicious emails. In: Proceedings of International Conference on Science of Cyber Security (SciSec'2024) (2024)
32. Longtchi, T., Xu, S.: Characterizing the evolution of psychological tactics and techniques exploited by malicious emails. In: Proceedings of International Conference on Science of Cyber Security (SciSec'2024) (2024)
33. Ma, B., Chu, T., Jin, Z.: Frequency and severity estimation of cyber attacks using spatial clustering analysis. Insur. Math. Econ. **106**, 33–45 (2022)
34. Marikyan, D., Papagiannidis, S., Alamanos, E.: A systematic review of the smart home literature: a user perspective. Technol. Forecast. Soc. Chang. **138**, 139–154 (2019)
35. Montañez, R., Golob, E., Xu, S.: Human cognition through the lens of social engineering cyberattacks. Front. Psychol. **11**, 1755 (2020)
36. Pendleton, M., Garcia-Lebron, R., Cho, J.H., Xu, S.: A survey on systems security metrics. ACM Comput. Surv. **49**(4), 62:1–62:35 (2016)
37. Peng, C., Xu, M., Xu, S., Hu, T.: Modeling multivariate cybersecurity risks. J. Appl. Stat. **0**(0), 1–23 (2018)
38. Poolsappasit, N., Dewri, R., Ray, I.: Dynamic security risk management using bayesian attack graphs. IEEE Trans. Depend. Secure Comput. **9**(1), 61–74 (2011)
39. Sun, H., Xu, M., Zhao, P.: Modeling malicious hacking data breach risks. North Am. Actuarial J. **25**(4), 484–502 (2021)
40. Sun, H., Xu, M., Zhao, P.: A multivariate frequency-severity framework for healthcare data breaches. Ann. Appl. Stat. **17**(1), 240–268 (2023)
41. Tasche, D.: Expected shortfall and beyond. J. Bank. Finance **26**(7), 1519–1533 (2002)
42. Xia, Q., Chen, Q., Xu, S.: Near-ultrasound inaudible trojan (nuit): Exploiting your speaker to attack your microphone. In: Calandrino, J.A., Troncoso, C. (eds.) 32nd USENIX Security Symposium, USENIX Security 2023, Anaheim, CA, USA, 9–11 August 2023. USENIX Association (2023)
43. Xu, M., Schweitzer, K.M., Bateman, R.M., Xu, S.: Modeling and predicting cyber hacking breaches. IEEE T-IFS **13**(11), 2856–2871 (2018)
44. Xu, M., Hua, L.: Cybersecurity insurance: modeling and pricing. North Am. Actuarial J. **23**(2), 220–249 (2019)
45. Xu, S.: The cybersecurity dynamics way of thinking and landscape (invited paper). In: ACM Workshop on Moving Target Defense (2020)
46. Xu, S.: Cybersecurity dynamics. In: Proceedings of the 2014 Symposium and Bootcamp on the Science of Security (HotSoS 2014), p. 14. ACM (2014)
47. Xu, S.: Cybersecurity dynamics: A foundation for the science of cybersecurity. In: Proactive and Dynamic Network Defense, vol. 74, pp. 1–31. Springer (2019)
48. Xu, S.: Sarr: a cybersecurity metrics and quantification framework. In: Third International Conference on Science of Cyber Security (SciSec'2021), pp. 3–17 (2021)

Exhaustive Exploratory Analysis of Low Degree Maximum Period NLFSRs By Graph Analysis

Eric Filiol[1]([✉]) and Pierre Filiol[2]

[1] Thales Digital Factory, Thales Group, Paris, France
eric.filiol@thalesgroup.com
[2] Lab-STICC, ENSTA Bretagne, Brest, France
pierre.filiol@ensta-bretagne.fr

Abstract. NonLinear Feedback Shift Registers (NLFSRs) are key primitives to design pseudorandom generators in modern stream ciphers and random sequence generation especially when the feedback function is of low degree which are of high interest for hardware implementation. Finding a systematic procedure of acceptable complexity for constructing NLFSRs with maximum period is still a general open problem and only a few results have been obtained so far.

In this paper, we present the final results of an exhaustive exploratory search and analysis of NLFSRs of low degree of the form $\sum_{i=0}^{n-1} c_i.x^i + x^n + x_j.x_k$ initiated in [11]. We first eliminate all polynomials producing short cycles very easily with a systematic approach. Then by modelling NLFSRs as graphs and considering the associated formal incidence matrix we express the maximum period property as graph properties. From them, we can generate equations (constraints) to be satisfied. This two-step approach enables to reduce the number of possible candidates greatly. They can then be tested finally for the maximum period property by HPC on GPGPUs and Massively Parallel Processor Array (MPPA). From those results, we have already identified new properties that should help to reduce the initial step further.

Keywords: NLFSR · Stream Cipher · Binary Sequence · Maximum Period · Graph Representation · Incidence Matrix

1 Introduction

Binary sequences produced by *feedback shift registers* (FSRs) are widely used in stream ciphers and random generators. These registers are the key primitives used in these cryptographic systems.

A binary n-stage feedback shift register is defined as a mapping from $\mathbb{F}_2^n$ to $\mathbb{F}_2$

$$(x_{n-1}, \ldots, x_0) \mapsto (f(x_{n-1}, \ldots, x_0), x_{n-1}, \ldots, x_1) \tag{1}$$

This research work was supported by Ops4Sec through the HPC resources it supplied.

where f is a Boolean function, called feedback function, $\mathbb{F}_2$ denotes the binary field and $\mathbb{F}_2^n$ the n-dimensional vector space over $\mathbb{F}_2$ consisting of the n-tuples of elements of $\mathbb{F}_2$. Whenever f is a linear transformation, we deal with a *Linear Feedback Shift Register* (LFSR) otherwise (f is nonlinear) with a *Nonlinear Feedback Shift Register* (NLFSR). In this paper, we focus on NLFSRs defined by a bijective mapping (nonsingular mapping).

Consider a binary sequence $\sigma = (\sigma_i)_{i=0}^{\infty}$. From the n first fixed terms $\sigma_0, \sigma_1, \ldots, \sigma_{n-1}$ (called the initial state vector), we derived the register output sequence uniquely defined by the recurrence relation for all $i > 0$:

$$\sigma_{n+i} = f(\sigma_i, \sigma_{i+1}, \ldots, \sigma_{i+n-1}) \qquad (2)$$

If there exists an integer $p > 0$ such that $\sigma_{i+p} = \sigma_i$, $\forall i \in \mathbb{F}_2^n \setminus \{(0, \ldots, 0)\}$, the sequence is called *periodic* of period p. The most desirable property for NLFSRs (as well as for LFSRs) is to have a maximal possible period length of $2^n - 1$. If we iterate f over $\mathbb{F}_2^n$, we then have two cycles (see Fig. 1): one of length 1 (the loop over the single point $(0, 0, \ldots, 0)$ and a cycle of length $2^n - 1$. In this case, NLFSRs generate maximal length sequences or *m-sequences* [12]. For cryptographic applications, the register initial value is part of the secret key. The maximum period property thus ensures that when clocking the register during the encryption, we come back to the initial value once the register first went through all possible non-null register states exactly once [22]. In other words, for any initial state of maximum period NLFSRs, any possible state during encryption by a NLFSR-based stream cipher has the same uniform probability to appear at any time instant.

From the cryptographic or random number generation perspective, it is strongly desirable that NLFSRs' feedback function fulfils the following conditions [1]:

– The number of feedback function's linear and nonlinear terms should remain as small as possible. It is especially a desirable property for hardware implementation (number of logic gates).
– The algebraic degree of the feedback function should be the lowest possible (at least 2 however).

At last, NLFSR-based stream ciphers are proven to be more secure that LFSR-based stream ciphers since most known cryptanalysis techniques either do not apply or are far more complex [14].

The goal is somehow that NLFSRs are as close as possible to all the main advantages of LFSRs such as low power consumption, easy implementation and high efficiency while providing a better resistance against known attacks [16, page 5].

The paper hence focuses on nonlinear feedback functions whose Algebraic Normal Form (ANF) is given by:

$$f(x_{n-1}, \ldots, x_0) = \sum_{i=0}^{n-1} c_i.x^i + x^n + x_j.x_k \qquad (3)$$

for all possible pairs $\{j, k\}$ such that $0 < j, k < n$. Here c_i are binary coefficients describing whether the i-th register cell is considered ($c_i = 1$) or not ($c_i = 0$) to compute $f(x_{n-1}, \ldots, x_0)$.

We have conducted an exhaustive exploratory analysis to find all feedback functions up to the degree $n = 28$. Thus binary coefficients c_i are considered as unknown variables.

The choice of considering NLFSRs instead of FSR generating de Bruin sequences of length 2^n (one single cycle containing also the null state) lies in the fact that the feedback function is far more complex and very dense. All register cells (or taps) are involved [12, Chap. VI, Theorems 7, 8 and 9] while maximum period sequences of length 2^n can be produced with as few as 2 taps (two non-zero coefficients c_i). Consequently, FSRs generating de Bruijn sequences are not suitable for hardware implementation (the gate complexity is far higher).

To reduce the computing time it has been necessary to find a new way of NLFSR modelling. For that purpose, we represent NLFSR as directed graphs whose incidence matrix exhibits specific properties to express the maximal period property (in this case we have a special instance of Hamiltonian circuit [6, Section 11.5]). It is worth mentioning that the present study can be easily applied to any other forms of feedback polynomials (for instance more quadratic terms).

It is the first exhaustive search to date whereas previous works only published a very few number of feedback polynomials due to the search complexity. In addition, complexity analysis is never neither precisely determined nor analysed. From our exhaustive results obtained, we have identified a number of news results that can be of high interest to explore further for $n > 28$.

In the rest of this article we will consider all operations in the finite field $(\mathbb{F}_2, +, .)$.

The paper is organized as follows. Section 2 analyses the overall complexity of searching of NLFSRs producing m-sequences and presents the related works with respect to this research topic. Section 3 presents first how to strongly reduces the exhaustive search by eliminating polynomials generating sequences with short period. Section 4 introduces our combinatorial model for NLFSRs and formalizes the maximal period property in terms of algebraic equations. Then Sect. 5 details the particular implementation aspects that have been used to perform an effective computation. Finally Sect. 6 presents the detailed results of our exhaustive search and identify a few new interesting properties before concluding in Sect. 7.

2   Preliminaries

2.1   Complexity Analysis of NLFSR Search

To date, there are no theoretical results that allow to easily find maximum period NLFSRs as is the case for LFSRs [12,18]. In this section we look at the approach currently being favoured in recent years. Let us rewrite (3) in a more simple way:

$$f(x_{n-1}, \ldots, x_0) = l(x_{n-1}, \ldots, x_0) + x_j . x_k \qquad (4)$$

where $l(x) = l(x_{n-1}, \ldots, x_0) = \sum_{i=0}^{n-1} c_i . x_i + x^n$ is the linear part of f.

The search for such NLFSRs of length n can be formalized according the two following steps:

1. Among the 2^n possible candidates corresponding to $l(x)$, for a given pair i, j fixing the degree-2 monomial, we retain those which validate a certain number of algebraic or combinatorial properties $I_1, I_2, \ldots, I_k$. If these properties are independent, with respective probability of being realized by a good candidate $P(I_i) = p_i$, then at the end of this stage we retain $N = 2^n \cdot \prod_{i=1}^{k} p_i$. This step has an incompressible complexity of 2^{n-1} (a symmetry property presented in Sect. 2.3 enables to cut search work in half).
2. For each valid candidate for $l(x)$, we check whether it is in the maximum period by calculating the cycle. Complexity is in 2^n in the worst case. The average complexity is 2^{n-1} from [12, Corollary 11, p. 183]. No result is known for this step that would allow us to reduce this complexity (except in certain very specific cases, see [12, Chapter VII]).

The overall worst-case complexity is therefore 2^{2n-1} and the average-case complexity is 2^{2n-2}. To reduce this complexity, the focus must be on the first stage to reduce the number of candidates to be tested in the second stage. Some of these properties are already known (see Sect. 2.3). In this paper we are going to add many others thanks to results from graph theory and matrix calculus on the associated incidence matrix. This significantly reduces the overall complexity of the exhaustive search. This has enabled us to perform an exploratory analysis up to $n = 28$.

2.2   Related and Previous Work

Since Jansen's seminal thesis in 1989 [15], the search for NLFSRs have been conducted according to two main approaches.

The first approach is iterative and constructive by nature. It starts with small NLFSRs and combines them using several techniques to obtain larger registers. In [20], the authors consider the connection polynomial built from the composition of two feedback polynomials $g(x_0, \ldots, x_n)$ and $f(x_0, \ldots, x_n)$ defined by $f * g = g(f(x_0, \ldots, x_n), f(x_1, \ldots, x_{n+1}), \ldots, f(x_m, \ldots, x_{n+m}))$. In [10], a NLFSR of length $k.n$ is produced by combining k NLFSRs of length n. Boolean functions (logic block) are added to the feedback polynomials in order to join composing cycles into a single maximal period cycle.

Szmidt [24] constructs feedback functions of Non-linear Feedback Shift Registers from feedback functions of Linear Feedback Shift Registers using the cross-join pairs method and Zech's logarithms in finite fields.

In [5], either trinomial primitive polynomials or quadratic feedback functions are considered to which monomials of degree $n-1$ are added. Other constructions and variants have been proposed (refer to the bibliography of the four above-mentioned main contributions).

Unfortunately, these polynomials are generally very dense and/or contain a large number of nonlinear terms, or highly nonlinear terms. In the case of hardware implementations, these candidates are less interesting (number of cycles, number of logic gates, etc.).

The second approach relates to more or less sophisticated exhaustive search. In [9], this approach has been initiated with parallel computing.

Later on in [21], the author studies NLFSRs implementation on FPGAs and discusses issues of their optimization. Search method of NLFSRs generating m-sequence was given. It was based on a practical synthesis and explores the possibility of NLFSR implementation of FPGA.

In [1], the authors consider a multi-stage hybrid algorithm which uses Graphics Processor Units (GPU) and developed for processing data-parallel throughput computation. They focus and give results for feedback polynomials of the form $l(x) + x_i.x_k + x_j.x_l$ (two quadratic monomials). Later the same authors [2] optimize their search methods by applying particular vector processor instructions. Their aim was to maximize the advantage of Single Instruction Multiple Data (SIMD) and Single Instruction Multiple Threads (SIMT) execution patterns. Their results are only partial and contains errors (especially regarding the number of sparse feedback polynomials, see Sect. 6).

Finally in 2022, the authors of [21] have extended their results given in [16], providing a bit more feedback polynomials. Unfortunately only a very few number of results are obtained due to the high computing complexity.

As a result, no exhaustive enumeration of maximum period NLFSR polynomials is yet available. Previous authors only give a limited number of solutions. They also did not give any complexity and rate reduction results (*i.e.* the number of candidates kept at the end of the first step). Exploratory exhaustive search is likely to allow identifying unsuspected properties which could help to find a general theory of construction of these NLFSRs as it is the case for their linear counterparts (LFSRs). This is the aim of the present article.

Studies considering the second approach by exhaustive search all try to reduce the first step of search described in Sect. 2 by using a very few number of algebraic properties satisfied by the feedback coefficients c_i. We recall them in Sect. 2.3. However, none of the authors really analysed the overall complexity of the exhaustive search, nor did they specify the rate of reduction that their approach made it possible to achieve on the first step.

2.3   Known Algebraic Properties

Golomb [12] in 1982 identifies a very important property that defines the condition for a shift register cycle to be branchless. In other words, the feedback function is bijective: each x has a unique successor $f(x)$ and any $f(x)$ has a unique predecessor (see Fig. 2).

Theorem 1. *[12]   The cycles generated by a feedback shift register have no branch points if and only if its feedback function can be decomposed as*

$$f(x_0, \ldots, x_{n-l}) = x_0 + g(x_l, \ldots, x_{n-1}) \tag{5}$$

It implies that any integer encoding the linear part $l(x)$ must be an odd value (monomial x_0 is always present). Later on Chan, Game and Rushanan [7] identified three more generic algebraic properties.

Theorem 2. *[7] Let $n_{L(g)}$ and $n_{NL(g)}$ denote respectively the number of linear and non-linear terms in g. If $x_0 + g(x_l, \dots, x_{n-1})$ generates a m-sequence, then*

$$n_{L(g)} + n_{NL(g)} \equiv 1 \mod 2. \tag{6}$$

Theorem 2 implies that the Hamming weight of the integer encoding $l(x)$ (the number of 1 in its binary representation) is even (we add the bit corresponding to monomial x_0 so $c_0 = 1$).

Theorem 3. *[7] If $x_0 + g(x_l, \dots, x_{n-1})$ generates a quadratic m-sequence, then $n_{L(g)} \neq O$, equivalently $g(x_l, \dots, x_{n-1})$ must contain some linear term.*

It implies that the Hamming weight of the integer encoding $l(x)$ must be at least equal to 2.

Theorem 4. *[7] If $x_0 + g(x_l, \dots, x_{n-1})$ generates a quadratic m-sequence, then $x_0 + g(x_{n-1}, \dots, x_1)$ generates a quadratic m-sequence as well.*

This Theorem enables to divide the search over half the pairs $\{i, j\}$ defining the degree 2 monomial in Eq. 4. Each time we have a solution, we generate the *conjugate* solution for replacing all indices i by $n - i$ in Eq. 4.

It is worth noting that the two first algebraic properties (Theorems 1 and 2) are statistically independent and as such we can eliminate 0.75 of candidates in Step 1 of the exhaustive search. In the next section, we present how to have more statistically independent such equations for a larger reduction.

3 Systematic and Exhaustive Elimination of Short Cycles

3.1 Short Cycles Theory

Another strong result by Golomb [12, Section VI.2.1, Theorem 13] asserts that the distribution of cycle lengths, for fixed inital state and variable feedback polynomials is a flat distribution for all lengths between 1 and $2^n - 1$. Consequently, we are just as likely to generate short cycles as long ones from any arbitrary initial state.

Our approach therefore consists of eliminating the maximum number of polynomials generating all short cycles of size at most λ. This latter parameter is chosen so as to find a good trade-off between low computational complexity and rapid elimination of the largest number of candidate polynomials for a given pair $\{j, k\}$.

For NLFSR of length n, Golomb [12, Section VII.2.2] provided a strict upper bound for the total number $z(n)$ of short cycles of length at most n:

$$z(n) < Z(n) = \frac{1}{n} \cdot \sum_{d|n} \mu(d) 2^{\frac{n}{d}}$$

where $\mu(.)$ is the Möbius function [19, Chap. 1].

Golomb [12, pp. 126–127 & Section VII.2.2] mentioned algebraic properties to quickly identify feedback polynomials producing short cycles. But the main issue lies in the complexity of establishing them:

- either we have to perform formal computing on the Algebraic Normal Form,
- or we must analyze the truth table of the feedback polynomial which has a complexity of 2^{n-1}.

In both cases, the algebraic properties is bound to be too complex to use and to evaluate for practical application. Incidentally, Golomb only gives formulae for cycle lengths 1 and 2 (cycle 1010101...). For instance, if we rewrite Eq. 3 as follows

$$f(x_{n-1},\ldots,x_0) = f_1(x_{n-1},\ldots,x_1) + x_0 \tag{7}$$

then the cycle $10101010\ldots$ occurs if $f_1(x_{n-1},\ldots,x_1) = f_1(x'_{n-1},\ldots,x'_1) \equiv n$ mod 2 where x' denotes the binary complementation.

3.2 Exhaustive Exploration of Short Cycles

The main issue in identifying short cycles lies in the fact that some cycles may be equivalent with respect to the monomial $x_j.x_k$ in Eq. 4. We have identified strong dependencies depending on the register length and indices $\{j,k\}$. In order to analyse these dependencies exhaustively and exploit them efficiently we designed Algorithm 1.

Algorithm 1. Exhaustive Exploration of Cycles of Length t

1: **procedure** SHCY(P, n, $f(x_{n-1},\ldots,x_0)$))   ▷ Pattern P of size t, NLFSR size,
      feedback polynomial f
2:     **for** all patterns P of length t **do**
3:          Init $= PPPPP...$ ▷ repeated pattern P up to length n
4:          reg $=$ Init   ▷ Initialize NLFSR with Init
5:          Clock NLFSR t times
6:          **if** reg $==$ Init **then** Output 1   ▷ P is a short cycle
7:          **else** Output 0   ▷ P is not a short cycle
8:          **end if**
9:     **end for**
10: **end procedure**

Algorithm 1 is applied to all linear part $l(x)$ in Eq. 4. From the results we can identify patterns that produce short cycles. For instance, for $n = 27$ and all pairs $\{j,k\}$ such that $|k-j| = 3q$ then only pattern 110 of size 3 produces a short cycle (in fact other pattern values are equivalent or do not produce short cycles at all). Whenever $|k-j| \neq 3q$ only pattern 001 produces a short cycle.

Due to lack of spaces we cannot give all patterns and relevant properties here but they will be made public very soon.

Each valid pattern (outputting a short cycle AND being not equivalent to previous patterns) provides an independent equation which eliminates linear part candidates $l(x)$ which probability $\frac{1}{2^t}$ where t is the pattern size. We have applied this procedure up to $\lambda = 7$. In average we reduce the number of feedback polynomials by a factor of 10^3. To go further in candidates elimination in the first step, we have now to consider a different approach based on graphs and on their associated incidence matrix.

4   NLFSR and Graph Incidence Matrix

Modelling NLFSRs using graph incidence matrices was first mentioned by Gonzalo, Ferrero and Soriano [13] in a rather imprecise and succinct manner. No results were given. No analysis of the independence of potential equations was presented. The computational and algorithmic aspects were not discussed, even though they are fundamental as soon as $n > 10$. Indeed, the size of the matrix grows exponentially with n, which limits their approach to small values of n. Our work, initiated in [11], is based on their approach, which we have effectively implemented and optimised.

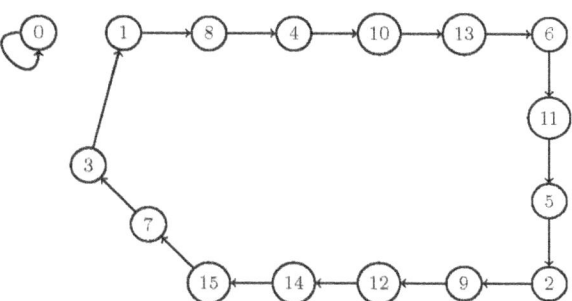

Fig. 1. Graph for $0, 1, 2, (1, 2)$.

4.1   Combinatorial Model for NLFSR

Let us consider a NLFSR of size n whose feedback polynomial has the general form $f(x) = x^n + \sum_{i=0}^{n-1} c_i.x_i + x_j.x_k$ for $0 < j, k < n$. For each possible pair $\{j, k\}$ we search for all n-uples $(c_0, \ldots, c_{n-1})$ for which the NLFSR fulfils the maximum period property. We can model any NLFSR by a directed graph of 2^n points. Any maximal period NLFSR more precisely is a $(2^n, 2^n)$-graph with two cycles: one cycle of length 1 (loop on the null point) and one cycle of length 2^{n-1}.

$$A = \begin{pmatrix} 0 & 0 & 0 & c_0 & 0 & 0 & 0 \\ 1+c_1 & 0 & 0 & 0 & c_1 & 0 & 0 \\ 1+c_0+c_1 & 0 & 0 & 0 & c_0+c_1 & 0 & 0 \\ 0 & 1+c_2 & 0 & 0 & 0 & c_2 & 0 \\ 0 & c_0+c_2+1 & 0 & 0 & 0 & c_0+c_2 & 0 \\ 0 & 0 & c_1+c_2 & 0 & 0 & 0 & 1+c_1+c_2 \\ 0 & 0 & c_0+c_1+c_2 & 0 & 0 & 0 & 0 \end{pmatrix} \tag{8}$$

To illustrate our approach, let us consider the feedback function $x_4 + x_2 + x_1 + x_0 + x_1.x_2$ denoted for short as $0, 1, 2, (1, 2)$. Figure 1 describes the two corresponding cycles.

To simplify notations for $i = (i_{n-1}, \ldots, i_1, i_0) \in \mathbb{F}_2^n$ let us note $i_0 = (0, i_{n-1}, \ldots, i_1)$ and $i_1 = (1, i_{n-1}, \ldots, i_1)$ the right-shifted versions i whose most significant bit is updated respectively with 0 and 1. This notation describes the state of an NLFSR, which changes from state i to state i_0 or i_1 depending on the feedback value $f(i)$ of the NLFSR calculated on state i.

Let us consider its incidence matrix whose entries are expressed as linear polynomials in the unknown c_i as follows: $\forall i = (i_{n-1}, i_{n-2}, \ldots, i_1, i_0) \in \mathbb{F}_2^n$ then $f(i)$ equals 0 or 1 and hence produce either i_0 or i_1 depending on the values of coefficients c_i. We can then define the formal incidence matrix $A = [a_{i,j}]$ where i and j are in $\mathbb{F}_2^n$

$$\forall i \in \mathbb{F}_2^n \setminus \{(0, 0, \ldots, 0)\} \begin{cases} a_{i,i_0} = 1 + f(i) \\ a_{i,i_1} = f(i) \end{cases} \tag{9}$$

For instance let us consider the formal feedback polynomials $f(x) = x_4 + c_3.x_3 + c_2.x_2 + c_1.x_1 + c_0 + x_1.x_2$ and $i = (0, 1, 1, 1)$. We have $a_{i,i_1} = c_2 + c_1$ and $a_{i,i_0} = 1 + c_2 + c_1$. If we consider the formal feedback polynomial $f'(x) = x_3 + c_2.x_2 + c_1.x_1 + x_0 + x_1.x_2$, the corresponding formal incidence matrix of order $2^n - 1 = 7$ is given in (8).

It is important to note that for $i = (0, 0, 1)$ and $i' = (1, 1, 1)$ have only one possible successor if we want the cycle to have maximal period (respectively i_1 and i'_1. Recall also from Theorem 1 that $c_0 = 1$. We can observe that the sum of matrix entries linewise is always equal to 1 as well as columnwise (in fact $c_0 = 1$).

4.2   Formalisation of Max-Period Property

The Max-Period property can be expressed in different ways (see Fig. 1).

If we discard the null loop, the graph is connected (one single cycle). Computing the graph spectrum from its Laplacian matrix L (if A is the adjacency matrix then $L = D - A$ where D is the diagonal matrix whose (i, i)-entry is the degree d_i of the i-th vertex; in our case $d_i = 2$ for all i) enables to check the graph connectivity. If the second smallest eigenvalue λ is equal to 0 whenever the graph is not connected [8, Section 1.1]. Unfortunately, computing the whole graph spectrum has intractable complexity for the matrix size order we deal with. Many properties of graphs can also expressed in terms of graph spectrum properties [3, 8].

It is a well known result that each entry $a_{i,j}^k$ of a power matrix A^k describes the number of paths of length l between i and j [4]. If we want a NLFSR be in maximal period then there must be no loop ($a_{i,i}^l = 0$) and there must exist a unique path between i and j (refer to Fig. 2) for any value of $l \geq 1$.

Exploiting this formalisation efficiently requires to manage two issues:

– A complexity issue. If almost all matrix computation have polynomial complexity, the actual complexity is exponential since the size of the matrix is in $\mathcal{O}(2^n)$. We then need to find ways of managing this explosion in complexity.
– The different equations have to be statistically independent in order to minimize their number. Optimally n statistically independent equations should drastically reduce the number of suitable candidates at the end of the first step of the exhaustive search.

To obtain several independent algebraic equations while limiting the computing effort, we calculated the successive powers $A^8 \ldots, A^{2^l}$ of the incidence matrix A (for $l < 8$ we already have eliminated all cycles of shorter sizes as described in Sect. 3). The results confirm that this approach does indeed yield statistically independent equations. The probability for a candidate to satisfy all of diagonal equations is indeed $\frac{1}{2^{l-2}}$. By restricting ourselves to $l \leq \frac{n}{4}$, the computational effort remains moderate while guaranteeing a very significant reduction in the first stage of the exhaustive search described in Sect. 2. It is worth noting that the statistical independence of the diagonal equations obtained is compliant with the fact that the cycle-length distribution is flat [12, Section 2.2]. Another point that is worth mentioning relates to the other nonnull entries that could also provide equations (see [4, Proposition 1.3.1]):

– for each such entry $a_{i,j}^l$ with $i \neq j$, the value must be equal to 1 (there is a unique walk from i to j),
– $tr(A^2) = 2^{n+1} - 2$.

We have evaluated the statistical independence of this additional equations and we have observed that they are always satisfied if the diagonal equation is. So they do not bring any new bit of information. It means that each squaring iteration provides only one significant equation.

5   Computational Approach

Modelling the maximum-period NLFSR search problem using graphs means that their incidence matrices are square matrices of size $2^n - 1$. As far as naive matrix multiplication is concerned, the computation of the diagonal equations is of cubic complexity but the size of the data is of exponential complexity in 2^{n-1}. It is therefore not possible to compute the matrix products directly (naive product of matrices) as soon as $n > 18$.

However these matrices are extremely sparse. The initial matrix defined in Eq. 8 has $2^{n+1} - 4$ entries out of $2^{2n} - 2^{n+1} + 1$ possible entries. The matrix sparsity is asymptotically defined by $\mathcal{S}(A) = \frac{1}{2^{n-1}}$ when $n \to \infty$.

We used a specific compact matrix representation. There are several possible forms of representation and we have opted for the form described in Table 1. The matrix entries are not integer or real values, but formal polynomials whose maximum number of monomials is 2^{2^n}. This maximum number is never reached in practice (for small values of l), which means that effective calculations can be carried out with limited memory requirements. This form consists of storing the (i, j) coordinates of the only non-zero inputs, together with their polynomials. We give this representation for the matrix (8) and in its general expression in Table 1.

Table 1. Compact Representation of (8) (left) - General Case (right)

1	4	c_0
2	1	$1 + c_1$
2	5	c_1
3	1	$1 + c_0 + c_1$
3	5	$c_0 + c_1$
4	2	$1 + c_2$
4	6	c_2
5	2	$c_0 + c_2 + 1$
5	6	$c_0 + c_2$
6	3	$c_1 + c_2$
6	7	$1 + c_1 + c_2$
7	3	$c_0 + c_1 + c_2$

...	...	...
i	i_0	...
i	i_1	...
$i + 1$	$(i + 1)_0$	...
$i + 1$	$(i + 1)_1$	...
...	...	...

This form of matrix product computation is fast, efficient and optimizes memory resources. The algorithm is described in Table 2. Note that for clarity of presentation, the form described in Table 1 has been implemented a bit differently so as to allow direct access to the various items without having to use local search in the table (thus eliminating tests). It is worth noting a few important points:

– Since the number of entries in Step 5 is constant, the overall complexity is that of main loop (Step 3). Hence the overall complexity is in $\mathcal{O}(N)$. However, for the initial matrix $N = 2^n - 1$. The number of non-zero entries of A roughly doubles with each squaring. So after l squaring, the number of non-zero entries is $2^l.N$.

Algorithm 2. Fast Large Sparse Matrix Square (LSMS)

1: **procedure** LSMS(A, N, n)   ▷ Matrix A to square, N size of the matrix
2:    Allocate result table *MatRes* of size 2.N
3:    **for** x from 1 to N **do**   ▷ For each nonzero entry of A
4:       $i = A[x][1]; j = A[x][2]; p_x = A[i][3];$ ▷ Coordinates (i, j) and corresponding polynomial
5:       **for** all y such that $A[y][1] == j$ **do**   ▷ Nonzero entries of the form (j, k)
6:          $p_y = A[j][y]$
7:          $MatRes[i][A[y][2]]+ = p_x * p_y$
8:       **end for**
9:    **end for**
10:   Return MatRes
11: **end procedure**

– At the end of the squaring procedure, result matrix *MatRes* is already ordered according to matrix line indices i. There is consequently no need of an additional sorting step.

6   Results and Discussion

6.1   Results

We have applied our method to search exhaustively all NLFSRs with feedback polynomials of the form given by (3). Until now we have completed this search up to $n = 27$ (for $n = 28$ search is under way). For small value of n ($n \leq 20$), we also performed a naive exhaustive search in order to validate our algebraic approach by comparing both results obtained.

This research work required four months of computing on an AMD Ryzen 32/64-core Linux machine with 256 Mb of RAM. We did not parallelize Algorithm 2, preferring to run threads on the different pairs $\{j, k\}$ for the monomial $x_j . x_k$.

Table 2 presents the definitive results for $15 \leq n \leq 27$. The (j, k) rate describes the proportion of pairs $\{j, k\}$ for which at least one solution has been found. The minimal weight of $l(x)$ is of high importance since it provides the most simple form for hardware implementation while maintaining excellent nonlinearity for cryptographic designs.

6.2   Analysis of Results

This exhaustive research enabled us first to compare our results with the few published ones. While the solutions we found systematically included and confirmed the few solutions already published, we have been able to disprove certain results. For instance Augustynowicz and K. Kanciak [2, p. 21, Table VII] claimed that no feedback polynomials of weight 7 does exist for $n \geq 26$ while we have found several ones. For instance

Table 2. Results of Exhaustive Search

n	Number of polynomials	(j, k) rate	Minimal weight of $l(x)$
15	204	0.7821	3
16	250	0.9428	3
17	302	0.9083	5
18	332	0.8235	5
19	404	0.8627	3
20	436	0.8596	5
21	554	0.8947	5
22	524	0.8666	5
23	568	0.8268	5
24	616	0.8650	5
25	756	0.8731	7
26	764	0.8933	7
27	782	0.8615	7
28	> 165	> 0.3781	≤ 7

- $n = 26$ $x_0 + x_1 + x_3 + x_5 + x_{10} + x_{14} + x_{18} + x_{16}.x_{17}$.
- $n = 27$ $x_0 + x_1 + x_2 + x_3 + x_7 + x_{10} + x_{18} + x_6.x_{14}$.
- $n = 28$ $x_0 + x_1 + x_2 + x_5 + x_{10} + x_{11} + x_{20} + x_9.x_{10}$.

A number of other of their results are also incomplete or wrong.

We have also initiated an in-depth analysis of the results to identify some interesting properties. For example, the number of solutions varies significantly according to the respective parities of the indices j and k of the monomial $x_j.x_k$. Thus, there are significantly more solutions when the parities are equal ($|k - j|$ even) than when they are different ($|k - j|$ odd).

Finally we can formulate the following conjecture concerning the number of polynomials as a function of n.

Conjecture 1. The number of feedback polynomials of the form (3), denoted $\#(n : 2)$, can be bounded by the following interval

$$\frac{4}{5}.(n^2 - 1) \leq \#(n : 2) \leq \frac{12}{5}.(n^2 - 1) \tag{10}$$

It is therefore in $\mathcal{O}(n^2)$.

This Conjecture is illustrated in Fig. 2.

6.3   Environmental Impact and Interest

We are aware that our research culminated in a four-month process of compute-time. This is an expensive process financially (electricity consumption) and envi-

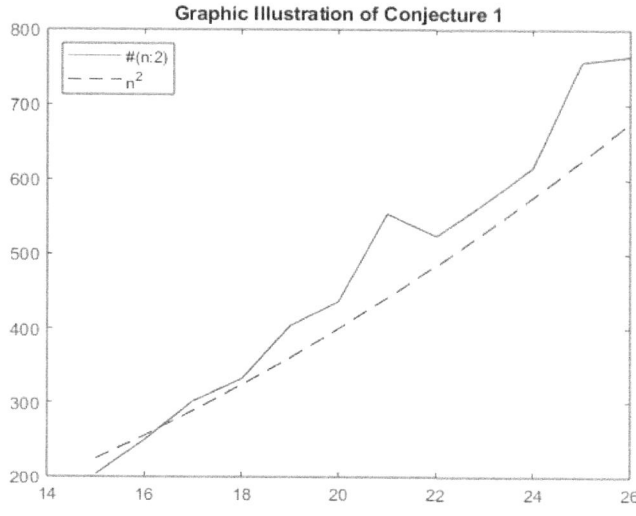

Fig. 2. Graphic Illustration of Conjecture 1

ronmentally. We tried to mitigate the environmental impact by choosing low consumption hardware (and optimized cooling systems like water cooling). Unfortunately, this is the price to pay when the only possible exploratory approach is intensive computing in the absence of general theoretical results.

Now, cryptography is a science that is carried out by numerous organisations around the world, often secret and often all doing the same research in parallel. This has a more considerable ecological impact. So by producing our results and making them public, we believe that all these calculations will no longer be necessary. What is more, having polynomials with low weight even for small lengths ($n < 29$) makes it possible to construct longer polynomials of acceptable complexity.

From all these results, we expect now to find theoretical properties and models that would cancel the need for extensive further computation.

7   Conclusion

In this paper we have presented how short cycles analysis and combinatorial modelling of an NLFSR via the incidence matrix of the state graph can help to significantly reduce the computational effort in the exhaustive search for feedback polynomials. This enabled us to carry out this search up to $n \leq 28$ for a minimal form that is very important in the design of encryption algorithms. Previous authors never performed such an exploratory and exhaustive search and only give computation time for the few instances they produced. Consequently we cannot compare our complexity results to any existing ones. We estimate that our approach is likely to reduce by an overall factor of at least 10^6. Future work should evaluate this reduction in a more precise way.

This exhaustive search will be carried out for $n > 28$. However, memory requirements quite soon exceeds current capacities ($n = 31$ requires a machine with 1Tb RAM). We are therefore considering two possible ways of exploiting our combinatorial approach:

- Emulate RAM with disk space and use MapReduce-type functions on Distributed File Systems (DFS) [17, Chap. 2]. The computation time will be longer, but this is of relative importance for an exhaustive one-time search (once and for all search).
- Use the l equations produced as a system of non-linear equations on $\mathbb{F}_2$ and solve them using Gröbner [23] bases. This will enable to generate valid candidates for step 2 directly without going through all possibles values.

The detailed results (polynomial lists, pattern properties) will be made public on the first author's webpage after the conference.

Acknowledgments. We would like to thanks reviewers for their useful comments and especially reviewer #1 for having suggested the content of Sect. 6.3.

References

1. Augustynowicz, P., Kanciak, K.: Scalable method of searching for full-period nonlinear feedback shift registers with GPGPU: new list of maximum period NLFSRs. Int. J. Electron. Telecommun. **64**, 167–71 (2018). https://doi.org/10.24425/119365
2. Augustynowicz, P., Kanciak, K.: The search of square m-sequences with maximum period via GPU and CPU. Infocommun. J. **11**, 17–22 (2019). https://doi.org/10.36244/ICJ.2019.4.3
3. Biggs, N.: Algebraic Graph Theory. 2nd ed., Cambridge University Press (1996)
4. Brouwer, A.E., Haemers, W.H.:. Spectra of Graphs. Universitext Serie, Springer Verlag (2012). https://doi.org/10.1007/978-1-4614-1939-6
5. Calik, C., Sonmez Turan, M., and Ozbudak, F.: On feedback functions of maximum length nonlinear feedback shift registers. In: Proceedings of the IEICE Transactions on Fundamentals of Electronics Communications and Computer Sciences, vol. E93.A, pp. 1226–1231 (2010). https://tsapps.nist.gov/publication/get_pdf.cfm?pub_id=905987, Accessed 2 Mar 2024
6. Cameron, P.J.: Combinatorics: Topics, Techniques. Algorithms, Cambridge University Press (1994)
7. Chan, A.H., Games, R.A., Rushanan, J.J.: On quadratic m-sequences. In: Proceedings of 1994 IEEE International Symposium on Information Theory, Trondheim, Norway, pp. 364–371 (1994). https://doi.org/10.1007/3-540-58108-1_20
8. Cvetkovic, D., Rowlinson, P., Simic, S.: Eigenspaces of graphs. Encyclopedia of Mathematics and Its Applications, vol. 66, Cambridge Universty Press (1997). https://doi.org/10.1017/CBO9781139086547
9. Dabrowski, P., Labuzek, G., Rachwalik, T., Szmidt, J.: Searching for nonlinear feedback shift registers with parallel computing. Inf. Process. Lett. **14**, 268–272 (2014)

10. Dubrova, E.: Generation of full cycles by a composition of NLFSRs. Des. Codes Cryptogr. **73**, 469–486 (2014). https://doi.org/10.1007/s10623-014-9947-3
11. Filiol, E., Filiol, P.: Graph-based modelling of maximum period property for nonlinear feedback shift registers. In: Secrypt 2024 Conference, Poster session, Sabrina De Capitani Di Vimercati and Pierangela Samarati (eds), SCITEPRES (2024). https://doi.org/10.5220/0000178800003767
12. Golomb, S.W.: Shift Register Sequences. Aegean Park Press, Laguna Hills, CA (1982)
13. Gonzalo, R.P., Ferrero, D., Soriano, M.: Nonlinear feedback shift registers with maximum period. In: Proceedings of the Sixth International Conference on Telecommunications Systems, Nashville, USA, pp. 13–16 (1998)
14. Hope4Sec Crypto Lab: Non-linear Shift Register-based Bent Combiner Cryptanalysis Results. Technical Report (2023). https://hope4sec.eu/Repository/Articles/h4scipherbent6.pdf, Accessed 29 May 2024
15. Jansen, C. J. A.: Investigations on nonlinear stream cipher systems: construction and evaluation methods. PhD. Thesis, Technical University of Delft (1989)
16. Kuznetsov, A.A., Potii, O.V., Poluyanenko, N.A., Gorbenko, Y.I., Kryvinska, N.: Stream ciphers in modern real-time it systems - analysis, design and comparative studies. Stud. Syst. Decis. Control **375** (2022). https://doi.org/10.1007/978-3-030-79770-6
17. Leskovec J., Rajaraman, A., Ullman, J.D.: Mining of massive datasets, 3rd ed. Cambridge University Press, Cambridge, United Kingdom (2020). https://doi.org/10.1017/CBO9781139924801
18. Lidl, R., Niederreiter, H.: Introduction to Finite Fields and Their Applications, 2nd ed. Cambridge University Press (1994). https://doi.org/10.1017/CBO9781139172769
19. McCarthy, P.J.: Introduction to Arithmetical Functions. Springer-Verlag (1986). https://doi.org/10.1007/978-1-4613-8620-9
20. Mykkeltveit, J., Siu, M.-K., Tong, P.: On the cycle structure of some nonlinear shift register sequences. Inf. Control **43**, 202–215 (1979). https://doi.org/10.1016/S0019-9958(79)90708-3
21. Poluyanenko N.: Development of the search method for nonlinear shift registers using hardware implemented on FPGA. EUREKA: Phys. Eng. **1**, 53–60 (2017). https://doi.org/10.21303/2461-4262.2017.00271
22. Rueppel, R.A.: Analysis and design of stream ciphers. Springer-Verlag (1986). https://doi.org/10.1007/978-3-642-82865-2
23. Sala, M.T., Mora, T., Perret, L., Sakata, S., Traverso, C.: Gröbner Bases, Coding, and Cryptography. Springer, Berlin, Germany (2009). https://doi.org/10.1007/978-3-540-93806-4
24. Szmidt, J.: Nonlinear feedback shift registers and Zech's logarithms. In: Proceeding of the 2019 International Conference on Military Communications and Information Systems (ICMCIS), Budva, Montenegro, pp. 1–4 (2019). https://doi.org/10.1109/ICMCIS.2019.8842713

Integrating Consortium Blockchain and Attribute-Based Searchable Encryption for Automotive Threat Intelligence Sharing Model

Tiange Xie[1,2], Feng Liu[1,2], Jiechao Gao[3], and Yinghui Wang[4(✉)]

[1] Institute of Information Engineering, Chinese Academy of Sciences, Beijing, China
{xietiange,liufeng}@iie.ac.cn
[2] School of Cyber Security, University of Chinese Academy of Sciences, Beijing, China
[3] Department of Computer Science, University of Virginia, Charlottesville, VA, USA
jg5ycn@virginia.edu
[4] School of Transportation Science and Engineering, Beihang University, Beijing, China
wangyinghui@buaa.edu.cn

Abstract. Sharing cyber threat intelligence (CTI) enhances organizations' abilities in threat detection and emergency response, fostering a proactive defense approach of "predict and prevent". Given the sensitive nature of automotive CTI and trust issues inherent in its sharing process, this paper presents an automotive CTI security sharing model that integrates blockchain and attribute-based searchable encryption. Leveraging consortium blockchain for CTI sharing, our model addresses the single point of failure and distrust in centralized systems through blockchain's decentralization and immutability. Furthermore, by combining attribute-based searchable encryption algorithms with smart contracts, we achieve fine-grained access control and ciphertext retrieval for automotive CTI data. This enables data users to independently search CTI on the blockchain, mitigating the risk of sensitive information disclosure. Additionally, we store encrypted CTI data off-chain in the Inter-Planetary File System (IPFS) to alleviate blockchain's storage burden. Finally, we develop an automotive CTI sharing prototype system to demonstrate the feasibility and effectiveness of our proposed model.

Keywords: Automotive CTI · Blockchain · Fine-Grained Access Control · Ciphertext Retrieval

1 Introduction

As vehicles become more intelligent and connected, cyber security incidents such as remote attacks and malicious control caused by cyber security issues are also increasing. Some original equipment manufacturers (OEMs) follow vulnerability management practices from the IT industry, responsibly handling and disclosing cyber security vulnerabilities in the automotive domain [5]. However, the automotive industry is a highly complex ecosystem spanning numerous fields, with a deep and massive supply system and a high degree of globalization in the industrial chain. Most of the heterogeneous and proprietary components provided by suppliers to OEMs have non-public sources, making it difficult for OEMs to obtain their complete source code and in-depth details. Also, most automotive components can only be analyzed as black boxes in terms of cyber security, unable to obtain detailed knowledge about cyber security vulnerabilities. A single organization can only obtain partial threat information, unable to defend against cyber attacks effectively. Therefore, sharing CTI between OEMs and their suppliers is of great significance for ensuring vehicle cybersecurity.

CTI sharing is an important step in building a cybersecurity defense system centered on threat intelligence [17]. It helps alleviate the automotive CTI information island problem and improve stakeholders' threat detection and emergency response capabilities. Existing automotive CTI sharing is based on centralized third-party data platforms, with unequal data and low sharing efficiency among participants [9]. More importantly, automotive CTI is highly sensitive, and due to industry competition, intellectual property rights and other factors, the lack of trust among stakeholders such as OEMs and suppliers is evident [18]. Furthermore, suppose CTI data is mishandled or leaked. In that case, the sharing organization may be subject to further attacks by attackers, resulting in economic or reputational losses [7,8,12,25,26,29,31], seriously hindering the willingness of stakeholders to participate in CTI sharing.

To address the above application issues and enable trusted and secure sharing of automotive CTI among stakeholders in the dynamic supply chain environment, this paper combines CTI sharing with blockchain and attribute-based searchable encryption technologies to ensure open and trusted CTI data sharing among users. It achieves one-to-many cyber security sharing and fine-grained access control within the automotive industry chain, thereby improving the credibility of automotive CTI sharing and effectively solving the issue of privacy and secret leakage in the CTI sharing process with the following main contributions:

(1) We combined the consortium blockchain with CP-ABKS and CP-ABE algorithms and achieved fine-grained access control and ciphertext retrieval for automotive CTI sharing in one-to-many scenarios. We also take the advantage of AES symmetric encryption, hash functions, and the IPFS system to ensure confidentiality and integrity throughout the sharing process, and the "off-chain" storage of IPFS alleviates the storage burden on the blockchain.

(2) Based on the consortium blockchain, two smart contracts are designed-a storage contract and a query contract-with the attribute-based searchable encryption algorithm deployed in the query contract, achieving trusted and secure sharing of automotive CTI without interaction between data owners and users.

(3) The CP-ABKS attribute-based searchable encryption algorithm's dependencies on the M4, bison, flex, GMP, and PBC cryptographic libraries are

compiled into the consortium blockchain docker, establishing a trusted and secure prototype system for automotive CTI sharing, verifying the feasibility and effectiveness of the model.

2 Related Work

Facing the complexity and diversity of cyber threats, scholars have been continuously exploring models and mechanisms capable of enabling CTI sharing [11,15,32,39]. Existing CTI sharing schemes are largely based on centralized management entities. However, excessive data centralization raises cyber security concerns and potential single points of failure, risking threat intelligence data leakage [9]. More importantly, CTI is inherently highly sensitive, and the automotive industry supply chain is profoundly complex, leading to intrinsic trust barriers in the intelligence sharing process [16]. These circumstances impede the willingness of stakeholders to share, diminishing the efficiency and security of CTI sharing. Trust barriers and sensitive information issues are significant factors constraining threat intelligence sharing. Blockchain technology integrates various key technologies such as distributed storage, consensus mechanisms, and smart contracts, exhibiting decentralized and tamper-proof characteristics [19]. It enables trustworthy CTI sharing among organizations without relying on a trusted authority [3,4,24].

For example, Badsha et al. [2] proposed a privacy-preserving cyeber security information sharing model based on the Ethereum blockchain, utilizing proxy re-encryption and CP-ABE encryption to achieve fine-grained access control for CTI. However, this scheme requires a centralized server and trusted entity to be responsible for storing encrypted data and managing keys, failing to achieve decentralization. Similarly, building on TATIS [21], Preuveneers et al. [22] leveraged blockchain technology and CP-ABE encryption to ensure confidentiality, reliability, and fine-grained access control for CTI sharing. This scheme also guarantees the source of exchanged data, increasing the overall trust in CTI sharing. However, the TATIS system uses the third-party centralized open-source identity and access management platform Keycloak, which may be subject to single points of failure, DoS attacks, and other cyber security vulnerabilities [37].

Searchable encryption (SE) technology can enable users to effectively retrieve encrypted keywords in an untrusted cloud server environment without leaking any information about the data [27]. Several works have focused on keyword searching on the blockchain, but existing schemes largely adopt symmetric encryption techniques, requiring data files and keyword trapdoors to be encrypted using the same key, failing to achieve efficient retrieval and access control, and potentially compromising privacy [6,14,23,30,35]. In other domains, researchers have proposed schemes combining blockchain and attribute-based searchable encryption to achieve fine-grained access control and ciphertext retrieval for data sharing [20,34,38]. For example, Yang et al. [36] proposed an attribute-based keyword search scheme that allows users to search for encrypted files on the blockchain based on attributes without any interaction with the data owner, avoiding potential privacy leakage. Scholars such as Gupta, Xiang, Hussien, and Gao [1, 10, 33] have also proposed similar schemes for data sharing fine-grained access control and searchability, but some schemes struggle to ensure the correctness of search results from the cloud server.

3 Automotive CTI Security Sharing Model

The consortium blockchain confines the maintenance nodes and visibility within the alliance, allowing trusted participants to privately disseminate highly sensitive data through member management services and channel mechanisms. To address the constraints of sensitive information and trust barriers in automotive CTI sharing, we construct a secure sharing model for automotive CTI by integrating consortium blockchain and attribute-based searchable encryption algorithms. This aims to achieve trustworthy automotive CTI sharing with support for fine-grained access control and ciphertext retrieval capabilities. A secure sharing solution for automotive CTI should meet the following security requirements: **Trustworthiness:** The system relied upon for automotive CTI sharing, as well as the sharing organizations themselves, should be trustworthy. **Privacy:** Processes such as storage, search, and transmission in automotive CTI sharing should ensure intelligence data remains undisclosed. Access permissions should be set for intelligence data to prevent unauthorized access. **Reliability:** Automotive CTI sharing system should be able to withstand security threats such as single point of failure and Denial-of-Service attacks. **Integrity:** Automotive CTI should be protected against tampering during storage and transmission processes to ensure integrity of intelligence data.

3.1 CP-ABKS Based Attribute Searchable Encryption Algorithm

This model adopts the CP-ABKS attribute-based searchable encryption scheme from reference [40]. This scheme combines CP-ABE (Conjunctive Policy Attribute-Based Encryption) and SE (Searchable Encryption) algorithms, consisting of five polynomial-time algorithms:

System Initialization (*Setup*): This algorithm is executed by the attribute authority. It takes the security parameter λ as input and outputs the system public parameters pm and the master key mk. It selects a bilinear mapping $e : G \times G \rightarrow G_T$, where G and G_T are cyclic groups of prime order p. It defines a random oracle hash function $H_1 : \{0,1\}^* \rightarrow G$ and a one-way hash function $H_2 : \{0,1\}^* \rightarrow Z_p$. It selects a, b, c from Z_p and g from G, and outputs the system public parameters $pm = (H_1, H_2, e, g, p, g^a, g^b, g^c, G, G_T)$, and the system master key $mk = (a, b, c)$.

Key Generation (*KeyGen*): This algorithm is executed by the attribute authority. It takes the attribute set $Atts$ of the data user and the master key mk as input and outputs the user's private key sk. It selects $r \in Z_p$, calculates $A = g^{((ac-r)/b)}$. For every attribute $at_j \in Atts$, it selects $r_j \in Z_p$, computes $A_j = g^r H_1(at_j)^{r_j}$, and $B_j = g^{r_j}$. The output is $sk = (Atts, A, \{(A_j, B_j)|at_j \in Atts\})$.

Encryption (*Encrypt*): The data owner executes the encryption algorithm. It takes the system public parameters pm, the keyword w of the data, and the access structure T as input, and generates keyword ciphertext I_w. It selects $r_1, r_2 \in Z_p$, calculates $W = g^{(cr_1)}$, $W_0 = g^a(r_1 + r_2)g^{bH_2(w)r_1}$, $W' = g^{br_2}$. For each node in access tree T, it computes secret shares of r_2, $\{q_v(0)|v \in lvs(T)\} \leftarrow Share(T, r_2)$, and for every $v \in lvs(T)$, it calculates $W_v = g^{q_v(0)}$, $D_v = H_1(att(v))^{q_v(0)}$. The output is $I_w = (T, W, W_0, W', \{(W_v, D_v)|v \in lvs(T)\})$.

Trapdoor Generation (*Trapdoor*): The data user executes the keyword trapdoor generation algorithm. It takes the system public parameters pm, the user's private key sk, and the interested keyword ω as input, and outputs the keyword trapdoor T_ω. It selects $s \in Z_p$, calculates $tok_1 = (g^a g^{bH_2(\omega)})^s$, $tok_2 = g^{cs}$, $tok_3 = A^s = g^{((acs-rs)/b)}$. For every $at_j \in Atts$, it computes $A'_j = (A_j)^s$ and $B'_j = (B_j)^s$. The output is $T_\omega = (Atts, tok_1, tok_2, tok_3, \{(A'_j, B'_j)|at_j \in Atts\})$.

Search (*Search*): The search algorithm is executed by nodes on the blockchain. It takes the keyword ciphertext I_w and the keyword trapdoor T_ω as input. If the attribute set $Atts$ of the data user satisfies the access tree T embedded in I_w and w matches ω, the search is successful and returns the stored information to the data user; otherwise, it fails. First, it checks if there exists a subset S of attributes in $Atts$ given by T_ω that satisfies the access tree T specified by the keyword ciphertext I_w. If S does not exist, it returns 0; otherwise, for each $att(v) = at_j \in S$ (when $v \in lvs(T)$, i.e., when v is a leaf node), it computes $E_v = e(A'_j, W_v)/e(B'_j, D_v) = e(g, g)^{rsq_v(0)}$. Combining $(T, \{E_v|att(v) \in S\})$ and $E_{root} = e(g, g)^{rsr_2}$, it computes $e(g, g)^{rsq_{root}(0)}$. If $e(W_0, tok_2) = e(W, tok_1) \cdot E_{root} \cdot e(tok_3, W')$, it returns 1; otherwise, it returns 0.

The encryption algorithm in the CP-ABKS scheme combines the one-way hash function H_2, making it impossible to use this algorithm to encrypt the symmetric key provided to the data user for intelligence data. Therefore, this model uses the classical CP-ABE algorithm proposed by Bethencourt et al. [3] to encrypt the symmetric key used for encrypting intelligence data.

3.2   System Model

The system model consists of five entities: the intelligence data owner, the intelligence data user, the third-party authorization server, the InterPlanetary File System (IPFS), and the blockchain. Under the premise of ensuring the trustworthiness and privacy of intelligence sharing using consortium member management services and channel mechanisms, this paper selects CP-ABKS and CP-ABE algorithms to achieve fine-grained access control and ciphertext retrieval for CTI. The system model is depicted in Fig. 1.

The authorization server is responsible for generating and transmitting the public key (PK) and master key (MK) used in the initialization algorithms of CP-ABKS and CP-ABE, as well as the private key (SK) used in the key generation algorithm. The InterPlanetary File System (IPFS), a distributed file system, eliminates the single point of failure issue. The data owner encrypts CTI data and stores it in IPFS to ensure the confidentiality of intelligence data and alleviate

Table 1. Symbol Explanation

Symbol	Meaning
data	Threat intelligence data
key	Key for AES algorithm
encdata	Ciphertext of threat intelligence data
hashipfs	Content hash value generated by IPFS
hashdata	Hash value of threat intelligence data
enckey	Ciphertext of AES key
deckey	Decryption key (key)
decdata	Decrypted threat intelligence data
dechash	Hash value of decrypted threat intelligence data
StoreCont	Data storage smart contract
QueryCont	Data query smart contract

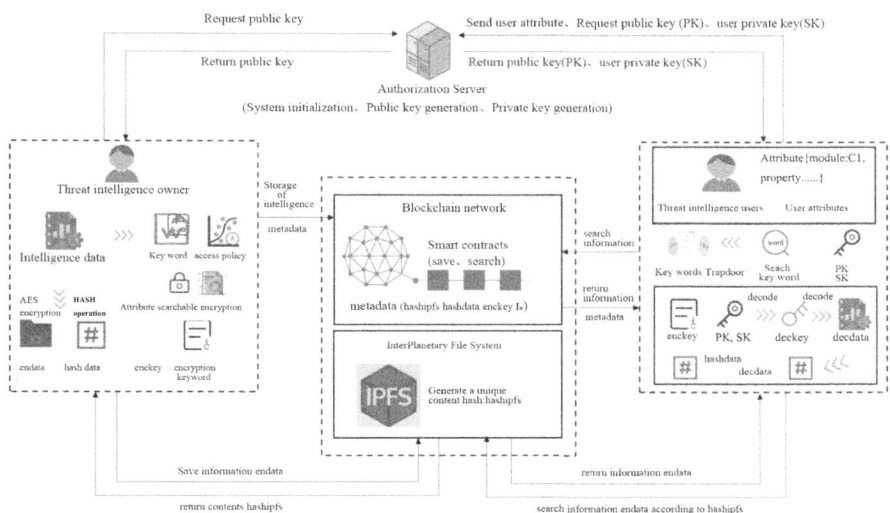

Fig. 1. Automotive CTI Security Sharing Model

the storage burden on the blockchain. The blockchain is used to store the content hash values, keyword ciphertexts, and other information of IPFS. The data owner encrypts keywords and encryption keys using their defined access policies, and then calls the smart contract to upload the content hash values and keyword ciphertexts of IPFS to the blockchain network. The data user generates a search trapdoor using their private key and keywords, submits the trapdoor information to the blockchain network, and invokes the data query smart contract to perform the search operation. If the search is successful, the blockchain nodes return the corresponding IPFS-generated hash values, keyword ciphertexts, and other information to the data user. The data user then invokes CP-ABE algorithm, hash functions, etc., to perform ciphertext decryption, data consistency, and integrity verification, and obtain the required CTI data. Table 1 provides symbols used in the model. The detailed mechanisms for how the CP-ABKS and CP-ABE algorithms integrate and function within the blockchain framework will be introduced in Sect. 3.3.

3.3 Automotive CTI Security Sharing

Automotive CTI security sharing involves two aspects: threat intelligence owners (OEMs, suppliers, testing organizations, etc.) upload encrypted intelligence data and its metadata, storing the intelligence metadata in the consortium blockchain through smart contracts; threat intelligence users (other OEMs, suppliers, etc.) conduct ciphertext retrieval on the consortium blockchain based on their attributes and interested keywords to obtain corresponding threat intelligence data. The specific implementation process is as follows:

Threat Intelligence Data Storage: The process of data storage involves four entities: the intelligence data owner, IPFS, the blockchain, and a third-party authorization server. The specific process of data storage is as follows:

The intelligence data owner calls the AES encryption algorithm to encrypt the intelligence data, obtaining the ciphertext *encdata*. Then, *encdata* is stored in IPFS, and IPFS returns the generated content hash value, *hashipfs*.

$$encdata = AES.Enc(key,data) \tag{1}$$

$$hashipfs = IPFS.Store(encdata) \tag{2}$$

The intelligence data owner calls the SHA256 algorithm to hash the intelligence data, obtaining the data hash value, *hashdata*.

$$hashdata = SHA256.Hash(data) \tag{3}$$

The intelligence data owner requests the public key PM from the authorization server, and the server calls the Setup algorithm to generate and store the public key PM and master key MK, then sends PM to the data owner. The data owner calls the Encrypt encryption algorithm to encrypt the keywords w and the AES encryption key according to the access structure T set by themselves, obtaining the keyword ciphertext I_w and ciphertext *enckey*.

$$PM, MK = CP\text{-}ABE.Setup(\gamma) \tag{4}$$

$$enckey = CP\text{-}ABE.Encrypt(PM, T, key) \tag{5}$$

$$I_w = CP\text{-}ABKS.Encrypt(PM, T, w) \tag{6}$$

The data owner uploads *hashipfs*, *hashdata*, *enckey*, and I_w to the blockchain network. The blockchain network receives the data storage request and triggers the storage smart contract *StoreCont*.

$$\text{StoreCont (hashipfs, hashdata, enckey, } I_w) \tag{7}$$

Threat Intelligence Data Access: The process of accessing threat intelligence data involves four entities: intelligence data user, IPFS, blockchain, and a third-party authorization server. The specific process of data access is as follows:

The intelligence data user sends the attribute set to the authorization server to request the public key PM and private key SK. The server utilizes PM, MK, and the data user's attribute set Atts, and calls the KeyGen key generation algorithm to generate SK_1 and SK_2, then sends PM, SK_1, SK_2 to the intelligence data user.

$$\text{SK}_1 = \text{CP-ABKS.KeyGen(PM, MK, Atts)} \tag{8}$$

$$\text{SK}_2 = \text{CP-ABE.KeyGen(MK, Atts)} \tag{9}$$

The intelligence data user generates a trapdoor T_ω using the public key PM, private key SK_1, and the keyword of interest ω. They then upload the keyword trapdoor T_ω and the public key PM to the blockchain network, where the query smart contract *QueryCont* executes the Search algorithm. If the search is successful, it returns *hashdata*, *hashipfs*, and *enckey* to the intelligence data user.

$$T_\omega = \text{Trapdoor(PM, SK}_1, \omega) \tag{10}$$

$$hashdata, hashipfs, enckey = \text{QueryCont.Search(T}_\omega, I_w) \tag{11}$$

The intelligence data user then calls the IPFS query algorithm using *hashipfs* obtained from the blockchain to retrieve the encrypted intelligence data *encdata* from the IPFS network.

$$encdata = \text{IPFS.Query(hashipfs)} \tag{12}$$

Next, the intelligence data user decrypts the AES key ciphertext *enckey* using the CP-ABE decryption algorithm to obtain the decryption key *deckey*.

$$deckey = \text{CP-ABE.Decrypt(PM, SK}_2, enckey) \tag{13}$$

The intelligence data user decrypts *encdata* using the AES decryption algorithm with the decryption key *deckey* to obtain the decrypted file *decdata*.

$$decdata = \text{AES.Dec(deckey, encdata)} \tag{14}$$

Finally, the intelligence data user computes the hash value *dechash* of *decdata* using the SHA256 algorithm. If *hashdata* and *dechash* are the same, the access is successful.

$$dechash = \text{SHA256.Hash(decdata)} \tag{15}$$

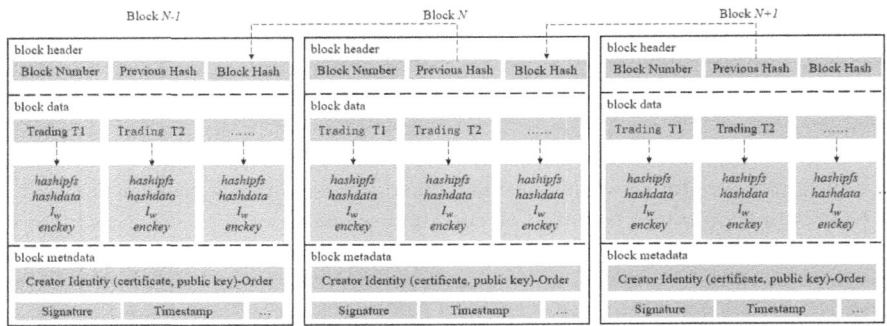

Fig. 2. Block Structure Design

The block structure of the consortium blockchain in the scheme is shown in Fig. 2 The block header contains three fields: block number, hash value of the previous block header, and hash value of the current block. The block data consists of an ordered list of transactions, where each transaction includes transaction header, transaction signature, transaction proposal, and a series of endorsements. Transactions such as T1, T2 store the hash value of automotive CTI data $hashdata$, the hash value generated by IPFS $hashipfs$, the ciphertext of the AES encryption key $enckey$, and the ciphertext of the keyword I_w. The block metadata includes a timestamp, certificate of the block writer, signature, etc.

Algorithm 1. StoreCont

Require: I_w, $hashdata$, $hashipfs$, $enckey$
Ensure: Null
1: **if** $len(args) \neq 4$ **then**
2:     **return** "incorrect arguments"
3: **else**
4:     Set $cti = \text{CTI}(\text{Hashdata} = args[1], \text{Hashipfs} = args[2], \text{Enckey} = args[3])$
5:     $ctiAsBytes, \_ := \text{json.Marshal}(cti)$ {Convert variable to byte form}
6:     Add/update key-value pair in ledger: $err := \text{PutState}(args[0], ctiAsBytes)$
7:     **if** $err \neq$ nil **then**
8:         **return** "failed to upload"
9:     **else**
10:         **return** nil
11:     **end if**
12: **end if**

3.4 Smart Contract Design

The smart contract for secure sharing of automotive CTI consists of two parts: data storage (*StoreCont*) and data querying (*QueryCont*). It integrates the consortium blockchain with the CP-ABKS searchable encryption algorithm to support fine-grained access control and ciphertext retrieval for automotive CTI sharing. Specifically, *StoreCont* is primarily responsible for storing metadata such as *hashdata, hashipfs, enckey,* and I_w of automotive CTI data. *QueryCont* is mainly responsible for implementing the CP-ABKS attribute searchable encryption algorithm. It searches for each ciphertext keyword I_w based on the keyword trapdoor T_ω, checks if the attribute set *Atts* of the data user satisfies the access tree T embedded in I_w, and verifies whether the keyword w set by the data owner matches the keyword of interest ω set by the data user. The specific algorithmic processes of the smart contract are depicted in the pseudocode shown in Algorithm 1 and Algorithm 2.

Algorithm 2. QueryCont

Require: T_w

Ensure: *hashdata, hashipfs, enckey*

 1: **if** len(*args*) $\neq$ 1 **then**

 2:     **return** "incorrect arguments"

 3: **else**

 4:     GetStateByRange(startKey, endKey)

 5:     *queryResultsIterator*.Next() {Return key-value query results}

 6:     **if** cpabks.Check(queryResponse.Key, T_ω) **then**

 7:         **return** 0

 8:     **else**

 9:         Compute $E_v = \frac{e(A'_j, W_v)}{e(B'_j, D_v)} = e(g, g)^{rsq_v(0)}$

10:         Compute $e(g, g)^{rsq_{root}(0)}$

11:         **if** $e(W_0, \text{tok}_2) == e(W, \text{tok}_1) E_{root} e(\text{tok}_3, W')$ **then**

12:             **return** 1

13:             fmt.Println(Value) {Output corresponding value if condition is met}

14:         **else**

15:             **return** 0

16:             **continue**

17:         **end if**

18:     **end if**

19: **end if**

---

4   Analysis of the Solution

4.1   Correctness Analysis

If the attribute set *Atts* of the intelligence data user satisfies the access tree T embedded in I_w and w matches ω, the search succeeds and returns the stored information to the data user; otherwise, it fails. The correctness of the CP-ABKS scheme is as follows:

$$e(W_0, \text{tok}_2) = e(g^a(r_1 + r_2)g^{bH_2(w)r_1}, g^{cs})$$
$$= e(g, g)^{acs(r_1+r_2)}e(g, g)^{bcsH_2(w)r_1}$$

$$e(W, \text{tok}_1)E_{\text{root}}e(\text{tok}_3, W') = e(g^{cr_1}, (g^a g^{bH_2(\omega)})^s)E_{\text{root}}e(g^{\frac{acs-rs}{b}}, g^{br_2})$$
$$= e(g, g)^{acs(r_1+r_2)}e(g, g)^{bcsH_2(w)r_1}$$

When the threat data user obtains the symmetric key ciphertext enckey, they use their own private key SK_2 to check if it satisfies the access structure. If the private key satisfies the access structure, then the symmetric key deckey can be recovered. The correctness of the CP-ABE scheme is as follows, where the specific meanings of the symbols can be found in the reference [3]:

$$\text{deckey} = \frac{C}{(e(C, D)/A)} = \left(\frac{Ce(g, g)^{rs}}{e(h^s, g^{((\alpha+\gamma)/\beta)})}\right) = \text{enckey}$$

4.2   Security Analysis

Trustworthiness: All the organizations participating in the intelligence sharing in the alliance chain have undergone identity verification before joining. In addition, uploading automotive CTI information requires verification by trusted nodes in the alliance chain, and only trusted organizations can participate in the verification process. These mechanisms largely ensure the trustworthiness of the automotive CTI sharing process.

Privacy: The permissioned nature and private data of Hyperledger Fabric [13,28] provide flexible identity systems and unique access control policies, ensuring that private information stored on the chain is not accessed by unauthorized organizations. Based on this, this paper combines AES, CP-ABE algorithm to encrypt plaintext intelligence into encdata, enckey, and uses the CP-ABKS algorithm to formulate access control policies by the intelligence data owner, realizing fine-grained access control and ciphertext retrieval of CTI sharing. Among them, the keywords are extracted by the intelligence data owner and encrypted through $I_w = (T, W, W_0, W', \{(W_v, D_v) \mid v \in \text{lvs}(T)\})$. The keywords are hashed by the H_2 one-way hash function, making it impossible for attackers to obtain public keywords, defending against replay attacks by attackers. In the process of keyword query, this model generates keyword trapdoor $T_w = \left(\text{Atts}, \text{tok}_1, \text{tok}_2, \text{tok}_3, \{(A'_j, B'_j) \mid a_j \in \text{Atts}\}\right)$ through the trapdoor algorithm, if the private key SK_1 of the intelligence data user is not public, attackers cannot construct a trapdoor T_w that meets the conditions to search for corresponding intelligence data. In the process of storing and querying automotive CTI, threat intelligence data has always been in an encrypted state encdata, unauthorized visitors cannot access threat intelligence, ensuring the sensitivity of threat intelligence. In addition, when data users perform keyword queries, they submit them to the blockchain network through keyword trapdoors, without any interaction with the intelligence data owner, avoiding potential privacy leaks by users.

Reliability: Traditional centralized CTI sharing platforms are prone to single points of failure. Once they encounter network attacks, it may lead to data loss or leakage, causing irreparable losses to intelligence sharing organizations. This paper uses Hyperledger Fabric blockchain for automotive CTI sharing. Multiple nodes on the chain back up the data, which can avoid single points of failure and DoS attack issues. Moreover, the model stores automotive threat intelligence data ciphertext in IPFS, which also avoids the threat of single points of failure, and enhances the reliability of automotive threat intelligence sharing. At the same time, IPFS alleviates the storage pressure of the blockchain and enhances the scalability of the blockchain network.

Integrity: This scheme stores encrypted automotive threat intelligence data in the IPFS network, and retrieves the unique hash value from the file content through $hashipfs = IPFS.Store(encdata)$ to retrieve the file. Once the stored content changes, the retrieval address $hashipfs$ will change accordingly, thereby achieving data storage integrity. In addition, using hash functions $hashdata = SHA256.Hash(data)$, $dechash = SHA256.Hash(decdata)$ to hash the original intelligence data and the intelligence data obtained by the visitor, verifying the integrity of shared intelligence data.

Traceability: On the blockchain, related parameters are passed through transactions, and any authorization records are immutable access transactions. Any node on the chain can trace the entire process. At the same time, every piece of automotive CTI information on the blockchain has the digital signature of the uploader, which can achieve the non-repudiation of automotive CTI sharing.

4.3   Feature Comparison

Table 2 presents a comparison between the proposed model and other methods proposed in the literature. From the comparison results, most studies are based on consortium chain's channel mechanism or attribute encryption algorithms to achieve privacy protection for data sharing. Some schemes also consider the issue of access control for shared data. In terms of specific scheme constructions, studies such as [22] and [14] not only implement fine-grained access control privacy protection but also consider the storage bottleneck of the blockchain by integrating the IPFS distributed network with the blockchain, thereby alleviating the burden of blockchain storage.

Similar to our research, [36] ensures privacy protection for fine-grained access control of shared data through an attribute-based searchable encryption scheme, supporting keyword ciphertext retrieval functionality. This scheme considers ciphertext retrieval when keywords are the same for shared data, but overlooks potential differences in access control policies needed for data with identical keywords, resulting in weak capabilities for fine-grained access control of shared data and low feasibility for practical applications. Additionally, the authors did not consider the storage bottleneck of blockchain systems, resulting in poor scalability.

Table 2. Comparison of Different Schemes

Scheme	Blockchain	Privacy Protection	Access Control	Ciphertext Retrieval	No User Interaction	Integrity Verification	Scalability	Smart Contract
[21]	Ethereum	✗	✗	✗	✗	✓	✓	✓
[22]	Ethereum	✓	Weak	✗	✗	✗	✓	✗
[14]	Consortium	✓	Strong	✗	✓	✓	✓	✗
[30]	Consortium	✓	✗	✓	-	✗	✗	✓
[36]	Consortium	✓	Weak	✓	✓	✗	✗	✓
Our method	Consortium	✓	Strong	✓	✓	✓	✓	✓

Through comparative analysis, the CTI sharing scheme proposed in this paper based on consortium chain and CP-ABE searchable encryption can achieve privacy protection and fine-grained access control for intelligence data, ensuring secure transmission of intelligence data to multiple parties. In this model, initialization, encryption, decryption, key generation, and keyword trapdoor generation are all computed "off-chain", and the ciphertext of CTI is stored in the IPFS network, while on-chain storage and query of intelligence metadata are performed through smart contracts, alleviating the computational and storage overhead of the blockchain and improving the scalability of the intelligence sharing system. At the same time, data users generate trapdoors to access intelligence data based on their own attributes and interested keywords without the need for interaction with intelligence data owners, reducing the corresponding communication overhead. Additionally, the model also employs AES encryption and hashing operations to ensure the confidentiality and integrity of threat intelligence information.

4.4   Analysis of Computational and Storage Costs

Tables 3 and 4 present the comparison of the computational cost and storage overhead between our proposed approach and the solutions proposed in references [36] and [40]. Here, T_e represents the time for exponentiation in the cyclic group G, T_{e1} represents the time for exponentiation in the cyclic group G_T, T_p represents the time for pairing operation, and H_t represents the time for hash operation in G; $|G|$ and $|G_T|$ represent the lengths of elements in cyclic groups G and G_T respectively. In addition, N, X, and S denote the user's attribute set, the set of leaf nodes in the access tree, and the minimum attribute set that satisfies the access tree, respectively.

Table 3. Comparison of Computational Cost

Algorithms	Strategy		
	Yang [36] (CP-ABKS)	Zheng [40] (CP-ABKS)	Ours (CP-ABKS + CP-ABE)
KeyGen	$(3N+1)T_e + T_p + T_{e1} + NH_t$	$(3N+1)T_e + NH_t$	$(6N+2)T_e + 2NH_t$
Encrypt	$(2X+4)T_e + 2T_{e1} + XH_t$	$(2X+4)T_e + (X+1)H_t$	$(4X+6)T_e + T_p + (2X+1)H_t$
Trapdoor	$(2N+5)T_e$	$(2N+5)T_e + H_t$	$(2N+5)T_e + H_t$
Search	$(2S+3)T_p + ST_{e1}$	$(2S+3)T_p + ST_{e1}$	$(2S+3)T_p + ST_{e1}$
Decrypt	T_e	–	$(2S+2)T_p + ST_{e1}$

Table 4. Comparison of Storage Cost

Strategy	KeyGen	Encrypt	Trapdoor
Yang (CP-ABKS) [36]	$(2N+1)|G| + |G_T|$	$(2X+3)|G| + 2|G_T|$	$(2N+3)|G|$
Zheng (CP-ABKS) [40]	$(2N+1)|G|$	$(2X+3)|G|$	$(2N+3)|G|$
Ours (CP-ABKS + CP-ABE)	$(4N+2)|G|$	$(4X+4)|G| + |G_T|$	$(2N+3)|G|$

4.5 Simulation

To verify the actual performance of the proposed scheme, we conduct simulation experiments using the PBC cryptography library under the Charm framework and test the actual running time and storage overhead of various algorithms using the SS1024 elliptic curve. Considering that the key generation, encryption, and decryption phases in this scheme are executed 'off-chain,' the experiments mainly compare the algorithm running time and storage overhead in the search phase and trapdoor generation phase for Yang [36], Zheng [40], and the proposed scheme. In the simulation experiments, the number of attributes is set to 5, 10, 15, 20, and 25 respectively, and the comparison results of the schemes are shown in Figs. 3.

From the comparison results of algorithm running time (Fig. 3a) and storage overhead (Fig. 3b), the computational and storage overhead of the 'on-chain' search algorithm and trapdoor generation algorithm in this scheme are consistent with the overhead of the other two schemes, meaning that the efficiency of ciphertext retrieval on the 'on-chain' is generally the same. Compared to the scheme proposed by Yang [36] and Zheng [40], our scheme uses CP-ABKS and CP-ABE algorithms to encrypt the threat intelligence keywords and the symmetric keys for threat intelligence encryption, respectively, achieving dual fine-grained access control for automotive network threat intelligence. This not only ensures the efficiency of network threat intelligence sharing but also further enhances the flexibility and strength of fine-grained access control of network threat intelligence.

5 Prototype System for Automotive CTI Sharing

To verify the feasibility of the automotive CTI sharing model combining consortium blockchain and CP-ABKS encryption technology, we constructed a prototype system for automotive CTI sharing involving stakeholders such as OEMs

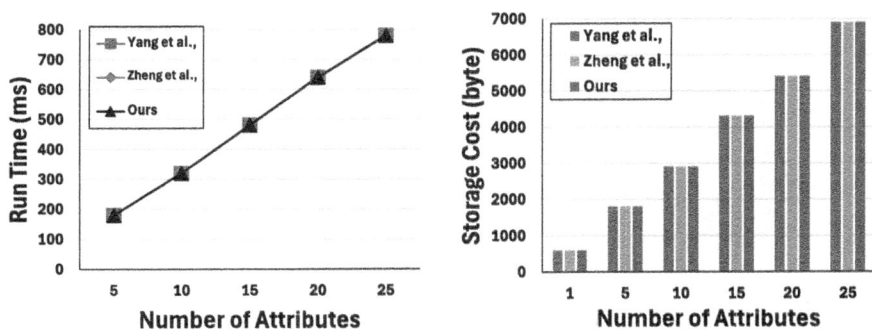

(a) Algorithm Running Time Comparison   (b) Algorithm Storage Cost Comparison

Fig. 3. Simulation Results

and suppliers. The prototype system includes environment configurations such as consortium blockchain, IPFS, Goland editor, and Pairing-based cryptography (PBC). For this experiment, CentOS 7 was selected as the operating environment for the prototype system. We utilized HyperLedger Fabric 2.4 and Docker to build the consortium blockchain network for system deployment and operation. Additionally, we used docker-compose technology to set up the IPFS network for storing encrypted automotive CTI data. In the CentOS virtual machine, Goland and the PBC cryptography library were installed and configured for performing "off-chain" AES symmetric encryption, SHA256 hash functions, and CP-ABE algorithm operations.

The blockchain network in the prototype system consists of an ordering service provider called Orderer and consortium members with two Peer nodes each (such as OEM 1, Supplier A, or Testing Organization B). Initially, the script for launching the consortium blockchain network is executed, and channels are created between consortium members such as OEM 1 and Supplier A. The system initialization results are shown in Fig. 4 (Channel 'mychannel' joined). After creating the mychannel channel, commands are executed in the consortium blockchain network for packaging chaincode, installing chaincode, approving chaincode definitions, and submitting chaincode. This deploys the StoreCont and QueryCont smart contracts on the consortium blockchain nodes. Subsequently, when sharing organizations (e.g., OEM 1, Supplier A, or Testing Organization B) upload shared data or keyword traps to the consortium blockchain network, the StoreCont or QueryCont smart contracts are called accordingly. The results of these operations are shown in Fig. 5. For example, when the data owner (such as Supplier A or Testing Organization B) invokes the StoreCont chaincode to store hashdata, hashipfs, enckey, and I_w metadata of automotive CTI data in the consortium blockchain, the prototype system returns "chaincodeInvokeOrQuery -> Chaincode invoke successful," indicating a successful chaincode invocation. Similarly, when the intelligence data user (such as OEM 1) constructs keyword traps based on certain attributes of vehicle models or components and the key-

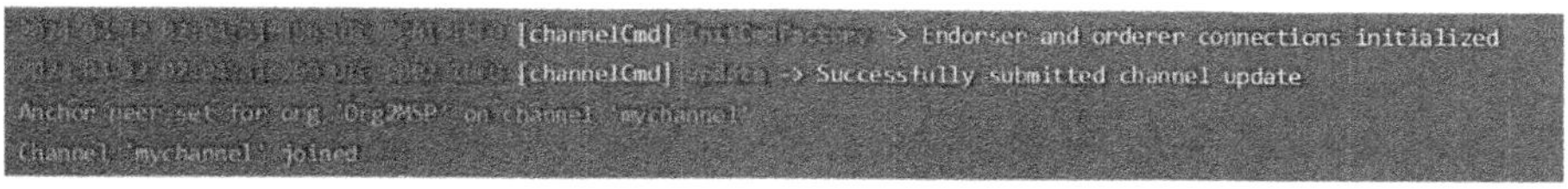

Fig. 4. Starting Blockchain Network and Channel Creation

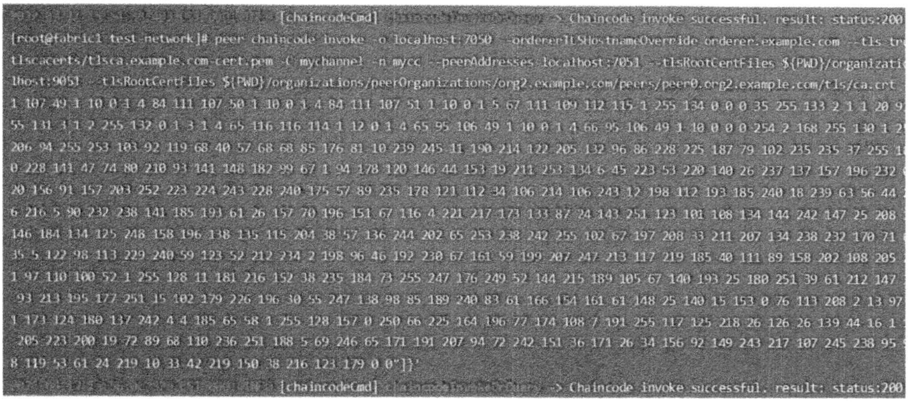

Fig. 5. Data Storage and Query Results

words of interest, the QueryCont smart contract is invoked to retrieve relevant automotive CTI data. If the QueryCont chaincode invocation is successful, the prototype system returns the corresponding intelligence data along with the statement indicating successful chaincode invocation.

In the prototype system, the execution times for invoking StoreCont and QueryCont smart contracts are 12.68 ms and 16.59 ms respectively. We built a blockchain system based on the Hyperledger Fabric open-source framework and CP-ABE, CP-ABKS encryption using Go language code. We developed Store-Cont and QueryCont smart contracts using the Chaincode on the Hyperledger Fabric platform, achieving a prototype system for secure sharing of automotive CTI with fine-grained access control and support for ciphertext retrieval. Additionally, Hyperledger Fabric does not compile the PBC cryptographic library into the Docker container, causing QueryCont chaincode instantiation to fail due to missing PBC library compilation dependencies. Therefore, in this experiment, we rewrote the Dockerfile for the ccenv container to compile the M4, bison, flex, GMP, and PBC cryptographic library dependencies required for executing the CP-ABKS searchable encryption algorithm into the container, thus creating an environment that integrates consortium blockchain with attribute-based search encryption mechanisms.

6 Conclusion

The sensitivity of automotive CTI has spurred significant interest in secure and reliable intelligence-sharing platforms. Our proposal addresses trust concerns and safeguards sensitive data by combining consortium blockchain with CP-ABE and CP-ABKS algorithms to create a fine-grained access control model for automotive CTI. Blockchain decentralization establishes a trusted environment for participants to share threat intelligence data equitably, mitigating trust issues and averting single points of failure found in centralized systems. Additionally, our model employs CP-ABKS encryption to encrypt keywords, storing their ciphertext in the consortium blockchain and enabling keyword ciphertext search through smart contracts. This approach grants sharing organizations flexible access control over threat intelligence users and supports keyword ciphertext retrieval, ensuring the confidentiality of shared automotive CTI. Furthermore, we propose a decentralized "off-chain" storage solution to ease storage burdens on the consortium blockchain and enhance system scalability. Moreover, hash operations on intelligence data uphold the integrity of shared automotive threat intelligence throughout the process. Overall, our proposed automotive CTI sharing model guarantees trustworthiness, confidentiality, reliability, and scalability in threat intelligence sharing. Future work could explore efficient access structure designs and updates to user access control policies to further bolster fine-grained access control for automotive CTI.

References

1. Alobadh, H., Yasin, S.M., Udzir, N.I.: Blockchain-based access control scheme for secure shared personal health records over decentralised storage. Sensors (2021)
2. Badsha, S., Vakilinia, I., Sengupta, S.: Blocynfo-share: Blockchain based cybersecurity information sharing with fine grained access control. In: 2020 10th Annual Computing and Communication Workshop and Conference, IEEE (2020)
3. Bethencourt, J., Sahai, A., Waters, B.: Ciphertext-policy attribute-based encryption. In: IEEE symposium on security and privacy (SP 2007), IEEE (2007)
4. Bhushan, B., Sinha, P., Sagayam, K.M., Andrew, J.: Untangling blockchain technology: a survey on state of the art, security threats, privacy services, applications and future research directions. Comput. Electr. Eng. (2021)
5. Bolz, R., Kriesten, R.: Automotive vulnerability disclosure: stakeholders, opportunities, challenges. J. Cybersecurity Priv. 1(2), 274–288 (2021)
6. Chen, L., Lee, W.K., Chang, C.C., Choo, K.K.R., Zhang, N.: Blockchain based searchable encryption for electronic health record sharing. In: FGCS (2019)
7. Gao, J., Wang, W., Nikseresht, F., Govinda Rajan, V., Campbell, B.: Pfdrl: personalized federated deep reinforcement learning for residential energy management. In: Proceedings of the 52nd ICPP (2023)
8. Goodwin, C., et al.: A framework for cybersecurity information sharing and risk reduction. Microsoft (2015)
9. Guo, X.: Network Threat Intelligence Sharing Mechanism and Implementation Based on Blockchain Smart Contracts. Master's thesis, BUPT (2019)

10. Gupta, B.B., Li, K.C., Leung, V.C., Psannis, K.E., Yamaguchi, S., et al.: Blockchain-assisted secure fine-grained searchable encryption for a cloud-based healthcare cyber-physical system. IEEE/CAA J. Automatica Sinica (2021)

11. Haass, J.C., Ahn, G.J., Grimmelmann, F.: Actra: a case study for threat information sharing. In: Proceedings of the 2nd ACM Workshop on Information Sharing and Collaborative Security, pp. 23–26 (2015)

12. Homan, D., Shiel, I., Thorpe, C.: A new network model for cyber threat intelligence sharing using blockchain technology. In: 2019 10th IFIP International Conference on New Technologies, Mobility and Security (NTMS), pp. 1–6. IEEE (2019)

13. Iftekhar, A., Cui, X., Tao, Q., Zheng, C.: Hyperledger fabric access control system for internet of things layer in blockchain-based applications. Entropy (2021)

14. Jiang, S., Liu, J., Wang, L., Yoo, S.M.: Verifiable search meets blockchain: a privacy-preserving framework for outsourced encrypted data. In: ICC (2009)

15. Kampanakis, P.: Security automation and threat information-sharing options. IEEE Secur. Priv. **12**(5), 42–51 (2014)

16. Lin, Y., Liu, P., Wang, H., Wang, W., Zhang, Y.: A survey of research on cyber-security threat intelligence sharing and exchange. J. Comput. Res. Dev. **57**(10), 2052–2065 (2020)

17. Mavroeidis, V., Bromander, S.: Cyber threat intelligence model: an evaluation of taxonomies, sharing standards, and ontologies within cyber threat intelligence. In: 2017 European Intelligence and Security Informatics Conference (EISIC), pp. 91–98. IEEE (2017)

18. Morris, D., Madzudzo, G., Garcia-Perez, A.: Cybersecurity threats in the auto industry: tensions in the knowledge environment. Technol. Forecast. Soc. Chang. **157**, 120102 (2020)

19. Nakamoto, S.: Bitcoin: A peer-to-peer electronic cash system. A peer-to-peer electronic cash system, Bitcoin (2008)

20. Niu, S., Xie, Y., Yang, P., Du, X.: Cloud-assisted attribute-based searchable encryption scheme on blockchain. In: JCRD (2021)

21. Preuveneers, D., Joosen, W.: Tatis: trustworthy apis for threat intelligence sharing with uma and cp-abe. In: Foundations and Practice of Security: 12th International Symposium, pp. 172–188. Springer (2020)

22. Preuveneers, D., Joosen, W., et al.: Distributed security framework for reliable threat intelligence sharing. Secur. Commun. Netw. **2020** (2020)

23. Qin, Z., Xu, J., Nie, X., Xiong, H.: A survey of public key searchable encryption schemes. J. Cyber Secur. **2**(3) (2017)

24. Riesco, R., Larriva-Novo, X., Villagrá, V.A.: Cybersecurity threat intelligence knowledge exchange based on blockchain: proposal of a new incentive model based on blockchain and smart contracts to foster the cyber threat and risk intelligence exchange of information. Telecommun. Syst. **73**(2), 259–288 (2020)

25. Siddiqui, Z., Gao, J., Khan, M.K.: An improved lightweight puf-pki digital certificate authentication scheme for the internet of things. IEEE Internet Things J. **9**(20), 19744–19756 (2022)

26. Skopik, F., Settanni, G., Fiedler, R.: A problem shared is a problem halved: a survey on the dimensions of collective cyber defense through security information sharing. Comput. Secur. **60**, 154–176 (2016)

27. Song, D.X., Wagner, D., Perrig, A.: Practical techniques for searches on encrypted data. In: Proceeding IEEE S&P (2000)

28. Stamatellis, C., Papadopoulos, P., Pitropakis, N., Katsikas, S., Buchanan, W.J.: A privacy-preserving healthcare framework using hyperledger fabric. Sensors (2020)

29. Stupka, V., Horák, M., Husák, M.: Protection of personal data in security alert sharing platforms. In: Proceedings of the 12th ICARS (2017)
30. Tahir, S., Rajarajan, M.: Privacy-preserving searchable encryption framework for permissioned blockchain networks. In: 2018 IEEE iThings (2018)
31. Tounsi, W., Rais, H.: A survey on technical threat intelligence in the age of sophisticated cyber attacks. Comput. Secur. **72**, 212–233 (2018)
32. Vázquez, D.F., Acosta, O.P., Spirito, C., Brown, S.: Conceptual framework for cyber defense information sharing within trust relationships. In: CYCON (2012)
33. Xiang, X., Zhao, X.: Blockchain-assisted searchable attribute-based encryption for e-health systems. J. Syst. Archit. (2022)
34. Yan, X., Yuan, X.: Blockchain-based and verifiable attribute-based searchable encryption scheme. J. Commun./Tongxin Xuebao (2020)
35. Yang, Y., Lin, H., Liu, X., Guo, W., Zheng, X.: Blockchain-based verifiable multi-keyword ranked search on encrypted cloud with fair payment. IEEE Access (2019)
36. Yang, Z., Zhang, H., Yu, H., Li, Z., Zhu, B., Sinnott, R.O.: Attribute-based keyword search over the encrypted blockchain. Comput. Model. Eng. Sci. **128**(1), 269–282 (2021)
37. Zhang, W., Bai, Y., Feng, J.: Tiia: a blockchain-enabled threat intelligence integrity audit scheme for IIoT. Futur. Gener. Comput. Syst. **132**, 254–265 (2022)
38. Zhang, Y., Zhu, T., Zheng, D.: Blockchain-based fine-grained multi-keyword searchable encryption scheme. Inf. Netw. Secur. (2021)
39. Zhao, W., White, G.: A collaborative information sharing framework for community cyber security. In: IEEE HST (2012)
40. Zheng, Q., Xu, S., Ateniese, G.: Vabks: verifiable attribute-based keyword search over outsourced encrypted data. In: IEEE INFOCOM (2014)

Multi-modal Multi-task Tiered Expert (M3TTE): An Effective Method for CDN Website Classification

Yulong Zhan[1,2,3(✉)], Yang Cai[3], Gang Xiong[1,2], Gaopeng Gou[1,2], and Xiaoqian Li[3]

[1] Institute of Information Enginnering, Chinese Academy of Science, Beijing, China
{zhanyulong,xionggang,gougaopeng}@iie.ac.cn
[2] School of Cyber Security, University of Chinese Academy of Sciences, Beijing, China
[3] CNCERT/CC, Beijing, China
{caiyang,lixiaoqian}@cert.org.cn

Abstract. In response to the escalating demand for Content Delivery Network (CDN) website classification, this paper introduces Multi-Modal Multi-Task Tiered Expert(M3TTE), a revolutionary method designed to enhance the efficiency and accuracy of different classification tasks of CDN websites. Facing the limitations of existing models, M3TTE leverages a tiered expert network, employing specialized experts for processing different modal features. This tiered structure ensures personalized treatment, addressing the challenges posed by diverse modalities. The framework is anchored in four key principles: personalized handling of features from different modalities, retention of each modality's specificity, integration of expert outputs for multi-task learning, and preservation of both global and task-specific features. M3TTE's architecture achieves a delicate balance between global and task-specific information, showing remarkable performance in CDN website classification.

Experiments, conducted on a diverse dataset collected from various devices and operating systems, underscore M3TTE's superiority over single-task and multi-task models. Impressively, M3TTE attains 74.93% accuracy and 74.89% F1 score for website category classification and 97.48% accuracy and F1 score for CDN classification. This paper presents M3TTE as an excellent solution, effectively tackling the complexity of CDN website classification through rigorous experiments and evaluations.

Keywords: Multi-Modal Learning · Multi-Task Learning · CDN Website Classification · Tiered Expert Network

1 Introduction

With the continuous development of the Internet, the personalization of user devices and the cloudification of Internet infrastructure have gradually become trends. CDNs have become an increasingly important and widely used infrastructure for Internet users. On the one hand, some users host their websites on

J. Zhao and W. Meng (Eds.): SciSec 2024, LNCS 15441, pp. 369–386, 2025.
https://doi.org/10.1007/978-981-96-2417-1_20

CDNs to address their lack of server resources or enhance security protection. On the other hand, CDNs bear the responsibility of accelerating and distributing content of certain Internet services. The rise of short videos, online gaming, and e-commerce further reinforces the application of CDNs. As stated in the reports [12,17], the growth rate of the CDN market over the past few years has averaged more than 25%, and even exceeded 30% in some countries such as China. In recent years, CDN traffic has grown significantly faster than total traffic. It is reported that by 2022, CDN traffic has accounted for 72% of total traffic.

The current application of CDN is extensive, with a promising outlook for development. Several CDN vendors are investing in and conducting research in this field, leading to intense competition for market share among multiple companies. The competitive factors for CDN vendors primarily revolve around user experience, including product content, after-sales service and cost efficiency. In an environment where there is little difference in commercial prices and services, the core of user experience lies in the real-time nature of CDN transmission.

With established investment and service frameworks, user experience needs to be further optimized, and the real-time performance of CDN transmission should be enhanced. There are two aspects of work that can strengthen the related user experience. For one thing, a straightforward method involves classifying traffic and setting different priorities for transmission [21]. Different transmission priorities are assigned to various flows, and higher-priority flows are transmitted preferentially. For another thing, strategies can be implemented to prioritize the transmission of flows belonging to one's own CDN brand within the transmission nodes of one's organization or collaborative partners. To protect user privacy, encrypted traffic transmission is usually employed. Therefore, the mentioned strategies need to conduct online classification based on various features of the received CDN traffic, such as traffic load, without decrypting it. This classification involves determining the type of content carried by the traffic and its correspondence to the CDN brand in two dimensions.

Indeed, historical efforts in CDN-related research primarily centered around the design, mechanism, and security of CDNs [5,10,13,22]. There is limited exploration into the classification of websites hosted on CDNs or the traffic transmitted through CDNs.

The existing papers on encrypted traffic predominantly revolve around the classification of encrypted traffic, addressing aspects such as whether the traffic is encrypted, the protocol employed, whether it pertains to a specific type of service, its manufacturer, or whether it corresponds to the visitation of a particular page, among other factors [1,6,7,14,15,19,25,26,28]. However, a noticeable gap exists with regards to papers that specifically focus on website classification or traffic within a CDN environment. Additionally, recent studies in this field have predominantly utilized machine learning and deep learning methods. These experiments typically focus on single-category classification in each iteration. While classification frameworks may be applicable to tasks involving the classification of traffic into multiple categories [1,26], to our knowledge, there are few papers on traffic classification that leverage multiple modalities of features in a single classification, providing results across multiple dimensions through multi-task learning.

In the context of multi-task learning, classical frameworks with non-large models generally use single-type features [16,20,23], which do not align with the multi-modal features of traffic classification for CDN websites. This paper mainly studies the method to solve the above problems—performing multi-task classification on the encrypted traffic of CDN based on the multi-modal features. Our main contributions are as follows.

1. By utilizing multi-modal features for multi-task classification of CDN website encrypted traffic, we solve the gaps in the previous literature on CDN, encrypted traffic classification and multi-task learning. The significance and application value of this research direction are expounded.
2. We reflect on the shortcomings of previous frameworks and models, proposing a learning model M3TTE for multi-task and multi-modal learning that is flexible, adaptable, and personalized, accommodating varying feature modalities and numbers of tasks.
3. We apply this model to two tasks: classification of CDN websites and brand recognition, employing a dataset collected and open-sourced by us, and design an experimental framework. We compare our model with previous single-task models for encrypted traffic classification, classical multi-task models, and optimized multi-task models. The experimental results show that the new model has better performance and higher efficiency. In addition, we conduct ablation studies to demonstrate the necessity of each component of the framework and explore the impact of relevant parameters on the experiments.

The remainder of the paper is structured as follows. Section 2 provides an overview of related work in the field. Section 3 delves into our approach, introducing our thoughts, the new method employed, details about the dataset, and the overall framework. In Sect. 4, we present comparative experiments, ablation studies, and hyperparameter experiments with discussion. Finally, Sect. 5 concludes the paper and suggests potential directions for future research.

2   Related Work

Encrypted traffic classification has found widespread applications in various domains, including capacity planning, traffic engineering, performance monitoring, Quality of Service(Qos), and information development. The evolution of encrypted traffic classification methods [9,18] can be broadly categorized into three main approaches: port and deep packet inspection, traditional machine learning, and deep learning.

Due to the rise of dynamic ports, the increase of encrypted traffic, and the increasing diversity of network traffic, port-based and deep packet inspection methods are gradually becoming less effective. Traditional machine learning methods often utilize statistical features, such as average packet size, average packet interval, median packet size, and so on. These features combined with algorithms like SVM, DT, K-means [2–4], have historically achieved good results. However, traditional machine learning methods face two main challenges. Firstly, they tend to rely on statistical features that cannot be obtained until the flow is complete, making them unsuitable for real-time online classification scenarios. Secondly, these methods typically involve two stages: feature extraction (requiring substantial expert knowledge) and learning, making it difficult to form an end-to-end structure. The local optimal results achieved in the feature extraction and learning stages may not necessarily lead to the final global optimal results.

At present, the most popular way to solve the above problems is based on deep learning. Most of them employ end-to-end structures with different types of neural networks that automatically learn discriminant features with better generalization ability, eliminating the need for expert-crafted features. This implies that, by leveraging deep learning algorithms, the global optimal solution can be obtained without requiring expert knowledge or significant manual effort. Many deep learning methods primarily utilize payload features, making online classification feasible.

DataNet [24] is proposed to handle the task using multilayer perceptron (MLP), stacked autoencoders (SAE), and convolutional neural network (CNN). In 2017, [19,25] employs 1DCNN and 2DCNN to process packet payloads, abstracting traffic data into common image data. Time-series features and related models are also utilized in this direction. A Bi-LSTM [28] model with different types of features is employed to handle specific tasks. Several papers have classified specific websites by encrypted traffic using website fingerprinting [6,7], combined with basic machine learning or deep learning models. Some basic deep learning models are combined for serial processing or integrated learning to realize encrypted traffic classification tasks. [1] concatenates CNN and Long short-term memory(LSTM) for processing payload features and flow temporal features extracted from handshake packets. [14] first perform flow representation, then use integrated learning based on CNN and LSTM for the representation vector. [26] treats data traffic as textual information and combines attention mechanisms for processing. They propose the SAM model, which is a great improvement over previous models due to the attention mechanism, making it currently the state-of-the-art online classification model for encrypted traffic classification.

The mentioned papers on encrypted traffic classification mainly focus on single-task classification and are unable to accomplish multiple learning tasks in a single iteration. Most of them use either a single feature or a combination of multiple features using the same processing approach. The paper [11] achieves multi-task learning for traffic, but only employs a simple model to address basic tasks such as malicious detection, encryption detection, and Trojan classification. This model is not suitable for more complex multi-task classification learning.

In the realm of multi-task learning, classic methods include shared bottoms, L2-constraints, cross-stitch and tensor decomposition [20] methods. In 2018, Google proposed the Multi-gate Mixture-of-Experts(MMOE) method [16] for multi-task learning. This method outperforms others in related or unrelated multi-task learning scenarios and is currently the widely used optimal multi-task framework [8,27]. However, it has not yet been applied to encrypted traffic classification, especially CDN-related traffic, and currently does not support multi-modal features.

3   Methods, Dataset and Framework

3.1   Methods

Before proposing a multi-modal multi-task model suitable for our CDN website classification task, we ground our considerations and experiments on the widely used MMOE model, as depicted in Fig. 1. Initially, we consider directly utilizing the widely applied MMOE model for multi-task learning to handle multi-modal data—treating the combination of multi-modal data as features in the MMOE(MMOE-Base). However, this approach doesn't take into account the specificity of each modality within the multi-modal data. Subsequently, we consider extending MMOE by assigning a dedicated expert for each modality. However, the original single gate proves challenging to adapt to the requirements of multiple modality experts. We explore two solutions to address the gate issue: one involves merging all modality inputs as side inputs to the gate (MMOE-all), and the other includes setting multiple gates, directing selected features as side inputs to different task outputs (MMOE-Selected). The corresponding schemes are illustrated in Fig. 2, and their experimental results can be seen in Sect. 4. Nevertheless, these approaches above mainly encounter three issues:

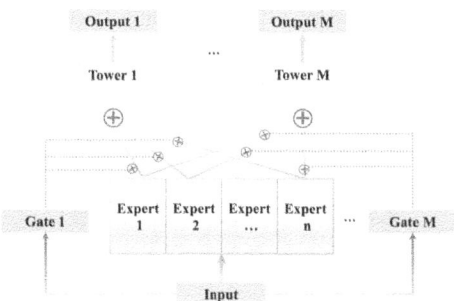

Fig. 1. MMOE model

1) The lack of consideration for the specificity of different modality features, compromises the crucial advantage of multi-modal learning.
2) Directly merging features from different modalities for processing is akin to adding data with different dimensions, likely leading to suboptimal classification results.
3) Selecting only some features for the gate side path may omit some effective features, resulting in poor learning performance.

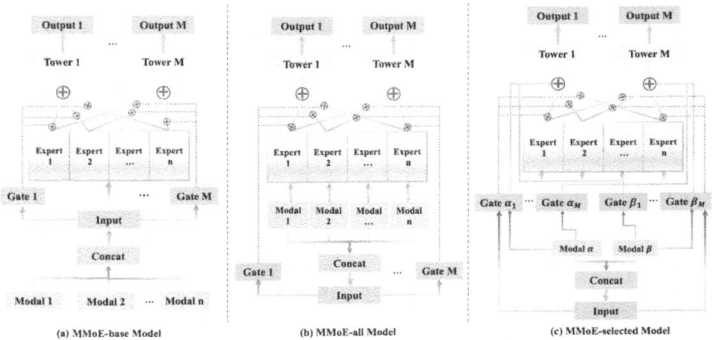

Fig. 2. MMOE-Base, MMOE-all, and MMOE-selected model

After careful consideration, we believe that an appropriate multi-modal multi-task learning model should achieve the following four points:

1) Ensure personalized processing of features from different modalities, avoiding the use of the same expert network for multiple modalities.
2) Preserve the specificity of each modality's features, recognizing that unabstracted modalities cannot be directly generalized as a whole.
3) Integrated expert handles multiple modalities to accomplish multi-task learning.
4) Model retains both the global characteristics of multi-modal features and personalized features suitable for each task.

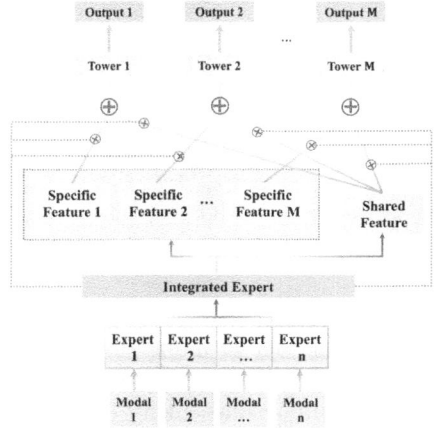

Fig. 3. M3TTE model

In light of these considerations, we designed the Multi-Modal Multi-Task Tiered Expert (M3TTE) method, as shown in Fig. 3. This method leverages the joint learning structure of multi-modal learning to ensure the specificity of various modality features. Throughout the process of multi-task learning, M3TTE not only utilizes both the global characteristics of multi-modal features, but also uses the personalized features of each task to achieve effective learning outcomes.

In the workflow of this method, we consider an input x with n modal features, denoted as x_1, x_2, ..., x_n. Through the network F_i of each modality expert, each modality feature x_i is transformed into an abstracted feature y_i.

$$y_i = F_i(x_i) \quad (i = 1, 2, ..., n) \tag{1}$$

By combining the abstracted features from each modality, the integrated expert network I is utilized to obtain the overall feature $feature_{all}$ corresponding to the input:

$$feature_{all} = I(y_1, y_2, ..., y_n) \tag{2}$$

The shared features and specific features corresponding to each learning task are calculated and input into the respective learning task's tower, resulting in the final output:

$$\begin{aligned} feature_{share} &= I_{sh}(feature_{all}) \\ feature_{specific_i} &= I_{sp_i}(feature_{all}) \quad (i = 1, 2, .., m) \end{aligned} \tag{3}$$

The M3TTE model will be applied in subsequent framework.

3.2 Dataset

Datasets in the field of encrypted traffic classification are not publicly available or relevant to CDNs, making them unsuitable for this study. Thus we collected data from various devices (mobile phones and personal computers) and multiple operating systems (Mac, Windows 7, 8, 10, 11). The dataset consists of traffic content collected using Wireshark while randomly accessing pages on websites hosted on CDNs, including both well-known and self-built sites. We exclude interference from access methods and device categories on the classification model. The general process of traffic collection is illustrated in Fig. 4.

The website categories mainly consist of five types: blog (18.9%), picture (20.6%), video (20.8%), Bulletin Board System (BBS) (20.6%), and Social Networking Services (SNS) (19.1%). CDN brands include Alibaba (12.1%), Tencent (12.1%), Baidu (14.2%), Qiniu (10.0%), Cloudflare (16.5%), Cloudfront (10.1%), Fastly (2.0%), self-built (2.1%), and non-CDN data (20.9%). There are 16,236,558 packets in 12,195 pcapng files, where each file represents a particular network flow corresponding to a website category and a CDN brand.

After purification by network five-tuple, the dataset named WebT2023 and the documentation are available on Google Drive https://drive.google.com/file/d/1vDYR5WeSEZQzNMsJbib0n3FEUxYN-Rdj/view?usp=share_link.

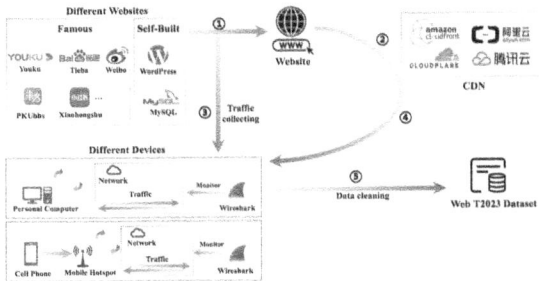

Fig. 4. Steps to build the dataset:①Select specific categories of websites, identify corresponding well-known sites for each category, and build websites of those categories with regularly updated content. ②Choose different well-known and widely used CDNs for experiments. ③Collect traffic using Wireshark while randomly visiting the websites through different devices with different operating systems without using any specific access methods. ④Successively change websites to be hosted on the CDNs chosen in step ② and repeat step ③. ⑤Purify the data by network five-tuple.

3.3   Framework

Combining the aforementioned dataset and methods, we present the framework for multi-modal multi-task CDN website classification, as shown in Fig. 5. The framework can be divided into three main components: preprocessing module, flow sampling and representation module, and the crucial M3TTE module.

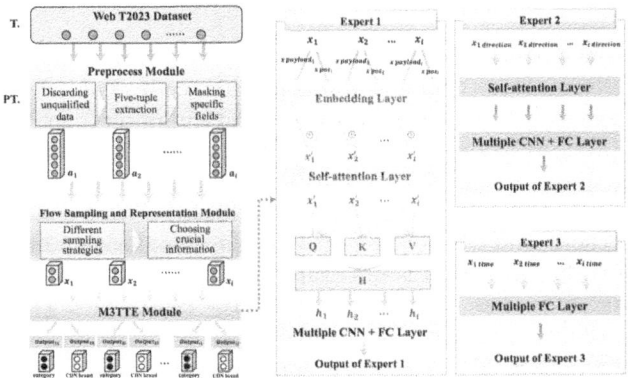

Fig. 5. Framework of the CDN website classification

Preprocessing Module. In this module, the original WebT2023 dataset undergoes preprocessing. Unqualified data is discarded through data cleaning based on five-tuple and flow truncation. To prevent the framework from relying on specific features, latent intricate features are removed. Multiple fields within a flow are masked with 0, including IP addresses, TCP/UDP ports, sequence numbers, and checksum. Data at the physical and link layers is also eliminated.

Flow sampling and Representation Module. The Flow Sampling and Representation Module is employed for online classification and system application. N packets in one flow are selected for representation, and various sampling strategies can be applied to process crucial information for representation. Different sampling methods include first N method, random N method, middle N method, and random continuous N method. Flow sampling enables online traffic classification with a small portion of traffic. For a packet, the first B bytes are selected for representation. Given the original dataset T, we choose a flow a from the preprocessed dataset, perform flow sampling and representation, and obtain input x suitable by subsequent module, denoted as FR:

$$x = FR(a) \quad a \in Preprocess(T) \tag{4}$$

M3TTE Module. For an input x in the module, we extract $x_{payload} + x_{pos}, x_{direction}$, and x_{time} as three modal features, which are then input into three different modal experts. $x_{payload}$ and x_{pos} represent the payload information and corresponding position information in the traffic. Each byte in x is converted to an integer in the range $[0, 255]$, and the corresponding position vector is $[0,1,...]$. $x_{direction}$ indicates the information about the direction and size of the packets. Its value is the number of bytes for N packets, multiplied by -1 if the direction is from client to server. x_{time} represents the time-related information of the traffic, with its value being the arrival time of the corresponding data packet relative to the start of the flow.

Combining the features from the three modalities mentioned above, we can complete the classification of CDN website categories and CDN brands.

In the framework, we have designed specific experts for three modals. The structure of the most complex modal expert1 is elaborated. In modal expert1, we handle payload and position information. As shown in Eq. 5, modal expert1 first embeds payload and position information separately and then adds them to obtain x', which contains information about both payload and sequences.

$$x'_i = Embedding_B(x_i^{byte}) \oplus Embedding_P(x_i^{pos}), \quad i \in (1, 2, ..., d) \tag{5}$$

Self-attention mechanism is employed, inspired by the analogy of traffic to text. The input $X' = (x'_1, x'_2, ..., x'_d) \in R^{D_x \times d}$ is mapped to three different spaces: query space (Q), key space (K), and value space (V), as expressed in Eq. 6.

$$\begin{aligned} Q &= W_q X' \in R^{D_k \times d} \quad (W_q \in R^{D_k \times D_x}) \\ K &= W_k X' \in R^{D_k \times d} \quad (W_k \in R^{D_k \times D_x}) \\ V &= W_v X' \in R^{D_v \times d} \quad (W_v \in R^{D_k \times D_x}) \end{aligned} \tag{6}$$

Any dimension in the query vector corresponds to one dimension in x', and it can be used to calculate the output h_i in the attention layer using Eq. 7. The calculation process is accelerated by GPU using Eq. 8.

$$
\begin{aligned}
h_i &= attention((K, V), q_i) \\
&= \sum_{j=1}^{d} softmax(Linear(k_j, q_n))v_j, \quad i \in (1, 2, ..., d)
\end{aligned}
\tag{7}
$$

The output H contains both information and contextual features.

$$
H = V softmax(\frac{K^T Q}{\sqrt{D_k}})
\tag{8}
$$

In modal expert1, CNN with different kernel sizes is used to handle H. Features of different granularity are extracted, followed by dimension reduction using Fully-Connected(FC) layers. In modal expert2, we apply self-attention and CNN to process the packet length and direction information. There is no complex mechanism used for handling the temporal information contained in expert 3. Subsequent processes follow the methods described earlier. Ultimately, for an input object x, we obtain output1 and output2 corresponding to the classification results of website category and CDN brand, respectively. Cross-entropy loss and the Adaptive Moment Estimation (Adam) optimizer are employed for optimization. The loss of the entire framework can be calculated as Eq. 9:

$$
Loss = \alpha_1 * CEloss(l_{category}, output1) + \alpha_2 * CEloss(l_{brand}, output2)
\tag{9}
$$

4   Experiments Evaluation

To assess the performance of the M3TTE model and the importance of components in this framework, we conducted comprehensive experiments and ablation study in our framework. Section 4.1 outlines the experiment setup, including the comparison of one-task and multi-task schemes, the experiment settings, the ablation strategy, and the evaluation metrics. Section 4.2 presents the results of the comparative experiment with other schemes, while the ablation study is conducted in Sect. 4.3. Finally, the influences of several hyperparameter settings are analyzed in Sect. 4.4. Analysis and discussion are provided in the corresponding sub-sections.

4.1   Experiment Setup

Compared Schemes. The M3TTE model is compared with both one-task methods in Sect. 2 (the DatanetMLP model, the 1D-CNN model, the 2DCNN model, the BiLSTM model, the Deep Packet model, the TSCRNN model, the website fingerprinting (WF) method [6], the SAM model) and multi-task methods in Sect. 3 (MMOE-Base, MMOE-all, MMOE-selected models). The code for the entire experiment is available at https://github.com/zhanyl12/M3TTE.

Experiment Setting. In the experiments, features from three modalities are utilized: the payload content of sampled data packets, the direction and length of sampled data packets, and the relative transmission time of sampled data packets, represented by features 1, 2, and 3 respectively. The experiments focused on two tasks: classification of websites hosted on CDN and classification of CDN brands, represented by task1 and task2.

The experiments are conducted using an Intel(R) Core(TM) i7-13700K CPU @ 3.40 GHz and an NVIDIA GPU 4070Ti. Python 3.7.4 and PyTorch are used to create the framework. Each scheme is evaluated with a 10-fold cross-validation.

The parameters in the compared methods (including one-task and multi-task methods) are mostly consistent with those in the corresponding papers. All models are applied to the sampling data. Concerning default parameters, the sampling strategy is set to First N, selecting 10 packets and sampling 50 bytes from each packet. The batch size is configured as 128, with 15 epochs, and a dropout rate of 0.1. For detailed parameters, please refer to the provided code.

Ablation Strategy. Our ablation study is conducted from two perspectives. On the one hand, we aim to verify the importance of the three modalities of data, assessing the performance of M3TTE-feature1, M3TTE-feature2, and M3TTE-feature3. On the other hand, we examine the importance of the specificity structure in M3TTE (modality experts and specific features), by M3TTE-specific.

Evaluation Metrics. We use four metrics and total training time to evaluate the classification performances: AC (Accuracy) evaluates the overall performance of the classifier on the whole dataset. Precision (PR), recall (RC), and F1-score (F1) are used to assess the quality of the classification for each category. The weighted metrics are calculated as follows:

$$\begin{aligned}
Accuracy(AC) &= \frac{TP+TN}{TP+TN+FP+FN} \\
Precision(PR) &= \frac{TP}{TP+FP} \\
Recall(RC) &= \frac{TP}{TP+FN} \\
F1-score(F1) &= \frac{2*PR*RC}{PR+RC}
\end{aligned} \tag{10}$$

TP, FP, TN, FN refer to true positive, false positive, true negative, false negative respectively.

4.2   Comparative Experiment

Table 1 and Fig. 6 show the performance of different models in two classification tasks. The results indicate that the M3TTE model outperforms various single-task models and the multi-task models based on MMOE in both task1 and

Table 1. Average performance with standard deviation over the 10 folds for different models in classification.

	AC(%)-task1	F1(%)-task1	AC(%)-task2	F1(%)-task2	Time(s)
DatanetMLP	34.08±9.89	29.20±15.37	85.75±1.22	85.60±1.28	413.2
1DCNN	20.37±1.21	6.91±0.74	29.17±19.36	17.59±23.78	628.4
2DCNN	42.64±1.93	42.58±2.00	86.91±1.40	86.82±1.41	730.2
BiLSTM	28.47±1.94	25.68±3.28	70.99±2.29	70.12±2.65	681.6
Deeppacket	35.72±9.52	31.90±15.37	88.90±0.84	88.72±0.87	640.0
TSCRNN	51.17±1.91	51.17±1.91	90.77±1.07	90.73±1.07	1300.5
SAM	70.56±1.11	70.61±1.06	97.27±0.83	97.27±0.83	1391.5
WF	32.64±1.49	32.18±1.51	17.19±1.36	18.56±1.39	633.1
M3TTE	**74.93±0.91**	**74.89±0.94**	**97.48±0.59**	**97.48±0.58**	986.4
MMOE-Base	39.20±1.95	39.38±1.36	78.94±14.32	78.73±14.28	343.2
MMOE-all	52.58±15.89	51.24±18.68	79.06±29.13	77.37±32.57	998.3
MMOE-selected	52.86±16.46	51.64±19.16	92.13±11.89	92.16±11.88	989.5

task2, achieving an accuracy of 74.93%, an F1 score of 74.89%, and an accuracy of 97.48% with an F1 score of 97.48%, respectively. Additionally, the M3TTE model requires the shortest total training time among the well-performing models, demonstrating the highest efficiency compared to the less optimal SAM model, reducing the training time by 29.1%.

Regarding single-task models, DatanetMLP, 1DCNN, 2DCNN, BiLSTM, Deeppacket methods based on MLP, CNN, LSTM, and SAE performed poorly in task1 and showed improvement in task2, except for 1DCNN, which still performed poorly. We attribute this to the 1DCNN model causing confusion in feature learning due to handling both packet headers and tails. The above models lag significantly behind our M3TTE model in both tasks. WF, applied to the identification of specific websites or pages, performed poorly in task1 and especially in task2, failing to accomplish website categorization and CDN brand classification. TSCRNN showed improved results in both task1 and task2 compared to the above methods, utilizing a time-series model for traffic processing and combining CNN networks to extract multi-granularity features. SAM achieved suboptimal results, stacking attention mechanism on top of the time-series model combined with CNN to extract multi-granularity features. However, SAM's accuracy and F1 values in task1 decreased by about 4.4% and 4.2%, respectively, compared to the M3TTE model, and both accuracy and F1 values in task2 lagged by about 0.2%. In our opinion, SAM's drawback lies in its inability to leverage multi-modal features and failure to exploit the correlation of multi-task learning for better performance in multiple tasks.

Regarding multi-task models, although the MMOE model is currently the best-performing multi-task classification model, its three improved models only yield satisfactory results in MMOE-selected in task2. The specific reasons for the poor performance of the MMOE models are explained in Sect. 3, primarily attributed to the incapacity of the current MMOE structure to adapt to multi-modal multi-task learning. The MMOE-selected model performs relatively better in task2 due to its abstraction, which can be seen as a ablation variant of the

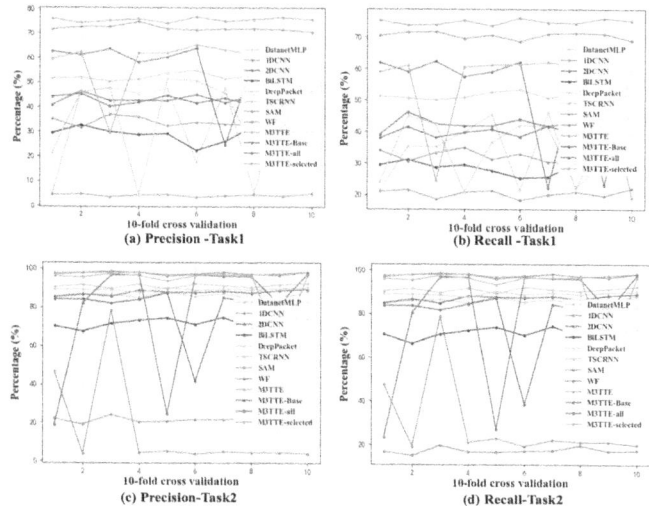

Fig. 6. Performance of classification by different models.

M3TTE method, preserving the specificity of different modal features to some extent. Although the MMOE-Base model has a shorter processing time, its performance is poor. The other two variant models show some improvements in performance, but they fall significantly behind M3TTE, and their efficiency is comparable.

As can be seen from the completion of the tasks, task1 involves website category classification, which is more challenging than task2, which focuses on CDN brand classification. In addition, the results of task 1 tend to exhibit larger variances.

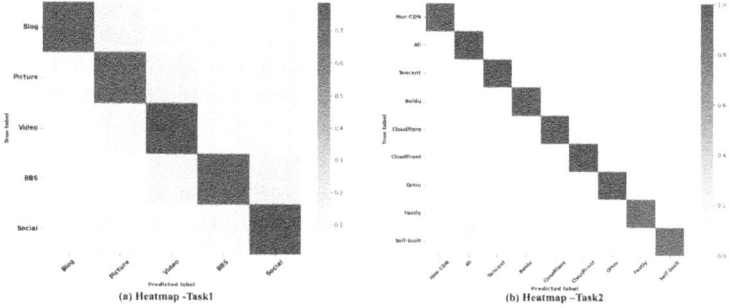

Fig. 7. Heatmaps of the classification results.

We select one fold of M3TTE's results and visualize the confusion matrix in the form of a heatmap, as shown in Fig. 7. In task1, the distribution of confusion cases is relatively uniform, with challenges in distinguishing between picture and video categories, as well as between BBS and social categories. The difficulty arises due to the presence of numerous common elements among these websites, posing a challenge for classification based on traffic. In task2, the classification performance improves compared to task1. But the classification of the Fastly and self-built categories is relatively poorer compared to other categories. This is attributed to the limited samples for these two categories, indicating an issue of data imbalance, which warrants further exploration in subsequent research.

4.3   Ablation Study

Table 2. Average performance with standard deviation over the 10 folds for ablation study.

	AC(%)-task1	F1(%)-task1	AC(%)-task2	F1(%)-task2	Time(s)
M3TTE	**74.93±0.91**	**74.89±0.94**	**97.48±0.59**	**97.48±0.58**	986.4
M3TTE-feature1	59.46±1.55	59.56±1.58	80.89±1.46	80.90±1.38	365.9
M3TTE-feature2	71.07±1.89	71.06±1.92	96.70±0.59	96.69±0.58	888.0
M3TTE-feature3	74.54±2.49	74.53±2.47	97.12±0.57	97.12±0.57	971.6
M3TTE-specific	29.66±2.24	27.84±3.19	73.83±2.04	73.66±2.26	254.1

The results of the ablation study are presented in Table 2 and Fig. 8, analyzing the impact of various modalities and specificity structures of the M3TTE model. The experimental findings reveal that each modality feature and specificity structure play a positive and effective role in model classification, with the absence of any element resulting in a deterioration of model classification performance. Among them, modal feature1 has the most significant impact among the required modal features for classification. Removing feature1 leads to a decrease of over 15% in accuracy and F1 in task1, and a drop of approximately 16.6% in task2. Distilling feature2 results in a decrease of about 3% and 0.8% in accuracy and F1 in both tasks, while feature3 decreases by around 0.6% and 0.3%. This indicates that the importance of packet payload, packet size and direction, and relative packet timing features decreases in turn. The M3TTE model can essentially complete the classification task as long as it receives packet data, with relatively minor impacts from network latency and other factors.

4.4   Hyperparameter Experiment

From Table 3, we can observe the impact of the number of bytes selected in one packet on the classification results in the M3TTE model. As the number of bytes increases, the training time of the model becomes longer, and the classification results of task1 and task2 gradually improve. The accuracy and F1 for task1

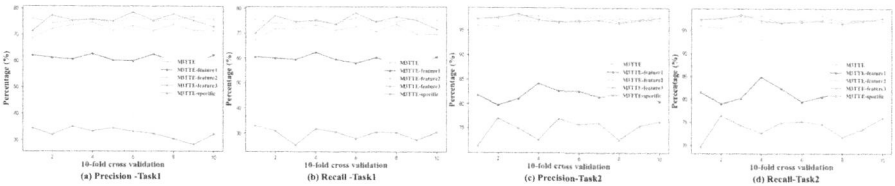

Fig. 8. Performance of ablation study.

Table 3. Average performance of classification by different byte numbers selected.

	AC(%)-task1	F1(%)-task1	AC(%)-task2	F1(%)-task2	Time(s)
M3TTE-30b	68.62±1.38	68.51±1.46	94.29±0.59	94.16±0.57	663.5
M3TTE-40b	70.15±1.25	70.17±1.19	96.75±0.45	96.75±0.45	798.2
M3TTE-50b	**74.93±0.91**	**74.89±0.94**	**97.48±0.59**	**97.48±0.58**	986.4
M3TTE-60b	73.99±1.63	73.99±1.66	96.75±0.61	96.75±0.61	1142.6
M3TTE-70b	73.65±1.36	73.66±1.40	96.75±0.80	96.75±0.80	1323.7

increase from 68.62% and 68.51% to 74.93% and 74.89%, respectively. Similarly, for task2, the metrics rise from 94.29% and 94.16% to 97.48% and 97.48%. However, after the byte value exceeds 50, the impact on the results gradually diminishes. Consistent with previous comparative experiments, the results of task1 exhibit smaller fluctuations compared to task2. Regarding the selection of the number of bytes, 50 appears to be a good choice in the experiment, as it provides a balanced training time and yields the best performance in the above experiments.

Table 4. Average performance of classification by different packet numbers for representation.

	AC(%)-task1	F1(%)-task1	AC(%)-task2	F1(%)-task2	Time(s)
M3TTE-6p	66.99±1.13	67.05±1.18	96.07±0.95	96.06±0.96	664.5
M3TTE-8p	71.39±1.37	71.39±1.38	96.54±0.78	96.54±0.79	804.5
M3TTE-10p	74.93±0.91	74.89±0.94	**97.48±0.59**	**97.48±0.58**	986.4
M3TTE-12p	75.63±0.99	75.67±1.00	96.88±0.53	96.87±0.53	1132.0
M3TTE-15p	79.94±0.96	79.99±0.99	97.21±0.35	97.21±0.35	1404.9
M3TTE-20p	**82.25±1.83**	**82.27±1.85**	97.30±0.64	97.30±0.64	1910.4

Table 4 illustrates the impact of the number of packets used in the M3TTE model on the classification results. The training time of the model increases with the growth of the number of packets. In task1, the classification performance improves as the number of groups increases. The accuracy and F1 scores increase from 66.99% accuracy and 67.05% F1 to 82.25% accuracy and 82.27% F1 with a balanced consideration of performance and training time. In task2, classification performance is relatively stable, with accuracy and F1 scores remaining around 96% and 97%, respectively. The metrics in task2 show little variation, indicating that the number of packets used in flow sampling has a minimal impact on CDN brand classification. Choosing to sample 10 packets can obtain relatively optimal and efficient results.

Table 5. Average performance of classification by different sampling strategy selection.

	AC(%)-task1	F1(%)-task1	AC(%)-task2	F1(%)-task2	Time(s)
First N	**74.93±0.91**	**74.89±0.94**	**97.48±0.59**	**97.48±0.58**	986.4
Random N	57.50±1.47	57.60±1.58	95.41±0.82	95.32±0.96	955.1
Random continous N	57.09±2.44	57.16±2.46	93.84±0.52	93.82±0.51	968.6
Mid N	60.10±1.79	60.02±1.80	93.26±1.04	93.21±1.09	958.3

We can understand the impact of flow sampling strategy on the results of the M3TTE model application from Table 5. When the number of sampled packets and bytes is constant, the flow sampling strategy has a negligible effect on the model's training time. The sampling strategy essentially analyzes the content and importance of packets at different positions in the flow. Among the four flow sampling strategies, the First N strategy performs the best, especially in task1, where it outperforms the other three strategies with a large margin. The accuracy and F1 scores improve by approximately 15%. In task2, the four strategies exhibit relatively similar performance, with the First N strategy leading by about 2% in accuracy and F1 over the second-ranked Random strategy. This indicates that the initial packets of a flow carry more important information, particularly regarding website category information. In contrast, the information about CDN brand can be obtained from packets at different positions in the flow, but the information in the initial packets is more conducive to discrimination.

5   Conclusion and Future Work

Taking CDN as a critical infrastructure of the current Internet, this paper proposes a method to optimize the user experience and Quality of Service (QoS) of CDN by classifying CDN website traffic into categories and brands, setting transmission priorities along two dimensions. To address this issue, we reflect on the shortcomings of previous studies and introduce a scalable classification model, M3TTE. We explain the design principles, specific architecture, and computational processes of this model in detail.

To evaluate the effectiveness of M3TTE and the importance of its components, we conduct complex experiments on our proprietary dataset, WebT2023.

The results of comparative experiments indicate that M3TTE has better experimental performance than the existing single-task and multi-task models. When compared with models that exhibit suboptimal performance, M3TTE proves to be more time-efficient. The ablation study highlights the necessity of both features and structure within the M3TTE model for the task of CDN website classification. The impacts of hyperparameters are also analyzed.

In future work, we aim to further optimize the details of the model to enhance its performance and efficiency. Additionally, classes with fewer samples may experience classification challenges. Exploring solutions to solve this problem is an avenue worth investigating in our future research. We plan to conduct corresponding research and experiments in the next stage.

References

1. Akbari, I., et al.: A look behind the curtain: traffic classification in an increasingly encrypted web. Proc. ACM Measur. Analy. Comput. Syst. **5**(1), 1–26 (2021)
2. Alshammari, R., Zincir-Heywood, A.N.: A flow based approach for SSH traffic detection. In: 2007 IEEE International Conference on Systems, Man and Cybernetics, pp. 296–301. IEEE (2007)
3. Alshammari, R., Zincir-Heywood, A.N.: Machine learning based encrypted traffic classification: identifying SSH and skype. In: 2009 IEEE Symposium on Computational Intelligence for Security and Defense Applications, pp. 1–8. IEEE (2009)
4. Moore, A.W., Zuev, D.: Internet traffic classification using Bayesian analysis techniques. ACM SIGMETRICS Perform. Eval. Rev. (2005)
5. Calder, M., Flavel, A., Katz-Bassett, E., Mahajan, R., Padhye, J.: Analyzing the performance of an anycast CDN. In: Proceedings of the 2015 Internet Measurement Conference, IMC 2015, pp. 531–537. Association for Computing Machinery, New York (2015). https://doi.org/10.1145/2815675.2815717
6. Cherubin, G., Jansen, R., Troncoso, C.: Online website fingerprinting: evaluating website fingerprinting attacks on tor in the real world. In: 31st USENIX Security Symposium (USENIX Security 22), pp. 753–770 (2022)
7. Cruz, M., Ocampo, R., Montes, I., Atienza, R.: Fingerprinting BitTorrent traffic in encrypted tunnels using recurrent deep learning. In: 2017 Fifth International Symposium on Computing and Networking (CANDAR), pp. 434–438. IEEE (2017)
8. Du, N., Huang, Y.: GLaM: efficient scaling of language models with mixture-of-experts (2022)
9. Gang, X., Jiao, M., Zigang, C., Yong, W., Li, G., Binxing, F.: Research progress and prospects of network traffic classification. Integr. Technol. **1**(1), 32–42 (2012)
10. Guo, R., et al.: CDN Judo: breaking the CDN DoS protection with itself. In: Proceedings 2020 Network and Distributed System Security Symposium (2020). https://api.semanticscholar.org/CorpusID:211263624
11. Huang, H., Deng, H., Chen, J., Han, L., Wang, W.: Automatic multi-task learning system for abnormal network traffic detection. Int. J. Emerg. Technol. Learn. **13**, 4–20 (2018). https://api.semanticscholar.org/CorpusID:55018932
12. Insight and Info: 2021 China CDN Service Market Research Report-Analysis of Market Scale Current Situation and Development Trends. Insight and Info (2021)
13. Li, W., et al.: CDN backfired: amplification attacks based on http range requests. In: 2020 50th Annual IEEE/IFIP International Conference on Dependable Systems

and Networks (DSN), pp. 14–25 (2020). https://doi.org/10.1109/DSN48063.2020.00022

14. Lin, K., Xu, X., Gao, H.: TSCRNN: a novel classification scheme of encrypted traffic based on flow spatiotemporal features for efficient management of iiot. Comput. Netw. **190**, 107974 (2021)
15. Liu, X., et al.: Attention-based bidirectional GRU networks for efficient https traffic classification. Inf. Sci. **541**, 297–315 (2020)
16. Ma, J., Zhao, Z., Yi, X., Chen, J., Hong, L., Chi, E.H.: Modeling task relationships in multi-task learning with multi-gate mixture-of-experts. In: Proceedings of the 24th ACM SIGKDD International Conference on Knowledge Discovery & Data Mining, KDD 2018, pp. 1930–1939. Association for Computing Machinery, New York (2018). https://doi.org/10.1145/3219819.3220007
17. Mordor Intelligence: Content Delivery Network (CDN) Security Market Size and Share Analysis - Growth Trends and Forecasts (2023–2028). Mordor Intelligence (2023)
18. Papadogiannaki, E., Ioannidis, S.: A survey on encrypted network traffic analysis applications, techniques, and countermeasures. ACM Comput. Surv. (CSUR) **54**(6), 1–35 (2021)
19. Rezaei, S., Kroencke, B., Liu, X.: Large-scale mobile app identification using deep learning. IEEE Access **8**, 348–362 (2019)
20. Ruder, S.: An overview of multi-task learning in deep neural networks (2017)
21. Shen, M., et al.: Machine learning-powered encrypted network traffic analysis: a comprehensive survey. IEEE Commun. Surv. Tutorials (2022)
22. Sun, J., et al.: iSwift: fast and accurate impact identification for large-scale CDNs. In: 2022 IEEE/ACM 30th International Symposium on Quality of Service (IWQoS), pp. 1–10 (2022). https://doi.org/10.1109/IWQoS54832.2022.9812890
23. Sun, X., Panda, R., Feris, R., Saenko, K.: AdaShare: learning what to share for efficient deep multi-task learning (2020)
24. Wang, P., Ye, F., Chen, X., Qian, Y.: DataNet: deep learning based encrypted network traffic classification in SDN home gateway. IEEE Access **6**, 55380–55391 (2018)
25. Wang, W., Zhu, M., Wang, J., Zeng, X., Yang, Z.: End-to-end encrypted traffic classification with one-dimensional convolution neural networks. In: 2017 IEEE International Conference on Intelligence and Security Informatics (ISI), pp. 43–48. IEEE (2017)
26. Xie, G., Li, Q., Jiang, Y.: Self-attentive deep learning method for online traffic classification and its interpretability. Comput. Netw. **196**, 108267 (2021)
27. Xue, F., Shi, Z., Wei, F., Lou, Y., Liu, Y., You, Y.: Go wider instead of deeper (2021)
28. Yao, H., Liu, C., Zhang, P., Wu, S., Jiang, C., Yu, S.: Identification of encrypted traffic through attention mechanism based long short term memory. IEEE Trans. Big Data **8**(1), 241–252 (2019)

Malware Variant Detection Based on Knowledge Transfer and Ensemble Learning

Yu Ding[1], Haoliang Sun[2(✉)], Binbin Li[1,3], Zisen Qi[1,3], Siyu Jia[1,3], Haiping Wang[1,3], and XingBang Tan[1]

[1] Institute of Information Engineering, Chinese Academy of Sciences, Beijing, China
dingyu@iie.ac.cn
[2] National Computer Network Emergency Response Technical Team/Coordination Center of China, Beijing, China
sunhl@cert.org.cn
[3] School of Cyber Security, University of Chinese Academy of Sciences, Beijing, China

Abstract. In the realm of cybersecurity, malware adopts various evasion tactics, such as obfuscation and code rewriting, to evade detection by network security protection systems, perpetually evolving in sophistication. This presents a formidable challenge for traditional signature-based and machine learning-based malware detection methods, rendering them ineffective against emerging malware variants. To mitigate this problem, security analysts strive to detect malware variant samples at the earliest possible stage. This typically involves labor-intensive manual analysis, where analysts meticulously scrutinize network activities, system calls, and other suspicious behaviors. However, given the vast volume of potentially malicious samples identified daily by network security systems, discerning the 'high-value' samples worthy of further investigation becomes a daunting task. To address this challenge, we propose a novel malware variant detection method rooted in knowledge transfer and ensemble learning. This method aims to detect and identify malware variant samples, streamlining the process of screening out highly suspected files for security analysts. By doing so, it alleviates the burden of manual analysis and judgment. This study leveraged the Microsoft Kaggle dataset to simulate real-world scenarios of malware variant detection. The results demonstrate a recall rate of 61.9% for malware variant samples, with an accuracy rate of 56.52%. Notably, security analysts can pinpoint a malware variant sample by analyzing an average of 1.77 samples, significantly reducing the manual analysis workload. These experimental findings underscore the effectiveness and robustness of our proposed method.

Keywords: Malware Variant Detection · Knowledge Transfer · Ensemble Learning

J. Zhao and W. Meng (Eds.): SciSec 2024, LNCS 15441, pp. 387–401, 2025.
https://doi.org/10.1007/978-981-96-2417-1_21

1   Introduction

In order to avoid detection by network security protection systems, malware usually uses methods such as obfuscation and rewriting to modify the code for continuous iterative evolution [5]. According to Kaspersky Lab's 2023 annual report *Kaspersky Security Bulletin 2023. Statistics* [1], 23,364 code modifications were detected for ransomware alone, and 43 new malware family variants were discovered. Especially in the context of targeted attacks by attackers with advanced technical means like state-sponsored attack groups, malware attacks have become more complex and the situation has become increasingly severe. Under this challenge, traditional signature-based and machine learning-based malware detection methods cannot detect new malware variants.

Traditional signature-based methods primarily rely on the distinctive signatures of malicious samples to identify malware files [6,14]. These signatures are byte sequences unique to each malware sample. While effective for known malware samples, signature-based methods struggle to identify new malware variants not present in the signature database. Furthermore, they require extensive manual analysis and judgment to detect new malware families, often after the malware has already caused significant damage. Similarly, traditional machine learning-based methods are limited to training on known malicious samples, and learning the characteristics of established malware families. While these methods demonstrate high accuracy in detecting known malware families, they fail to identify the unique features of malware variants due to the absence of corresponding samples during training. Consequently, they are ill-equipped to detect and classify emerging malware variants.

In order to discover malware variant samples as early as possible, security analysts usually use sandboxing and other technologies to conduct detailed manual analysis and judgment on network activities, system calls and other behaviors suspected to be malware variant samples. This kind of manual analysis and judgment is time-consuming and labor-intensive. Due to the immense number of malicious samples identified daily by network security protection systems, and the low percentage of these samples that are genuine variants of malicious software, it is crucial to screen out high-value malicious variants from the vast pool of malware samples for further manual analysis and research. Traditional malware family classification focuses more on which malware family a malicious sample belongs to, while malware variant sample detection focuses more on whether the sample is a malware variant. The challenge of this problem is that there is no malware variant sample data, which is similar to the detection of unknown category samples. Because the classification problem requires model training based on sample data with category labels, this problem cannot be solved simply with a classification algorithm.

In light of this background, We propose a novel malware variant detection method based on knowledge transfer and ensemble learning. The method analyzes the malware's disassembled ASM file, extracts the opcodes, and converts them into vectors. These vectors are used to train multiple models to identify known malicious families. Ensemble learning is then used to determine whether the malware is a variant sample. The main contributions are as follows:

- We propose a novel malware variant detection method based on knowledge transfer and ensemble learning, which can identify variant samples of malware. This method aids security analysts in screening out sample files highly suspected to be malware variants, thereby reducing the workload of manual analysis and judgment. The effectiveness and robustness of the proposed method are verified by our experiments.
- The model training of the method proposed in this paper is divided into two stages: pre-training and transfer training. The pre-training stage could use all known malware samples to train a multi-classification Transformer Block, and the transfer training a binary classifier for each malware family based on the Transformer Block.
- On the open-source Microsoft Kaggle dataset, this method achieved a recall rate of 61.9% and an accuracy rate of 56.52%. Security analysts can identify a malware variant sample by analyzing an average of only 1.77 samples, significantly reducing the manual analysis burden.

2 Related Work

Malware detection and classification have been a research hotspot in academia for a long time. Classic techniques for it can be categorized into three main types: Static, Dynamic, and Hybrid analysis [12]. Static analysis refers to analyzing the Portable Executable files (PE files) without running them. Dynamic analysis executes infected files in a simulated environment using virtual machines or emulators to analyze their malicious functionality [9]. Hybrid analysis combines both static and dynamic methods for analysis improvement. [4]

Since the evolution of malware causes the performance of ML classification models to significantly degrade over time [15], this problem is defined as model aging or similar concepts such as time decay [10], degradation [2] and model degradation [8]. Hammad et al. [7] evaluated the effectiveness of top Android anti-malware products against code obfuscation. They used a large dataset containing 3,000 benign applications, 3,000 malicious applications, and 73,362 obfuscated applications. The research found that all anti-malware products succumb to code obfuscation, and most products are susceptible to trivial transformations, such as simple changes to Android manifest files. Zhang et al. [15] They proposed a framework named APIGraph, which utilizes the similarity information among Android malware in terms of semantically equivalent or similar API usages to enhance state-of-the-art malware classifiers, thus naturally slowing down classifier aging. Cordonsky et al. [3] introduced an innovative method for discriminating between known and unknown malware families.

Their approach involved training a neural network using sandbox behavior data from malware samples. Following this, they eliminated the pre-softmax layer of the network to craft a signature generation model capable of generating unique signatures for each sample. These signatures were then represented as 30-dimensional vectors. Subsequently, the researchers analyzed the output vector space to establish a decision boundary, determined by the Euclidean distance of each point from the origin. This approach enabled the method to discern whether a given malware instance was a variant or not.

3 Methodology

In this section, we introduce the malware variant detection method as illustrated in Fig. 1. The goal of this method is to determine whether malicious software samples are variants. To achieve this objective, we first pre-process the disassembled ASM files of malicious samples, extracting opcodes and converting them into an opcode vector. This vector serves as input for the malware variant detection model. Subsequently, we train models to identify multiple known malicious families. we conduct model training to recognize multiple known malicious families, which consists of two stages: pre-training and transfer training. Finally, the predicted results of these models are integrated to determine whether the malware is a variant sample. We will provide a detailed introduction of the method in the following subsections.

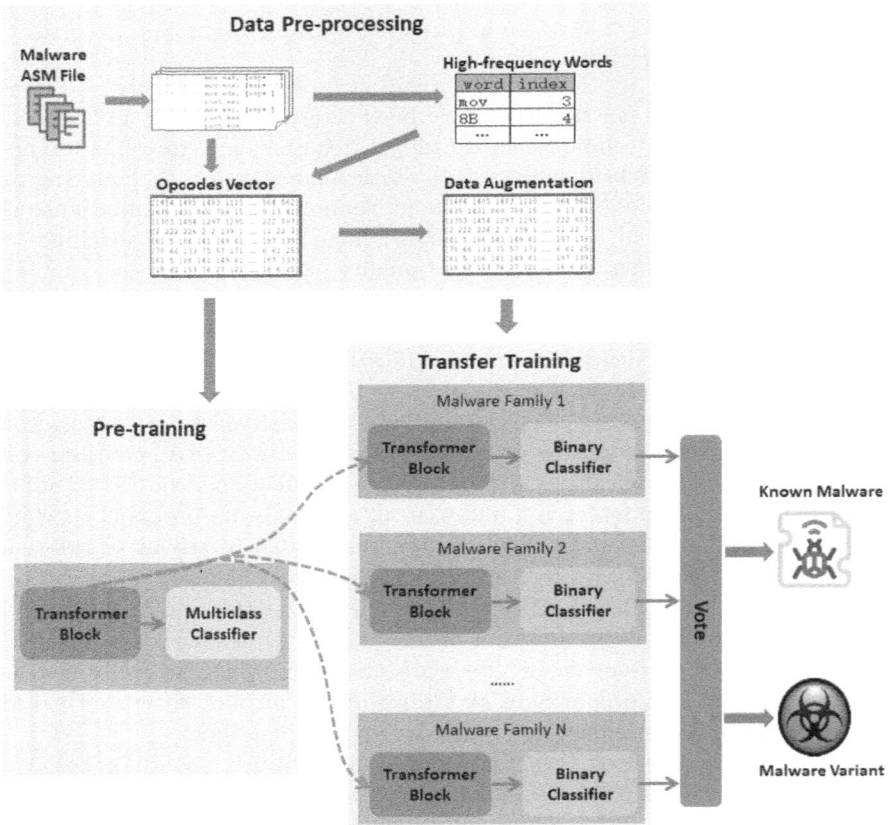

Fig. 1. Overview of our method.

3.1   Data Source

In this study, we detect variants of malware samples based on the disassembled binary files in the ASM format. The ASM file is a source file in assembly language composed of a series of statements, each of which ends with a line feed or semicolon. Figure 2 shows the format of the ASM file in our dataset, containing hexadecimal code and corresponding assembler code as well as metadata information about the binary file, such as function calls and parameters. Behavioral patterns differ between different malware families, which are controlled by code. Therefore, identifying features of malware families can be obtained from the ASM files of malware samples.

```
.text:00401060  8B 44 24 08     mov eax, [esp+8]
.text:00401064  8A 08           mov cl, [eax]
.text:00401066  8B 54 24 04     mov edx, [esp+4]
.text:0040106A  88 0A           mov [edx], cl
.text:0040106C  C3              retn
.text:0040106C                                  ;  --------------------------
.text:0040106D  CC CC CC        align 10h
.text:00401070  8B 44 24 04     mov eax, [esp+4]
.text:00401074  8D 50 01        lea edx, [eax+1]
.text:00401077
.text:00401077                  loc_401077:      ; CODE XREF: .text:0040107Cj
.text:00401077  8A 08           mov cl, [eax]
.text:00401079  40              inc eax
.text:0040107A  84 C9           test    cl, cl
.text:0040107C  75 F9       jnz short loc_401077
.text:0040107E  2B C2           sub eax, edx
.text:00401080  C3              retn
```

Fig. 2. The format of the ASM file in our dataset. It contains the hexadecimal opcode (the left decimal digit) and the corresponding assembler code (the right assembly language).

3.2   Data Pre-processing

The ASM files of each malware are textual and necessitate preprocessing for digital representation before training. Our approach involves transforming the ASM file into a fixed-length integer vector, accomplished through a three-step conversion process:

- **Extracting Opcodes**: The contents within the red box in Fig. 2 are referred to as **opcodes**. We extract the hexadecimal numbers for each line of the ASM file and their corresponding disassembly opcodes while removing the '**.text**', '**.rdata**', and other section identifiers at the beginning of each line, as well as any comments at the end of each line.
- **Words Representation**: This step aims to count the number of occurrences of each word in all ASM files and sort the words in descending order according to the number of occurrences. Since the words in the files contain a large

number of variable values, the number of words has reached 80 million. To reduce the computational complexity, only the top **N** words with the highest occurrence times are selected as the high-frequency words. We represent high-frequency words using integer numbers, which are based on the descending order of word occurrences. All the Low-frequency word sequences are represented using a special non-zero integer, requiring the use of N+1 integer numbers to represent all words.

– **Conversion of Opcodes to Vectors**: Initially, each malware opcode sample transforms a vector based on the N+1 word index. However, due to the considerable length of these words, to expedite training, only the first **L** words of each file are chosen as the definitive representation of the sample. Consequently, this yields a 1-D vector of length L for each raw ASM file of a malware sample.

3.3 Malware Variant Detection

Model Training for Malware Variant Detection is divided into two stages: **pre-training** and **transfer training**. In the pre-training phase, the objective is to train a Transformer Block on the opcode vectors of all known malware families. This enables the Transformer Block to learn the characteristics of each known malware family. In the transfer training phase, each malware family is trained separately based on the pre-trained Transformer Block. This process yields a binary classifier for each specified malware family. Finally, the determination of whether a sample is a malicious variant is based on the voting results of N specified malware family identifiers, with the first L words of each file chosen as the definitive representation of the sample.

Pre-training. The pre-training phase involves training the Transformer Block and Classifier based on the sample data from all known malware families. Each malicious sample in the training dataset is labeled with its real malware family. Upon completing the model training in the pre-training phase, the trained Transformer Block is saved and subsequently utilized to train a specified malware family classifier in the transfer training phase.

In detail, the training model in the pre-training stage consists of two components: the Transformer Block and the Classifier. The input to the Transformer Block is an opcode vector constructed based on the ASM file of the malware sample, and its output is the vector representation of the malicious sample in the feature space. The Classifier in the pre-training stage is a multi-classifier for all known malware families. Both the Transformer Block and the Classifier are trained using the real labels of all known samples, aiming to enable the Transformer Block to learn the characteristics of all known malware families. The implementation of the Transformer Block can utilize various neural network structures. In the experiments of this paper, it was implemented based on the self-attention neural network [13]. The neural network structure of the Transformer Block is illustrated in Fig. 3.

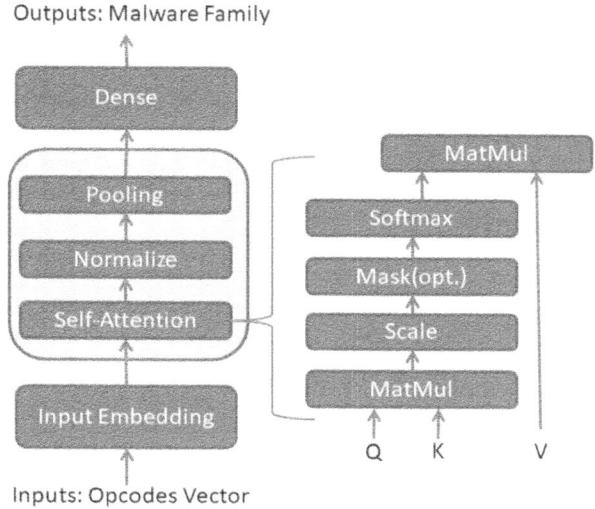

Fig. 3. Our proposed self-attention neural network model.

Transfer Training. The transfer training phase builds upon the Transformer Block from the pre-training phase. Transfer training is conducted for each known malware family to develop a two-classification model tailored to a specified malware family. The key idea is to transfer the classification feature knowledge acquired by the Transformer Block during pre-training to a dedicated malware family classification model.

The binary classification model for a specified malware family resembles the model from the pre-training stage and comprises a Transformer Block and a Classifier. The structure of the Transformer Block remains consistent with that of the pre-training stage, with weights initialized to those obtained during pre-training. The distinction lies in the Classifier of the N-specified malware family identification models in the transfer training phase, which is a binary classifier constructed based on the Sigmoid function. The output result is either 0 or 1, indicating whether the predicted sample belongs to the specified malware family. The training samples utilized in the transfer training phase remain consistent with the pre-trained training samples. However, during the training of the binary classification model for each malware family, the labels of samples not belonging to that family are designated as 0, while the labels of samples from the specified family are marked as 1. Due to the fewer number of samples for some malware families compared to samples from other families, there exists an imbalance in the training data, leading to lower recognition performance of classifiers for certain malware families. To address this issue, data augmentation operations are performed on the samples of each malware family during model identification. Specifically, a new training sample is constructed by flipping the feature vector of the malware family samples, thereby enhancing the training dataset and improving the classifier's performance for the specified malware family.

Through transfer training, N-designated malware family recognition models are obtained. When predicting samples, these N recognition models vote on whether the sample belongs to the known N malware families. Samples with a total vote count of 0 are malware variant samples and samples with a total vote count of ≥ 1 are known malware family samples. This method can be expressed by the following formula:

$$v = \sum_{i=1}^{N} max(\text{Sigmoid}_i(\text{Transformer}_i(\text{x}))) \tag{1}$$

$$y = \begin{cases} \text{Non-Variant if } v \geq 1 \\ \text{Variant} \quad \text{if } v = 0 \end{cases} \tag{2}$$

where N represents the total number of malware family identification models. $max(\text{Sigmoid}_i(\cdot))$ represents the voting result (0 or 1) of the i-th malware family identification model for this sample. Transformer represents the neural network used by the Transformer Block.

By summing the votes of N malware family identification models, the total number of votes v for the sample is obtained. When the total number of votes $v \geq 1$, the sample is determined to be a known malware family sample (Non-Variant); when the total number of votes $v = 0$, the sample is determined to be a malicious variant family sample (Variant).

4   Experiments and Results

In this study, two sets of experiments were conducted using *Simda* and *Kelihos_ver1* as malware variant samples, respectively. This approach simulated the occurrence of multiple types of malware variant samples in real scenarios and verified the feasibility and effectiveness of the detection model. Through comparative experiments, it has been demonstrated that incorporating pre-training and data augmentation can substantially enhance the detection accuracy of malware family samples.

4.1   Dataset

The dataset utilized in this study is sourced from the Microsoft Malware Classification Challenge (BIG 2015) [11], an open-source dataset. It comprises 10,868 malware ASM files, with each ASM file belonging to a specific malware family. The dataset encompasses 9 malware families, and the distribution of samples for each malware family is illustrated in Fig. 4.

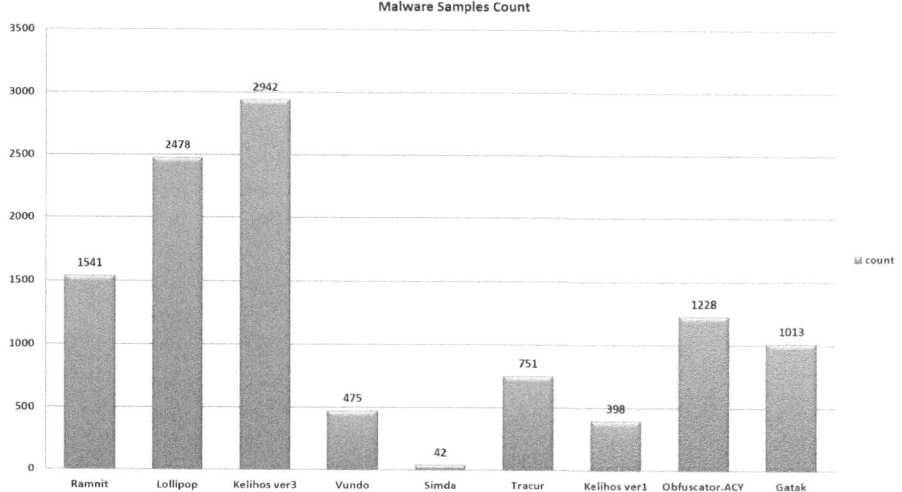

Fig. 4. The number of samples for each malware family.

4.2 Experiment Parameters

In the data preprocessing stage of this experiment, the opcodes of each ASM file are extracted. Subsequently, the top 10,000 most commonly used opcodes are selected as the top N high-frequency opcodes. These top N high-frequency opcodes are then mapped to N integers in descending order of occurrence, ranging from 3 to N + 2. Other low-frequency opcodes are mapped to a special integer, designated as 2. Each sample's opcode is converted into an integer vector based on the corresponding N integer sequence index. To streamline the model training process, the first L integers of each vector are used as training vectors, where L is set to 15,000 in this experiment.

In both the model pre-training and transfer training stages, the Transformer Block employed is a self-attention neural network structure. The input of the self-attention model is a 1D vector with a length of L. The parameter value of the Embedding layer in this experiment is set to 128, and the output dimension of the Self-Attention layer is configured as 192. Additionally, the epochs for both the pre-training and transfer training stages are set to 12. The min_delta of early stopping's $val\_accuracy$ is set at 0.00005, with a patience of 3.

4.3 Model Evaluation

The objective of this methodology is to identify malware variant samples, with a particular focus on effectively filtering out variant sample data. Initially, security researchers utilize this approach to sift through collected malware samples, isolating those that are highly suspected to be variants. Subsequently, researchers employ sandboxing and other technical methodologies to conduct a comprehensive analysis and assessment of these selected samples. The samples identified herein are highly suspected to be malware variants, consisting of true cases (TP) and false positives (FP). Thus, the effectiveness of the method is evaluated based on the presence of actual malicious variants among the samples detected as malware variants, with a higher count of true positives indicating superior performance. Furthermore, it is essential to minimize the inclusion of known malware family samples among the samples identified as malware variants, with the goal of reducing false positives to the minimum possible extent. To assess the performance of our method, we employ four key metrics: accuracy, precision, recall, and F1-score. These metrics provide a quantitative measure of the model's effectiveness in correctly classifying samples. The formulas for accuracy and F1-score are as follows:

$$\text{Accuracy} = \frac{TP + TN}{TP + TN + FP + FN} \tag{3}$$

$$\text{Precision} = \frac{TP}{TP + FP} \tag{4}$$

$$\text{Recall} = \frac{TP}{TP + FN} \tag{5}$$

$$\text{F1-score} = 2 \times \frac{\text{Precision} \times \text{Recall}}{\text{Precision} + \text{Recall}} \tag{6}$$

where TN represents true negatives and FN represents false negatives.

4.4   Results Analysis

We conducted two sets of experiments to validate the effectiveness of our proposed method. The first set of experiments used *Simda* as a malware variant and compared the model's performance when pre-training and data augmentation were used. The second set of experiments used both *Simda* and *Kelihos_ver1* as malware variants to simulate and evaluate the detection and identification performance of our method in scenarios where multiple malware variant types exist. The experimental results of the two sets of experiments are shown in Table 1. The following subsections will provide a detailed description and analysis of the experimental results.

Experiment 1: *Simda* as the Malware Variant This series of experiments focuses on using Simda as the malware variant. Simda family samples are not involved in the pre-training and transfer training stages. Instead, they are incorporated into the test set solely for evaluating the detection performance of the trained model. Specifically, 80% samples of the BIG 2015 dataset served as the training set, with the remaining 20% samples allocated to the test set. The test set comprises 2251 samples, including 42 malware variant samples (Simda family) and 2209 known malware family samples.

Table 1. Malware variant detection experiment results. 'DA' denotes Data Augmentation.

Exp.	Variant	Pretrain	DA	Accuracy	Precision	Recall	F1-Score
1	Simda	✗	✗	95.88%	24.21%	54.76%	33.58%
		✓	✓	**98.37%**	56.52%	**61.90%**	59.09%
		✓	✗	95.20%	21.93%	59.52%	32.05%
		✗	✓	97.74%	40.00%	38.10%	39.02%
2	Simda & Kelihos_ver1	✓	✓	88.88%	**89.95%**	40.68%	56.03%

This experimental series encompasses four experiments. The baseline model, which does not employ pre-training and data augmentation, serves as a reference. The effectiveness and advanced performance of employing pre-training and data augmentation are validated in another experiment. Additionally, experiments utilizing only pre-training or data augmentation individually assess the extent to which these techniques enhance model accuracy.

- **Using pre-training and data augmentation**

As depicted in Table 1, the model's accuracy using pre-training and data augmentation reached 98.40%, with a precision rate of 56.52%, recall rate of 61.90%, and F1-Score of 59.09%. The confusion matrix is presented in Table 2, indicating that out of the 46 malware variant samples detected, 26 were correctly identified, while 20 were false positives. Compared with the baseline model, the model employing pre-training and data augmentation exhibited improvements across various metrics: accuracy increased by 2.49%, precision rate by 32.31%, recall rate by 7.14%, and F1-Score by 25.51%. Notably, the average number of sample data analyzed for a malicious variant sample decreased from 4.13 to 1.77. This method effectively filters out known malware family samples, identifies highly suspected malware variant samples, and reduces the workload of manual sample analysis and judgment.

Table 2. Confusion matrices for pre-training and data augmentation were used

True	Predicted	
	Malware Variant	Known Malware Family
Malware Variant	26	16
Known Malware Family	20	2146

The Ramnit family exhibits the highest number of samples mistakenly detected as malware variants among known malware families, totaling 6 samples. This is followed by the Lollipop family and the Gatak family, each with 4 samples detected erroneously. The actual distribution of the detected samples among different families is illustrated in Fig. 5.

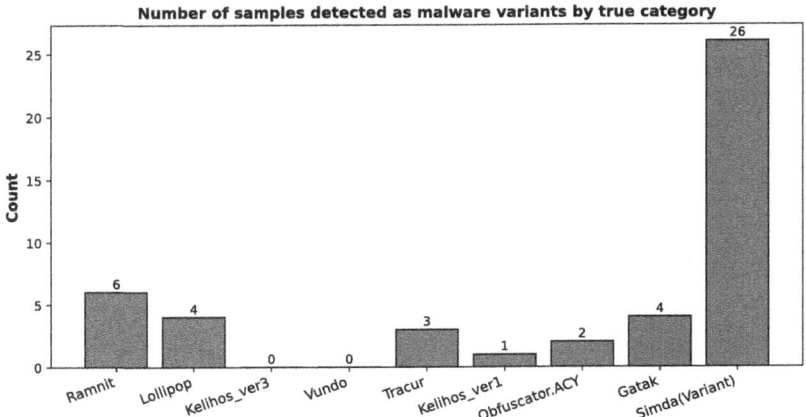

Fig. 5. Number of samples detected as malware variants by true category.

- **Solely relying on pre-training without data augmentation**

In this experiment, the model solely relies on pre-training without utilizing data augmentation. The results of the experiments are presented in Table 1. Comparing these results to the baseline model (which does not utilize pre-training or data augmentation), there is no significant improvement in the detection performance of malware variant samples. Moreover, compared to the model using both pre-training and data augmentation, using pre-training alone leads to marginal improvements, with an increase in accuracy of 0.63%, a precision rate increase of 16.52%, and a recall rate increase of 23.8%, resulting in an increase of 20.07% in the F1-Score. The confusion matrix for this experiment is presented in Table 3. Notably, 114 malware variant samples were detected, with only 25 correctly identified out of 42 actual malware variants, while 89 samples from known malware families were erroneously classified as malware variant samples. Security analysts would need to analyze an average of 4.56 samples to discover a malware variant. Compared to a model utilizing both pre-training and data enhancement, the absence of pre-training significantly increases the workload of manual analysis and judgment, approximately 2.58 times higher.

The experimental results indicate that pre-training alone does not improve model performance. However, when combined with data enhancement techniques, pre-training significantly boosts model performance. This difference is due to how pre-training interacts with the subsequent migration training process. When pre-training occurs without data enhancement, the model's parameters cannot be effectively updated during transfer learning because both pre-training and migration training use the same training data. In contrast, data enhancement introduces new, augmented samples during migration training, allowing the pre-trained model to better learn specific features of malicious families and thus improve performance.

Table 3. Confusion matrix when using only pre-training

True	Predicted	
	Malware Variant	Known Malware Family
Malware Variant	25	17
Known Malware Family	89	2077

– **Solely relying on data augmentation without pre-training**

In Table 1, the model achieves an accuracy rate of 97.74%, a precision rate of 40.00%, a recall rate of 38.10%, and an F1-Score of 39.02% without pre-training but with data augmentation. Compared to the baseline model, the detection accuracy of malware variant samples increases by 1.86%, precision rate by 15.79%, and F1-Score by 5.44%, despite a decrease in recall rate by 16.66%. Moreover, when utilizing pre-training alongside data augmentation, the model's accuracy improves by 0.63%, precision rate by 16.52%, and recall rate by 23.8%, resulting in a remarkable 20.07% enhancement in F1-Score. The confusion matrix depicted in Table 4 highlights that out of 40 detected malware variant samples, 16 were accurately identified, while 24 of the 42 actual malware variants were detected, indicating some misclassification. On average, security analysts investigate 2.5 samples to identify a malware variant. Notably, solely relying on data enhancement sans pre-training escalates the manual analysis and judgment workload by 1.4 times.

Overall, these results indicate that data enhancement can effectively improve the detection effect of malware variant samples and reduce the workload of manual analysis and judgment. Compared with pre-training, data augmentation improves model performance more significantly.

Table 4. Confusion matrix when using only data augmentation

True	Predicted	
	Malware Variant	Known Malware Family
Malware Variant	16	42
Known Malware Family	24	2142

Experiment 2: *Simda* & *Kelihos_ver1* as the malware variant This experiment involves utilizing multiple malicious family samples as test data and simulating a scenario where multiple malware variants coexist in the malicious sample data. Specifically, samples from the two malicious families, Simda and Kelihos_ver1, with the smallest sample data, were selected as malware variants. An 80% sample size was allocated for the training set, while the remaining 20% comprised the test set.

In Table 1, the model achieves an accuracy rate of 88.88%, a precision rate of 89.95%, a recall rate of 40.68%, and an F1-Score of 56.03%. The test set encompasses a total of 2,526 test samples, comprising 440 malware variant samples from Simda and Kelihos_ver1 families and 2,086 known malware family samples. The confusion matrix shown in Table 5 indicates that out of 199 detected malware variant samples, 179 were accurately identified from the 440 actual malware variant samples, while 20 samples from known malware families were misclassified as malware variants. Out of the 199 detected variants of malware samples, 179 were confirmed to be genuine instances of malware. On average, security analysts analyzed 1.11 samples to identify a variant of malicious software. The experimental results suggest that this method retains its robustness even when faced with multiple types of malware variants.

Table 5. Confusion matrix when Simda and Kelihos_ver1 are unknown malware family.

True	Predicted	
	Malware Variant	Known Malware Family
Malware Variant	179	261
Known Malware Family	20	2066

5   Conclusion

Traditional machine learning-based methods rely solely on known samples of malicious software families for training, limiting their ability to detect new variants. The emergence of malware variants poses a more serious threat compared to known families, emphasizing the importance of early detection to mitigate potential losses. To address this challenge, we proposed a knowledge transfer and ensemble decision-based detection technique for malware variants. Our approach processes the disassembled ASM files of malware samples, converting text into vector representations as inputs to the detection model. Training the model involves two stages: pre-training and transfer learning. In the pre-training stage, a multi-class model is trained on samples from all malware families to obtain a Transformer Block. Subsequently, in the transfer learning stage, we further train binary classification models for each malware family based on the pre-trained weights. Finally, the ensemble decision is made based on the voting results of N binary models to determine whether a sample is a malware variant. Experiment results on the Microsoft Kaggle dataset demonstrate the effectiveness of our approach. We achieved a recall rate of 61.9% and an accuracy rate of 56.52% for detecting malware variant samples. Security analysts can discover a malware variant sample by analyzing an average of 1.77 samples, significantly reducing the manual analysis workload. These results underscore the validity and robustness of our method.

References

1. AMR: Kaspersky security bulletin 2023. statistics. https://securelist.com/ksb-2023-statistics/111156/. Accessed 04 Dec 2023

2. Cai, H.: Assessing and improving malware detection sustainability through app evolution studies. ACM Trans. Softw. Eng. Methodol. (TOSEM) **29**(2), 1–28 (2020)

3. Cordonsky, I., Rosenberg, I., Sicard, G., David, E.O.: DeepOrigin: end-to-end deep learning for detection of new malware families. In: 2018 International Joint Conference on Neural Networks (IJCNN), pp. 1–7. IEEE (2018)

4. Ding, Y., et al.: Malware classification based on semi-supervised learning. In: International Conference on Science of Cyber Security, pp. 287–301. Springer (2022)

5. Geng, J., Wang, J., Fang, Z., Zhou, Y., Wu, D., Ge, W.: A survey of strategy-driven evasion methods for PE malware: transformation, concealment, and attack. Comput. Secur. **137**, 103595 (2024)

6. Gopinath, M., Sethuraman, S.C.: A comprehensive survey on deep learning based malware detection techniques. Comput. Sci. Rev. **47**, 100529 (2023)

7. Hammad, M., Garcia, J., Malek, S.: A large-scale empirical study on the effects of code obfuscations on android apps and anti-malware products. In: Proceedings of the 40th International Conference On Software Engineering, pp. 421–431 (2018)

8. Lei, T., Qin, Z., Wang, Z., Li, Q., Ye, D.: EveDroid: event-aware android malware detection against model degrading for IoT devices. IEEE Internet Things J. **6**(4), 6668–6680 (2019)

9. Mohamed, G.A.N., Ithnin, N.B.: Survey on representation techniques for malware detection system. Am. J. Appl. Sci. **14**(11), 1049–1069 (2017). https://doi.org/10.3844/ajassp.2017.1049.1069

10. Pendlebury, F., Pierazzi, F., Jordaney, R., Kinder, J., Cavallaro, L.: {TESSERACT}: eliminating experimental bias in malware classification across space and time. In: 28th USENIX Security Symposium (USENIX Security 2019), pp. 729–746 (2019)

11. Ronen, R., Radu, M., Feuerstein, C., Yom-Tov, E., Ahmadi, M.: Microsoft malware classification challenge. arXiv preprint: arXiv:1802.10135 (2018)

12. Sihwail, R., Omar, K., Ariffin, K.Z.: A survey on malware analysis techniques: static, dynamic, hybrid and memory analysis. Int. J. Adv. Sci. Eng. Inf. Technol. **8**(4–2), 1662–1671 (2018)

13. Vaswani, A., et al.: Attention is all you need (2017)

14. Ye, Y., Li, T., Adjeroh, D., Iyengar, S.S.: A survey on malware detection using data mining techniques. ACM Comput. Surv. (CSUR) **50**(3), 1–40 (2017)

15. Zhang, X., et al.: Enhancing state-of-the-art classifiers with API semantics to detect evolved android malware. In: Proceedings of the 2020 ACM SIGSAC Conference on Computer And Communications Security, pp. 757–770 (2020)

An Efficient IOC-Driven BigData Tracing and Backtracking Model for Emergency Response

Haiping Wang[1,2], Jianqiang Li[3], Binbin Li[1,2(✉)], Tianning Zang[1,2], Zisen Qi[1,2], Siyu Jia[1,2], Yu Ding[1], and Yifei Yang[1]

[1] Institute of Information Engineering, Chinese Academy of Sciences, Beijing, China
{wanghaiping,libinbin}@iie.ac.cn
[2] School of Cyber Security, University of Chinese Academy of Sciences, Beijing, China
wanghaiping20@mails.ucas.ac.cn
[3] National Computer Network Emergency Response Technical Team/Coordination Center of China, Beijing, China

Abstract. In the realm of intelligence-driven emergency response, utilizing Indicators of Compromise (IOCs) like IP addresses and domain names is crucial for the swift analysis of and response to cyber-attacks within vast network security logs. However, with daily log increases of hundreds of TBs and total volumes reaching into the hundreds of PBs, the tasks of tracking and backtracking critical data face significant challenges, including substantial computational resource consumption and extended processing times. In this paper, we introduce a data tracking and backtracking model based on the principles of data tiered organization and pre-tracking backtracking to enhance emergency response efficiency and reduce resource consumption. The model initially scores IOC clues based on dimensions such as activity and confidence levels, periodically filtering and caching those with higher scores. Based on these clues, it pre-tracks and backtracks associated data, thereby avoiding the computational overhead caused by multiple repetitive offline filtering operations. Furthermore, to improve the retrieval efficiency of threat intelligence clues in scenarios involving massive data volumes, we have designed a three-tier data pre-tracking and backtracking state index to manage the data involved in pre-tracking and backtracking. We also propose a data query task decomposition algorithm to optimize the data retrieval process and enhance emergency response efficiency. Experimental results demonstrate that our method reduces processing time by more than 85.6% compared to traditional direct raw data access, highlighting our model's real-world efficiency and practicality.

Keywords: IOC-Driven · BigData · Data Tracing · Data Backtracking · Emergency Response

© The Author(s), under exclusive license to Springer Nature Singapore Pte Ltd. 2025
J. Zhao and W. Meng (Eds.): SciSec 2024, LNCS 15441, pp. 402–419, 2025.
https://doi.org/10.1007/978-981-96-2417-1_22

1   Introduction

With the increasing complexity of cybersecurity threats and the continuous evolution of attack methods, traditional security emergency response mechanisms, which are based on rules or event-triggering, demonstrate certain limitations against novel, unknown, and advanced persistent threats (APTs). The emergent, covert, and rapidly spreading nature of cyber security incidents urgently calls for the development of a more precise and efficient response strategy. In this context, the concept of intelligence-driven emergency response [1,2] has emerged, offering a fresh perspective in tackling these challenges.

The concept of intelligence-driven emergency response, by integrating the stages of information collection, in-depth analysis, and judgment, along with precise action implementation, introduces a more dynamic and efficient mode of security protection. The essence of this mode lies in the utilization of key cyber indicators such as IP addresses and domain names, identified as Indicators of Compromise (IOC), enabling rapid identification and tracking of critical security threats within complex network security logs. In recent years, significant advancements have been achieved in intelligence-driven security emergency response and event tracking research [3,4], particularly in the detailed analysis of attack behavior characteristics and verification of attack techniques and tools [5–8]. However, facing the challenge of ever-increasing volumes of network security log data, with daily increases reaching hundreds of TBs and total volumes exceeding hundreds of PBs, existing research has not sufficiently explored or studied how to adapt to processing large-scale data. There is a need for enhancing multitasking and parallel processing efficiency, overcoming the constraints of limited computing resources, and optimizing model operational performance.

To achieve real-time tracking of newly added data and efficient retrospection of historical data in scenarios characterized by massive datasets, we introduce a novel intelligence-driven emergency response data tracking and retrospection model tailored for petabyte-scale network security log volumes. Specifically, we **first** employ statistical methods to utilize multidimensional threat intelligence clues, such as reputation scores and source reliability, for periodic evaluations to predict the likelihood of future potential threats. Only those threats assessed with a high probability of occurrence are selected as pre-tracking and retrospection query clues. This predictive and filtering mechanism ensures the focus and efficiency of pre-tracking and retrospection activities. **Next**, to optimize the data retrieval process based on pre-tracking and retrospection, a three-tiered data preservative state index is constructed. This index is designed to manage and maintain the partitioned description information of pre-tracking and retrospection task information and result data. **Lastly**, by subdividing the intelligence-driven emergency response tasks into real-time data tracking and historical data retrospection subtasks, based on associated objectives and time-frames, the three-tier index optimizes the decomposition and pruning process for historical data retrospection tasks. Meanwhile, for real-time data pre-tracking tasks, IOC-associated data results are directly distributed to corresponding real-time tracking tasks, reducing unnecessary queries and significantly enhancing the efficiency of emergency responses.

Our model effectively tackles the challenges of consolidating and extracting common data across tasks, managing refined data state indexes, and distributing data on-demand. Experimental results based on the public dataset CTU-13-Dataset indicate that, under equivalent data volumes and identical query conditions, the tiered data tracking and retrospection model proposed in this paper can reduce the time consumption of data extraction tasks by more than 85.6%. The main contributions can be summarized as follows:

- We propose an efficient data tracking and retrospection model for PB-scale data volumes in intelligence-driven emergency response data computation scenarios, based on the principles of data tiered organization and pre-tracking retrospection.
- We introduce a caching and replacement strategy for Indicators of Compromise (IOC) clues, which intelligently assesses and dynamically filters IOCs with high credibility, confidence, and activity levels, aiming to reduce the volume of original data in computation tasks, thereby improving the efficiency of emergency responses.
- We develop a three-tier indexing structure that efficiently manages pre-extraction task states and data partition descriptions, enabling millisecond-level retrieval of storage status and location for IOC-threat intelligence leads. This structure streamlines the consolidation of computational logic in data queries.
- Experimental results on the public dataset CTU-13-Dataset demonstrate that our approach effectively avoids the issue of repetitive reads of original data, achieving significant results: data partition operations are reduced by 56.7%, the amount of data processed is compressed to 2.8%, equating to a 97.2% reduction in data size; meanwhile, task execution time is significantly shortened by 85.6%.

The structure of the paper is as follows: Sect. 2 presents an overview of the related work. Section 3 introduces the framework employed in this study. Sections 4 to 6 provide a detailed explanation of our models, namely *IOC Caching and Replacement Strategies, Data Pre-Retention Status Index* and *Task-Driven Query Decomposition*. Section 7 discusses the experimental design and presents the results. Finally, Sect. 8 concludes the paper.

2   Related Work

In the domain of intelligence-guided cybersecurity event tracking and traceback studies, the development and deployment of data tracing and backtracking models have garnered significant attention from both industry practitioners and academic scholars. Contemporary research reflects a diversified progression, encompassing various aspects from standalone event analysis to the consolidation of global threat intelligence.

Earlier studies predominantly focused on cybersecurity event tracking via log analysis and intrusion detection techniques, excavating multiple data sources such as system logs and network traffic to discern anomalous behaviors and preliminarily establish event traceback routes. Pedro [10] provides an overview of anomaly-based network intrusion detection techniques and discusses the utilization of log analysis for cybersecurity event tracking. Chonka [11] examines real-time digital forensics, including the application of log analysis and intrusion detection methodologies in network event tracing. Robin Sommer [12] discusses the employment of machine learning in network intrusion detection within realistic environments, featuring extensive analysis of log data. Nevertheless, these methods can be limited in their accuracy and comprehensiveness in tracing and attribution when confronted with intricate and stealthy attacks, due to the insufficient integration of global threat intelligence.

With the continuous advancement of threat intelligence sharing mechanisms, exemplified by the proliferation of standards like STIX [13] and TAXII [14], researchers have begun to investigate the integration of threat intelligence into cybersecurity event tracking and traceback models. By real-time interfacing, analyzing, and aggregating multisource threat intelligence data, these efforts have significantly enhanced the capacity to pinpoint event origins and accurately delineate attack trajectories. Muhtadi [15] investigates how threat intelligence can be incorporated into intrusion detection systems to augment event tracing and attribution capabilities. Liu [16] proposes a sophisticated persistent threat detection framework built around threat intelligence, which includes optimized methods for event tracking and traceback. In paper [17], the authors explore the application of shared threat intelligence in predicting cyberattacks and its potential impact on the practice of event tracking and attribution. Further, Mohammadi [18] discusses the pivotal role of threat intelligence in security event tracking within the context of the Internet of Things (IoT) environment.

Moreover, machine learning and data mining techniques have been extensively employed in the construction of data tracing and backtracking models, aiming to automate the learning and prediction of network attack patterns, thereby enhancing the capability to trace and attribute novel and unknown threats. Park [19] explores machine learning-based IP traceback techniques utilizing flow correlations for data tracing. Yu [20] introduces an intelligent data traceback system grounded in association rule mining. Li [21] proposes a hybrid data traceback method combining an improved Apriori algorithm with support vector machines (SVMs). While Xu [22] primarily centers on anomaly detection, it demonstrates how deep learning methodologies can also be leveraged to build data tracing and backtracking models. Lastly, Srivastava [23] delves into the application of data mining techniques in predictive modeling, applicable to cybersecurity event tracking and traceback. Despite these advancements, these methods continue to confront challenges in processing vast amounts of data and achieving rapid response times.

In summary, despite the robust foundation laid by existing research for intelligence-driven cybersecurity event tracing and attribution, there remains room for improvement in the efficiency of current data tracing and backtracking models when dealing with petabyte-scale network event analysis. Against this backdrop, this study aims to propose and implement a novel intelligence-driven data tracing and backtracking model to support cybersecurity event analysis, aiming to further enhance the accuracy and real-time nature of tracking and attribution efforts while effectively countering contemporary cybersecurity challenges in cyberspace.

3   Model

3.1   Scenario Definition

Before delving into the proposed model solution in this paper, this section will provide a comprehensive exposition of the application scenarios.

The cybersecurity monitoring system continuously captures logs and security alerts encompassing protocols such as URL, DNS, TLS, UDP, and TCP, generating tens of billions of new records daily, with data storage requirements reaching hundreds of terabytes. Real-time logs are temporarily stored in Kafka [27] while historical data is partitioned by hour and stored within big data platforms like HIVE [25]. Due to storage constraints, Kafka buffers only the last three days' worth of real-time data, whereas HIVE tables retain approximately 90 days' worth of raw log information.

Figure 1 illustrates our simplification and abstraction of the problem, with the current time denoted as t_0. Under this assumption, the cybersecurity monitoring device continuously monitors all objects in the target network and generates real-time security audit logs data S_1. Due to insufficient disk space, historical data in partitions P_6 through P_{10} have been purged, while historical data in partitions P_1 through P_5 remains intact.

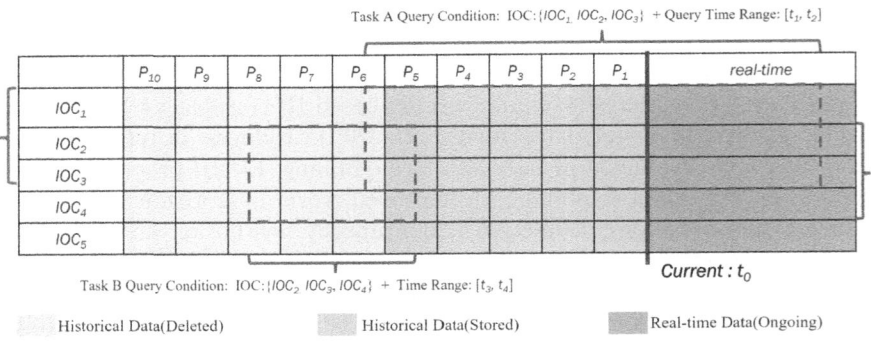

Fig. 1. Application Scenario Description Example.

There are two data tracing and backtracking tasks: Task A and Task B. Task A requires querying for leads related to IOC_1, IOC_2, and IOC_3 within a time range of $[t_1, t_2]$, where $t_1 < t_0 < t_2$. Conversely, Task B necessitates querying information pertaining to IOC_2, IOC_3, and IOC_4 with a query time window of $[t_3, t_4]$, where $t_3 < t_4 < t_0$. Currently, two challenges are encountered:

- Data partitions P_6 associated with Task A, and data partitions P_6 through P_8 linked to Task B have been deleted.
- The large scale of dataset S_1 leads to significant computational resource consumption, lengthy processing times, and low concurrency capabilities.

3.2   Model Framework

To address the aforementioned two issues, we adopt a hierarchical data extraction strategy where we pre-track and retrospectively filter data to selectively preserve and long-term store critical target-related information, thereby reducing data volume and extending the data retention period. This approach transforms complex tracing and backtracking tasks based on vast amounts of raw data into efficient query tasks operating on the pre-screened and retained subset of data. Consequently, it helps decrease resource consumption, accelerates processing speeds, and enhances concurrent processing capacity. Figure 2 depicts the model architecture proposed in this paper for intelligence-driven tracking and traceback in ultra-large-scale data environments, which can be summarized into the following five key steps.

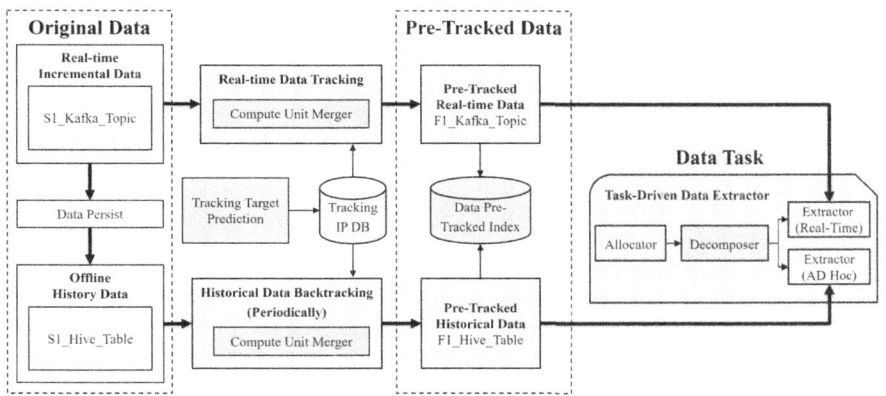

Fig. 2. Intelligence-Driven Massive Data Tracing and Backtracking Framework.

- **Step1 Tracking Target Prediction**: Combining multi-dimensional information, such as the credibility score, source reliability, association with known attacks, frequency, and hit rate trends of threat intelligence IOC clues, statistical models are utilized to predict the likelihood of their future occurrence. By setting a threshold, IOCs exceeding this threshold are identified as key tracking targets and are subsequently cached.

- **Step2 Real-time Data Pre-Tracking**: Based on predicted key tracking targets (e.g., critical IPs, domain names as IOC indicators), the Flink [24] real-time computing framework is leveraged to filter out relevant prioritized data from massive datasets. Given the cyclical nature of predicted tracking targets, multiple rounds of issued tracking tasks are automatically consolidated, ensuring one-pass reading and processing of real-time data.
- **Step3 Historical Data Pre-Backtracking**: Utilize Spark [26] and similar computing frameworks to schedule and initiate periodic historical data backtracking tasks in batches, thereby realizing the preprocessing and retrospective analysis of historical data.
- **Step4 Three-layer Data Pre-Tracking Status Index**: For pre-tracking and backtracking data associated with each IOC threat intelligence clue, a data pre-retention status index is constructed at the granularity of time slices. The index features a three-tier structure: the first layer comprises an IOC hash index node based on the Bloom filter; the second layer consists of a list index node for the IOC's pre-tracking and backtracking time range; and the third layer is an index node for partition information of pre-retained data.
- **Step5 Task-Driven Data Extraction**: Based on the targets and time-frames associated with intelligence-driven response tasks, these tasks are decomposed into two subtasks: real-time data tracking and historical data backtracking. Leveraging pre-indexed information, historical data backtracking tasks are optimized through decomposition and pruning, ultimately transforming them into query retrieval tasks for pre-tracked and retroactive data.

4   IOC Caching and Replacement Strategies

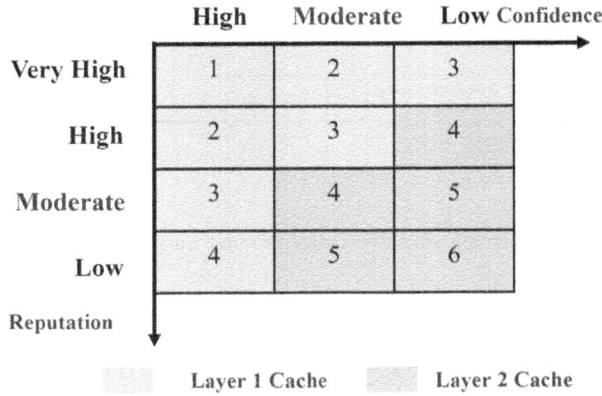

Fig. 3. Hierarchical IOC Caching Strategy.

During the preprocessing and retrospective analysis of IOC data, we initially employ a fixed-capacity cache to store prioritized IOC information. Subsequently, we utilize distributed real-time computing platforms like Flink [24] and Spark Streaming to perform live filtering and proactive tracking on network monitoring logs. Concurrently, we leverage distributed in-memory computing systems such as Spark for conducting associative pre-backtracking analysis on historical records.

The central idea of the proposed IOC clue caching and replacement algorithm is as follows:

- A dual-layer IOC clue caching architecture is constructed, where Layer 1 is designated to accommodate IOC information with higher levels of reputation and confidence, initially allocated with 20% of the total cache capacity. Conversely, Layer 2 holds other IOC data, occupying 80% of the available space.
- The caching regions are stratified according to the IOC's levels of reputation and confidence. Reputation is categorized into four tiers: low, moderate, high, and very high, while confidence is divided into three levels: low, moderate, and high. As depicted in the orange section of Fig. 3, IOCs with both high confidence and very high reputation, as well as those with moderate confidence but high reputation, are prioritized for caching within Layer1.
- Layer1's space is dynamically managed such that, when its occupancy ratio remains below 47.4% (equivalent to $18/38 \times 100\%$), an increase in the number of newly added IOCs with high confidence and reputation levels prompts a redistribution from Layer2 to Layer1. Once Layer1's proportion reaches or surpasses 47.4%, instead of further reducing Layer2's capacity, direct expansion of Layer1 occurs to accommodate the emerging demand.
- The replacement strategy for Layer2's cache space is predicated on a comprehensive evaluation of various dynamic IOC metrics, including *Acquisition Time, Latest Hit Time*, the number of hits in the past week, the number of hits in the last month, and the cumulative count of hit log records to date. Using entropy-based methods, the weights for each indicator are determined, and a weighted sum approach is employed to calculate a score for every IOC, which subsequently determines their retention order within the cache. Regularly, based on the capacity constraints of Layer 2, the top N IOCs with higher scores are maintained.

The detailed description of the scoring algorithm and ranking rules is as follows:

- Firstly, a min-max [28] normalization method is employed to standardize various evaluation metrics, encompassing *Acquisition Time, Latest Hit Time*, the number of logs hit within the last week, the number of logs hit within the last month, and the cumulative count of logged hits to date.
- Subsequently, the independence of feature coefficients is ascertained by calculating their covariance, which enables the selection of those with independent distribution characteristics.

- Based on the principle of the entropy method, the quantitative weights for each assessment indicator in the comprehensive scoring are determined and allocated.
- Following this, by combining the obtained feature coefficients and their respective weights, a weighted sum approach is used to compute the overall score for each IOC.
- Ultimately, the IOCs are ranked in descending order according to their computed comprehensive scores, thereby selecting the top N IOCs.

5   Data Pre-retention Status Index

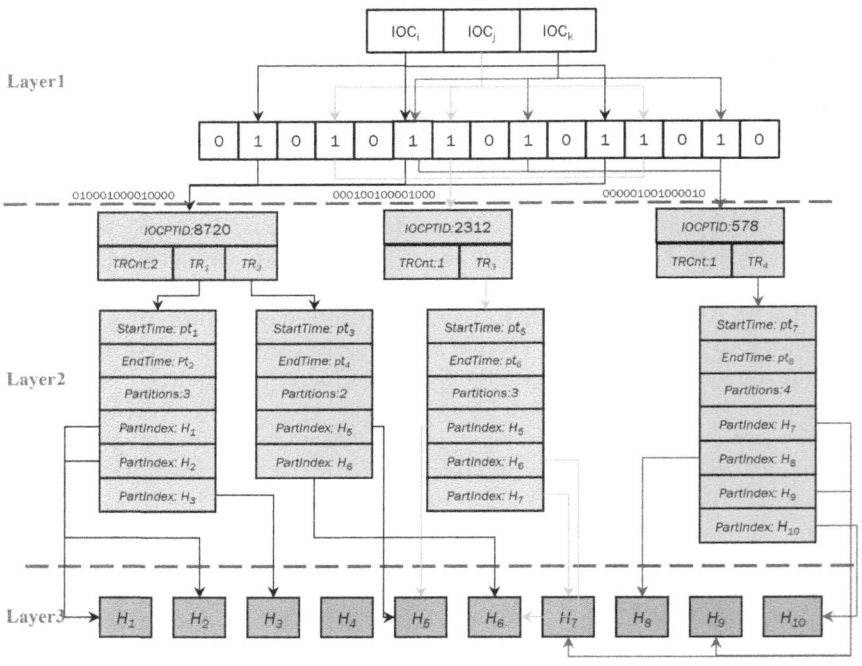

Fig. 4. Data Pre-Tracking Status Index Structure.

The above Fig. 4 illustrates the structure of the data pre-tracking and backtracking status index, which consists of a three-tier architecture.

- **Bloom Filter-based IOC Existence Hash Index Layer**: When the number of IOCs reaches millions to tens of millions, determining whether the associated data of a given IOC has already been traced and retraced, and locating the detailed description of its pre-tracing and backtracking tasks, becomes a challenging task. To tackle this issue, we employ a Bloom Filter as the top-level design for the data pre-tracing and backtracking state index, ensuring constant time querying for the presence of IOC pre-tracing and backtracking task information. Here, parameter m denotes the width of the bit array within the Bloom Filter (measured in bits), n represents the total count of unique keys inserted into the filter, corresponding to the number of threat intelligence IOCs in this study; and k refers to the number of independent hash functions used, set to 3 in this paper.

- **IOC Pre-Tracked Time Interval Index**: The structure is designed to maintain information about all pre-tracing and backtracking tasks associated with IOC-related data; it comprises Index IDs, counts of initiated data pre-tracing and backtracking time intervals, and pointers to the data within each interval. Each temporal interval index records the start time, end time, and the corresponding number of data partitions, dynamically maintaining pointers to the descriptive information of each data partition. By analyzing the start and end timestamps of these intervals in conjunction with the data partitioning algorithms of the storage engine, one can accurately determine the exact number of data partitions about the current time slot and their detailed contents.

- **Log Data Partition Information Descriptor Node**: The key contents of a data partition descriptor node include the dataset name, partition field identifier, and corresponding hourly timestamps. For instance, "2024022317" signifies a data partition for the time interval between 17:00 to 18:00 on February 23rd, 2024.

6   Task-Driven Query Decomposition

During threat intelligence IOC-driven emergency response, rapid querying and feedback of associated data are critical. By establishing a fine-grained IOC pre-tracking and backtracking index, it becomes possible to swiftly locate and return processed intermediate results, thereby enabling hierarchical data filtering within tasks and significantly enhancing the efficiency of emergency response.

In this section, we will discuss how to utilize pre-tracking and backtracking state indexes for partition pruning and dynamic aggregation in data query retrieval tasks under both single-task and multi-task contexts.

Figure 5 presents an instance of data pre-tracking and backtracking. Utilizing an IOC caching and replacement strategy algorithm, pre-backtracking of IOC_1 has been accomplished within the time span from H_6 to H_1. Similarly, pre-backtracking for IOC_2 and IOC_3 encompasses the data range from H_8 to H_1. As for IOC_4, backtracking for the period between H_8 and H_4 has been completed. All data obtained through backtracking is partitioned and stored on an hourly basis, as detailed in the green cell annotations within the chart. Furthermore, at present, continuous pre-tracking of newly emerging real-time data associated with IOC_1, IOC_2, and IOC_3 is underway.

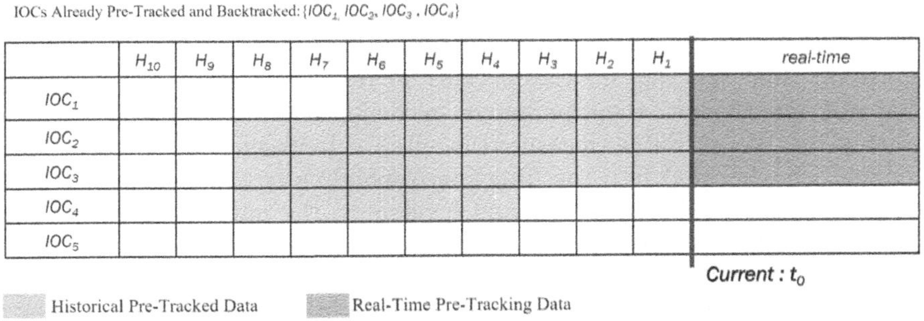

IOCs Already Pre-Tracked and Backtracked: {IOC_1, IOC_2, IOC_3, IOC_4}

	H_{10}	H_9	H_8	H_7	H_6	H_5	H_4	H_3	H_2	H_1	real-time
IOC_1											
IOC_2											
IOC_3											
IOC_4											
IOC_5											

Current : t_0

  Historical Pre-Tracked Data   Real-Time Pre-Tracking Data

Fig. 5. Data Pre-Tracking and Backtracking Example.

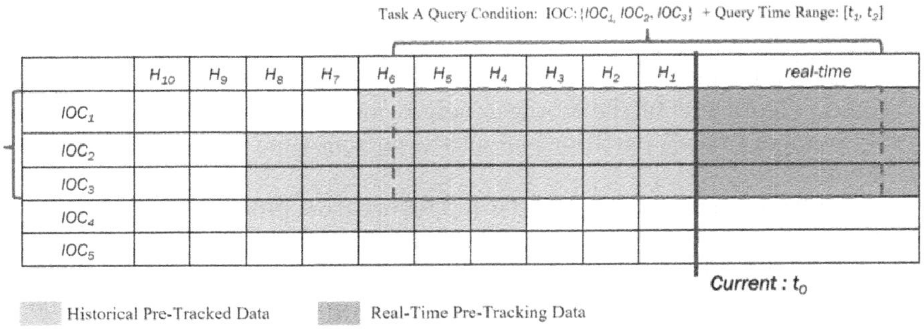

Task A Query Condition: IOC: {IOC_1, IOC_2, IOC_3} + Query Time Range: [t_1, t_2]

	H_{10}	H_9	H_8	H_7	H_6	H_5	H_4	H_3	H_2	H_1	real-time
IOC_1											
IOC_2											
IOC_3											
IOC_4											
IOC_5											

Current : t_0

  Historical Pre-Tracked Data   Real-Time Pre-Tracking Data

Fig. 6. Single Task Decomposition and Pruning.

Figure 6 reveals the refined data query strategy for Emergency Response Task A, encompassing task decomposition, partition pruning, and subtask optimization. Given the temporal relationship where $t_1 < t_0 < t_2$, t_0 serves as the temporal demarcation point, dividing the task into two distinct parts: data tracking and data backtracking.

Contrasting with traditional approaches that rely solely on querying raw data, this scheme harnesses the power of pre-tracked and backtracked data, manifesting three major advantages:

Within the time interval from t_0 to t_2, due to the completion of pre-tracking for IOC_1, IOC_2, and IOC_3, the method of obtaining their associated data shifts from exhaustive scanning of the full *S1_Kafka_Topic* data source to direct access of the KAFKA topic *F1_Kafka_Topic* housing the pre-tracking results. This shift effectively narrows the data subscription scope, reduces computational resource consumption, and thereby significantly boosts data acquisition efficiency.

In the time window between H_5 and t_0, considering that IOC_1, IOC_2, and IOC_3 have undergone pre-backtracking within the H_6 to H_1 range, the original requirement to perform offline data backtracking on all raw data partitions P_5, P_4, P_3, P_2, and P_1 is now adjusted to querying based on the pre-backtracked result data partitions H_5, H_4, H_3, H_2, and H_1. Given the significantly smaller data volume of partitions H_1 to H_5 compared to P_1 to P_5, this adjustment significantly alleviates the data processing burden.

In the time window between t_1 and H_5, since the associated data for IOC_1, IOC_2, and IOC_3 has been pre-filtered and stored at a lower cost in H6, compared to the original scheme, this proposal provides richer and more comprehensive result data.

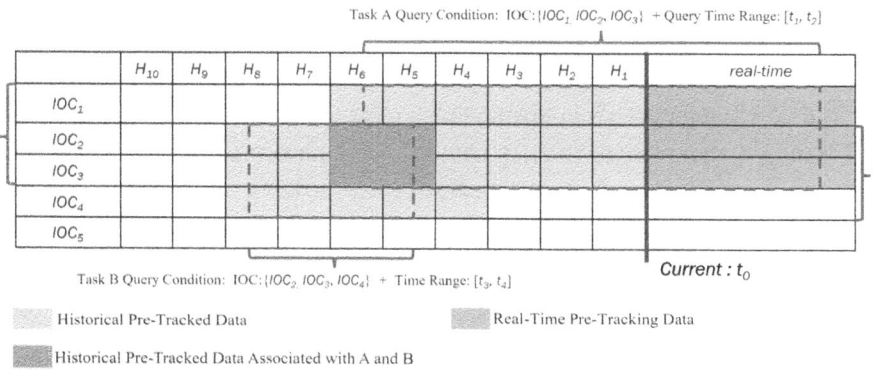

Fig. 7. Multiple Tasks Decomposition, Pruning, and Merging.

The above diagram Fig. 7 illustrates how, within a multi-task scenario encompassing tasks A and B among others, data pre-tracing technology is employed to trace state indices, thereby enabling task decomposition, precise pruning of data partitions, and the optimization of query strategies. Compared to handling a single task in isolation, the multi-task setting presents distinct advantages: It allows for the consolidation of disk access demands from multiple tasks targeting the same data partition, exemplified by the blue area in the diagram. This approach aims to eliminate redundant data re-reads, reduce disk I/O operations, consequently leading to substantial resource savings and enhanced overall efficiency.

7   Experiment

7.1   DataSets

In experimental settings, we rely on the publicly available CTU-13-Dataset of network traffic data from Czech Technical University (CTU) to simulate real-world intelligence-driven emergency response scenarios for validating the efficacy of the proposed algorithm. The dataset's essential characteristics are as follows (Table 1):

Table 1. Statistical Analysis and Overview of the CTU-13 Dataset

Scene ID	StartTime	Duaration(hrs)	NetFlows	Botnet Flows
1	2011/08/10 09:46:59	6.15	2,824,636	39,933(1.41%)
2	2011/08/11 09:53:40	4.21	1,808,122	18,839(1.04%)
3	2011/08/12 15:25:56	66.85	4,710,638	116,887(2.48%)
4	2011/08/15 11:00:30	4.21	1,121,076	1,719(0.15%)
5	2011/08/15 16:43:28	11.63	129,832	695(0.53%)
6	2011/08/16 10:04:03	2.18	558,919	4,431(0.79%)
7	2011/08/16 13:52:03	0.38	114,077	37(0.03%)
8	2011/08/16 14:23:17	19.5	2,954,230	5,502(0.17%)
9	2011/08/17 12:01:12	5.18	2,753,884	179,880(6.5%)
10	2011/08/18 10:21:46	4.75	1,309,791	106,315(8.11%)
11	2011/08/18 15:40:53	0.26	107,251	8,161(7.6%)
12	2011/08/19 10:32:31	1.21	325,471	2,143(0.65%)
13	2011/08/15 17:13:51	16.36	1,925,149	38,791(2.01%)

We integrated all 13 scenario NetFlow records from the CTU-13-Dataset, extracting five critical fields: *StartTime*, *SrcAddr*, *Sport*, *DstAddr*, and *Dport*. Further, for instances where the *Label* field indicates *Botnet Flows*, we filtered and de-duplicated *SrcAddr* and *DstAddr*, yielding a collection of 48,637 unique IP addresses as the IOC (Indicators of Compromise) set.

To simulate a massive data scenario more effectively, we replicated the 19,976,700 NetFlow records from the CTU-13-Dataset, amounting to 2.6 GB, expanding it by a factor of 150 to reach 2,996,505,000 records and 383.1 GB in size. The augmented dataset was then partitioned and stored at an hourly granularity, resulting in a total of 144 data partitions.

7.2   Scene Simulation

Prior to conducting the comparative experiments, we utilized all available IOC (Indicators of Compromise) information to pre-filter records from the complete *NetFlow* dataset that are associated with IOCs. This was done to provide a ioc-driven emergency response based on the data of preprocessing tracing and backtracking.

The raw *NetFlow* data spans from 09:46:59 on August 10, 2011, to 11:45:43 on August 19, 2011, covering approximately 218 h with data recorded for 144 h. To accurately simulate and evaluate the effectiveness of our methods across different scenarios, we designed four distinct backtracking durations in a laboratory environment:

– **6-H Backtracking**: equivalent to 2.4% of the total time span.
– **24-H Backtracking**: representing 11% of the overall duration.

- **72-H Backtracking**: 3 days, accounting for 33% of the total length.
- **168-H Backtracking**: 7 days, corresponding to 77% of the entire time period.

For each backtracking duration setting, we randomly selected 5 IP addresses as IOC leads and conducted 5 sets of experiments; in each set, the task start time was set using a random algorithm. For every task, we computed the netflow logs associated with these IOCs, grouping them by source IP *SrcAddr*, destination IP *DstAddr*, and *hour*, while summing up *sTos*, *dTos*, *TotPkts*, and *TotBytes* hourly.

7.3 Environment

The integrated test dataset was stored on a 3-node HIVE cluster, where each server features dual 24-core CPUs operating at 2.2 GHz, 256 GB of RAM, and 24 TB hard drives for ample storage capacity. The servers are equipped with 10 Gigabit Ethernet cards to ensure high-speed network communication. For data processing, we leverage the Spark distributed in-memory computing framework to perform efficient filtering operations, ensuring that each task draws from an equal share of computational resources from a dedicated queue during its execution.

7.4 Evaluation Metrics

During the experimental evaluation phase, we compared three methods:

- **FRD-BASED**: Tracing and backtracking using the full raw dataset.
- **PSD-BASED**: Tracing and backtracking based on pre-filtered data.
- **PSDI-BASED**: Optimizing the backtracking process by leveraging pre-stored data and associated indexing mechanisms.

We quantitatively assessed the performance of these different strategies under the same task using three metrics: number of read partitions **Partitions**, data size read from storage engine**DataSize(GB)** and task execution time **Time(s)**.

7.5 Results Analysis

Table 2, 3, 4 and 5 describes the results of five sets of comparative experiments conducted under four different backtracking durations.

Table 2 data shows that under a backtracking duration of 6 h, compared with directly querying based on raw data, our proposed method achieves a 56.7% reduction in the number of data partitions read, a 97.2% decrease in data size, and an 85.6% reduction in execution time.

Table 3 data reveals that, with a backtracking duration of 24 h, our proposed method achieves a 74.3% decrease in the number of data partitions accessed, a

Table 2. 6-h backtracking: mean comparison across 5 experiments

Alorithm	Partitions	DataSize(GB)	Time(s)
FRD-BASED	6	32.56	16.32
PSD-BASED	5.8	1.04	2.752
PSDI-BASED	**2.6**	**0.922**	**2.3486**
PSD vs FRD	-3.3%	-96.8%	-83.1%
PSDI vs PSD	-55.2%	-11.3%	-14.7%
PSDI vs FRD	**-56.7%**	**-97.2%**	**-85.6%**

Table 3. 24-h backtracking: mean comparison across 5 experiments

Alorithm	Partitions	DataSize(GB)	Time(s)
FRD-BASED	14	51.82	20.08
PSD-BASED	13.6	1.352	2.786
PSDI-BASED	**3.6**	**0.99**	**1.6918**
PSD vs FRD	-2.9%	-97.4%	-86.1%
PSDI vs PSD	-73.5%	-26.8%	-39.3%
PSDI vs FRD	**-74.3%**	**-98.1%**	**-91.6%**

Table 4. 72-h backtracking: mean comparison across 5 experiments

Alorithm	Partitions	DataSize(GB)	Time(s)
FRD-BASED	28.8	93.4	29.44
PSD-BASED	27.6	1.734	3.8158
PSDI-BASED	**4.2**	**1.062**	**2.8168**
PSD vs FRD	-4.2%	-98.1%	-87.0%
PSDI vs PSD	-84.8%	-38.8%	26.2%
PSDI vs FRD	**-85.4%**	**-98.9%**	**-90.4%**

98.1% reduction in data volume, and a 91.6% decrease in execution time, when contrasted against direct querying based on raw data.

Table 4 data indicates that during a backtracking duration of 72 h, our proposed method realizes an 85.4% decrease in the count of data partitions read, a 98.9% diminution in data size, and a 90.4% reduction in execution time, when juxtaposed with direct querying using raw data.

Table 5. 168-h backtracking: mean comparison across 5 experiments

Alorithm	Partitions	DataSize(GB)	Time(s)
FRD-BASED	127.4	294.92	74.96
PSD-BASED	125.8	4.48	7.8136
PSDI-BASED	**9.8**	**3.12**	**4.6532**
PSD vs FRD	−1.3%	−98.5%	−89.6%
PSDI vs PSD	−92.2%	−30.4%	−40.4%
PSDI vs FRD	**−92.3%**	**−98.9%**	**−93.8%**

Table 5 demonstrates that for a backtracking duration of 168 h, our proposed method yields a 92.3% decrease in the quantity of data partitions accessed, a 98.9% contraction in data volume, and a 93.8% decline in execution time, relative to direct querying based on raw data.

8 Conclusion

This paper introduces a novel and efficient approach to data tracking and backtracking, designed to enhance the efficiency of emergency responses while significantly reducing resource consumption. Our method employs intelligent caching, and periodic evaluations, and prioritizes high-value Indicators of Compromise (IOCs), thus eliminating the need for repetitive offline filtering. Moreover, the adoption of a three-tier indexing architecture, combined with an innovative decomposition algorithm, optimizes data retrieval and greatly reduces computational overhead. The effectiveness of our model has been validated through experimental evaluations with the CTU-13-Dataset. Results demonstrate a significant reduction in the redundancy of raw data reading and a marked improvement in operational efficiency. Specifically, our approach led to a 56.7% decrease in data partition operations, reduced the processing volume to just 2.8% (representing a 97.2% reduction in data size), and cut task execution time by 85.6%.

References

1. Roberts, S.J., Brown, R.: Intelligence-Driven Incident Response: Outwitting the Adversary. O'Reilly Media, Inc., Newton (2017)
2. Schlette, D., Caselli, M., Pernul, G.: A comparative study on cyber threat intelligence: the security incident response perspective. IEEE Commun. Surv. Tutor. **23**(4), 2525–2556 (2021)
3. Gómez, J.A. (2011). The Targeting Process: D3A and F3EAD
4. Yang, D., Li, Q., Zhu, F., Cui, H., Yi, W., Qin, J.: Parallel emergency management of incidents by integrating OODA and PREA loops: the C2 mechanism and modes. IEEE Trans. Syst. Man Cybern. Syst. **53**(4), 2160–2172 (2023)

5. Emara, S.F., Abdelhady, S., Zaki, M.: A novel traceback model for DDoS attacks using modified Floyd-Warshall algorithm. Int. J. Inf. Comput. Secur. **20**(1/2), 84–103 (2023)
6. Xiao, F., Chen, E., Xu, Q., Zhang, X.: ICSTrace: a malicious IP traceback model for attacking data of the industrial control system. Secur. Commun. Netw. **2021**, 7525092:1-7525092:14 (2021)
7. Saurabh, S., Sairam, A.S.: ICMP based IP traceback with negligible overhead for highly distributed reflector attack using bloom filters. Comput. Commun. **42**, 60–69 (2014)
8. Jeong, E.-H., Lee, B.K.: An IP Traceback protocol using a compressed hash table, a sinkhole router and data mining based on network forensics against network attacks. Future Gener. Comput. Syst. **33**, 42–52 (2014)
9. https://mcfp.weebly.com/the-ctu-13-dataset-a-labeled-dataset-with-botnet-normal-and-background-traffic.html
10. Garcia-Teodoro, P., Verdejo, J.E.D., Maciá-Fernández, G., Vázquez, E.: Anomaly-based network intrusion detection: techniques, systems and challenges. Comput. Secur. **28**(1–2), 18–28 (2009)
11. Chonka, A., Skinner, N.: Real-time network forensics: building a back-end for full packet capture. Digit. Investig. **9**(1), S2–S14 (2012)
12. Sommer, R., Paxson, V.: Outside the closed world: on using machine learning for network intrusion detection. In: IEEE Symposium on Security and Privacy, pp. 305–316 (2010)
13. STIX. https://oasis-open.github.io/cti-documentation/stix/intro
14. TAXII. https://oasis-open.github.io/cti-documentation/taxii/intro.html
15. Al-Muhtadi, M., Tipper, H.: Integrating threat intelligence into intrusion detection systems. Int. J. Distrib. Sensor Netw. **14**(1), 1509835 (2018)
16. Liu, Y., Jin, X., Ni, Q.: A threat intelligence-driven framework for detecting advanced persistent threats. Futur. Gener. Comput. Syst. **112**, 445–457 (2020)
17. Al-Nemrat, A.M., Furnell, S.: Cyber threat intelligence sharing and prediction of cyber attacks. J. Netw. Comput. Appl. **67**, 20–32 (2016)
18. Al-Fuqaha, A., Guizani, M., Mohammadi, M., Aledhari, M., Ayyash, M.: Internet of things: a survey on enabling technologies, protocols, and applications. IEEE Commun. Surv. Tutor. **17**(4), 2347–2376 (2015)
19. Park, J., Son, S.: Machine learning-based IP traceback using flow correlation. IEEE Trans. Inf. Forensics Secur. **13**(1), 200–214 (2018)
20. Yu, H., He, D., Li, Y., Hu, X.: An intelligent data traceback system based on association rule mining. J. Netw. Comput. Appl. **67**, 49–60 (2016)
21. Li, Z., Chen, L., Zhang, Y., Song, W.: A hybrid data traceback approach based on improved Apriori algorithm and SVM in cloud computing environment. IEEE Trans. Serv. Comput. **7**(6), 533–546 (2014)
22. Xu, L., Wang, Y., Ren, K.: Anomaly-based intrusion detection using deep learning over network flows. IEEE Trans. Inf. Forensics Secur. **10**(12), 2456–2468 (2015)
23. Srivastava, J., Shrivastava, A., Singh, S.K.: Predictive modeling using data mining techniques for cyber crime investigation. Procedia Comput. Sci. **32**, 442–449 (2014)
24. Carbone, P., Katsifodimos, A., Ewen, S., Markl, V., Haridi, S., Tzoumas, K.: Apache Flink^{TM}: stream and batch processing in a single engine. IEEE Data Eng. Bull. **38**(4), 28–38 (2015)
25. Thusoo, A., et al: Hive - a petabyte scale data warehouse using Hadoop. In: ICDE 2010, pp. 996–1005 (2010)

26. Shanahan, J.G., Dai, L.: Large scale distributed data science using apache spark. In: KDD, pp. 2323–2324 (2015)
27. Wang, G., et al.: Building a replicated logging system with apache kafka. Proc. VLDB Endow. **8**(12), 1654–1655 (2015)
28. Zhao, Z., Kleinhans, A., Sandhu, G., Patel, I., Unnikrishnan, K.P.: Capsule networks with max-min normalization. CoRR arxiv:1903.09662 (2019)

Integrating CP-ABE and Device Fingerprint Into Federated Learning

Chunlu Chen[1]($\boxtimes$), Rodrigo Roman[2], Kevin I-Kai Wang[3], and Kouichi Sakurai[1]

[1] Kyushu University, Fukuoka, Japan
chen.chunlu.270@s.kyushu-u.ac.jp, sakurai@inf.kyushu-u.ac.jp
[2] University of Malaga, Malaga, Spain
rroman@uma.es
[3] The University of Auckland, Auckland, New Zealand
kevin.wang@auckland.ac.nz

Abstract. The rapid expansion of Internet of Things (IoT) devices and advancements in mobile computing have underscored the significance of edge computing in addressing data processing and analytics challenges at the network. Federated Learning (FL) emerges as a strategic response, enabling collaborative model training across multiple participants without necessitating the sharing of raw data, thus preserving privacy and effectively leveraging decentralized data. However, FL's architecture presents distinct security challenges, particularly in safeguarding data and models against unauthorized access and manipulation. This paper presents a FL framework that integrates Attribute-Based Encryption (ABE) to provide fine-grained data access control while ensuring data security. In addition, we employ device fingerprints to enhance user authentication and device security, ensuring that only verified devices can participate in the model training process, thereby minimizing the risks of malicious access and data breaches. This paper delineates our contributions towards enhancing FL security, including the deployment of encryption methodologies, access management, and a comprehensive discussion on security mechanisms.

Keywords: Federated Learning · Encryption · Privacy-Preserving · Device Fingerprint

1 Introduction

1.1 Background and Motivation

The proliferation of Internet of Things (IoT) devices and the growth in mobile computing power have significantly emphasized the importance of edge computing for managing the data processing and analytical challenges encountered at the network's edge [37]. This approach, which involves processing data either directly on the devices or at the network edge, offers a solution to the limitations faced by traditional cloud computing, including high latency and bandwidth

© The Author(s), under exclusive license to Springer Nature Singapore Pte Ltd. 2025
J. Zhao and W. Meng (Eds.): SciSec 2024, LNCS 15441, pp. 420–436, 2025.
https://doi.org/10.1007/978-981-96-2417-1_23

constraints. Despite these benefits, the adoption of edge computing introduces several risks, particularly concerning the security and privacy of data within edge devices [45]. In this context, Federated Learning (FL) has emerged as a promising strategy, enabling the collaborative training of models across multiple participants without necessitating the sharing of raw data, thereby preserving privacy and efficiently leveraging decentralized data sources [10,60,68]. Nonetheless, the federated architecture introduces unique security challenges, especially in terms of protecting the model and data against unauthorized access and manipulation.

1.2   Existing Works

The security of FL presents multifaceted challenges encompassing data protection, model integrity, and user privacy. A breach in any of these areas can significantly compromise the integrity of the system [31,80]. The potential integration of malicious entities heightens the susceptibility to information exploitation [47]. This section explores various defense strategies in the context of FL.

- **Encryption Techniques:** Homomorphic Encryption (HE) and Secret Sharing (SS) are central to current encryption practices. HE uniquely enables computations on encrypted data, preserving privacy throughout the process [4,16,21]. SS, conversely, distributes parts of the encryption key across several entities, allowing decryption only by authorized groups [8,17]. However, these encryption techniques may increase the computation and the complexity of the system. The primary challenge lies in balancing efficient training with model accuracy [16,43,76]. In addition, the lightweight approach is implemented through the addition of masks, which diminishes the computational requirements for encryption by overlaying data with these masks [59,78].
- **Attribute-Based Encryption:** Attribute-based encryption (ABE) achieves fine-grained access control over encrypted data through user attributes, unlike traditional public key encryption, which relies on a single key pair to control access [51]. The most common forms of ABE are Key-Policy ABE (KP-ABE) and Ciphertext-Policy ABE (CP-ABE). In KP-ABE, access control policies are stored with the user's private keys [19]. In CP-ABE, the access control policies are stored with the ciphertext [7]. CP-ABE is more suitable for scenarios where the sender needs to control access to different data by different users, while KP-ABE is better suited for scenarios that require a unified access policy for data. Besides these two methods, there are other forms such as Revocable ABE [33] and Lattice-Based ABE [30]. However, in KP-ABE, the complexity of key management is higher because the access control policy is stored with the user's private keys. The revocation mechanism in Revocable ABE schemes requires additional key management operations, which can impact system performance. Moreover, Lattice-Based ABE schemes require larger key sizes and more complex mathematical operations. Therefore, this study integrates federated learning (FL) with CP-ABE as the encryption method to ensure data security while providing flexible access control mechanisms. This means the system can tailor access permissions based on pre-

defined attributes or policies that may change over time or in response to specific scenarios.

- **Obfuscation Techniques:** Differential Privacy (DP) [79] is a key technique in this category, introducing intentional ambiguity to the data or certain features, preventing external parties from extracting individual raw data from the information transmitted by devices [3,24,71]. Current obfuscation methods often involve adding noise to data [26,32,67]. Techniques like augmenting training data or modifying label information can further enhance data privacy [29,40]. However, a critical trade-off exists: too much noise can reduce model accuracy, while too little may reveal training data [18,53].

- **Secure Multi-Party Computation Techniques:** Given the limitations of pure encryption or obfuscation methods, hybrid solutions like Secure Multi-Party Computation (MPC) have gained prominence [49]. Contemporary MPC protocols, often combined with DP technologies, strive to both preserve data privacy and reduce communicational overheads [20]. Adjusted noise addition is also used to improve model precision [62]. However, this approach is still limited due to its complex algorithms.

- **Blockchain Techniques:** Blockchain technology in FL provides a decentralized and secure framework, enhancing data integrity and auditability [6,48]. This integration helps mitigate single points of failure and potential data breaches, thus fostering a more robust and reliable FL ecosystem [50,74]. However, this approach increases computational and communication overhead, as blockchain transactions can be resource-intensive and slow, potentially hindering the efficiency of FL processes. Furthermore, the complexity of integrating blockchain with FL may result in increased system complexity.

1.3   Challenging Issues

Traditionally, Machine Learning (ML) systems rely on centralizing raw data on a server for model training, which, despite its effectiveness, introduces computational bottlenecks and raises concerns over legal, security, and privacy issues as data volumes expand [23,46]. FL presents a novel paradigm that mitigates these security and privacy challenges by creating global models without the need for direct data exchange between individual sources, instead sharing model parameters or intermediate results, such as gradients [13,14,34–36,75]. However, FL systems are still vulnerable to various security threats, with attackers potentially taking on different roles, from passive spectators to active participants, posing diverse threats to the privacy and security of the system [25,42,57,58].

In edge computing, the widespread distribution of devices complicates the management and security of nodes [47,70,77]. Moreover, the dynamism of device participation in the network and the heterogeneity of edge devices introduce additional complexities in continuously verifying device identities and ensuring compliance with security policies [9]. Verifying the integrity of data and model updates also remains a significant challenge, with attackers potentially modifying these updates to disrupt the training process or infer private information [41,52, 54].

1.4   Our Contributions

This study integrates the CP-ABE method with FL to provide fine-grained data access control mechanisms while ensuring data security [12]. Furthermore, we enhance user uniqueness by utilizing device fingerprints to strengthen authentication and reduce risks of deception. The contributions of this study are as follows:

- We introduce an ABE-FL system architecture that incorporates device fingerprints, enhancing the data access control mechanisms and protecting sensitive data within the FL environment.
- We implement foolproof fine-grained access control based on device fingerprints, thereby enhancing security and protecting data privacy and integrity across various edge computing environments.
- We adjust our approach based on the diversity of devices within FL, ensuring consistent authentication and adherence to security protocols. Through ABE-FL, we enhance the system's adaptability by adjusting access permissions based on device fingerprints and policies, which involves modifying access permissions to accommodate the diverse characteristics of devices within the FL environment.

1.5   Comparison with Related Works

While existing security mechanisms offer substantial layers of protection within FL environments, striking an optimal balance among model accuracy, system complexity, and operational efficiency emerges as a formidable challenge [1,42]. Traditional approaches, such as HE and SS, while effective in securing data, often introduce significant computational overhead and complexity, potentially impeding the fluidity of FL processes [17,39]. Similarly, while MPC and blockchain techniques fortify privacy and data integrity, they too grapple with the drawbacks of increased computational and communicational demands [44].

In recent years, ABE-FL has emerged as a promising paradigm [69]. In particular, ABE is mainly used to enforce access policies that determine which clients can participate in the federated learning process. For example, in the healthcare field, CP-ABE can used to encrypt a request for collaboration (e.g. healthcare data about a certain illness) that only those hospitals that comply with certain attributes (e.g. hospitals with specialized doctors) will be able to decrypt [38]. Also, in the area of collaborative smart cities, CP-ABE is also used to encrypt a single challenge with a policy set by the central authority that only authorized participants will be able to decrypt [73]. All these schemes define complex scenarios with multiple levels of participants, which require of blockchain technologies to implement various strategies such as client reputation management. In addition, these schemes do not take advantage of ABE to protect the federated learning phase, delegating this task to other solutions such as homomorphic encryption.

We assume that the server is trustworthy, hence we do not use methods that focus on the defense against malicious servers (e.g. homomorphic encryption), thus avoiding the associated overhead. We adopt constructs related to Attribute-Based Encryption (ABE), which not only simplify key management but also establish secure communication between clients and servers, and introduce additional access control policies to ensure that only users with specific attributes can access sensitive data. This results on an approach that reduces the complexity of the previously introduced state of the art, without requiring the integration of blockchain technologies to provide functional services. Moreover, we introduce device fingerprints to enhance user authentication and device security, ensuring that only verified devices can participate in the model training process, thereby reducing the risks of malicious access and data breaches.

1.6 Structure of the Paper

The structure of this paper is organized as follows: Sect. 2 provides an overview of the CP-ABE scheme. Section 3 provide details our proposed scheme. Section 4 outlines the threat model and security goals, and analyzing the security of our scheme. The future directions and paper concludes with Sect. 5.

2 Ciphertext-Policy Attribute-Based Encryption

In this section, we provide a detailed exposition of CP-ABE scheme.

2.1 Preliminaries

Bilinear Maps: This section introduces essential concepts regarding groups that possess efficiently computable bilinear maps. Consider two multiplicative cyclic groups, G_1 and G_2, both of prime order p. Assume g is a generator for G_1 and introduce a bilinear map $e : G_1 \times G_1 \rightarrow G_2$. The bilinear map e exhibits the properties below:

– Bilinearity: For any $u, v \in G_1$ and $a, b \in \mathbb{Z}_p$, it holds that $e(u^a, v^b) = e(u, v)^{ab}$.
– Non-degeneracy: $e(g, g) \neq 1$.

Access Structure: Given a set of attributes $P = \{P_1, P_2, \ldots, P_n\}$, a collection $\Gamma \subset 2^P$ is termed monotone if it is closed under superset operations. This means for any B, C, if $B \in \Gamma$ and $B \subset C$, then C must also be in Γ. An access structure (or, specifically, a monotone access structure) is a collection Γ consisting of non-empty subsets of P, excluding the empty set. Here, authorized sets are the members of Γ, and sets not in Γ are unauthorized sets.

Linear Secret Sharing Schemes (LSSS) [66]: An LSSS over a set of parties P is deemed linear (over $\mathbb{Z}_p$) if:

- The share belonging to each party is represented as a vector over $\mathbb{Z}_p$.
- An W matrix, known as the share-generating matrix, has ℓ rows and n columns. For row i of W, the function ρ assigns the corresponding party label $\rho(i)$. Considering a column vector $v = (s, r_2, \ldots, r_n)$—with s being the secret to share over $\mathbb{Z}_p$ and $r_2, \ldots, r_n$ as randomly chosen elements of $\mathbb{Z}_p$—the product Wv yields ℓ shares of the secret s according to the scheme. Each share $(Wv)_i$ is associated with party $\rho(i)$.

Here, LSSS is comprised within Γ. For any set of attributes s of an authenticated user, let $I \subset \{1, 2, \ldots, \ell\}$ denote $\{i; \rho(i) \in S\}$. Then, there exist constants $\{\omega_i \in \mathbb{Z}_p\}_{i \in I}$ such that, if $\{\lambda_i\}$ represents a collection of valid shares of a secret s under LSSS, then the equation $\sum_{i \in I} \omega_i \lambda_i = s$ holds true.

2.2   Device Fingerprint

The user's device serves as an identity verification and information processing center, typically storing a large amount of personal information [27]. For example, biometric data such as fingerprints or facial recognition can be used as locks for the device, ensuring that only authorized users can access the data on the device. Additionally, external hardware devices, such as security tokens, can be used. These devices generate random numbers to provide decryption keys, further enhancing security.

Besides biometric information and hardware tokens, the user's regular usage habits also contain important personal information, such as browser fingerprinting emerges as a sophisticated stateless identification technique [28], utilizing the aggregation and analysis of diverse data from a user's browser and device, such as browser type and version, operating system, screen resolution, supported fonts and plugins, language settings, and additional browser. This amalgamation of data constructs a unique 'fingerprint' for devices (or users), enabling websites or online service providers to identify or even track the behavior of these devices without the need for explicit identifiers.

As a unique identifier for devices, device fingerprinting introduces novel opportunities in this domain. Its applications are extensive, encompassing tracking [63], security enhancement [2,15], anomaly behavior detection [61], and bot detection [64]. For instance, the FedFingerprinting [5] framework aims to enhance the efficiency and accuracy of detecting anonymous users accessing illegal websites via the Tor network. Concurrently, FL4IoT [65] is dedicated to applying FL to tasks related to device fingerprinting and identification based on network traffic. In combating device spoofing, research [55,56] efforts have amalgamated behavioral fingerprinting with ML technologies to discern device models and types.

The device fingerprints are digital traces of information carried by computer network devices that can be used to uniquely identify devices based on unique patterns of features generated by remote devices [11,22,72]. This paper discusses and leverages the fingerprint information of devices to ascertain their uniqueness. Even in the face of collaborative attacks, the proposed scheme ensures the device fingerprint information of users, serving as a cryptographic key, is never disclosed externally. Unless divulged by the owner, the security of the key is assured, thereby enhancing key security through this methodology. This study is based on the Waters' CP-ABE [7] scheme. Our scheme leverages the uniqueness of device fingerprints to identify devices, thus enhancing security by ensuring that only authenticated devices can access the system.

2.3   Fingerprint-Based ABE Scheme

Our proposed method is based on CP-ABE, and we utilize inherent device fingerprints, to enhance key protection. By employing constructs related to ABE, we not only simplify key management and establish secrets between clients and servers but also incorporate additional constraint policies that clients must adhere to. Specifically, these policies require clients to adhere to certain criteria, including their geographical location and computational capabilities. These criteria are not only for unique identification (authentication) but also for ensuring adherence to predefined access policies (authorization). This framework was developed to address the need for more stringent access controls that are tailored to the capabilities and responsibilities of each client within the system. Furthermore, device fingerprints are unique, immutable characteristics derived from each participating device—an essential component of the asymmetric encryption process. This ensures that encryption keys are intrinsically linked to device identities, significantly reducing the risk of key cloning and deception.

This encryption scheme mainly includes the following functions:

- **Key Generation:** A bilinear group G_1 of prime order p with a generator g is selected, along with a bilinear map $e : G_1 \times G_1 \rightarrow G_2$. Random exponents $a, b \in Z_p$ are chosen, in addition to a hash function $H : \{0,1\}^* \rightarrow G$. The public key PK and system master key MK is published as:

$$PK = g, g^b, e(g,g)^a, \quad MK = g^a$$

Utilizing the master secret key MK and a specific attribute set S, the system selects a random value $t \in Z_p$ for each participant. This leads to the creation of a personalized secret key:

$$SK = \left(g^{a+bt}, g^t, (K_X)_{X \in S} \right), \quad \forall_{X \in S} K_X = H(X)^t$$

- **Encryption:** The encryption function employs the public key PK, a text M, and an access structure (W, ρ) that covers all attributes to encrypt the message. The function ρ links each row of W to specific attributes, with W being an $\ell \times n$ matrix. Initially, generated a random vector $\gamma = (s, y_2, \cdots, y_n) \in Z_p^n$ and random values $r_1, r_2, \cdots, r_\ell \in Z_p$. This vector is integral to sharing the encryption exponent s. For each row i of W, the algorithm calculates $\lambda_i = \gamma \cdot W_i$. The output is a ciphertext CT as follows:

$$\text{CT} = \left(Me(g,g)^{as}, g^s, \widehat{CS} \right),$$

$$\widehat{CS} = \left(g^{b\lambda_1} H \left(X_{\rho_1} \right)^{r_1}, g^{r_1} \right), \left(g^{b\lambda_2} H \left(X_{\rho_2} \right)^{r_2}, g^{r_2} \right), , , \left(g^{b\lambda_\ell} H \left(X_{\rho_\ell} \right)^{r_\ell}, g^{r_\ell} \right).$$

– **Decryption:** The decryption process requires the ciphertext CT and the secret key SK. Assuming the attribute set S meets the criteria of the access structure (W, ρ), a subset I is defined as $\{i = \rho(i) \in s\}, I \in \{1, 2, \ldots, \ell\}$. A specific structure $\{\omega_i \in Z_p\}$ ensures that valid shares $\{\lambda_i\}$ of a secret s allow the decryption algorithm to reconstruct the text M. The decryption calculation is as follows:

$$\frac{e\left(g^s, g^{a+bt} \right)}{\prod_{i \in I} \left(e \left(g^{b\lambda_i} H \left(X_{\rho_i} \right)^{r_i}, g^t \right) e \left(H \left(X_{\rho_i} \right)^t, g^{r_i} \right) \right)^{\omega_i}} = \frac{e(g,g)^{as} e(g,g)^{bts}}{\prod_{i \in I} e(g,g)^{bt\omega_i \lambda_i}} = e(g,g)^{as}$$

$$\frac{Me(g,g)^{as}}{e(g,g)^{as}} = M$$

3 Our Method

In this section, we provide a detailed exposition of our proposed scheme. The integration of CP-ABE into FL environments enables flexible governance over participants' interactions with the model by using device fingerprint to establish fine-grained access controls.

In the envisioned FL framework, we assume that the central server within the system is trustworthy. The central server can honestly access and process the attribute information and model parameters collected from clients to ensure the effective progression of the learning process. Additionally, our system includes N edge devices (clients), each utilizing its local data to train the model. During the initial phase of the system's startup, the server collects attribute information from each device, such as hardware or software details. This step is aimed at authorizing devices to participate in the process and verifying their identity. By using the inherent device information, including the additional constraint policies when encrypting the data or model parameters (gradients), we ensure that this information remains confidential from external parties, preventing unauthorized third parties or external attackers from gaining direct access.

In addition, although we assume that the communication channel is trustworthy, we can make use of encrypted communication channels such as Transport Layer Security (TLS) to protect it. TLS offers several advantages, including robust encryption, data integrity, and endpoint authentication, which safeguard against eavesdropping and data tampering. Moreover, this proposal can be adapted to specific usage environments by integrating other secure communication strategies, such as secure multi-party computation and other advanced cryptographic techniques, to further enhance the security framework.

The implementation approach is outlined as follows:

- **Server Side:**
  - **Initialization:** The server initializes the global model Θ. It collects device fingerprints (such as hardware or software information) from each client to authorize their participation and verify their identity. Based on the collected client attribute information, a policy matrix P is constructed, where each row represents an access policy, and columns correspond to client attribute information.
  - **Model Encryption:** Using the collected device fingerprints, the initialized global model is encrypted based on the Fingerprint-Based ABE Scheme and distributed to clients C_i, where $i \in \{1, 2, \ldots, n\}$. The model updates are encrypted as follows:

    $$Enc(\theta) = (\theta e(g, g)^{as}, g^s, \widehat{CS})$$

    where $Enc(\theta)$ represents the encrypted global model update, and $\widehat{CS}$ denotes the attribute-based encryption function associated with client C_i's attributes, thus facilitating access control based on user attributes.
  - **Model Aggregation:** After receiving updates from clients, the server conducts an evaluation to determine the impact of each update on the global model's performance. This involves assessing whether the contributions from clients enhance or degrade the model's effectiveness. Positive contributions lead to the client being classified as normal, while negative contributions prompt concerns about potential threats, resulting in adjustments to the policy matrix P. The server then selectively aggregates updates, incorporating only those that demonstrably improve the model's performance.
  - **Feedback of Results:** Based on the latest status of the policy matrix P, the participation weight of each client in future training rounds is determined. After the model aggregation is complete, the server cycles the updated model back to the clients for further training. This iterative process continues until the model converges or the predefined number of iterations is completed.
- **Client Side:**
  - **Local Decryption:** Upon receiving the encrypted model, it is decrypted based on the local device fingerprint. This process verifies whether the client's C_i local attribute information matches the attribute information collected by the server. If it meets the set of ABE policies, the global model θ can be decrypted:

$$Dec_i(Enc(\theta)) = \frac{e\left(g^s, g^{a+bt_i}\right)}{\prod_{i \in I}\left(e\left(g^{b\lambda_i} H(X_{\rho_i})^{r_i}, g^{t_i}\right) e\left(H(X_{\rho_i})^{t_i}, g^{r_i}\right)\right)^{\omega_i}} \oplus Enc(\theta)$$

This formula assumes that the client has successfully generated an attribute key that matches the policy matrix P, and this decryption restores the original model parameters θ_i by reversing the encryption mechanism. X_{ρ_1} represents a specific attribute, for example, 'device type' = 'smartphone', and $H(\cdot)$ represents mapping through a hash function.

- **Local Model Training:** Each client trains the global model Θ using their dataset D_i, thus updating the local model parameters θ_i.
- **Model Upload:** After local training is completed, the client uploads the local model parameters θ_i to the server.

4 Security Analysis

In this section, we analyze the security and demonstrate that our scheme satisfies the security goals.

4.1 Security Goals

The overarching security goal for integrating FBE scheme with FL is to establish a secure, privacy-preserving, and resilient collaborative learning framework. This goal encompasses ensuring the confidentiality, integrity, and availability of data and models across distributed devices. Through the employment of device fingerprints for enhanced encryption and authentication, alongside access control mechanisms, this approach aims to protect against unauthorized access, and manipulation, thereby maintaining the trustworthiness and reliability of the FL system.

4.2 Threat Model

The threat model considers potential adversaries aiming to compromise the confidentiality, integrity, or availability of the device fingerprint-based ABE-FL system, specifically those with legitimate access to the system but malicious intent. This includes unauthorized client attempts to access or decrypt data not intended for them.

4.3 Security Analysis

To safeguard against issues such as unauthorized access or decryption, our framework employs LSSS, ensuring that only attributes meeting a predefined access structure can decrypt. The resilience against data falsification hinges on the difficulty of reconstructing the LSSS without an exact match of attributes. This is underpinned by the equation $\sum_{i \in I} \omega_i \lambda_i = s$, where I denotes the set of indices for which the function $\rho(i)$ is included in the set of attributes S, thereby guaranteeing that only authorized data are processed.

   In addition, the encryption mechanism's security is predicated on the impossibility of decrypting the ciphertext without the specific secret key SK, which is derived from the user's attributes and the system's master key MK. The decryption process illustrating that without access to $SK = \left(g^{a+bt}, g^t, (K_X)_{X \in S}\right)$, it is infeasible for an adversary to reconstruct $e(g, g)^{as}$ and access the encrypted message M. This dual-layered approach, combining the LSSS with a secure key management protocol, effectively mitigates the risk posed by internal threats, ensuring the integrity and confidentiality of the FL process.

4.4   Privacy Protection Mechanism

Privacy protection serves as a cornerstone within our framework, aimed at safeguarding sensitive user information and the intricacies of learning models amidst a broad spectrum of security threats. By introducing and implementing the strategies outlined below, we aim to establish a secure and efficient protection mechanism:

Device Fingerprint-Based ABE Encryption Scheme: This scheme is foundational to our privacy protection mechanism. Utilizing unique device fingerprints, we ensure encryption keys are deeply tied to a device's identity, significantly mitigating risks associated with key cloning and unauthorized data access. In the key generation phase, a bilinear group G_1 based on bilinear maps is selected, with random exponents $a, b \in Z_p$ set, and a hash function H chosen. This leads to the construction of the system's public key PK and master key MK, ensuring each participant receives a personalized key based on their attribute set S. During the encryption and decryption processes, information is encrypted using the public key PK, the text M, and an access structure (W, ρ) covering all attributes. The decryption process necessitates a key matching the attribute set S, thereby ensuring that only users with specific attributes can decrypt and access the information.

End-to-End Encryption. Our framework capitalizes on the unique identity of device fingerprints, fortifying the bond between encryption keys and the device's identity to substantially reduce the risks of key replication and unauthorized data breaches. During the key generation phase, we utilize a bilinear group G_1 with chosen random exponents $a, b \in Z_p$ and select a hash function H, forming the basis for our system's public key PK and master key MK. This ensures personalized key allocation for each participant, derived from their attribute set S. The encryption employs the public key PK, targeting text M, and adheres to an access structure (W, ρ) that spans all attributes. For decryption, a key aligning with the attribute set S is required, guaranteeing that only users possessing the matching attributes can decrypt and access the transmitted data, thereby enhancing end-to-end encryption strategy within the FL system.

5   Future Directions and Conclusion

In this section, we explore future directions and provide a conclusion.

5.1   Future Directions

Looking towards potential future research directions, we are focused on two directions.

Trade-Offs between Security, Privacy, and Efficiency : We focus on the trade-off between security, privacy, and computational efficiency. These are crucial for the effective deployment and functioning of ML systems, each playing a vital role in ensuring the technology's trustworthiness and usability. To achieve an optimal balance among these aspects, we are ready to undertake an extensive investigation of further security mechanisms, including MPC and DP. MPC presents valuable opportunities to improve data privacy and security by facilitating computations on encrypted data, thus offering an additional safeguard against data breaches and unauthorized access. Similarly, DP introduces a mathematical framework for quantifying and managing privacy risks, allowing for the analysis of datasets in a way that protects the information of individual participants. These approach not only strengthens the security framework but also aligns with the increasing demand for privacy-preserving computational techniques in the realm of ML.

Decentralized Architectures: The decentralized nature of FL presents unique challenges and opportunities in scalability, and communication overhead, especially in distributed environments that span across various geographic locations and computing infrastructures. Scalability is a critical concern as the number of participating nodes increases, necessitating efficient algorithms and protocols that can handle the growth without compromising performance. Communication overhead, particularly in the context of exchanging model updates, requires innovative compression techniques and network protocols to minimize latency and bandwidth usage. Lastly, the aspect of consensus mechanisms in FL involves ensuring that all participating nodes agree on the model updates in a secure and verifiable manner, which is crucial for the integrity and reliability of the learning process.

To address these challenges, our research will delve into cryptographic techniques, network protocols, and innovative consensus algorithms tailored for FL environments. We aim to develop solutions that not only enhance the security and privacy aspects of FL but also improve its efficiency, making it more adaptable to real-world applications. Additionally, we will explore the potential of leveraging blockchain technology as a means to secure model exchanges and enforce consensus, thereby adding an extra layer of transparency and trust to the FL process.

5.2 Conclusion

By utilizing unique device fingerprints for enhanced encryption, our ABE-FL framework establishes a secure, privacy-preserving, and resilient collaborative learning environment. The integration of device fingerprints in key generation significantly reduces the risk of unauthorized data access, ensuring that only verified and compliant devices can participate in the FL process. However, achieving a fully secure and efficient federated learning system continues to pose significant challenges. Our ongoing efforts are directed towards integrating advanced security features and refining the FL system architecture to improve collaborative learning in diverse and distributed settings.

Acknowledgments. The research of the first author is partially supported by the Japan Science and Technology Agency, Support for Pioneering Research Initiated by the Next Generation (JST SPRING) under Grant JPMJSP2136. The second author is supported by the Spanish Ministry of Universities and the European Union - NextGeneration EU funds through the Grants for the Requalification of the Spanish University System 2021–2023 (University of Malaga, Resolution July 1st 2021). The third author is partially supported by the International Exchange, Foreign Researcher Invitation Program of National Institute of Information and Communications Technology (NICT), Japan. The fourth authors are supported by KAKENHI-PROJECT- JP 24K02932.

Disclosure of Interests. The authors have no competing interests to declare that are relevant to the content of this article.

References

1. Aledhari, M., Razzak, R., Parizi, R.M., Saeed, F.: Federated learning: a survey on enabling technologies, protocols, and applications. IEEE Access **8**, 140699–140725 (2020)
2. Annamalai, M.S.M.S., Bilogrevic, I., De Cristofaro, E.: Fp-fed: privacy-preserving federated detection of browser fingerprinting. arXiv preprint arXiv:2311.16940 (2023)
3. Asoodeh, S., Liao, J., Calmon, F.P., Kosut, O., Sankar, L.: Three variants of differential privacy: lossless conversion and applications. IEEE J. Sel. Areas Inf. Theory **2**(1), 208–222 (2021)
4. Aziz, R., Banerjee, S., Bouzefrane, S., Le Vinh, T.: Exploring homomorphic encryption and differential privacy techniques towards secure federated learning paradigm. Future Internet **15**(9), 310 (2023)
5. Bang, J., Jeong, J., Lee, J.: Fedfingerprinting: a federated learning approach to website fingerprinting attacks in tor networks. IEEE Access (2023)
6. Bao, X., Su, C., Xiong, Y., Huang, W., Hu, Y.: Flchain: a blockchain for auditable federated learning with trust and incentive. In: International Conference on Big Data Computing and Communications (BIGCOM), pp. 151–159. IEEE (2019)
7. Bethencourt, J., Sahai, A., Waters, B.: Ciphertext-policy attribute-based encryption. In: 2007 IEEE Symposium on Security and Privacy (SP 2007), pp. 321–334. IEEE (2007)
8. Bonawitz, K., et al.: Practical secure aggregation for privacy-preserving machine learning. In: ACM SIGSAC Conference on Computer and Communications Security, pp. 1175–1191 (2017)
9. Cai, R., et al.: Many-task federated learning: a new problem setting and a simple baseline. In: Proceedings of the IEEE/CVF Conference on Computer Vision and Pattern Recognition, pp. 5036–5044 (2023)
10. Cao, Z., et al.: Privacy matters: vertical federated linear contextual bandits for privacy protected recommendation. In: Proceedings of the 29th ACM SIGKDD Conference on Knowledge Discovery and Data Mining, pp. 154–166 (2023)
11. Charyyev, B., Gunes, M.H.: Locality-sensitive iot network traffic fingerprinting for device identification. IEEE Internet Things J. **8**(3), 1272–1281 (2020)

12. Chen, C., Anada, H., Kawamoto, J., Sakurai, K.: A hybrid encryption scheme with key-cloning protection: user/terminal double authentication via attributes and fingerprints. J. Internet Serv. Inf. Secur. **6**(2), 23–36 (2016)

13. Chiu, T.C., Shih, Y.Y., Pang, A.C., Wang, C.S., Weng, W., Chou, C.T.: Semisupervised distributed learning with non-iid data for aiot service platform. IEEE Internet Things J. **7**(10), 9266–9277 (2020)

14. De Oliveira, D.C., Liu, J., Pacitti, E.: Data-intensive workflow management: for clouds and data-intensive and scalable computing environments. Synth. Lect. Data Manag. **14**(4), 1–179 (2019)

15. Durey, A., Laperdrix, P., Rudametkin, W., Rouvoy, R.: FP-redemption: studying browser fingerprinting adoption for the sake of web security. In: Bilge, L., Cavallaro, L., Pellegrino, G., Neves, N. (eds.) DIMVA 2021. LNCS, vol. 12756, pp. 237–257. Springer, Cham (2021). https://doi.org/10.1007/978-3-030-80825-9_12

16. Fang, H., Qian, Q.: Privacy preserving machine learning with homomorphic encryption and federated learning. Future Internet **13**(4), 94 (2021)

17. Fazli Khojir, H., Alhadidi, D., Rouhani, S., Mohammed, N.: Fedshare: secure aggregation based on additive secret sharing in federated learning. In: Proceedings of the 27th International Database Engineered Applications Symposium, pp. 25–33 (2023)

18. Geyer, R.C., Klein, T., Nabi, M.: Differentially private federated learning: a client level perspective. arXiv preprint arXiv:1712.07557 (2017)

19. Goyal, V., Pandey, O., Sahai, A., Waters, B.: Attribute-based encryption for fine-grained access control of encrypted data. In: Proceedings of the 13th ACM Conference on Computer and Communications Security, pp. 89–98 (2006)

20. Hao, M., Li, H., Xu, G., Liu, S., Yang, H.: Towards efficient and privacy-preserving federated deep learning. In: IEEE International Conference on Communications (ICC), pp. 1–6. IEEE (2019)

21. Hardy, S., et al.: Private federated learning on vertically partitioned data via entity resolution and additively homomorphic encryption. arXiv preprint arXiv:1711.10677 (2017)

22. Humayed, A., Lin, J., Li, F., Luo, B.: Cyber-physical systems security–a survey. IEEE Internet Things J. **4**(6), 1802–1831 (2017)

23. IEEE: Ieee approved draft guide for architectural framework and application of federated machine learning (2020). https://ieeexplore.ieee.org/document/9154804

24. Jia, J., Salem, A., Backes, M., Zhang, Y., Gong, N.Z.: Memguard: defending against black-box membership inference attacks via adversarial examples. In: ACM SIGSAC Conference on Computer and Communications Security, pp. 259–274 (2019)

25. Kairouz, P., et al.: Advances and open problems in federated learning. Found. Trends Mach. Learn. **14**(1–2), 1–210 (2021)

26. Kariyappa, S., Qureshi, M.K.: Gradient inversion attack: leaking private labels in two-party split learning. arXiv preprint arXiv:2112.01299 (2021)

27. Kumar, V., Paul, K.: Device fingerprinting for cyber-physical systems: a survey. ACM Comput. Surv. **55**(14s) (2023)

28. Laperdrix, P., Bielova, N., Baudry, B., Avoine, G.: Browser fingerprinting: a survey. ACM Trans. Web (TWEB) **14**(2), 1–33 (2020)

29. Lee, H., Kim, J., Ahn, S., Hussain, R., Cho, S., Son, J.: Digestive neural networks: a novel defense strategy against inference attacks in federated learning. Comput. Secur. **109**, 102378 (2021)

30. Li, J., Ma, C., Zhang, K.: A novel lattice-based ciphertext-policy attribute-based proxy re-encryption for cloud sharing. In: Meng, W., Furnell, S. (eds.) SocialSec 2019. CCIS, vol. 1095, pp. 32–46. Springer, Singapore (2019). https://doi.org/10.1007/978-981-15-0758-8_3

31. Li, Z., Huang, Z., Chen, C., Hong, C.: Quantification of the leakage in federated learning. arXiv preprint arXiv:1910.05467 (2019)

32. Liang, Z., Wang, B., Gu, Q., Osher, S., Yao, Y.: Differentially private federated learning with laplacian smoothing. arXiv preprint arXiv:2005.00218 (2020)

33. Liu, C.W., Hsien, W.F., Yang, C.C., Hwang, M.S.: A survey of attribute-based access control with user revocation in cloud data storage. Int. J. Netw. Secur. 18(5), 900–916 (2016)

34. Liu, J., Pacitti, E., Valduriez, P., De Oliveira, D., Mattoso, M.: Multi-objective scheduling of scientific workflows in multisite clouds. Futur. Gener. Comput. Syst. 63, 76–95 (2016)

35. Liu, J., Pacitti, E., Valduriez, P., Mattoso, M.: A survey of data-intensive scientific workflow management. J. Grid Comput. 13(4), 457–493 (2015)

36. Liu, J., et al.: Efficient scheduling of scientific workflows using hot metadata in a multisite cloud. IEEE Trans. Knowl. Data Eng. (TKDE) 31(10), 1940–1953 (2018)

37. Liu, L., Chen, C., Pei, Q., Maharjan, S., Zhang, Y.: Vehicular edge computing and networking: a survey. Mob. Netw. Appl. 26(3), 1145–1168 (2021)

38. Liu, W., Zhang, Y.H., Li, Y.F., Zheng, D.: A fine-grained medical data sharing scheme based on federated learning. Concurr. Comput. Pract. Exp. 35(20), e6847 (2023)

39. Liu, X., Li, H., Xu, G., Chen, Z., Huang, X., Lu, R.: Privacy-enhanced federated learning against poisoning adversaries. IEEE Trans. Inf. Forensics Secur. 16, 4574–4588 (2021)

40. Liu, Y., et al.: Defending label inference and backdoor attacks in vertical federated learning. arXiv preprint arXiv:2112.05409 (2021)

41. Lyu, L., Yu, H., Yang, Q.: Threats to federated learning: a survey. arXiv Cryptography and Security (2020)

42. Ma, C., et al.: On safeguarding privacy and security in the framework of federated learning. IEEE Netw. 34(4), 242–248 (2020)

43. Ma, J., Naas, S.A., Sigg, S., Lyu, X.: Privacy-preserving federated learning based on multi-key homomorphic encryption. Int. J. Intell. Syst. (2022)

44. Malekzadeh, M., Hasircioglu, B., Mital, N., Katarya, K., Ozfatura, M.E., Gunduz, D.: Dopamine: differentially private federated learning on medical data. arXiv: Learning (2021)

45. Mao, Y., You, C., Zhang, J., Huang, K., Letaief, K.B.: A survey on mobile edge computing: the communication perspective. IEEE Commun. Surv. Tutor. 19(4), 2322–2358 (2017)

46. McMahan, B., Moore, E., Ramage, D., Hampson, S., Arcas, B.A.: Communication-efficient learning of deep networks from decentralized data. In: International Conference on Artificial Intelligence and Statistics (AISTATS), pp. 1273–1282 (2017)

47. Mothukuri, V., Parizi, R.M., Pouriyeh, S., Huang, Y., Dehghantanha, A., Srivastava, G.: A survey on security and privacy of federated learning. Futur. Gener. Comput. Syst. 115, 619–640 (2021)

48. Nguyen, D.C., et al.: Federated learning meets blockchain in edge computing: opportunities and challenges. IEEE Internet Things J. 8(16), 12806–12825 (2021)

49. Pettai, M., Laud, P.: Combining differential privacy and secure multiparty computation. In: Annual Computer Security Applications Conference, pp. 421–430 (2015)

50. Qammar, A., Karim, A., Ning, H., Ding, J.: Securing federated learning with blockchain: a systematic literature review. Artif. Intell. Rev. **56**(5), 3951–3985 (2023)
51. Rasori, M., La Manna, M., Perazzo, P., Dini, G.: A survey on attribute-based encryption schemes suitable for the internet of things. IEEE Internet Things J. **9**(11), 8269–8290 (2022)
52. Ratnayake, H., Chen, L., Ding, X.: A review of federated learning: taxonomy, privacy and future directions. J. Intell. Inf. Syst. ,1–27 (2023)
53. Ren, H., Deng, J., Xie, X.: Grnn: generative regression neural network–a data leakage attack for federated learning. ACM Trans. Intell. Syst. Technol. (TIST) **13**(4), 1–24 (2022)
54. Rodríguez-Barroso, N., Jiménez-López, D., Luzón, M.V., Herrera, F., Martínez-Cámara, E.: Survey on federated learning threats: concepts, taxonomy on attacks and defences, experimental study and challenges. Inf. Fusion **90**, 148–173 (2023)
55. Sánchez, P.M.S., Valero, J.M.J., Celdrán, A.H., Bovet, G., Pérez, M.G., Pérez, G.M.: A survey on device behavior fingerprinting: Data sources, techniques, application scenarios, and datasets. IEEE Commun. Surv. Tutor. **23**(2), 1048–1077 (2021)
56. Sánchez Sánchez, P.M., Huertas Celdran, A., Buendía Rubio, J.R., Bovet, G., Martínez Pérez, G.: Robust federated learning for execution time-based device model identification under label-flipping attack. Clust. Comput., 1–12 (2023)
57. Shejwalkar, V., Houmansadr, A., Kairouz, P., Ramage, D.: Back to the drawing board: a critical evaluation of poisoning attacks on production federated learning. In: 2022 IEEE Symposium on Security and Privacy (SP), pp. 1354–1371. IEEE (2022)
58. Shen, S., Zhu, T., Wu, D., Wang, W., Zhou, W.: From distributed machine learning to federated learning: in the view of data privacy and security. Concurr. Comput. Pract. Exp. **34**(16), e6002 (2022)
59. So, J., et al.: Lightsecagg: a lightweight and versatile design for secure aggregation in federated learning. Proc. Mach. Learn. Syst. **4**, 694–720 (2022)
60. Supriya, Y., Gadekallu, T.R.: A survey on soft computing techniques for federated learning-applications, challenges and future directions. ACM J. Data Inf. Qual. (2023)
61. Thom, J., Thom, N., Sengupta, S., Hand, E.: Smart recon: network traffic fingerprinting for iot device identification. In: 2022 IEEE 12th Annual Computing and Communication Workshop and Conference (CCWC), pp. 0072–0079. IEEE (2022)
62. Truex, S., et al.: A hybrid approach to privacy-preserving federated learning. In: ACM Workshop on Artificial Intelligence and Security, pp. 1–11 (2019)
63. Vastel, A., Laperdrix, P., Rudametkin, W., Rouvoy, R.: Fp-stalker: tracking browser fingerprint evolutions. In: 2018 IEEE Symposium on Security and Privacy (SP), pp. 728–741. IEEE (2018)
64. Vastel, A., Rudametkin, W., Rouvoy, R., Blanc, X.: Fp-crawlers: studying the resilience of browser fingerprinting to block crawlers. In: MADWeb'20-NDSS Workshop on Measurements, Attacks, and Defenses for the Web (2020)
65. Wang, H., Eklund, D., Oprea, A., Raza, S.: Fl4iot: Iot device fingerprinting and identification using federated learning. ACM Trans. Internet Things **4**(3), 1–24 (2023)
66. Waters, B.: Ciphertext-policy attribute-based encryption: an expressive, efficient, and provably secure realization. In: Catalano, D., Fazio, N., Gennaro, R., Nicolosi, A. (eds.) PKC 2011. LNCS, vol. 6571, pp. 53–70. Springer, Heidelberg (2011). https://doi.org/10.1007/978-3-642-19379-8_4

67. Wei, K., et al.: Federated learning with differential privacy: algorithms and performance analysis. IEEE Trans. Inf. Forensics Secur. **15**, 3454–3469 (2020)

68. Wen, J., Zhang, Z., Lan, Y., Cui, Z., Cai, J., Zhang, W.: A survey on federated learning: challenges and applications. Int. J. Mach. Learn. Cybern. **14**(2), 513–535 (2023)

69. Wu, D., Pan, M., Xu, Z., Zhang, Y., Han, Z.: Towards efficient secure aggregation for model update in federated learning. In: GLOBECOM 2020 - 2020 IEEE Global Communications Conference, pp. 1–6 (2020). https://doi.org/10.1109/GLOBECOM42002.2020.9347960

70. Xia, Q., Ye, W., Tao, Z., Wu, J., Li, Q.: A survey of federated learning for edge computing: research problems and solutions. High-Conf. Comput. **1**(1), 100008 (2021)

71. Xie, Y., Chen, B., Zhang, J., Wu, D.: Defending against membership inference attacks in federated learning via adversarial example. In: International Conference on Mobility, Sensing and Networking (MSN), pp. 153–160. IEEE (2021)

72. Xu, Q., Zheng, R., Saad, W., Han, Z.: Device fingerprinting in wireless networks: challenges and opportunities. IEEE Commun. Surv. Tutor. **18**(1), 94–104 (2015)

73. Yin, X., Qiu, H., Wu, X., Zhang, X.: An efficient attribute-based participant selecting scheme with blockchain for federated learning in smart cities. Computers **13**(5), 118 (2024)

74. Yuan, S., Cao, B., Sun, Y., Peng, M.: Secure and efficient federated learning through layering and sharding blockchain. arXiv preprint arXiv:2104.13130 (2021)

75. Zhang, C., Xie, Y., Bai, H., Yu, B., Li, W., Gao, Y.: A survey on federated learning. Knowl.-Based Syst. **216**, 106775 (2021)

76. Zhang, C., Li, S., Xia, J., Wang, W., Yan, F., Liu, Y.: {BatchCrypt}: efficient homomorphic encryption for {Cross-Silo} federated learning. In: USENIX Annual Technical Conference, pp. 493–506 (2020)

77. Zhang, Y., Zeng, D., Luo, J., Xu, Z., King, I.: A survey of trustworthy federated learning with perspectives on security, robustness, and privacy. arXiv preprint arXiv:2302.10637 (2023)

78. Zhang, Z., et al.: Lsfl: a lightweight and secure federated learning scheme for edge computing. IEEE Trans. Inf. Forensics Secur. **18**, 365–379 (2022)

79. Zhao, J., Huang, C., Wang, W., Xie, R., Dong, R., Matwin, S.: Local differentially private federated learning with homomorphic encryption. J. Supercomput. **79**(17), 19365–19395 (2023)

80. Zhu, L., Liu, Z., Han, S.: Deep leakage from gradients. In: Neural Information Processing Systems (NIPS), vol. 32 (2019)

Automatic Alert Categories Standardization for Heterogeneous Devices with Incomplete Semantic Knowledge Based on LSTM

Haiping Wang[1,2], Jianqiang Li[3], Binbin Li[1,2(✉)], Tianning Zang[1,2], Yifei Yang[1], Siyu Jia[1,2], Zisen Qi[1,2], and Yu Ding[1]

[1] Institute of Information Engineering, Chinese Academy of Sciences, Beijing, China
{wanghaiping,libinbin}@iie.ac.cn
[2] School of Cyber Security, University of Chinese Academy of Sciences, Beijing, China
wanghaiping20@mails.ucas.ac.cn
[3] National Computer Network Emergency Response Technical Team/Coordination Center of China, Beijing, China

Abstract. In the realm of cybersecurity situational awareness, systems compile alert logs from diverse network monitors. These logs exhibit large-scale dimensions, numerous alert types, and autonomous alert categories. Ensuring precise and efficacious analysis necessitates the standardization and incorporation of these logs into a unified alert type categorization. Conventionally, security experts can manually categorize and standardize alert types by scrutinizing their semantic descriptions. However, this approach is inherently subjective and limited to alerts with comprehensive descriptions already known to the experts. To confront these challenges, we pioneer a groundbreaking method that automates alert type standardization without dependence on semantic information, harnessing solely the inherent attack behavior attributes and attacker-victim ratios extracted from the logs themselves. Specifically, we commence by encapsulating each alert type within a 27-dimensional feature vector, derived from its corresponding alert logs over a defined time frame. Subsequently, we utilize Long Short-Term Memory (LSTM) neural networks to homogenize these alert types. We undertake experiments on online alert logs, achieving an outstanding 92.308% accuracy in automatically categorizing 223 distinct alert types into 12 standardized classifications. These outcomes attest to the significant enhancement in efficiency and precision engendered by our method in alert categorization. Given that the alert logs encompass data from multiple NDR (Network Detection and Response Device) devices, this experimentation further substantiates the robustness of the proposed methodology.

Keywords: Alert Categories · Automatic Standardization · Incomplete Semantic · LSTM

J. Zhao and W. Meng (Eds.): SciSec 2024, LNCS 15441, pp. 437–456, 2025.
https://doi.org/10.1007/978-981-96-2417-1_24

1   Introduction

In the context of cyber security situational awareness for backbone network traffic [1,2], enhancing attack event detection is paramount. Probe-generated alerts serve as a key resource for event scrutiny [3–5], offering deep insights into network security posture, thereby bolstering situational awareness and enabling swift response strategies.

In order to boost cyber-attack detection, traffic is duplicated across multiple vendor-specific network probes generating independent alert types. These alerts are diverse, complex, and inconsistent, varying greatly across devices. Moreover, they often lack sufficient semantic clarity and exhibit complex interdependencies. The analytical challenges posed by these alerts can be summarized as follows:

- **Diverse Alert Types**: Data sources spawn myriad alert types; combined from several sources, their count can escalate into hundreds of thousands or millions.
- **Rapid Expansion**: Evolving security operations, equipment updates, and system upgrades drive a dynamic growth in alert rules, thus increasing device-generated alert varieties.
- **Semantic Knowledge Gap**: Independent management of alerts and detection rules by disparate teams leads to ambiguous rule semantics, causing confusion about alert types and standardization errors.
- **Intricate Semantic Interconnections**: Alerts/events often feature vague semantics, while varied alert rules detect at different granularities. Moreover, similarity and inclusion relationships exist among these rules.

Before aggregating and correlating alert logs, unifying alert type categories is essential for effective analysis. Example: Map "TCP FLOOD DDOS" and "TCP FLOOD" to a consistent category - "Network Attack Event> DDoS Attack>TCP Flood". Traditional methods involve security experts manually analyzing, categorizing, and standardizing alerts based on their cybersecurity expertise. However, this reliance on expert judgment can introduce subjectivity, threatening the accuracy and comprehensiveness of alert analysis. Moreover, it presents difficulties in upkeep, demands considerable time, and lacks consistency and efficiency.

In response to these challenges, we present a model designed to automatically standardize alert types from various sources lacking complete semantic context into a unified taxonomy, which amplifies situational awareness and expedites timely countermeasure deployment. Initially, using arrival time interval sequences from alert logs, we devise four algorithms to capture time-series attributes—*Heartbeat, Density, LongTail,* and *Random*. For every alert type, we compute their respective coefficients across four prospective attack linkages, culminating in a 16-dimensional *attack pattern* vector. Subsequently, we examine the proportional attributes between attacker IPs (*aIPs*), source Ports (*sPorts*), victim IPs (*vIPs*), and destination ports (*dPorts*), generating an 11-dimensional *attacker-victim ratio* feature vector. Ultimately, we harness a Long Short-Term Memory (LSTM [6–8]) network to categorize each alert type accurately, leveraging both the attack behavior features and the attacker-victim ratio data. The key contributions can be encapsulated as:

- We introduce a pioneering model for the automatic standardization of alert categories based on incomplete semantic knowledge. This model enables the automatic sorting and normalization of alert types from multiple sources into an unified category.
- We leverages LSTM neural networks to make comprehensive decisions by considering *attack behavior* and *attackers victims ratio*, allowing us to effectively characterize the relationship between attackers and victims.
- We conduct extensive experiments on real network environment alert logs from online NDRs, achieving an impressive 92.308% accuracy. This high accuracy demonstrates the efficacy and reliability of our model in practical applications.

The paper's structure is outlined as follows: Sect. 2 reviews pertinent literature. Section 3 delineates the framework adopted in this research. Sections 4 to 5 delve into the two models central to our approach, named *Attack Behavior Character* and *Attackers Victims Ratio Character*, detailing the extraction of behavior feature vectors from alert logs. Section 6 outlines the LSTM-based neural network utilized for alert classification standardization. Section 7 covers the experimental setup, encompassing the training and testing phases, and unveils the outcomes. The paper culminates with conclusions in Sect. 8.

2   Related Work

Recently, academia and industry have converged their efforts on core research areas in cybersecurity incident standardization, including the *Cybersecurity Standard Framework*, *Standardized Alert Formats*, and *Attack Type Classification*.

2.1   CyberSecurity Standard Framework

In the sphere of standard specification formulation, The MITRE Corporation has introduced the notably influential ATT&CK model [9–11], which methodically portrays and categorizes well-known tactics, techniques, and procedures (TTPs) leveraged in cyberattacks. This model fosters enhanced comprehension among stakeholders and capacitates them to proficiently confront adversarial maneuvers within the digital ecosystem. It segments the multifaceted phases of an attack into eleven discrete tactical classifications: Initial Compromise, Execution, Persistence, Privilege Escalation, Defense Evasion, Credential Access, Reconnaissance, Lateral Movement, Collection, Command and Control, culminating in Impact.

Simultaneously, the U.S. National Institute of Standards and Technology (NIST) has developed the Common Cybersecurity Framework (CSF) [12–17], providing the industry with overarching guiding principles. Furthermore, the International Organization for Standardization (ISO), in cooperation with the International Electrotechnical Commission (IEC), has co-published the ISO/IEC 27001 series of standards [18–20], setting up a harmonized international benchmark for Information Security Management Systems (ISMS).

Of significance, the Chinese National Standard GB/T 22239-2019 [21–23] also reflects the scholarly contributions and practical applications within the cybersecurity domain in the People's Republic of China.

2.2   Standardized Alert Formats

To foster consistent representation of security incidents and guarantee the seamless and efficient circulation and sharing of security alerts and incident data across disparate systems, scholars and professionals are actively engaged in devising standardized guidelines for articulating the contents of security events. Among these initiatives, STIX [24,25]—Structured Threat Information eXpression—acts as a standardized linguistic medium and architectural framework, purposefully crafted to express and distribute cybersecurity threat intelligence including cyber attack events. It encapsulates comprehensive depictions of attacker behavioral patterns, motivations, methodologies, and related network threat components, and by offering a universal data format, facilitates efficacious communication of threat intelligence amongst various security entities, institutions, and governmental departments within a shared infrastructure.

Supplementing this, TAXII (Trusted Automated eXchange of Indicator Information) constitutes a protocol explicitly designed for automated and secure interchange of threat intelligence conforming to STIX standards across diverse organizational and systemic borders. TAXII outlines message formats and transmission protocols for information sharing, substantively bolstering real-time and efficient dissemination of threat intelligence within the cybersecurity community.

Meanwhile, IODEF [26,27]—Incident Object Description Exchange Format—is a standardized format for describing and reporting cybersecurity incidents, established by the Internet Engineering Task Force (IETF). Utilizing IODEF, incident responders and administrators can systematically document and convey critical details about security incidents, encompassing but not restricted to event types, timestamps, affected systems and network resources, severity levels, and corresponding response strategies or remediation plans.

Furthermore, VERIS [28]—Vocabulary for Event Recording and Incident Sharing—is a standardized lexicon developed by the Verizon Data Breach Investigations Report team, intended to accurately capture and disseminate the intricacies of cybersecurity incidents. VERIS adopts a structured methodology that meticulously describes the contextual backdrop of security events, the attributes of attackers' maneuvers, specifics regarding compromised assets, the extent of harm inflicted, and the entire incident response cycle. By employing VERIS, organizations can uniformly and impartially document and compare a variety of security incidents, thereby enriching their comprehension of threat environments and refining risk management strategies correspondingly.

2.3   Attack Type Classification

In the forefront of scientific inquiry pertaining to the identification and categorization of security event types, a principal endeavor involves the creation of diverse algorithmic and modeling approaches geared toward accurate detection and classification of various security events derived from monitoring systems. A case in point is the study presented by LOCoCAT [29], which introduces a resource-efficient intrusion detection scheme custom-designed for Controller Area Network (CAN bus) ecosystems, adeptly categorizing detected malevolent activities into pre-established attack classes. By contrast, research documented in [30] utilizes a Multi-Layer Perceptron (MLP) grounded in deep learning techniques to devise an efficacious phishing attack detection methodology, particularly apt for implementation on edge devices with constrained memory and power supplies.

Another scholarly endeavor [31] investigates phishing email attacks by breaking them down into three core categories: falsified emails, exploitation of vulnerabilities within email bodies, and attachment deceptions. This meticulous analysis scrutinizes the common attributes and distinguishing characteristics of these attacks, thereby illuminating the operational characteristics of organizations perpetrating phishing email campaigns. Following this, the article [32] advances a methodology for classifying security incidents within the Internet of Things (IoT) landscape, striving to efficiently discern security threats amidst a highly interconnected milieu of smart devices. Conclusively, the research published under [33] concentrates on elevating cybersecurity standards in cyberspace by leveraging URL-based attributes to precisely recognize malicious URLs in a multi-classification context.

Moreover, some researchers employ machine learning methods to perform data technology categorization through automated semantic analysis of text. For instance, Paper [34] presents the hybrid machine learning model Sato, which excels in automatically discerning semantic types in relational table columns by ingeniously integrating contextual cues within the table and column value data. Meanwhile, Paper [35] delves into the application of text auto-categorization technologies, examining key aspects such as document representation, classifier construction, and classifier evaluation. However, it's crucial to note that these methods rely heavily on the completeness and definiteness of the text content, potentially struggling to cope with scenarios where descriptions of security event types are missing or ambiguous.

Distinct from previous investigations, the current research zeroes in on scenarios where the semantic definitions of alert categories remain indistinct and where alert typologies adapt dynamically to environmental changes. Its central aim is to explore a mechanism that can automatically standardize alert categories emanating from third-party sources.

3 Model

3.1 Definition

In this paper, the proposed solution leverages various attributes of alert logs, including alert ID aID, alert type AT, occurrence time t, source ip sIP, source port $sPort$, destination ip dIP and destination port $dPort$. Before delving into the specifics, it is necessary to introduce some definitions.

- **aIP**: The Attacker IP, often denoted as *sIP*, sometimes may be equal to *dIP* or take on other values, depending on the specific alert type *AT*.
- **vIP**: The Victim IP, typically represented as *dIP*, may occasionally be equal to *sIP* or assume other values, depending on the specific alert type *AT*.
- **sIPC**: Class C IP address that *sIP* belongs to.
- **dIPC**: Class C IP address that *dIP* belongs to.
- **aIPC**: Class C IP address that *aIP* belongs to.
- **vIPC**: Class C IP address that *vIP* belongs to.

In addition, we define the following indicators related to alert type *AT*;

- **BV(AT)**: 16-dimensional *attack behavior* vector for *AT*;
- **RV(AT)**: 11-dimensional *attackers victims ratio* vector for *AT*;
- **Dist(aIP)**: Number of distinct attack IP *aIP*;
- **Dist(vIP)**: Number of distinct victim IP *vIP*;
- **Dist(aIPC)**: Number of distinct *aIPC*;
- **Dist(vIPC)**: Number of distinct *vIPC*;
- **DistP(sPort)**: Number of distinct *sPort*;
- **DistP(dPort)**: Number of distinct *dPort*;
- **DistP(aIP)**: Number of distinct *sPort* for attacker IP *aIP*;
- **DistP(vIP)**: Number of distinct *dPort* for victim IP *vIP*;
- **DistP(aIPC)**: Number of distinct *sPort* for attacker IPs belongs to *aIPC*;
- **DistP(vIPC)**: Number of distinct *dPort* for victim IPs belongs to *vIPC*;

3.2   Framework

Standardizing alert type categories from disparate devices requires a two-step classification approach: feature extraction and vector-based classification. Conventional text mining fails due to the absence of semantic context; hence, this study derives meaningful alert traits from raw logs. Amidst inconsistent log formats and missing attributes, consistent data points like timestamps, IP addresses, and port details help form insightful characteristics.

Our method groups alerts within set time intervals, creating behavior profiles that encapsulate attack patterns and attacker-to-victim ratios by analyzing temporal trends, IP ratios, and port usage patterns.

For the classification stage, given the sequential nature of features across multiple periods, we deploy an LSTM neural network to manage this temporal data complexity. It adeptly processes time-series feature vectors for standardized classification purposes. The strategy involves chronologically extracting and feeding these vectors into the LSTM over multiple periods, allowing the model to concurrently recognize attack behaviors, attacker-victim ratios, and temporal attributes across various alert types.

Figure 1 illustrates the framework of our automatic alert categories standardization solution, which can be summarized in the following seven main steps.

- **Step1: Determine Standardized Category**. Choose an unique alert type category, referred to as **UATC** (Unique Alert Type Category, shown in Table 1), as the ultimate standard category to which all alert types detected by multi-source heterogeneous cyber attack monitoring devices or systems should be standardized.

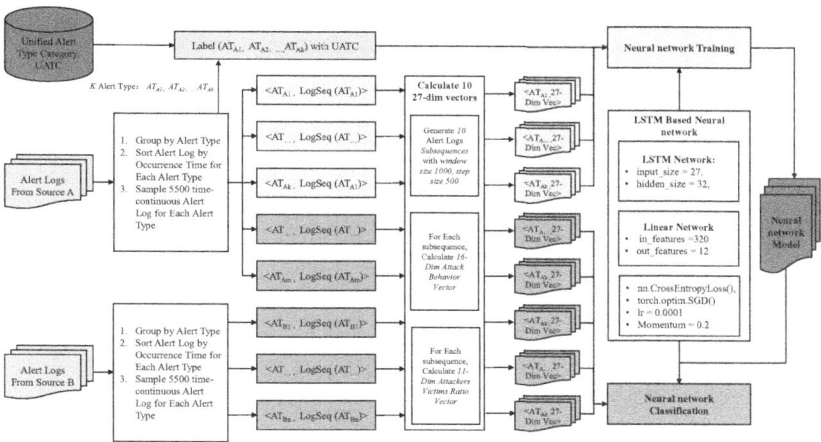

Fig. 1. Automatic Alert Categories Standardization Framework.

- **Step2: Assign *UATC* Labels to Training Data**. Tag k alert types from one or multiple IDSs/IPSs/probes with the UATC label.
- **Step3: Organize&Sample Alerts**. Categorize alert logs by type and source, sort chronologically, then select 5500 continuous logs. Divide these logs into 10 sequences, each with a 1000-sample window and 500-sample step.
- **Step4: Compute Attack Behavior Vectors**. Derive a 16-dimensional attack behavior vector for each alert type and log subsequence (explained in Sect. 4).
- **Step5: Calculate Attacker-Victim Ratio Vectors**. Generate an 11-dimensional attacker-victim ratio vector for each type and subsequence (details in Sect. 5).
- **Step6: Train LSTM-Neural Net**. Train an LSTM-based network using k labeled types and their corresponding 27-dimensional vectors (network structure detailed in Sect. 6).
- **Step7: LSTM-Based Classification**. For non-standardized alert types, produce their 10 27-dimensional vectors and classify them into the UATC using the trained LSTM model.

4   Attack Behavior Character

This section outlines our method for modeling time interval series properties related to cyber attack events. We represent alert attack behavior with a concise 16-dimensional vector, constructed by multiplying four unique attack patterns by four characteristic weights. These attack patterns include:

- **one-to-one**: Attacks by a single attacker (aIP) targeting a specific victim (vIP) share identical characteristics.

Table 1. UATC Categories

Primary Category	Secondary Category	IncidentID
Malicious Program	Mining Virus	IN01001
Malicious Program	Trojan Horse	IN01002
Malicious Program	Botnet	IN01003
Malicious Program	Ransomware	IN01004
Malicious Program	Worm	IN01005
Malicious Program	Network Scan	IN01006
Cyber Attack	Exploit	IN02001
Cyber Attack	Backdoor	IN02002
Cyber Attack	Login Attempt	IN02003
Cyber Attack	Denial of Service	IN02004
Cyber Attack	Phishing	IN02005
Others	–	IN03001

- **many-to-one**: Multiple attackers within the same *aIPC* targeting the same *vIP* exhibit consistent attack traits.
- **one-to-many**: An attacker (*aIP*) attacking multiple victims within a single *vIPC* displays uniform attack characteristics.
- **many-to-many**: Coordinated attacks from multiple attackers in a single *aIPC* against multiple targets in a single *vIPC* share homogeneous features.

Observing cyber attack alert logs, we've discerned distinctive time series patterns among various alert types. Key characteristics include: Botnet attacks show a rhythmic pattern akin to a heartbeat, with bots or C&C servers regularly communicating with the botmaster, resulting in recurring alerts. During DDOS attacks, numerous high-frequency alerts emerge densely over time intervals. Exploitation and APT events display low-frequency, random occurrences, leading to irregularly distributed alerts across intervals. Typically, alerts cluster closely in time within a single attack, but intervals between different attacks tend to be larger, reflecting a long-tail distribution.

Based on the time pattern characteristics associated with attack behaviors in security events as described above, we have designed four coefficients *heartbeat*, *density*, *random*, and *longtail* to characterize alert behavior. Alerts of the same type often share similar distributions in these coefficients across their time interval sequences. Thus, each attack type AT produces a 16-dimensional vector $BV(AT)$ representing its attack behavior. The details are showed in Algorithm 1. Subsequent sections detail the equations for calculating these four coefficients.

4.1   HeartBeat Coefficient

Cyber attacks often display periodic behavior, evident in recurrent communications between attackers and victims, generating alerts with periodic patterns

---

Data: Alert logs $logSet(AT)$
Result: 16-dimensional attack behavior feature vector
Vector v = new Vector(16);
M = Dist(aIP); N = Dist(vIP);
L = Dist(aIPC); K = Dist(vIPC);
Group $logSet(AT)$ by $\langle aIP, vIP \rangle$ and sort by ts;
foreach $logSeq(AT, aIP_m, vIP_n)$ **do**
  $\quad$ $sumH$ += Heartbeat($logSeq(AT, aIP_m, vIP_n)$) ;
  $\quad$ $sumD$ += Density($logSeq(AT, aIP_m, vIP_n)$) ;
  $\quad$ $sumR$ += Random($logSeq(AT, aIP_m, vIP_n)$) ;
  $\quad$ $sumL$ += LongTail($logSeq(AT, aIP_m, vIP_n)$) ;
end
v[0]= $sumH \div (M \times N)$; v[1]= $sumD \div (M \times N)$;
v[2]= $sumR \div (M \times N)$; v[3]= $sumL \div (M \times N)$;
Group $logSet(AT)$ by $\langle aIPC, vIP \rangle$ and sort by ts;
foreach $logSeq(AT, aIPC_l, vIP_n)$ **do**
  $\quad$ $sumH$ += Heartbeat($logSeq(AT, aIPC_l, vIP_n)$) ;
  $\quad$ $sumD$ += Density($logSeq(AT, aIPC_l, vIP_n)$) ;
  $\quad$ $sumR$ += Random($logSeq(AT, aIPC_l, vIP_n)$) ;
  $\quad$ $sumL$ += LongTail($logSeq(AT, aIPC_l, vIP_n)$) ;
end
v[4]= $sumH \div (L \times N)$; v[5]= $sumD \div (L \times N)$;
v[6]= $sumR \div (L \times N)$; v[7]= $sumL \div (L \times N)$;
Group $logSet(AT)$ by $\langle aIP, vIPC \rangle$ and sort by ts;
foreach $logSeq(AT, aIP_m, vIPC_k)$ **do**
  $\quad$ $sumH$+= Heartbeat($logSeq(AT, aIP_m, vIPC_k)$) ;
  $\quad$ $sumD$ += Density($logSeq(AT, aIP_m, vIPC_k)$) ;
  $\quad$ $sumR$ += Random($logSeq(AT, aIP_m, vIPC_k)$) ;
  $\quad$ $sumL$ += LongTail($logSeq(AT, aIP_m, vIPC_k)$) ;
end
v[8]= $sumH \div (M \times K)$; v[9]= $sumD \div (M \times K)$;
v[10]= $sumR \div (M \times K)$; v[11]= $sumL \div (M \times K)$;
Group $logSet(AT)$ by $\langle aIPC, vIPC \rangle$ and sort by ts;
foreach $logSeq(AT, aIPC_l, vIPC_k)$ **do**
  $\quad$ $sumH$ += Heartbeat($logSeq(AT, aIPC_l, vIPC_k)$) ;
  $\quad$ $sumD$ += Density($logSeq(AT, aIPC_l, vIPC_k)$) ;
  $\quad$ $sumR$ += Random($logSeq(AT, aIPC_l, vIPC_k)$) ;
  $\quad$ $sumL$ += LongTail($logSeq(AT, aIPC_l, vIPC_k)$) ;
end
v[12]= $sumH \div (L \times K)$; v[13]= $sumD \div (L \times K)$;
v[14]= $sumR \div (L \times K)$; v[15]= $sumL \div (L \times K)$;
return v ;

Algorithm 1: Attack Behavior Calculation Alogrithm

in time intervals. This study employs the ARMA [36] method to quantify this periodicity as the *Heartbeat* coefficient. Given a time interval sequence S of length N, the *Heartbeat Coefficient* is computed through ARMA autocorrelation coefficients of S. Below is a summary of the calculation procedure.

Firstly, calculate $r(k)$ using Eq. 1.

$$r(k) = \sum_{t=1}^{N} f(t)f(t+k) \tag{1}$$

$f(t)$ means the t th element of S. $r(k)$ represents the ACF(Auto-Correlation Function) or autocovariance. For a lag of length k, $r(k)$ computes the autocovariance between the sequence and itself lagged k units. If $t+k \geq N$, then $f(t+k) = f(t+k-N)$, where $k=0,1,...,N-1$.

Secondly, find the max value of all $r(k)$ and marked it as $Max(r)$.

Thirdly, calculate $\alpha(k)$, where

$$0 \leq \alpha(k) = r(k) \div Max(r) \leq 1 \tag{2}$$

Finally, calculate the ARMA autocorrelation coefficients of S respectively $\alpha(0)$, $\alpha(1)$, $\alpha(2)$,...,$\alpha(N-1)$, set *Heartbeat Coefficient* with Eq. 3:

$$Heartbeat(S) = \max_{k=0 \to N-1} \alpha(k) \tag{3}$$

4.2 Density Coefficient

Assuming the alert log sequence S spans N alerts from start time t_1 to end time t_2, its *density coefficient* can be defined as:

$$D(S) = \frac{N}{t_2 - t_1} \tag{4}$$

Given m alert log sequences $S_1, S_2, ..., S_m$ with corresponding $D(S)$ values $D(S_1), D(S_2), ..., D(S_m)$, where D_{max} and D_{min} denote the maximum and minimum $D(S)$ respectively, normalize each $Density(S_i)$ using min-max normalization [37] as shown in Eq. 5:

$$Density(S_i) = \frac{D(S_i) - D_{min}}{D_{max} - D_{min}} \tag{5}$$

Lastly, we set *Density(S)* as the weighted average of all *$Density(S_i)$* values.

4.3 Random Coefficient

Drawing inspiration from entropy theory, the random coefficient gauges unpredictability of time intervals in alert sequence S. With S having N elements and m unique intervals (Δt_0, Δt_1, Δt_2,..., $\Delta t_{(m-1)}$) affected by variable time scales, noise, etc.), these intervals usually differ. After statistical analysis of alert intervals and acceleration ratios, we categorize them into six tiered groups based on

1-minute, 5-minute, 30-minute, and 1-hour windows, calculating the probabilities P_0, P_1, P_2, P_3, P_4, P_5 for each interval. Finally, the random coefficient for alert sequence S is derived thus:

$$Random(S) = \sum_{i=0}^{5} P_i \times \log_2 P_i \qquad (6)$$

4.4   LongTail Coefficient

Observing the time intervals in alert sequence S, we find similarity with economic *long-tail distribution*. This aligns with the tendency for intervals between separate attack stages to exceed those between consecutive alerts in the same attack. To measure this variation across alert types, we define the LongTail coefficient as follows: after sorting all intervals in S ascendingly, we calculate the average of the top 80% values as μ_1 and the average of the bottom 20% as μ_2.

$$LongTail(S) = \mu_1 \div \mu_2 \qquad (7)$$

5   Attackers Victims Ratio Character

Comparing attacker IPs (aIP) to victim IPs (vIP) and source ports ($sPort$) to destination ports ($dPort$), we observe unique *attacker-victim ratio* attributes for different cyber attack types. This part elucidates modeling these ratios with 6 coefficients for aIP to vIP, and 5 for $sPort$ to $dPort$.

5.1   IP Ratio

Our research reveals unique attacker-victim IP distribution patterns for diverse alert types, characterized as *one-to-one*, *one-to-many* (e.g., vulnerability scans), *many-to-one* (e.g., DDoS attacks), and *many-to-many*. To measure these IP address interaction patterns, we devised six *IP ratio* coefficients.

$$aIP\_vIP\_ratio = Dist(aIP) \div Dist(vIP) \qquad (8)$$

aIP_vIP_ratio represents the coefficient of the ratio between attacker and victim IP counts.

$$aIP\_aIPC\_ratio = Dist(aIP) \div Dist(aIPC) \qquad (9)$$

The aIP_aIPC_ratio coefficient signifies the level of concentration among attacker IP addresses. A higher aIP_aIPC_ratio suggests increased concentration, meaning more attacker IPs stem from the same Class C IP address and participate in attacks. If attackers use multiple IPs from the same Class C address in security events, the aIP_aIPC_ratio will notably exceed 1. Conversely, when no such concentration is observed, this ratio's value will hover around 1.

$$vIP\_vIPC\_ratio = Dist(vIP) \div Dist(vIPC) \tag{10}$$

vIP_vIPC_ratio is employed to gauge the frequency of coordinated attacks targeting multiple victim IPs within the same Class C IP address.

$$aIP\_vIPC\_ratio = Dist(aIP) \div Dist(vIPC) \tag{11}$$

aIP_vIPC_ratio is used to evaluate whether an attacker IP frequently attacks multiple victims of the same class C IP address.

$$aIPC\_vIP\_ratio = Dist(aIPC) \div Dist(vIP) \tag{12}$$

$aIPC\_vIP\_ratio$ is used to characterize the attack behaviors that use multiple IPs in the same class C IP address to attack a specific victim IP address.

$$aIPC\_vIPC\_ratio = Dist(aIPC) \div Dist(vIPC) \tag{13}$$

$aIPC\_vIPC\_ratio$ is used to characterize the attack behavior that uses multiple IP of the same class C IP address to attack multiple victim IP addresses in another class C IP address.

5.2 Port Ratio

Attackers' source port to victim's target port ratio coefficients also reveal consistent trends. Specifically, these attack patterns are notable:

- Fixed source ports targeting specific destination ports on victims' systems.
- Random source ports aiming for specific destinations on victims' systems.
- Specific source ports attacking random destination ports on victims' systems.
- Random source ports attacking random destination ports on victims' systems.

To measure the distribution properties between source and destination ports, we devised five distinct port ratio coefficients.

$$dPort\_sPort\_ratio = \frac{DistP(dPort)}{DistP(sPort)} \tag{14}$$

$dPort\_sPort\_ratio$ signifies the ratio of destination to source ports. It approximates 0 when attackers' source ports vary while the destination port is constant. However, it is often much greater than 1 when various attackers use the same source port to target ever-changing victim destination ports.

$$sPort\_aIP\_ratio = \frac{DistP(sPort)}{\sum_{i=0}^{Dist(aIP)-1} DistP(aIP_i)} \tag{15}$$

$sPort\_aIP\_ratio$ indicates how the source port changes dynamically with the attacker's IP address and during attacks. This metric involves $DistP(aIP_i)$, which counts the duplicates of source ports used by attacker aIP_i.

$$dPort\_vIP\_ratio = \frac{DistP(dPort)}{\sum_{j=0}^{Dist(vIP)-1} DistP(vIP_i)} \tag{16}$$

$dPort\_dIP\_ratio$ is used to characterize whether the destination port dynamically changes with the victim IP address and attack process. Where $DistP(vIP_i)$ equals to the number of duplicate $dPort$ for victim vIP_i.

$$sPort\_aIPC\_ratio = \frac{DistP(sPort)}{\sum_{j=0}^{Dist(aIPC)-1} DistP(aIPC_i)} \tag{17}$$

$sPort\_aIPC\_ratio$ is used to characterize the attack behavior that uses multiple IPs in the same class C IP address with the same source port to launch different attack process. Where $DistP(aIPC_i)$ equals to the number of duplicate $sPort$ for attacker IPs belong to $aIPC_i$.

$$dPort\_vIPC\_ratio = \frac{DistP(dPort)}{\sum_{j=0}^{Dist(vIPC)-1} DistP(vIPC_i)} \tag{18}$$

$dPort\_vIPC\_ratio$ is used to characterize the attack behavior that victims within the same class C IP address be attacked frequently through the same destination port. Where $DistP(vIPC_i)$ equals to the number of duplicate $dPort$ for victim IPs belong to $vIPC_i$.

6   LSTM-Based Neural Network

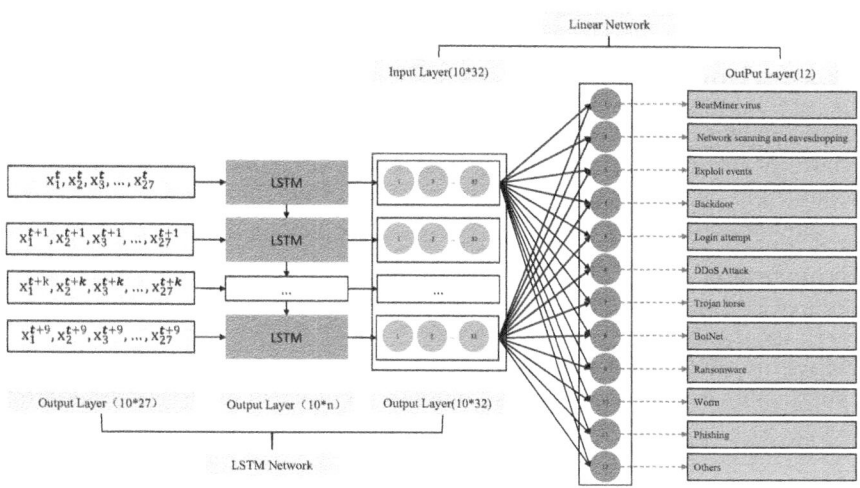

Fig. 2. LSTM-Based Neural Network.

In Sects. 4 and 5, we outline a method to extract features of *attacker behavior* and *attackers victims ratio*. In Sect. 4, we categorize alert logs within a given time frame by their one-to-one, one-to-many, many-to-one, and many-to-many patterns, computing *Heartbeat, Density, Random*, and *LongTail* coefficients for each cluster, resulting in 16-dimensional *attack behavior* feature vectors via weighted averaging. In Sect. 5, we derive 11-dimensional *attackers victims ratio* vectors. Combining both processes yields 27-dimensional vectors per alert type. Herein, we detail the architecture of deep learning networks that build standardized classification models using these vectors.

Given the temporal dependency of the proposed *attack behavior* and *attackers victims ratio* coefficients on alert log timestamps, we treat the 27-dimensional vectors as temporally sequenced and contextually relevant, aligning with LSTM neural network design. Alert logs are organized chronologically and by type, forming groups with corresponding 27-dimensional feature vectors, which establish a contextually linked sequence.

This section delves into constructing deep learning networks based on 27-dimensional feature vectors extracted from *attack behavior* and *attackers victims ratio*. Illustrated in Fig. 2, we introduce an LSTM-based neural network model that handles these vectors as time-sequential and context-aware.

Multi-time range 27-dimensional feature vectors are input to an LSTM network, which processes and converts them into 32-dimensional representations. These 32-dimensional vectors are subsequently mapped onto a single alert type category, UATC, through a linear layer. As depicted in Fig. 2, our proposed LSTM-based model takes 10 sets of 27-dimensional feature vectors per alert type, transforming them into 32-dimensional vectors. These vectors then pass through a fully-connected linear network, which refines them into 12 distinct categories, as outlined in Table 1.

7   Experiment

This section assesses the effectiveness of the proposed automated alert category standardization method. Two main experiments are conducted to validate the reasonableness of the constructed feature vector and the efficiency of our alert type standardization model:

- **Experiment 1**: Verifies the validity of our feature vector using the Chi-squared [38] distribution-based normality test and peak coefficient theory to check its stationarity.
- **Experiment 2**: Evaluates the model's efficiency by conducting eight sets of experiments concentrating on the model's accuracy in alert type standardization.

7.1   DataSets

To evaluate the efficacy of our solution, we gathered real-time online alerts from two distinct NDR device manufacturers. Over one week, we processed alert logs

Table 2. Experimental Data Statistics.

Source	DataSet	AlertLogs	Types	UATC
NDR A	TrainSet A	1,738,000	43	11
NDR A	TestSet A	242,000	29	11
NDR B	TrainSet B	4,812,500	177	9
NDR B	TestSet B	473,000	63	6

from both NDR systems for the experiment. Utilizing Flink, we produced 10 batches of 27-dimensional feature vectors for each alert type, initiating extraction when the alert log count hit 5500 per type. Chronologically sorted logs were segmented into groups using a sliding window technique with a 1000-length window and a 500-step stride, generating 10 distinct time intervals. Within each of these intervals, we extracted 27-dimensional feature vectors, obtaining 10 sets, or 'batches', of such vectors. The window size was chosen to encapsulate entire attack sequences. Sliding windows enabled the extraction of numerous feature vector groups.

Table 2 shows that we compiled 1,321 batches from a pool of 7,265,500 alerts provided by two security vendors combined. Each batch includes 10 27-dimensional vectors capturing the essence of a specific alert type across a unique time span. Ensuring impartial assessment, we randomly split the 1,321 batches into training (1,191) and test (130) sets. These 1,321 batches represent 223 diverse alert types, all manually classified into 12 standard alert categories by expert professionals as detailed in Table 1.

7.2 Verification of Coefficient Stationarity

Our method's core idea centers on distinguishing alert types by extracting behavioral features from corresponding logs. Therefore, the method's efficacy depends on correctly identifying these traits. A well-defined behavior feature should aptly represent a specific alert type, showing a level of consistency and predictable fluctuations within a set range across different time periods, rather than randomness.

To guarantee that each characteristic coefficient reliably corresponds to at least one alert type and maintains stable values across diverse time frames, we executed Experiment 1 (Table 3). This involved calculating 100 coefficient values for various alert types over 100 distinct time intervals. We utilized a normal distribution test, based on the Chi-squared distribution [39], to examine the stationarity of these coefficients. Additionally, we used kurtosis to assess the density distribution, where higher kurtosis values indicate greater concentration, per peak coefficient theory [40]. The normality test relied on the Chi-squared distribution's 3-sigma rule, suggesting that the probabilities for intervals $[\mu - \delta, \mu + \delta], [\mu - 2\delta, \mu + 2\delta]$, and $[\mu - 3\delta, \mu + 3\delta]$ should be approximately 0.6827, 0.9545, and 0.9973, respectively.

Table 3. Experiment 1: Analysis of Coefficient Stationarity Across Alert Types

coefficient	Representative Alert Type	Kurtosis	$\mu \pm \delta$	$\mu \pm 2\delta$	$\mu \pm 3\delta$
Heartbeat(H):aIP-vIP	Socks proxy Access	9.68	0.85	0.96	0.98
Density(D):aIP-vIP	ICMP Scanning	37.89	0.95	0.98	0.98
Random(R):aIP-vIP	Protocol NULL Flood	82.67	0.98	0.98	0.99
LongTail(L):aIP-vIP	Zgrab Scanner	8.85	0.9	0.95	0.97
Heartbeat(H):aIP-vIPC	OSAMiner LinkBack Domain	8.75	0.83	0.94	0.98
Density(D):aIP-vIPC	Command Injection Scanner	40.29	0.96	0.96	0.97
Random(R):aIP-vIPC	IP Scanner	10.69	0.87	0.96	0.97
LongTail(L):aIP-vIPC	C&C Malicious Domain	22.00	0.91	0.97	0.99
Heartbeat(H):aIPC-vIP	Shadow Command Injection	10.64	0.87	0.95	0.98
Density(D):aIPC-vIP	Sysrv-hello Linkback Domain	19.85	0.86	0.97	0.99
Random(R):aIPC-vIP	CVE-2021-35394 Vulnerability	7.37	0.9	0.94	0.98
LongTail(L):aIPC-vIP	Cacti File Upload Vulnerability	77.82	0.99	0.99	0.99
Heartbeat(H):aIPC-vIPC	Weak Crossing Login	9.06	0.78	0.98	0.98
Density(D):aIPC-vIPC	Path Traversal Attack	18.42	0.89	0.94	0.99
Random(R):aIPC-vIPC	Webshell Delivery	6.90	0.77	0.96	0.97
LongTail(L):aIPC-vIPC	CVE-2020-9757 Vulnerability	61.76	0.95	0.99	0.99
aIP-vIP-ratio	Character SQL Injection	19.48	0.96	0.96	0.97
aIP-aIPC-ratio	Parameter Contamination	18.30	0.92	0.96	0.98
aIPC-vIP-ratio	Shiro Remote Code Execution	11.87	0.86	0.96	0.96
vIP-vIPC-ratio	Chinad Malicious Domain	31.36	0.97	0.97	0.97
aIP-vIPC-ratio	Sensitive File or Path Spy	6.02	0.85	0.93	0.99
aIPC-vIPC-ratio	Unauthorized Access and Permission Bypass	10.14	0.86	0.94	0.96
aIP-sPort-ratio	Alien Malicious Domain	26.28	0.95	0.96	0.97
vIP-dPort-ratio	Jsctrl Malicious Domain	98.01	0.99	0.99	0.99
dPort-sPort-ratio	EternalBlue Linkback Domain	20.41	0.96	0.97	0.98
aIPC-sPort-ratio	Directory Traversal Attack	33.34	0.88	0.99	0.99
vIPC-dPort-ratio	WannaCry Malicious Domains	66.74	0.92	0.99	0.99

According to Table 3, all 27 characteristic coefficients have at least one corresponding alert type with a probability exceeding 0.77 within the $[\mu - \delta, \mu + \delta]$ range. For the $[\mu - 2\delta, \mu + 2\delta]$ interval, 22 coefficients attain or surpass 0.95, while the remaining 5 lie between 0.93 and 0.94. In the $[\mu - 3\delta, \mu + 3\delta]$ interval, all coefficients display consistently high distribution probabilities above 0.96, with most ranging from 0.98 to 0.99. Moreover, each of the 27 coefficients is associated with at least one alert type featuring peak coefficients over 6. The above experimental findings provide substantial evidence that the proposed 27-dimensional feature vectors exhibit stationary properties, indicating their capability to effectively capture and describe alert type behaviors.

7.3   Evaluation Accuracy of Alert Type Standardization

In order to identify the optimal LSTM network design and tune model parameters, as well as validate the proposed method's effectiveness, we conducted

Experiment 2, depicted in Table 4. This experiment encompasses eight sub-experiments that comprehensively assess the accuracy of our automated alert type standardization approach using various configurations:

- **LSTM-32**: LSTM neural network with 32 hidden layers.
- **LSTM-64**: LSTM neural network with 64 hidden layers.
- **LSTM-32-L2**: LSTM neural network with 32 hidden layers and L2-regularization.
- **LSTM-64-L2**: LSTM neural network with 64 hidden layers and L2-regularization.
- **BiLSTM-32**: Bi-directional LSTM neural network with 32 hidden layers.
- **BiLSTM-64**: Bi-directional LSTM neural network with 64 hidden layers.
- **BiLSTM-32-L2**: Bi-directional LSTM neural network with 32 hidden layers and L2-regularization.
- **BiLSTM-64-L2**: Bi-directional LSTM neural network with 64 hidden layers and L2-regularization.

The experimental process was divided into training and testing phases. The manually annotated dataset was split into a training set and a test set at an 8:2 ratio.

Table 4. Experiments 2: Evaluation of Alert Type Standardization Accuracy Using Various LSTM-Based Neural Networks. The 'Epoch' reported refers to the epoch with the highest accuracy achieved during the training phase across 2000 epochs. Best results are highlighted in **bold**.

Group Name	Regularization	Epoch	Train Loss	Train Accuracy	Test Loss	Test Accuracy
LSTM-32	-	1687	0.0375	0.98405	0.65292	0.90769
LSTM-64	-	808	0.0858	0.96809	0.54165	0.87692
LSTM-32-L2	L2(weight = 1e-4)	1274	0.17791	0.93787	0.3809	0.9
LSTM-64-L2	L2(weight = 1e-4)	567	0.21572	0.93283	0.40363	0.89231
BiLSTM-32	-	1764	0.02467	0.98657	0.57134	0.88462
BiLSTM-64	-	1343	0.08232	0.98489	0.58431	0.89231
BiLSTM-32-L2	L2(weight = 1e-4)	1809	0.03265	0.98657	0.5174	0.88462
BiLSTM-64-L2	L2(weight = 1e-4)	1163	0.04986	0.98741	**0.35415**	**0.92308**

Experimental Setup. The experimental LSTM network was constructed using the PyTorch [41] deep learning framework, a widely-used open-source library. For the optimization algorithm, we chose Stochastic Gradient Descent (SGD), configuring it with a learning rate of 0.0001 and a momentum of 0.2. Additionally, we implemented the Cross-Entropy loss function [42] for measuring the discrepancy between the actual and predicted outputs. Each model underwent training over 2000 epochs. Following each epoch, the model was set to evaluation mode for predictions, using the training datasets to calculate the *train accuracy*. This process allowed us to monitor both *train loss* and *train accuracy* across the epochs. The optimal model was selected based on the highest *train accuracy*, ensuring careful consideration to prevent overfitting.

Results Analysis. As shown in Table 4, we assessed the accuracy of alert type standardization across eight distinct experiments, each employing a unique LSTM-based network configuration. In every experiment, the model demonstrating superior performance during the training phase was chosen. The outcomes reveal that all experimental models achieved commendable classification accuracy rates, with particular emphasis on models like **BiLSTM-64-L2**, which attained an impressive 92.308% accuracy on the testing dataset. These results highlight the effectiveness of our proposed approach, especially when utilizing a bi-directional LSTM network featuring 64 hidden layers and L2 regularization, showcasing its proficiency in automatically standardizing alert types.

8 Conclusion

This paper presents a solution for standardizing alert categories from heterogeneous devices with incomplete semantic knowledge for the first time. Our solution incorporates the characteristics of *attack behavior* and *attackers victims ratio*, which are quantified into a 27-dimensional feature vector model. By utilizing a sliding window counting mechanism for alert logs, multiple time-sequence-based and context-dependent 27-dimensional feature vectors are extracted for each alert type. The proposed standardization model is trained using an LSTM-based neural network. We conducted extentsive experiments from two perspectives based on online NDR alert logs: the stationarity of individual feature coefficients, the separability of all features, and the accuracy of the automatic standardization of alert types. The results demonstrate that our approach can effectively identify common features among different alert types under incomplete information and extract them using quantitative models for the standardization process.

References

1. Endsley, M.R.: Design and evaluation for situation awareness enhancement. In: Proceedings of the Human Factors Society Annual Meeting, pp. 97–101 (1988)
2. Wellens, A.R.: Group situation awareness and distributed decision making: from military to civilian application. In: Individual and Group Decision Making: Current Issues, pp. 267–291 (1993)
3. Stiawan, D., et al.: An improved LSTM-PCA ensemble classifier for SQL injection and XSS attack detection. Comput. Syst. Sci. Eng. **46**(2), 1759–1774 (2023)
4. Wang, C., et al.: State prediction using LSTM with optimized PMU deployment against DoS attacks. J. Intell. Fuzzy Syst. **42**(6), 5957–5971 (2022)
5. Cheng, Z., Sun, D., Wang, L., Lv, Q., Wang, Y.: MMSP: a LSTM based framework for multi-step attack prediction in mixed scenarios. In: ISCC 2022, pp. 1–6 (2022)

6. Poornima, R., Elangovan, M., Nagarajan, G.: Network attack classification using LSTM with XGBoost feature selection. J. Intell. Fuzzy Syst. **43**(1), 971–984 (2022)
7. Kulikov, D.A., Platonov, V.V.: Adversarial attacks on intrusion detection systems using the LSTM classifier. Autom. Control Comput. Sci. **55**(8), 1080–1086 (2021)
8. Srinivasan, S., Deepalakshmi, P.: An innovative malware detection methodology employing the amalgamation of stacked BiLSTM and CNN+LSTM-based classification networks with the assistance of Mayfly metaheuristic optimization algorithm in cyber-attack. Concurr. Comput. Pract. Exp. **35**(10) (2023)
9. MITRE. Adversarial Tactics, Techniques & Common Knowledge (ATT&CK) (2024). https://attack.mitre.org/
10. Al-Sada, B., Sadighian, A., Oligeri, G.: Analysis and characterization of cyber threats leveraging the MITRE ATT&CK database. IEEE Access **12**, 1217–1234 (2024)
11. Kim, Y., Lee, I., Kwon, H., Lee, K., Yoon, J.: BAN: predicting APT attack based on Bayesian network with MITRE ATT&CK framework. IEEE Access **11**, 91949–91968 (2023)
12. Amazon Web Services: NIST Cybersecurity Framework-Aligning to the NIST CSF in the AWS Cloud (2017)
13. Ibrahim, A., Valli, C., McAteer, I., Chaudhry, J.: A security review of local government using NIST CSF: a case study. J. Supercomput. **74**(10), 5171–5186 (2018)
14. Roy, P.P.: A high-level comparison between the NIST cyber security framework and the ISO 27001 information security standard. In: 2020 National Conference on Emerging Trends on Sustainable Technology and Engineering Applications, NCETSTEA 2020, vol. 53, pp. 27001–27003 (2020)
15. Benz, M., Chatterjee, D.: Calculated risk? A cybersecurity evaluation tool for SMEs. Bus. Horiz. **63**(4), 531–540 (2020)
16. Kohnke, A., Sigler, K., Shoemaker, D.: Implementing Cybersecurity: A Guide to the National Institute of Standards and Technology Risk Management Framework. CRC Press (2017)
17. National Institute of Standards and Technology. Framework for Improving Critical Infrastructure Cybersecurity [NIST Cybersecurity Framework]. National Institute of Standards and Technology, Gaithersburg, MD (2018). https://nvlpubs.nist.gov/nistpubs/CSWP/NIST.CSWP.04162018.pdf
18. Carvalho, C., Marques, E.: Adapting ISO 27001 to a public institution. In: Iberian Conference on Information Systems and Technologies, CIST, vol. 2019-June, no. June, pp. 19–22 (2019)
19. Phirke, A., Ghorpade-Aher, J.: Best practices of auditing in an organization using ISO 27001 standard. Int. J. Recent Technol. Eng. **8**(2), 691–695 (2019). Special Issue 3
20. Disterer, G.: ISO/IEC 27000, 27001 and 27002 for information security management. J. Inf. Secur. **04**(02), 92–100 (2013)
21. Leszczyna, R.: A review of standards with cybersecurity requirements for smart grid. Comput. Secur. **77**, 262–276 (2018)
22. Leszczyna, R.: Cybersecurity and privacy in standards for smart grids – a comprehensive survey. Comput. Stand. Interfaces **56**(2017), 62–73 (2018)
23. Hao, X., Zhou, F., Chen, X.: Analysis on security standards for industrial control system and enlightenment on relevant Chinese standards. In: Proceedings of 2016 IEEE 11th Conference on Industrial Electronics and Applications, ICIEA 2016, pp. 1967–1971 (2016)
24. https://oasis-open.org/committees/tc_home.php?wg_abbrev=cti

25. Apoorva, M., Eswarawaka, R., Reddy, P.V.B.: A latest comprehensive study on structured threat information expression (STIX) and trusted automated exchange of indicator information (TAXII). In: Satapathy, S.C., Bhateja, V., Udgata, S.K., Pattnaik, P.K. (eds.) Proceedings of the 5th International Conference on Frontiers in Intelligent Computing: Theory and Applications. AISC, vol. 516, pp. 477–482. Springer, Singapore (2017). https://doi.org/10.1007/978-981-10-3156-4_49
26. https://tools.ietf.org/html/rfc7970
27. Montville, A.W., Black, D.: Enumeration reference format for the incident object description exchange format (IODEF). RFC **7495**, 1–10 (2015)
28. https://www.verizon.com/about/news/verizon-dbir
29. de Melo, C.B., Dutt, N.D.: LOCoCAT: low-overhead classification of CAN bus attack types. IEEE Embed. Syst. Lett. **15**(4), 178–181 (2023)
30. Chang, P.: Multi-Layer Perceptron Neural Network for Improving Detection Performance of Malicious Phishing URLs Without Affecting Other Attack Types Classification. CoRR abs/2203.00774 (2022)
31. Lee, J., Lee, Y., Lee, D., Kwon, H., Shin, D.: Classification of attack types and analysis of attack methods for profiling phishing mail attack groups. IEEE Access **9**, 80866–80872 (2021)
32. Erfani, M.: A feature exploration approach for IoT attack type classification. DASC/PiCom/CBDCom/CyberSciTech, pp. 582–588 (2021)
33. Manyumwa, T., Chapita, P. F., Wu, H., Ji, S.: Towards fighting cybercrime: malicious URL attack type detection using multiclass classification. In: IEEE BigData, pp. 1813–1822 (2020)
34. Zhang, D., Suhara, Y., Li, J., Hulsebos, M., Demiralp, Ç., Tan, W.-C.: Sato: contextual semantic type detection in tables. Proc. VLDB Endow. **13**(11), 1835–1848 (2020)
35. Sebastiani, F.: Machine learning in automated text categorization. ACM Comput. Surv. **34**(1), 1–47 (2002)
36. Kurt Barbé, Gosselin, H.: An ARMA time series approach for analyzing long memory dynamics in measurements. In: I2MTC, pp. 1–6 (2016)
37. Zhao, Z., Kleinhans, A., Sandhu, G., Patel, I., Unnikrishnan, K.P.: Capsule networks with max-min normalization, CoRR, abs/1903.09662 (2019)
38. Best, D.J., Rayner, J.C.W.: Chi-squared components for tests of fit and improved models for the grouped exponential distribution. Comput. Stat. Data Anal. **51**(8), 3946–3954 (2007)
39. Cui, H., Yan, G., Song, H.: A novel curvelet thresholding denoising method based on chi-squared distribution. Signal Image Video Process. **9**(2), 491–498 (2015)
40. Bono, R., Arnau, J., Alarcón, R., Blanca, M.J.: Bias, precision, and accuracy of skewness and kurtosis estimators for frequently used continuous distributions. Symmetry **12**(1), 19 (2020)
41. https://github.com/pytorch/pytorch
42. Mao, A., Mohri, M., Zhong, Y.: Cross-Entropy Loss Functions: Theoretical Analysis and Applications. CoRR abs/2304.07288 (2023)

Family Similarity-Enhanced Implicit Data Augmentation for Malware Classification

Zisen Qi[1,2], Yijing Wang[1], Binbin Li[1,2(✉)], Tianning Zang[1,2], Hao Cui[2], Xingbang Tan[1], and Yu Ding[1]

[1] Institute of Information Engineering, Chinese Academy of Sciences, Beijing, China
{qizisen,libinbin}@iie.ac.cn
[2] School of Cyber Security, University of Chinese Academy of Sciences, Beijing, China

Abstract. Malware poses a persistent threat to network security. Machine learning models, owing to their remarkable detection efficiency and generalization prowess, are prevalently utilized for malware detection and classification. Nonetheless, the prevalent class imbalance in malware datasets poses a significant hurdle, affecting the training accuracy of such models. To address this challenge, this paper introduces a novel data augmentation method for malware classification, named Family Similarity-Enhanced Data Augmentation (FSDA) (Code is available at https://github.com/jasonqzs/FSDA), which leverages the concept of family similarity to achieve implicit data augmentation techniques. FSDA introduces pertinent family-specific features into long-tailed classes, enabling effective classification of malware belonging to such families. Specifically, the methodology initially leverages long-tailed distributed data to train the model's backbone and classifier. Subsequently, it estimates the covariance matrix for each class and constructs a knowledge graph that captures the intricate relationships between any two classes. Finally, the approach adaptively enriches tail samples by propagating information from all similar classes within the knowledge graph. The experiments conducted on two prevalent datasets, Malimg and MaleVis, demonstrate the efficacy of the proposed FSDA method.

Keywords: Malware Classification · Data Augmentation · Family Similarity

1 Introduction

To mitigate the risks posed by malicious code, security experts have deployed diverse detection strategies, ranging from sample-based features and heuristic rules to machine learning approaches. In recent years, machine learning-driven detection methods have garnered significant attention and been extensively explored in both academic and industrial settings. However, the imbalance in malware classes and the long-tailed distribution of data have posed significant challenges to the training and detection efficacy of machine learning models.

J. Zhao and W. Meng (Eds.): SciSec 2024, LNCS 15441, pp. 457–472, 2025.
https://doi.org/10.1007/978-981-96-2417-1_25

However, this pivotal issue has garnered relatively scant attention in prior malware detection and classification research. Training deep models on long-tail distributed data poses a great challenge. Specifically, the separating hyperplane tends to be heavily skewed to tail classes due to their weak statistical ability. Additionally, tail classes are prone to overfitting as their limited samples fail to adequately capture their intra-class diversity.

To enhance the influence of tail classes during training, a common and intuitive approach is re-balancing, which typically involves data resampling and loss re-weighting. Data resampling strives for training parity by balancing the data distribution through selective sampling, while loss re-weighting prioritizes tail categories by assigning heavier penalties to the losses incurred by tail classes. Although these methods can mitigate the bias in the separating hyperplane, overfitting on tail categories remains a considerable challenge due to their limited intra-class diversity. Data augmentation is an efficient way for enriching tail categories, leveraging techniques such as cropping, rotation, mixup, and generative adversarial networks to generate novel samples [1–4]. However, as the volume of augmented examples grows, it can significantly slow down the training speed. Fortunately, ISDA [5] offers an instance-based implicit data augmentation approach that mitigates this issue. This method generates new instances by transforming the original instances along semantic directions sampled from the feature covariance matrix, thereby reducing computational costs. Unfortunately, for tail classes, estimating a diverse covariance matrix can be challenging [13].

In this paper, we introduce a novel method named Family Similarity-Enhanced Data Augmentation (FSDA) for Malware Classification. FSDA effectively enhances tail instances by leveraging semantic transformation directions from analogous categories, thereby broadening the intra-class diversity. Initially, we train a network using long-tail distributed data. Subsequently, we estimate a covariance matrix for each category, encapsulating all possible transformation paths within that category. For tail categories with limited covariance matrices, we enrich them by incorporating similar categories. To identify these similar categories, we construct a knowledge graph where non-diagonal elements signify the degree of similarity between two categories. By propagating transformation directions from analogous categories, we generate sufficient instances to augment tail instances. Moreover, to guarantee that the tail instance incorporates all necessary features for transfer, we complement it with features from similar categories. Consequently, our approach generates instances that are distinctly different from the original training data. The key contributions of this paper are outlined below.

- We introduce FSDA, a novel Family Similarity-Enhanced Data Augmentation technique tailored for Malware Classification. This method innovatively leverages semantic transformation directions from related classes to enhance tail classes, significantly enhancing the intra-class diversity for tail categories.
- We devise a learnable knowledge graph that enables our method to dynamically identify and select appropriate similar categories for various samples, thereby facilitating rational transformations.

– The proposed FSDA method exhibits superior performance compared to current state-of-the-art techniques on long-tailed datasets Malimg and MaleVis-LT.

2 Background and Related Work

With the advent of advancements in machine learning, specifically deep learning, large models, and multimodal learning, researchers have explored the potential of multimodal machine learning approaches for malware classification. Among these, the image-based processing method stands out as a promising direction. This approach involves converting malware files into grayscale or RGB images and subsequently utilizing image-based machine learning models to classify the malware. This method offers the potential to streamline the classification process and reduce the reliance on manual analysis, making it a viable alternative to traditional techniques [6–10]. In 2011, Nataraj et al. [6] pioneered the research on image processing for malware feature analysis. They converted malware into grayscale images and classified malware families using advanced computer vision techniques. Subsequently, Yuan et al. [7] introduced a byte-level malware classification method based on Markov images and deep learning. Their approach converts malware binaries into Markov images using byte transfer probability matrices, and then employs a deep convolutional neural network for classification. Aslan et al. [8] presented a novel hybrid architecture that integrates two extensively pre-trained network models in an optimized manner. This architecture comprises four key stages: data acquisition, deep neural network architecture design, training of the proposed architecture, and evaluation of the trained network. Ghahramani et al. [9] proposed Deep Image, which comprises three malware detection methods rooted in visualization techniques: the clustering approach, the probabilistic approach, and the deep learning approach. Chaganti et al. [10] introduced an efficient neural network model, EfficientNetB1, for malware family classification. They utilized a malware byte-level image representation technique to perform this classification. Jia et al. [11] proposed an image-based malware classification method called IMCSCL, which leverages self-supervised and contrastive learning. They visualize malware using opcode semantic features and detect malware using a contrastive learning method with an enhanced feature encoder network.

The above image-based malware classification models often overlook the intricate long-tail challenge posed by malware data. This challenge manifests in high classification accuracy for head classes with abundant samples, but limited performance for tail classes due to their significantly smaller sample sizes and diverse intra-class features. As a result, the models' generalization ability is constrained, resulting in comparatively lower classification accuracy for tail classes. To tackle this long-tail distribution problem, researchers typically employ data rebalancing and augmentation techniques. Resampling is a popular rebalancing strategy that aims to achieve a balanced data distribution through methods such as over-sampling [14] and under-sampling [15]. However, oversampling can introduce overfitting by duplicating tail data, while under-sampling can compromise feature representation by discarding head data. Re-weighting is another rebalancing strategy that focuses on giving tail classes more attention. A simple approach involves re-weighting the loss of each class inversely proportional to its sample size [16]. Additionally, the class-balance loss [17] emphasizes the effective number of samples. While rebalancing methods have achieved notable improvements, the intra-class diversity for tail categories remains limited, rendering them prone to overfitting. To enhance the intra-class diversity for tail categories, researchers have explored various data augmentation methods. One straightforward approach involves linearly combining samples from the same tail class to generate new ones [18,19]. However, this combination method can sometimes yield meaningless samples. To address this limitation, Generative Adversarial Networks (GANs) [20] have been introduced. For instance, Sharma et al. [12] proposed MIGAN, which can efficiently generate high-quality synthetic malware images that aid in classifying malware samples into families. MIGAN augments and resamples datasets to mitigate the class imbalance issue and improve classification accuracy for tail classes. These explicit data augmentation methods, while effective, can slow down convergence speed as the number of samples increases. ISDA [5] offers an alternative solution, utilizing class-wise covariance matrices and instance-wise features to formulate Gaussian distributions, enabling the generation of infinite samples. However, ISDA encounters challenges when applied to tail classes due to the inability to estimate a diversified covariance matrix from insufficient samples. To address this limitation, MetaAug [21] endeavors to identify an optimal covariance matrix that can generalize effectively to unseen data samples. However, due to its reliance on available samples for covariance matrix computation, it still encounters limited diversity for tail classes. Recognizing that head categories are often more diversified, certain approaches have been proposed to transfer knowledge from head to tail categories. For example, Liu et al. [22] conceptualize each category as a "feature cloud" and augment tail categories by transplanting the intra-class angular distribution from head categories. Chu et al. [23] extend this concept by synthesizing new high-level features for tail categories through a fusion of class-specific features from head categories with class-generic features from tail categories. These advancements aid in enriching the intra-class diversity of tail categories, ultimately leading to an enhancement in classification performance.

Drawing inspiration from implicit data augmentation techniques in image processing, this paper introduces a malware classification model that leverages family-similar implicit data augmentation. The proposed model transfers the features of samples from related classes to tail classes, achieving effective data augmentation for tail classes without the need for generating new samples. This approach ultimately enhances the accuracy of the classification model.

3   The Proposed Method

3.1   Framework Overview

Typically, classifiers tend to exhibit lower performance for tail classes. To enrich their diversity and improve classification accuracy, we introduce a family similarity-enhanced data augmentation approach specifically tailored for malware classification. The framework of this approach is depicted in Fig. 1.

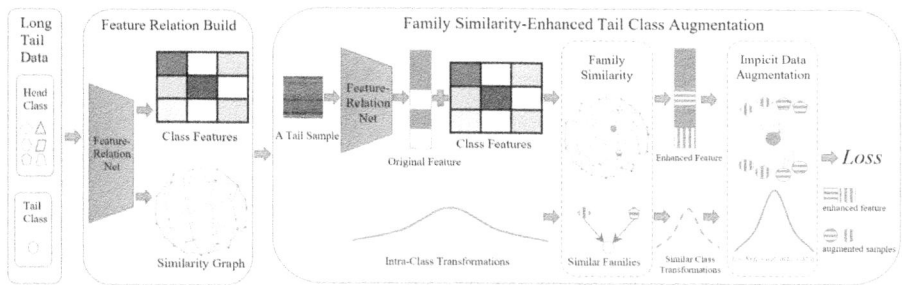

Fig. 1. Overview of the proposed malware classification framework FSDA.

We initially leverage all available samples to train a feature-relation network. Following this, we extract the features and compute both the covariance matrix and the class feature matrix for each category. The covariance matrix encapsulates all the semantic transformation directions specific to each category, while the class feature matrix encapsulates its distinctive features. We then construct a similarity graph that captures the similarity between any two categories. To enrich tail samples, we first propagate them through the feature relation net to obtain their semantic features. Subsequently, we aim to augment these tail samples by sampling semantic transformations from their respective category's covariance matrix. However, the covariance matrix for tail categories is often limited in scope; therefore, we refine it by leveraging similar categories defined in the similarity graph. Furthermore, tail instances may lack certain features for transfer. To ensure completeness, we complement these instances with features from similar families. As a result, an unlimited number of instances can be generated by altering a single instance with infinite semantic transformation directions derived from the refined distribution.

3.2   Implicit Data Augmentation

Given a long-tail distributed dataset $\{s_i, t_i\}_{i=1}^{K}$ where $t_i \in L = \{l_1, l_2, \ldots, l_{Nc}\}$ and Nc represents the total number of categories, we were inspired by the observation that instance-balanced sampling leads to more generalizable representations. Based on this insight, we employed the original dataset distribution to train a feature relation network.

For a sample s_i belonging to a tail class l_c, we first extract its feature representation as $f_i = F(s_i)$. Drawing inspiration from ISDA, we randomly sample along the distribution $T(0, \alpha\Sigma_c)$ to generate features with varying semantic transformations. Specifically, we obtain these transformed features as $\tilde{f}_i \sim T(f_i, \alpha\Sigma_c)$.

Given that the semantic transformation directions vary across different categories, it is essential to compute a unique covariance matrix for each category. To obtain these covariance matrices, we first calculate the class feature for each category by averaging the features of all samples within that class, as defined in Eq. (1):

$$\varphi_c = \frac{1}{N_c} \sum_{i=1}^{N_c} f_i \tag{1}$$

Here, c ranges from 1 to C, N_c represents the number of samples in class l_c, and f_i denotes the feature representation of the i-th sample in class l_c.

Next, we compute each element of the covariance matrix using the following formula:

$$\Sigma_c(p, q) = \frac{\sum_{i=1}^{N_c} (f_i^p - \varphi_c^p)(f_i^q - \varphi_c^q)}{N_c - 1} \tag{2}$$

Subsequently, we obtain the covariance matrices for all classes, denoted as $\Sigma = \{\Sigma_1, \Sigma_2, \ldots, \Sigma_C\}$. However, a key challenge arises due to the limited variations within covariance matrices of tail classes. To address this issue, we introduce a novel reasoning module that effectively transfers rich variations from other classes.

3.3   Family Similarity-Enhanced Implicit Data Augmentation

In this section, we delve into the process of transferring knowledge to enhance a tail sample. Leveraging the constructed similarity graph, we enrich the covariance matrices of tail classes through familial similarity, thereby augmenting their representational capacity.

Similarity Graph Construction. Current methods typically rely on transferring variations from prevalent head classes to underrepresented tail classes. Conversely, our approach seeks to transfer knowledge from similar categories. To accomplish this, we employ the feature relation network similarity constructor C to construct a similarity graph, which is based on the confusion matrix derived from the training data. This graph is essentially a category-to-category directed graph $< V, E >$, where V represents the category nodes and $E \in R^{C \times C}$ denotes the adjacency matrix. Each element E_{ij} captures the similarity between l_i and l_j, which is calculated as follows:

$$E_{ij} = \frac{\sum_{i=1}^{N_i} S(C(f_i) = l_j \wedge t_i = l_i)}{N_i} \tag{3}$$

where S serves as an indicator function, and E_{ij} signifies the proportion of samples from category l_i that are predicted to belong to category l_j.

Similarity-Based Semantic Transformation. We propose a methodology to refine the covariance matrix of each tail class by leveraging the covariance matrices of its similar classes. Intuitively, categories are deemed similar when they share a common software framework, originate from the same developers, or exhibit analogous behaviors. By transferring semantic transformations from these similar classes, we can derive a more accurate intra-class covariance for tail classes, thereby enhancing their representational capacity.

$$\Sigma_c^r = \sum_{i=1,i\neq c}^{C} E_{c,i}\Sigma_i \tag{4}$$

Ideally, computing the family-similar class-conditional covariance matrix for the entire training dataset in each epoch would be ideal but highly resource-intensive. To address this, we employ an online approach that updates and transfers the covariance matrix in a batch-wise manner.

Utilizing the reasoning-based covariance matrix, we can manipulate an instance in a more semantically diverse direction. However, an instance may lack certain features crucial for a similarity-based transformation. For instance, only when a specific compiler is detected in the samples can we adjust its optimization level. To guarantee that a tail instance possesses all the necessary features for transformation, we complement it with features from similar categories.

$$\varphi_c^r = \sum_{i=1,i\neq c}^{C} E_{c,i}\varphi_i \tag{5}$$

After refining the feature distribution of each tail sample using a similar-class feature matrix and a similar-class covariance matrix, f_i can undergo diverse semantic transformations along random directions sampled from $N(\alpha\varphi_c^r, \beta(\Sigma_c + \Sigma_c^r))$. During training, these transformations generate the augmented feature $\tilde{f}_i$ according to Eq. 6:

$$\tilde{f}_i \sim N(f_i + \alpha\varphi_c^r, \beta(\Sigma_c + \Sigma_c^r)) \tag{6}$$

where α and β are positive coefficients that regulate the intensity of family similarity-driven semantic data augmentation. In our experiments, we implement a decay mechanism for both α and β using the ratio $\frac{t}{T}$, where t represents the current number of epochs and T denotes the total number of epochs. This decay ensures that the influence of similarity-enhanced augmentation gradually diminishes as training progresses.

Loss Function. Utilizing our data augmentation technique, a straightforward approach to training a classifier is to generate samples for tail classes using

Eq. 6 until they reach a comparable number of samples as the head classes. Assuming that each sample in the tail class l_c is augmented M times, we obtain a new dataset $\{(f_i^1), (f_i^2), \ldots, (f_i^M)\}_{i=1}^{N_c}$, where f_i^k represents the k-th augmented feature of f_i, sampled from $\tilde{f}_i$. This process effectively enriches the representation of tail classes, improving the overall performance of the classifier.

Subsequently, we can employ the traditional cross-entropy loss function to train the classifier:

$$Loss(\delta_F, W, B) = \sum_{c \in T_c} \frac{1}{N_c} \sum_{i=1}^{N_c} \frac{1}{M} \sum_{k=1}^{M} -log(\frac{e^{w_{t_i}^T f_i^k + b_{t_i}}}{\Sigma_{j=1}^C e^{w_j^T f_i^k + b_j}})$$
$$+ \sum_{c \in H_c} \frac{1}{N_c} \sum_{i=1}^{N_c} -log(\frac{e^{w_{t_i}^T f_i + b_{t_i}}}{\Sigma_{j=1}^C e^{w_j^T f_i + b_j}}) \tag{7}$$

Where T_c denotes the set of tail classes and H_c represents the set of head classes, $W = [w_1, w_2, \ldots, w_C]^T$ represents the weight matrix, and $\boldsymbol{B} = [b_1, b_2, \ldots, b_C]^T$ corresponds to the biases of the final fully connected layer.

3.4   Algorithm Model Design

In this section, we present the proposed malware classification model, which leverages family-similar data augmentation. The model's algorithm comprises five steps, detailed in Algorithm 1.

4   Experiments

4.1   Datasets

The Malimg dataset comprises 25 distinct families, encompassing a grand total of 9,339 malware image samples. Notably, it exhibits an imbalance in its categorical distribution. The malware classes within this dataset are diverse, including Adialer.C, Agent.FYI, Allaple.A, Allaple.L, Alueron.gen!J, Autorun.K, Benign, C2LOP.P, C2LOP.gen!g, Dialplatform.B, Dontovo.A, Fakerean, Instantaccess, Lolyda.AA1, Lolyda.AA2, Lolyda.AA3, Lolyda.AT, Malex.gen!J, Obfuscator.AD, Rbot!gen, Skintrim.N, Swizzor.gen!E, VB.AT, Wintrim.BX, and Yuner.A.

The MaleVis dataset encompasses 26 classes, comprising 25 distinct types of malware along with benign samples. It consists of a total of 9100 training RGB images and 5126 validation RGB images. All training classes involve a uniform distribution of 350 image samples, whereas the validation set varies in the number of images per class. The malware classes include Adposhel, Agent-fyi, Allaple.A, Amonetize, Androm, AutoRun-PU, BrowseFox, Dinwod!rfn, Elex, Expiro-H, Fasong, HackKMS.A, Hlux!IK, Injector, InstallCore.C, MultiPlug, Neoreklami, Neshta, Regrun.A, Sality, Snarasite.D!tr, Stantinko, VBA/Hilium.A, VBKrypt, and Vilsel.

Algorithm 1. Pseudocode of FSDA

Input: The long tail dataset D, D=$\{s_i, t_i\}_{i=1}^{K}$, where $t_i \in L = \{l_1, l_2, \ldots, l_C\}$
Output: *Classifier(Loss(FDSA))*
1: **for** each sample s_i in l_c **do**
2: 　　get sample feature: $f_i = F(s_i)$
3: **end for**
4: **for** for each class l_c in L **do**
5: 　　get class feature: $\varphi_c = \frac{1}{N_c} \sum_{i=1}^{N_c} f_i$
6: **end for**
7: **for** for all classes in L **do**
8: 　　get class similarity graph:$< V, E >$
9: 　　V is l_i a class in L, $l_i \in L = \{l_1, l_2, \ldots, l_C\}$
10: 　　E_{ij} is a similarity edge in graph,$E_{ij} = \frac{\sum_{i=1}^{N_i} S(C(f_i) \wedge y_i = l_i)}{N_i}$
11: **end for**
12: **for** for each class l_c in L **do**
13: 　　get class covariance matrix $\Sigma_c(p, q) = \frac{\sum_{i=1}^{N_c}(f_i^p - \varphi_c^p)(f_i^q - \varphi_c^q)}{N_c - 1}$
14: **end for**
15: **for** for tail class l_c in L **do**
16: 　　transfer features from similar class $\varphi_c^r = \sum_{i=1, i \neq c}^{C} E_{c,i} \varphi_i$
17: 　　tail sample re-fined by similar-classes $\tilde{f}_i \sim N(f_i + \alpha \varphi_c^r, \beta(\Sigma_c + \Sigma_c^r))$
18: **end for**
19: train a classfier *Classifier(Loss(FDSA))* use the cross-entrory loss, assume each sample in tail class l_c is augmented M times, the loss is:

$$Loss(\delta_F, W, B) = \sum_{c \in T_c} \frac{1}{N_c} \sum_{i=1}^{N_c} \frac{1}{M} \sum_{k=1}^{M} -log(\frac{e^{w_{y_i}^T f_i^k + b_{y_i}}}{\sum_{j=1}^{C} e^{w_j^T f_i^k + b_j}}) +$$

$$\sum_{c \in H_c} \frac{1}{N_c} \sum_{i=1}^{N_c} -log(\frac{e^{w_{y_i}^T f_i + b_{y_i}}}{\sum_{j=1}^{C} e^{w_j^T f_i + b_j}})$$

4.2　Evaluation Metrics

The performance of a detection model is evaluated using various metrics, such as Accuracy, Precision, Recall, and F1-Score. These metrics are derived from four fundamental classifications: True Positives (TP), True Negatives (TN), False Positives (FP), and False Negatives (FN). Specifically, TP signifies positive samples accurately identified as such, TN indicates negative samples correctly labeled as negative, FP represents negative samples mistakenly classified as positive, and FN denotes positive samples incorrectly labeled as negative.

Accuracy quantifies the percentage of samples that are accurately classified by the detection model relative to the total number of samples. It is computed using the following formula:

$$Accuracy = \frac{TP + TN}{TP + TN + FP + FN} \tag{8}$$

Precision measures the proportion of positive samples that are correctly identified by the detection model among all samples labeled as positive. It assesses the model's precision rate and is calculated using the following formula:

$$Precision = \frac{TP}{TP + FP} \qquad (9)$$

Recall quantifies the ability of the detection model to correctly identify positive samples, thereby evaluating the completeness rate of the model. It is calculated using the formula:

$$Recall = \frac{TP}{TP + FN} \qquad (10)$$

F1-score serves as a comprehensive metric that balances both precision and recall of the detection model. It is calculated as the harmonic mean of the two and offers an insight into the overall performance of the model. The formula for calculating F1-score is:

$$F1 = \frac{2 \cdot Precision \cdot Recall}{Precision + Recall} \qquad (11)$$

4.3 Experimental Environment

The experiment was conducted in an environment utilizing a CentOS 7.9.2009 system, equipped with a 32-core Hygon C86 7390 Processor clocked at 2700.296 MHz, an NVIDIA A40 graphics card, and 128GB of memory. The programming language chosen for the experiment was Python 3.10.14, and the Pytorch 2.0.1 deep learning framework was adopted for the implementation.

4.4 Results on Malimg

We conducted two types of control experiments on the Malimg dataset. Firstly, we compared models designed by other researchers to assess their performance. Secondly, we analyzed the impact of data augmentation under different backbone architectures. To evaluate the effectiveness of these models and augmentation techniques, we employed metrics such as accuracy, recall, precision, and the F1 score.

Comparison with Different Models. To validate the effectiveness of our approach, we evaluated the performance of various models on the Malimg dataset. Specifically, we randomly split the dataset into 80% for training and the remaining 20% for testing. The results of this comparison are presented in Table 1, where "*" denotes the results reported in IMCSCL [11].

Our models demonstrate superior performance compared to other approaches, achieving a 2.85% improvement in accuracy over previous work. These models are built upon the ResNet32, ResNet50, and ResNet50v2 backbones, with the introduction of the Family Similarity-enhanced Data Augmentation Function (FSDAF). Specifically, we set the hyperparameters $\alpha = 0.01$

Table 1. Comparison results on the Malimg dataset.

Model	Accuracy(%)	Recall(%)	Precision(%)	F1(%)
Kalash et al.*	87.22	87.22	87.38	87.21
Wong et al.*	86.98	96.98	87.33	85.37
Zeng et al.*	91.15	91.15	90.89	90.96
Khan et al.*	94.70	94.70	94.61	94.58
Lo et al.*	94.74	94.74	94.61	94.62
Jia et al.*	96.77	96.77	96.78	96.76
FSDA-R32	**98.78**	**97.08**	**97.01**	**96.93**
FSDA-R50	**99.62**	**99.36**	**99.35**	**99.35**
FSDA-R50v2	**99.52**	**98.78**	**98.69**	**98.90**

and $\beta = 0.25$, and designate the number of header classes as 3. During training, the model undergoes 100 epochs using stochastic gradient descent (SGD) with a momentum of 0.9 and a weight decay of 2×10^{-4}. We initialize the learning rate at 0.01 and adjust it to 0.1 after 40 epochs.

Results with Different Backbones. We further investigated the performance of various backbones, both with and without the Family Similarity-enhanced Data Augmentation Function, on the Malimg dataset. Following the same procedure, we randomly split the dataset into 80% for training and 20% for testing. The results, presented in Table 2, demonstrate the impact of different backbones and the effectiveness of the FSDAF in improving performance.

Table 2. Results with different backbones on the Malimg dataset.

Model	Accuracy(%)	Recall(%)	Precision(%)	F1(%)
ResNet-32	97.87	93.73	93.37	93.29
FDSA-R32	**98.78**	**97.08**	**97.01**	**96.93**
ResNet-50	99.62	99.36	99.35	99.35
FDSA-R50	99.62	99.36	99.35	99.35
AlexNet+ResNet-50	99.14	98.05	98.03	98.04
FDSA-AR50	**99.36**	**98.64**	**98.71**	**98.65**
ResNet-50v2	99.47	98.67	98.61	98.63
FDSA-R50v2	**99.52**	**98.78**	**98.69**	**98.90**

The experiments reveal that the FSDA, when integrated with different backbones, can enhance the performance to varying degrees. Specifically, for

ResNet32, the accuracy improves significantly by almost 1%, indicating the effectiveness of the FSDAF in this case. However, for ResNet50, the FSDA has almost no effect on the accuracy. We hypothesize that ResNet50, with its stronger representational capacity, is already able to perform well on the Malimg dataset, even with its imbalanced distribution. Therefore, the additional benefits of the FSDA may be less pronounced in this case.

4.5   Results on MaleVis-LT

The MaleVis dataset is balanced, comprising 9100 training and 5126 validation RGB images. Each training class contains exactly 350 image samples, while the validation set varies in the number of images per class. In our experimental setup, we utilize an imbalance factor λ to generate diverse training sets for the MaleVis-LT dataset. Specifically, λ represents the ratio between the number of samples in the most frequent class and the least frequent class. In this section, we compare the performance of original backbones with our proposed FDSA on the MaleVis-LT dataset and examine the effectiveness of FDSA across various imbalance factors λ.

We conducted experiments to compare the performance of various original backbones, including ResNet32, ResNet50, ResNet50v2, and AlexNet+ResNet50, with our proposed FDSA on the MaleVis-LT dataset. The results of these comparisons are presented in Table 3, highlighting the performance of each backbone.

Table 3. Comparison results on the MaleVis-LT dataset.

Model	Accuracy(%)	Recall(%)	Precision(%)	F1(%)
ResNet-32	80.71	85.08	81.79	81.17
FDSA-R32	**81.33**	**86.03**	**82.21**	**82.69**
ResNet-50	82.68	87.70	85.60	84.91
FDSA-R50	**82.69**	**87.76**	**85.92**	**85.08**
AlexNet+ResNet-50	82.62	87.62	85.60	84.91
FDSA-AR50	**83.28**	**87.84**	**85.04**	**85.42**
ResNet-50v2	82.81	87.82	84.27	84.69
FDSA-R50v2	**82.97**	**87.83**	**84.62**	**84.64**

The experimental results on the MaleVis-LT dataset are analogous to those observed on the Malimg dataset. Specifically, FSDA is able to enhance the performance of the original backbones to varying degrees. For ResNet32, we noticed a significant improvement in accuracy, approximately 0.5%, while for ResNet50, the effect was nearly negligible. In this section, we set the hyperparameters α to 0.75 and β to 1, and determined the number of header classes as 5. The imbalance parameter λ was set to 50 to generate the imbalanced MaleVis-LT dataset. The model was trained for 100 epochs using stochastic gradient descent (SGD) with a momentum of 0.9 and a weight decay of $2 * 10^{-4}$. We initialized the learning rate at 0.01 and reduced it to 0.1 after the 40th epoch.

Results with Different Imbalance Parameters λ. In this section, we conduct a series of experiments to validate the effectiveness of FDSA under varying imbalance parameters λ, specifically with values ranging from 10 to 100. We utilize ResNet50 as the backbone and present the performance results for different imbalance factors λ in Table 4.

Table 4. Results with different imbalance parameters on the MaleVis-LT dataset.

λ	Model	Accuracy(%)	Recall(%)	Precision(%)	F1(%)
10	ResNet-50	86.33	93.20	88.64	82.10
10	**FDSA-R50**	**86.48**	**93.24**	**88.85**	**90.23**
50	ResNet-50	82.68	87.70	85.60	84.91
50	**FDSA-R50**	**82.69**	**87.76**	**85.92**	**85.08**
100	ResNet-50	81.17	84.98	84.69	82.10
100	**FDSA-R50**	**81.53**	**85.20**	**85.36**	**82.05**

The results demonstrate that as the imbalance parameter λ increases, the effectiveness of FDSA becomes increasingly prominent. This indicates that the more severe the long-tail effect in the dataset, the more impactful the family similarity-enhanced implicit data augmentation provided by FDSA becomes. For these experiments, we maintained $\alpha = 0.75$ and $\beta = 1$, with the number of header classes set to 5, and imbalance parameters λ ranging from 10 to 100. The model was trained for 100 epochs using stochastic gradient descent (SGD) with a momentum of 0.9 and a weight decay of $2 * 10^{-4}$. The initial learning rate was set to 0.01 and adjusted to 0.1 after the 40th epoch.

Visualization of Implicit Data Augmentation. In the MaleVis-LT dataset, the sample *ce0d7ed9a7bd40515f27efa5c6cab6341d4d03d8resized_image.png* in the VBKrypt family cannot be correctly classified before augmentation. After implicit data augmentation, the classification model can correctly classify the samples into the VBKrypt family (Fig. 2).

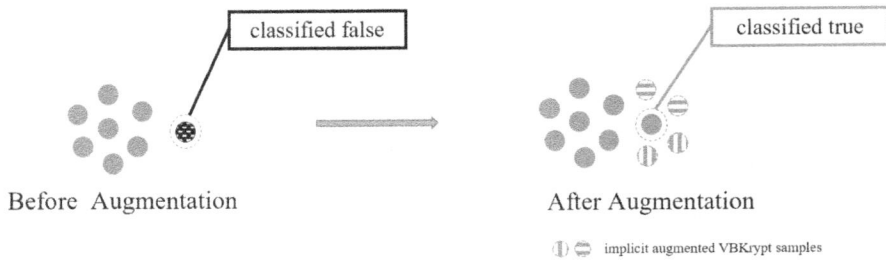

Before Augmentation   After Augmentation

implicit augmented VBKrypt samples

Fig. 2. Visualization of implicit data augmentation on the MaleVis-LT dataset.

5   Conclusion

In this paper, we introduce FSDA, a Family Similarity-Enhanced Data Augmentation method specifically designed for malware classification. Our approach leverages transformation directions from other classes to augment instances from tail categories with distinct semantics. The experimental results validate that FSDA can effectively enhance the performance of long-tail classification. In future work, we aim to explore transformations based on file structural semantics and behavioral semantics to achieve even more reasonable data augmentation, thereby further improving the classification accuracy for malware, especially those belonging to tail categories.

References

1. Simonyan, K., Zisserman, A.: Very deep convolutional networks for large-scale image recognition. In: 3rd International Conference on Learning Representations (ICLR 2015), pp. 1—14. Computational and Biological Learning Society, San Diego (2015)
2. He, K., Zhang, X., Ren, S., Sun, J.: Deep residual learning for image recognition. In: 2016 IEEE Conference on Computer Vision and Pattern Recognition (CVPR), Las Vegas, pp. 770–778. IEEE (2016)
3. Ratner, A.J., Ehrenberg, H.R., Hussain, Z., Dunnmon, J., Ré, C.: Learning to compose domain-specific transformations for data augmentation. In: Proceedings of the 31st International Conference on Neural Information Processing Systems (NIPS 2017), pp. 3239–3249. Curran Associates Inc., Red Hook, NY, USA (2017)
4. Bowles C., et al.: GAN Augmentation: Augmenting Training Data using Generative Adversarial Networks. arXiv, abs/1810.10863 (2018)
5. Wang, Y., Pan, X., Song, S., Zhang, H., Wu, C., Huang, G.: Implicit semantic data augmentation for deep networks. In: Neural Information Processing Systems, pp. 12635–12644. Curran Associates, Inc., Canada (2019)

6. Nataraj, L., Karthikeyan, S., Jacob, G., Manjunath, B.: Malware images: visualization and automatic classification. In: VizSec 2011: 2011 International Symposium on Visualization for Cyber Security, Pittsburgh, Pennsylvania, USA, pp. 1–7. Association for Computing Machinery (2011)

7. Yuan, B.G., Wang, J.F., Liu, D., Guo, W., Wu, P., Bao, X.H.: Byte-level malware classification based on Markov images and deep learning. Comput. Secur. **92**(13), 101740–101752 (2020)

8. Aslan, Ö., Yilmaz, A.A.: A new malware classification framework based on deep learning algorithms. IEEE Access **9**(16), 87936–87951 (2021)

9. Ghahramani, M., Taheri, R., Shojafar, M., Javidan, R., Wan, S.: Deep Image: A precious image-based deep learning method for online malware detection in IoT Environment. arXiv preprint arXiv:2204.01690 (2022)

10. Chaganti, R., Ravi, V., Pham, T.D.: Image-based malware representation approach with EfficientNet convolutional neural networks for effective malware classification. J. Inf. Secur. Appl. **69**(23), 103306–103328 (2022)

11. Jia, Y., Meng, Y., Zhuang, H.: IMCSCL: image-based malware classification using self-supervised and contrastive learning. In: 2023 IEEE 23rd International Conference on Software Quality, Reliability, and Security (QRS), Chiang Mai, pp. 672–683. IEEE (2023)

12. Sharma, O., Sharma, A., Kalia, A.: MIGAN: GAN for facilitating malware image synthesis with improved malware classification on novel dataset. Expert Syst. Appl. **241**(22), 122678–122700 (2024)

13. Li, S., Gong, K., Liu, C.H., Wang, Y., Qiao, F., Cheng, X.: MetaSAug: meta semantic augmentation for long-tailed visual recognition. In: 2021 IEEE/CVF Conference on Computer Vision and Pattern Recognition (CVPR), Nashville, TN, USA, pp. 5208–5217. IEEE (2021)

14. Byrd, J., Lipton, Z.C.: What is the effect of importance weighting in deep learning. In: International Conference on Machine Learning, Stockholm, pp. 872–881. Curran Associates, Inc. (2019)

15. Buda, M., Maki, A., Mazurowski, M.A.: A systematic study of the class imbalance problem in convolutional neural networks. Neural Netw. **106**(10), 249–259 (2018)

16. Huang, C., Li, Y., Loy, C.C., Tang, X.: Learning deep representation for imbalanced classification. In: 2016 IEEE Conference on Computer Vision and Pattern Recognition (CVPR), Las Vegas, pp. 5375–5384. IEEE (2016)

17. Cui, Y., Jia, M., Lin, T.Y., Song, Y., Belongie, S.: Class-balanced loss based on effective number of samples. In: 2019 IEEE/CVF Conference on Computer Vision and Pattern Recognition (CVPR), Long Beach, CA, USA, pp. 9260–9269. IEEE (2019)

18. He, H., Bai, Y., Garcia, E.A., Li, S.: ADASYN: adaptive synthetic sampling approach for imbalanced learning. In: 2008 IEEE International Joint Conference on Neural Networks (IEEE World Congress on Computational Intelligence), Hong Kong, pp. 1322–1328. IEEE (2008)

19. Zhang, H., Cisse, M., Dauphin, Y.N., Lopez-Paz, D.: mixup: beyond empirical risk minimization. In: 6th International Conference on Learning Representations, Vancouver, BC, Canada. OpenReview.net (2018)

20. Bowles, C., et al.: GAN Augmentation: Augmenting Training Data using Generative Adversarial Networks. arXiv abs/1810.10863 (2018)

21. Li, S., Gong, K., Liu, C.H., Wang, Y., Qiao, F., Cheng, X.: MetaSAug: meta semantic augmentation for long-tailed visual recognition. In: 2021 IEEE/CVF Conference on Computer Vision and Pattern Recognition (CVPR), Nashville, TN, USA, pp. 5208–5217. IEEE (2021)

22. Liu, J., Sun, Y., Han, C., Dou, Z., Li, W.: Deep representation learning on long-tailed data: a learnable embedding augmentation perspective. In: 2020 IEEE/CVF Conference on Computer Vision and Pattern Recognition (CVPR), Seattle, WA, USA, pp. 2970–2979. IEEE (2020)
23. Chu, P., Bian, X., Liu, S., Ling, H.: Feature space augmentation for long-tailed data. In: Computer Vision–ECCV 2020: 16th European Conference, Glasgow, United Kingdom, pp. 694–710. Springer, Cham (2020)

Author Index

J. Zhao and W. Meng (Eds.): SciSec 2024, LNCS 15441, pp. 473–474, 2025.
https://doi.org/10.1007/978-981-96-2417-1

The manufacturer's authorised representative in the EU is Springer
Nature Customer Service Centre GmbH, Europaplatz 3, 69115 Heidelberg,
Germany. If you have any concerns regarding our products, please
contact ProductSafety@springernature.com

Printed and bound by CPI Group (UK) Ltd, Croydon, CR0 4YY
05/05/2026
02103581-0004